Braking of Road Vehicles

Braking of Road Vehicles

Andrew Day

AMSTERDAM • BOSTON • HEIDELBERG • LONDON • NEW YORK • OXFORD
PARIS • SAN DIEGO • SAN FRANCISCO • SINGAPORE • SYDNEY • TOKYO

Butterworth-Heinemann is an imprint of Elsevier

Butterworth-Heinemann is an imprint of Elsevier
The Boulevard, Langford Lane, Kidlington, Oxford, OX5 1GB, UK
225 Wyman Street, Waltham, MA 02451, USA

First published 2014

British Library Cataloguing in Publication Data
A catalogue record for this book is available from the British Library

Library of Congress Cataloging in Publication Data
A catalog record for this book is available from the Library of Congress

ISBN: 978-0-12-397314-6

For information on all Butterworth-Heinemann publications visit our website at store.elsevier.com

Printed and bound in the United States
14 15 16 17 10 9 8 7 6 5 4 3 2 1

Contents

Preface

This book is intended to be an introduction to the science and engineering of road vehicle braking, and is based on lecture courses related to automotive chassis engineering, brakes and braking. Most significantly, the short course for industry on the 'Braking of Road Vehicles', which I have organised as part of my contribution to knowledge transfer and exchange between industry and academia since 1996 (the course itself started at Loughborough University in 1966 and I gave my first lecture on it in 1979), has brought me into contact with many expert practitioners who have given freely of their knowledge and time, and hundreds of delegates with their own questions and thirst for knowledge. I extend my thanks to all colleagues, companies and organisations who have helped me extend my knowledge in this way, and have named a few of them below.

Over the years road vehicle braking has become an increasingly broad and complicated field as its importance in road vehicle safety has been increasingly recognised and developed, and I have observed that whilst many engineers have very deep knowledge of detailed and highly specialised aspects of braking, the more general principles and practice of braking can be under-appreciated. The purpose of this book is therefore to provide this basic knowledge in a formalised way, present the principles and theory, explain the analyses and applications, and provide some interpretation and discussion of the principles and practice, while leaving the advanced topics to the specialists who are represented in the wealth of research literature that is available in the public domain. The first seven chapters set out the basic engineering theory and analysis for automotive brake and braking system design. The subsequent chapters present a closer look at some of the 'application-oriented' aspects of braking, including legislation and safety, testing, brake noise and judder, electronic braking, and finally a few case studies from my 37 years of endeavour in the field of automotive braking. The braking industry has its own way of analysing and presenting designs and data; wherever possible I have tried to adopt a generic approach, avoiding approaches, nomenclature and presentations that may be familiar to some but unknown to others. Readers may try to follow what I hope is the logical development of the subject matter in the first seven chapters, and then 'dip into' the other five. Alternatively they may wish to search for information relevant to their particular interest, but I recommend that at least they read Chapters 1 and 2, which contain some fundamentally important observations.

The book is dedicated to two Peters, Peter Newcomb and Peter Harding, who mentored and guided me many years ago in my introduction to road vehicle braking. The original *Braking of Road Vehicles* book, co-authored by Peter Newcomb and Bob Spurr, and published by Chapman & Hall in 1967, was the classic definitive introductory textbook on the subject. Peter Newcomb advised me throughout my PhD research and we worked together for many years on all aspects of braking, including the 'Braking of Road Vehicles' short course. Peter Harding was a gifted engineer and manager (and rock-climber) at Mintex Ltd., manufacturers of friction materials, and had the most remarkable knowledge of braking and friction materials gained from a lifetime in the industry. Much of the knowledge presented in this book started from them and has been accumulated by me over 37 years. Although I cannot always remember the original sources, where possible I have attempted to reference them.

My acknowledgements and thanks go to the following people and organisations:

Federal Mogul (Ferodo) and in particular Ray Rashid for advice on Chapter 2.
John Baggs and Peter Marshall (both formerly of Ford) Eddie Curry (MIRA), and Dave Barton (Leeds University) for sharing their knowledge of passenger car braking system design and brake design analysis on which Chapters 3 and 5 are based.
Brian Shilton and Colin Ross (formerly of Wabco and Knorr-Bremse respectively) and Neil Williams and Paul Thomas (Meritor) for sharing their knowledge of commercial vehicle braking system design and brake design analysis on which Chapters 4 and 5 are based.
Ludwig Fein and Thomas Svensson of Ford, and Mike MacDonald (and many colleagues at Jaguar Land Rover) of Jaguar Land Rover, Neil Williams of Meritor, Colin Ross and Brian Shilton for sharing their knowledge of brake system layout on which Chapter 6 is based.
Marko Tirovic of Cranfield University for sharing his vast knowledge of the thermal analysis of brakes on which Chapter 7 is based.
Colin Ross, and Winfried Gaupp formerly of TUV Nord for sharing their vast knowledge of braking legislation and especially their help and advice in writing Chapter 8.
Rod McLellan for sharing his vast knowledge of brake and vehicle testing on which Chapter 9 is based.
The authors of the research reviews and many papers on which Chapter 10 is based, especially John Fieldhouse for sharing his knowledge of brake noise and judder, and Awsse Alasadi and David Bryant for analyses, diagrams, help and advice in this chapter.
Ian Moore and Ludwig Fein of Ford, and Colin Ross for sharing their knowledge of electronic braking systems on which Chapter 11 is based.
Jos Klaps (formerly of Ford) for sharing his wealth of knowledge about braking systems generally, and specifically steering drift as presented in Chapter 12, and help with many diagrams and figures. George Rosala for his help and advice on brakes over many years.
Jaguar Land Rover, Valx, and Les Price for the cover pictures.

Federal Mogul (Ferodo) for information on friction materials.
Meritor, Link Engineering, and Arfesan for diagrams, images and information relating to their products.
To many other people and organisations for information and diagrams that I have picked up over the years.

Andrew Day
January 2014

CHAPTER 1

Introduction

'Never start anything you can't stop' applies to many aspects of modern life but nowhere does this maxim apply more appropriately than to transport. For road vehicles, whether intended for personal or commercial use, it is surprising how performance data still appears to concentrate on the capability of the engine and powertrain to accelerate the vehicle, or to provide an attractive power-to-weight ratio to maintain speed, with scarcely a mention of the ability of the braking system to decelerate it quickly and safely. The conventional view of vehicle braking systems, even in the technologically advanced twenty-first century world of road transport, is that brakes are 'straightforward'; what could be simpler than pushing one material against another to create a friction force to absorb the energy of motion and slow the vehicle down?

Yet the braking system of a modern road vehicle is a triumph of technological advances in three distinct scientific and engineering disciplines. Firstly, materials science and engineering has delivered technologically advanced friction couples that form the heart of any road vehicle braking system. These advanced friction couples provide reliable, durable and consistent friction forces under the most arduous conditions of mechanical and thermal loading in operating environments where temperatures may exceed 800°C. The materials that make up these friction couples are in many ways quite environmentally sustainable; e.g. the cast iron for brake discs or drums is a relatively straightforward formulation which utilises a high proportion of scrap iron. The friction material includes in its formulation naturally occurring materials such as mineral fibres and friction modifiers, together with recycled components such as rubber in the form of tyre crumb.

Secondly, advanced mechanical engineering design has enabled high-strength braking system components to be optimised to deliver consistent and controllable braking torques and forces over a huge range of operational and environmental conditions. The use of computer-aided design and analysis methods has enabled stress concentrations to be identified and avoided, with the result that structural failures of brake components are unusual in any aspect of modern braking systems. The modern 'foundation brake' (i.e. the wheel brake rotor/stator unit) has been designed to dissipate the heat converted from the kinetic energy of the moving vehicle through the process of friction to the environment as quickly and effectively as possible. Design advances such as ventilated brake discs and sliding calipers have only been possible through the use of modern modelling and simulation techniques so that the underlying scientific principles can be applied effectively.

Thirdly, close and accurate control of braking systems and components through electronics and software engineering has moved braking firmly into the area of active vehicle safety. Forty years ago, Antilock Braking Systems (ABS) demonstrated the safety benefits of maintaining directional control while braking under high deceleration and/or low adhesion conditions. It quickly became clear that 'intelligent' control of the braking system had much more to offer, ranging from traction control where the brake on a spinning wheel could be applied to match the wheel speed to the available road speed or traction availability, through electronic braking distribution to maximise the brake torque depending on the adhesion conditions at each tyre/road interface, and most recently to stability control (ESC) where judicious application of individual wheel brakes according to carefully developed and extremely sophisticated control algorithms could help mitigate the effect of potentially hazardous manoeuvres. It is worth noting that this required a change in legislation, in the sense that non-driver-initiated brake application, or 'intervention' as it is known, had to be permitted before such active safety could be legally incorporated in production vehicles.

However, alongside the remarkable technology advances that have emanated from these three areas of endeavour, it should be noted that the road vehicle braking system is still a remarkably low-cost part of the overall vehicle, and the reliability and maintainability of the braking system on any modern road vehicle is extremely high. Despite the complexity and sophistication of the braking system, and the often environmentally challenging conditions under which the brakes have to operate, routine maintenance is mostly all that is required, and when replacement of, for example, the brake pads or discs is required, the correct parts can be obtained and fitted quickly almost anywhere in the world. The need for regular and appropriate maintenance of any vehicle's braking system must not be underestimated; 1.7% of road accidents in Germany in 2009 were attributed to faulty brakes (ECE, 2012).

Almost since the dawn of wheeled road transport, friction between a rotor (attached to the wheel) and a stator (attached to the vehicle body, chassis or axle) has been utilised in some form to provide controlled vehicle retardation. Other methods have historically been employed, e.g. dragging a heavy object on the road behind the vehicle, or simply steering the vehicle into a conveniently positioned obstacle, but these do not offer much in the way of sustainability, consistency or reliability. Using the vehicle's engine to provide a retarding torque (engine braking) is standard practice in commercial vehicles as a form of 'retarder' in the transmission to generate braking torque. Aerodynamically designed 'air brakes' are found to be effective in taking on some of the duty of the friction brakes at high speeds in high-performance cars. But the most significant recent development in non-friction-based road vehicle braking is regenerative braking. For many years it has been accepted that the kinetic energy of a moving vehicle is dissipated into the environment during braking. It is only with the world's current concerns over CO_2 emissions and limited fossil fuel reserves that this has been challenged.

The braking system of any road vehicle is subject to extensive legislative standards and requirements in many regions of the world. In this book the legislative framework considered is the European Union (UN) Legislation and Regulations, although comparison with the US legislation is made where appropriate. EU law states that all road vehicles are required to have a working braking system that meets the legislative requirements. Included in the braking system requirements are 'service' and 'secondary' braking systems so that the vehicle can be safely brought to rest even in the event of the failure of one part of the system, and a 'parking brake' that can hold the vehicle safely on a specified incline. In Europe, vehicle manufacturers have to demonstrate that their vehicle meets the design and performance standards specified in the UN Regulations through a process of Type Approval. Once a vehicle is sold, the responsibility passes to the owner or user of the vehicle to ensure that the vehicle's braking system continues to meet legal requirements; usually this takes the form of a regular compulsory examination of the vehicle. The design and performance standards associated with Type Approval are regarded as minimum standards, and most vehicle manufacturers have their own 'in-house' standards that exceed the 'legal requirements', often by a considerable margin. For example, UN Regulation 13H (UN, Feb 2014) states that the minimum service braking performance defined by the 'Type-0 test with engine disconnected' for a passenger car (category M_1) is a mean deceleration of 6.43 m/s^2 for a driver pedal effort (brake pedal force) of between 6.5 and 50 daN. Car manufacturers would typically design for substantially more vehicle deceleration for this level of pedal effort, but have to bear in mind the requirement for the secondary braking system to provide a deceleration of not less than 2.44 m/s^2 within the same range of pedal effort. Pedal effort is important because of the large range of physical capability of different drivers. Likewise, the parking brake is covered by a set of legislative requirements and standards, including operating force.

Fundamental to the design of a braking system for a road vehicle (under UN regulations) is that a brake is required at every road wheel. The only exception is light trailers (Category O_1: trailers with a maximum mass not exceeding 0.75 tonnes), which do not need to be fitted with wheel brakes, relying instead upon the brakes of the towing vehicle. In commercial vehicle parlance, the brake unit at the wheel is known as the 'foundation brake'. This term, which is applied exclusively to friction brakes, is used throughout this book to define the wheel brake unit for all vehicles: commercial vehicles, passenger cars, trailers, etc. The function of the foundation brake is to generate a retarding torque (i.e. one that opposes the direction of rotation of the wheel to which it is attached), which is proportional to the actuation force applied. There are two distinct types of automotive foundation brake in common use today, namely the 'drum' brake (Figure 1.1(a)), where the stators are brake shoes fitted with friction material linings that are expanded outwards to press against the inner surface of a rotor in the form of a

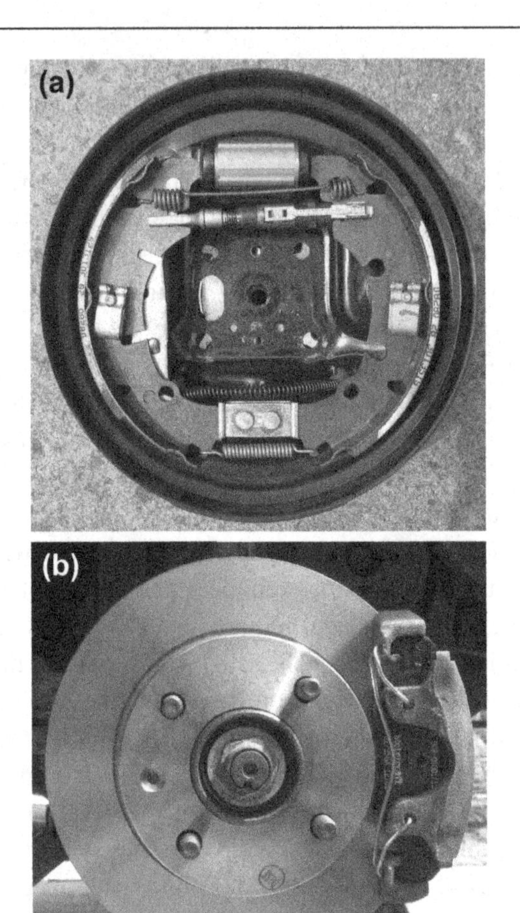

Figure 1.1: (a) Drum Brake. (b) Disc Brake.

brake drum, and the 'disc' brake (Figure 1.1(b)), where the stators are brake pads that are clamped against the outer surfaces of a rotor in the form of a brake disc. Included in the definition of 'foundation brake' are the mounting fixtures such as the 'anchor plate', (also termed 'torque plate', 'spider', or 'reaction frame'), which is firmly bolted to the axle or steering knuckle. The mechanism by which the force provided by the actuation system is applied to the stator elements (pads or shoes) is also considered as being part of the foundation brake.

The brake actuation system comprises the mechanical, electrical and electronic components, which recognise and interpret the 'driver demand', e.g. from the movement of the brake pedal and/or the force applied by the driver to it, and convert it into forces applied to the individual foundation brakes to generate the required retarding torque. The basic brake actuation system transmits the force on the brake pedal through various mechanical connections to the foundation brake. These mechanical connections may take the form of cables, rods and linkages, hydrostatic, hydraulic, or pneumatic systems. They fall into two distinct categories, namely those that rely upon the 'muscular' energy of the driver, and those that rely upon a separate energy source to provide the actuation force which is under the control of the driver. The former usually has a 'servo' or 'booster' in the system to provide power assistance in order to reduce pedal efforts (termed 'power brakes' in the USA), and tends to be used on lighter vehicles such as passenger cars. The latter type of system is used on heavy commercial vehicles in the form of pneumatic or 'air brake' systems, although power hydraulic braking systems are also fitted to some types of commercial vehicle, sometimes in the form of combined 'air over hydraulic' systems. Over the last 25 years the 'mechanical' basis of brake actuation systems has been increasingly augmented by electromechanical technology and electronic control to provide the 'intelligent' control that is such a valuable safety-enhancing feature of modern road vehicle braking systems. This can range from electromechanical handbrake actuation through Electronic Brakeforce Distribution (EBD) at each wheel to full Electronic Stability Control (ESC).

The fundamental scientific principles of the design and analysis of foundation brakes and actuation systems were established many years ago. The required basic performance of the braking system for any road vehicle is always specified in terms of the required brake force at each wheel. This depends upon the design specification of the vehicle, so this is always the starting point for road vehicle braking system design. The design of the foundation brake and the actuation system components, although included in this book, is usually completed in detail by the specialist, and from the vehicle manufacturer's point of view braking system design has tended to become a process of specification and selection. Many vehicle manufacturers have in the past contracted the braking system design out to 'full service suppliers' who have the specialist skills and knowledge to design and deliver a vehicle braking system that meets the vehicle manufacturer's requirements. But increasingly, the importance of the braking system to the overall safety of the vehicle, the need for close integration of the braking system with other vehicle control and management systems, and the sensitivity of the customer to the braking system performance have encouraged most vehicle manufacturers to retain a substantial interest in the braking system design. This has meant that a detailed knowledge of brakes and braking systems is valuable to the vehicle manufacturer and it is the purpose of this book to address this.

The end user of the braking system on any vehicle is the driver, whose expectations are quite straightforward; he or she should be able to apply the brakes in a smooth and controllable manner to generate an equally smooth and controllable vehicle deceleration that is consistent throughout all conditions of vehicle operation and environments. In the foundation brake, this requires remarkable stability in the frictional performance of the brake friction pair, namely the friction material and the rotor, over a wide range of operational and environmental conditions. Most drivers are very sensitive to changes in the braking response of the vehicle, so 'brake pedal feel' is a major attribute in a successful road vehicle to the extent that poor brake response can adversely affect vehicle sales. Drivers (and passengers and other road users) generally do not like their brakes to make a noise, or create uncomfortable vibration while applied, so attention to the noise and vibration aspects of a brake installation on a vehicle is very important for the vehicle manufacturer to avoid customer dissatisfaction.

Because of the concerns over the sustainability of road transport, friction braking is now increasingly seen as wasteful, and strategies for the speed control of road vehicles without using the foundation brakes are increasingly being developed and employed in the form of regenerative braking. Regenerative braking requires technology by which kinetic energy can be converted to and from a more easily storable form of energy, and a device to which that energy can be transferred, stored and recuperated. Some systems are relatively easily included in existing powertrain technology, e.g. alternators that charge only on engine overrun, but the technology to achieve significant deceleration by regenerative braking is expensive and adds mass to the vehicle. 'Hybrid' vehicle technology has shown over the last 20 years how regenerative braking can be combined with friction braking to create effective 'mixed-mode' braking systems. Mostly these are based on electric motor/generator and battery technology, but some hydraulic systems and mechanical (flywheel) systems have also been explored. Because the foundation brake and regenerative braking systems have to work together, their integrated operation or 'blending' has to be transparent to the driver because of the sensitivity previously noted. This presents a major challenge to the vehicle chassis system engineers. Nevertheless, mixed-mode braking technology is progressing and already European legislation has been rewritten (UN, Mar 2014, Feb 2014) to accommodate developments in mixed-mode braking systems for road vehicles.

This book covers the design, implementation, and operation of brakes and braking systems for cars and commercial vehicles with associated trailers, which represent the majority of road vehicles. The principles described do apply to other types of road vehicle, though for some other types (e.g. motorcycles) specific aspects may be significantly different to those presented here. Examples of analyses and calculations are included, together with some examples of 'things that can go wrong' and likely causes. The book starts with a consideration of the science and technology of friction as applied to friction materials and vehicle foundation brakes; this is because an understanding of friction is considered to be fundamentally important in effective road vehicle braking system design. The decelerating road vehicle, including the specific configurations of two-axled rigid vehicles and multi-axle vehicle and trailer combinations, is then analysed to establish an understanding of the requirements of the braking system to achieve the levels of vehicle deceleration, stability, driver effort and performance that are needed to achieve safe braking under all operational conditions. The dynamic distribution of brake force at each axle (and wheel) is then analysed taking account of longitudinal and lateral weight transfer, and parameters such as adhesion utilisation and vehicle braking efficiency are defined and developed so that safe and legally compliant braking system designs for different types of road vehicles can be generated. Friction in tyre/road contact is also considered and the importance of the tyre/road 'grip' (adhesion) is explained and analysed.

Foundation brake design focuses on automotive disc brakes and drum brakes. After developing the basic mechanical theory of these two types of brake, the performance characteristics of each are explained and discussed. Brake torque calculation enables the actuation system to be designed to provide the required distribution of braking forces at each axle. The mechanical loads in the brakes lead to one of the most important operational challenges in foundation brake design, namely that which relates to frictional energy transformation, heat transfer and the temperatures generated during braking. The thermal analysis of brakes is explained, including the calculation of temperatures reached in the brake components during braking, the effect of vehicle speed, load and deceleration, and the importance of sizing the foundation brakes to withstand thermal and mechanical loads. Aspects of braking legislation that influence vehicle braking system design are explained and discussed. Because European legislation is based on the system of Type Approval, legal requirements, design analysis and validation test procedures are explained. Since all designs ultimately require experimental verification, component, system and vehicle testing principles, procedures and technologies are explained for all parts of the braking system.

Brake-related noise, vibration and judder are major concerns in all road vehicles, ranging from brake squeal to brake judder in passenger cars through to heavy commercial vehicles. The phenomena of brake noise and judder are introduced, and experimental and theoretical approaches to minimising the propensity for brake noise, vibration and judder are reviewed. Electronic braking methods and practice are described, which include the application of advanced technologies in modern road vehicle braking systems, in particular foundation brake control and well-established features such as antilock braking (ABS). Rapidly developing technologies of Electronic Brake Distribution (EBD), Electronic Stability Control (ESC), autonomous braking and regenerative braking for road vehicles are explained in terms of the basic principles and the technological implementation. The underlying theory associated with these is explained together with a consideration of how the overall safety of the vehicle is enhanced. The book ends with a selection of case studies, which are intended to illustrate the experimental verification of braking system design and how and why the actual performance of brakes and braking systems can vary from the design.

References

UN, Mar 2014. UNECE Regulation 13. Uniform provisions concerning the approval of vehicles of categories M, N and O with regard to braking, E/ECE/324/Rev.1/Add.12/Rev.8, March 2014.

UN, Feb 2014. UNECE Regulation 13H. Uniform provisions concerning the approval of passenger cars with regard to braking, E/ECE/324/Rev.2/Add.12H/Rev.3, February 2014.

ECE, 2012. Final report of contributions to impact assessment of policy options to improve the EU systems of PTI and of roadside vehicle testing.

CHAPTER 2

Friction and Friction Materials

Introduction

The brakes of road vehicles have relied upon friction for hundreds of years. The use of frictional forces generated between two bodies in sliding contact to provide a retardation mechanism for moving bodies can be traced back in history almost to the origins of human endeavour (Dowson, 1979). Wheel bearings were first noted 5000 years ago, but the use of wheel brakes cannot be traced back this far; it is almost certain that frictional retardation was invoked by dragging, e.g. a log behind a horse-drawn cart when descending a steep hill, and parking while ascending a steep hill (for example, to allow the horse to rest) would be achieved by a 'sprag' to prevent its rolling back. A mechanism that pressed a friction pad against a rotating wheel was a subsequent technological advance, probably dating in England from the 1700s on horse-drawn carriages. Braking devices of this basic form were then utilised on railway carriages and trucks, and then on the first 'horseless carriage' road vehicles in Europe in the later 1800s (Newcomb and Spurr, 1989).

Fundamental to the operation of the friction brake is the dynamic or sliding coefficient of friction (μ) between the rotor and the stator components. (The symbol μ is universally used for the coefficient of friction; in this book it is used specifically to represent the dynamic or sliding coefficient of friction when the bodies in contact are moving relative to each other. The static coefficient of friction, when the bodies in contact have no relative motion between them, is represented here by μ_S.) The iron tyre on a wooden cart wheel worked well as a rotor surface against materials such as wood, leather and felt (Newcomb and Spurr, 1989), but these were sensitive to the environment (e.g. mud and rain), and as vehicle speed, size and weight increased, the amount of energy to be dissipated increased, and operating temperatures increased beyond the limits of these materials. Recognising the importance of temperature stability of the coefficient of friction, resin-bonded composite friction materials were invented to extend the operating range and durability of the stator material. Modern resin-bonded composite friction materials have developed so far that few car drivers nowadays give a moment's consideration to the work that the friction material has to do when they apply the brakes, or the environment in which they have to function.

Sliding friction between two dry contacting surfaces is often known as 'Coulomb friction' after Charles Coulomb (1736–1806), but despite its everyday nature the friction forces involved can usually only be estimated from previous experience and experimental evidence. In friction brakes the surfaces in sliding contact are often coated with 'transfer films' as a result of the sliding process so that the surfaces in contact are not just the bare metal or friction material. For example, a metal surface will have its surface disturbed by abrasion, adhesion or deformation during sliding to create metal fragments and other particles. Superimposed on these could be layers of oxide, which could in turn be covered by layers of organic material transferred from the friction material.

A friction surface, even if it appears to be geometrically smooth, is very rough on the microscopic scale with a distribution of asperities across it. One explanation of the genesis of friction is the interaction of microscopic asperities on the two surfaces, but when two such surfaces are forced together it is very difficult to say where and how contact between them occurs. Model systems have been studied in which the materials and surfaces were scientifically controlled in the expectation that once the friction of such systems was understood, more and more complicated systems could be examined. Although the scientific understanding of friction has benefited from such research, the extremely complicated nature of braking friction, involving high energy, high temperature, high speed and high pressure, remains an inexact science that relies upon specialist knowledge and understanding. Whilst from an engineering point of view a constant coefficient of friction between two sliding bodies may seem a reasonable assumption, in working with friction brakes it is vital to understand that the coefficient of friction is most likely to be variable, and also it is helpful to understand why.

The basic empirical laws of friction, which are known as Amontons' laws after Guillaume Amontons (1736–1806), are stated below. Coulomb introduced a fourth law, which stated that the friction force is independent of sliding speed, but whereas Amontons' laws of friction represent a good practical basis for brake friction pairs, the Coulomb law does not, for reasons explained later.

1. Friction force is independent of the nominal or apparent area of the surfaces in sliding contact.
2. Friction force F is proportional to the normal force N between two bodies in sliding contact, i.e. $F = \mu N$, where μ is the coefficient of friction.
3. The friction force always opposes the direction of sliding (i.e. the relative motion).

A simple physical explanation of the first two laws relates to the difference between the 'real' area of contact A_R (based on the total surface areas of the microscopic asperities in contact) and the 'apparent' (or 'nominal') area of contact A_N (indicated by the overall size of the contact interface) at any friction interface. The real area of contact A_R is very much less than and independent of the apparent area A_N, but is proportional to the normal force between the two bodies in sliding contact. A simplified model of the contacting surfaces based on the idea of contacting asperities is a series of elastic hemispheres (i = 1 − n) on one surface pressing against another perfectly flat, rigid surface. Each hemisphere can be considered to adhere to the flat surface, generating a shear force that is proportional to the area of contact between it and the rigid surface, and the sum of all the areas of the hemispheres in contact with the rigid surface equates to the real area of contact:

$$A_R = \Sigma A_{Ri} \tag{2.1}$$

If the constant f (N/m^2) denotes the specific friction force (i.e. the tangential friction force per unit real area of contact), then for any individual contact, i :

$$F_i = fA_{Ri} \tag{2.2}$$

Making the assumption that the area of contact between each hemisphere and the rigid surface is proportional to the normal force between them:

$$A_{Ri}/N_i = \text{constant and therefore } A_R/N = \text{constant} \tag{2.3}$$

Thus, since $\mu = F/N = fA_R/N$, μ must also be constant.

This simple theory assumes that the surfaces adhere to one another at the real areas of contact when pressed together by a normal force, and that no adhesion remains when the normal load is removed. Any variation in μ is attributed to variations in f arising from differing degrees of contamination of the surfaces. For a full understanding of the phenomenon of friction it would be necessary at the very least to be able to determine A_R and f, but the more this is investigated the more complicated it appears and there is no easy way of doing this for the types of friction material pairs used in modern automotive braking systems. Hence it remains necessary to measure rather than calculate the frictional properties of any friction material, although skilled formulators are able to predict approximate friction (and wear) behaviour based on their knowledge, expertise and experience.

'Tribology' is the generic name for the science of friction, lubrication and wear. Braking friction constitutes 'dry' friction and so the science of friction and wear are mostly of interest. There are three main mechanisms that are considered to contribute to the generation of friction and wear in a brake friction pair (Spurr, 1976):

1. Adhesion. Components of the friction material adhere to asperities on the mating surface, and the friction force is generated by the action of shearing those junctions.
2. Abrasion. Components of the friction material abrade the mating surface, actually removing parts of any transfer film and the mating surface material. Avoiding excessive wear of the mating surface can be achieved by carefully balancing those constituents of the friction material that contribute to transfer film generation.

3. Deformation. The friction material is deformed by the action of interfacial shearing. The high hysteresis of the friction material causes the work of deformation to appear as heat, which is then dissipated by conduction back into the friction material and transfer across the friction interface into the mating body. The proportion of heat that flows each way determines the effectiveness of heat energy transfer from the brake friction interface.

Research has also identified other processes and mechanisms that contribute to the tribological behaviour of particular friction pairs, but these three mechanisms are sufficient for a basic understanding of brake friction science.

There are many different types of friction materials in use today. Some of these are relatively humble (cast iron brake shoes are still used on railway stock) while others are exotic, e.g. carbon fibre composites that are used on aircraft and Formula 1 racing cars. Sintered metal friction materials are used in some industrial applications and in high-duty, high-performance applications such as rally cars and motorcycles, while cork- or paper-based friction materials are still used in low-duty applications, usually oil immersed. The majority of modern road vehicles, however, use resin-bonded composite friction materials that have been developed and refined to provide the friction and wear performance required for automotive brakes. Whichever type of friction material is chosen, for road vehicles the functional requirements of a modern friction material include:

- To provide a consistent and reliable frictional force that enables the vehicle to meet the legislative demands of braking when sliding in a controlled manner against a defined mating surface. This should encompass all aspects of vehicle usage, in particular speed, temperature, mechanical and thermal loadings, and environmental conditions such as humidity and moisture, dirt and dust.
- To be durable. The friction material is usually (though not necessarily) the sacrificial part of the friction pair. Thus it will wear out with time and usage, and its effective life must be in line with vehicle manufacturers' service targets. It must wear in a steady and controlled manner, and not suffer from degradation of performance over its life.
- To be mechanically and thermally strong enough to withstand the loads applied during use. This includes the attachment of the friction material by adhesive bonding or mechanical means (e.g. riveting) to the shoe or backplate components, and having an acceptable compressibility so that the brake pedal movement is not excessive.
- To be 'tribologically compatible' with the other part of the friction pair. The friction material may, for example, abrade the mating surface in order to maintain consistency of performance with time. Ultimately, therefore, the disc or drum may become worn. 'Tribologically compatible' in friction material terms therefore means that wear of the mating surface should be kept within accepted limits, and not that mating surface wear is zero.

- To minimise or preferably avoid, through formulation and design, frictionally initiated instabilities or vibrations in the brake, suspension or vehicle system. Typically this refers to noise and vibration.
- To be environmentally acceptable in use: no emission of hazardous fumes, debris or waste.
- To be cost-effective in design, manufacture and use.

It is important to understand that the friction material is only one-half of the brake friction pair. Vehicle brakes have for many years used cast iron or steel brake discs or drums, and most automotive friction materials have been designed and developed using cast iron as the mating surface. However, the type of cast iron (flake graphite, spheroidal graphite, etc.), as well as its precise metallurgical composition, is known to be important. For example, small (trace) amounts of elements such as titanium and vanadium have been found to drastically affect the friction and wear performance of certain types of friction materials, while much larger amounts of titanium in a cast iron not only affect the friction and wear performance but also can render it almost unmachineable (Chapman and Hatch, 1976). The surface finish of the rotor can also affect the friction and wear performance of the brake lining. An example of the dependence of the coefficient of friction of a brake friction material on the mating material is illustrated in Table 2.1. These data were generated on a small sample pin-on-disc test machine using 10 mm diameter friction material specimens (see Chapter 9).

From Table 2.1 it is clear that the selection of the mating material against which the friction material slides is of paramount importance. Cast iron is used in brake rotors for the great majority of cars and trucks, while stainless steel is increasingly used for motorcycle disc brakes, primarily for cosmetic reasons. One major problem with stainless steel for disc brake rotors is that the wet performance is substantially lower than the dry performance, sometimes by up to a factor of 3, i.e. from 0.6 in dry conditions to 0.2 in wet conditions. This inconsistency of braking performance is unacceptable and manufacturers provide slots and/or grooves to help clear water film from the friction surface. Special formulations of friction material have been developed for operation against stainless steel discs.

Table 2.1: Measured Coefficients of Friction for Different Mating Materials

Mating Material	Coefficient of Friction Measured on a 1 cm Diameter Friction Material Specimen Under Steady-State Conditions at 1 m/s Sliding Speed, 0.689 N/m^2, Interface Pressure Initial Temperature 80°C
Flake graphite cast iron	0.40
Mild steel	Very high erratic friction forces measured
Stainless steel	0.60

Friction Materials: Composition, Manufacture and Properties

Modern resin-bonded composite friction materials are specially formulated to give good friction and wear performance under the sliding contact conditions of braking. The basis of such formulations is usually a polymeric binder (resin) and a fibrous matrix that provides most of the mechanical strength necessary to withstand the generated frictional forces. Asbestos, once used universally in friction materials, has not been used in friction materials in Europe since the 1980s, and its place has been taken by a range of substitutes including cotton, mineral, metal and organic fibres. These fibres are considered to have fewer known health and environmental disadvantages. Fillers, friction modifiers, lubricants, etc. are included to tailor the mechanical, thermal, friction and wear properties, and often only a small amount of one constituent can have a substantial effect on the overall friction and wear performance of the friction material. The binder is a thermoset polymer most commonly based upon a phenolic resin, and may include other related organic compounds to achieve the desired chemical, processing and thermophysical properties. The following information about modern friction materials concentrates on friction materials for disc brake pads. The technology of formulation and manufacture of friction linings for drum brakes used on the rear axles of many mass-produced cars is similar to but less demanding than for disc brake pads, but large drum brakes on commercial vehicles tend to use friction materials that are more similar in terms of the technology of formulation and manufacture to the disc brake pads discussed here.

The generic type of friction material that is most commonly used in disc brakes on all classes of modern road vehicles is the 'resin-bonded composite' friction material. Within this generic type there are generally considered to be three classes, namely 'non-asbestos organic' (NAO), 'low steel' and 'semi-metallic' (precise nomenclature varies from manufacturer to manufacturer, and also in different countries), and each class includes many different formulations designed to match the operational duty and requirements specified for particular brakes and vehicles, usually to operate against a cast iron rotor. Of these three classes, NAO friction materials are usually the most expensive. They tend to have lower μ levels (typically 0.3–0.4), and have superior wear characteristics up to around 220°C, although wear can increase dramatically at higher μ levels and higher duty. They tend not to use significant quantities of iron or steel in the formulation, are relatively clean in operation, and have low noise propensity in terms of avoiding brake squeal, for example. Low steel friction materials, as the name suggests, use iron and/or steel in the formulation (the example formulation shown in Table 2.2 would fall into this class of friction material). They have higher μ levels than NAO materials (typically 0.35–0.5) and are considered to provide good 'pedal feel'. They have good fade and high-speed/duty performance, but noise propensity tends to be higher. Wear characteristics may not be as good as for NAO materials and they are not so clean in operation (wheel dust is discussed

Table 2.2: Typical Friction Material Formulation (Barton, 2013)

Constituent	% by Weight
Whiting (chalk)	38
Bronze powder	15
Graphite	10
Vermiculite	8
Phenolic resin	8
Steel fibres	6
Rubber particles	5
'Friction dust'	5
Sand	3
Aramid fibres	2

later). Semi-metallic friction materials tend to be simpler formulations compared with NAO and low steel materials, and have a ferrous metal content of up to 40% by weight. These friction materials are the lowest cost of the three classes and tend to have low μ levels (typically 0.25–0.35). High wear can be experienced with these materials at high speed and low temperatures, but wear and fade performance are better at higher temperatures. They tend to be prone to noise and judder compared to NAO materials and, because of the high metallic content, heat transfer through the disc brake pad can affect the actuation system, e.g. by causing high brake fluid temperatures in a hydraulic system. Passenger car braking system designs tend to emphasise frictional performance, while commercial vehicles tend to emphasise durability or wear life.

Other generic types of automotive friction materials exist that are usually specially designed to operate against different rotor materials, and these include the following:

- Sintered. Formulated and manufactured using powder metallurgy technology, this type of friction material can give high friction performance under extreme conditions of use, such as in motorsport applications. Sintered friction materials are often used in motorcycle disc brake pads, where they can provide more consistent frictional performance when operating against stainless steel discs, especially in wet conditions.
- Ceramic. Research has indicated that ceramic friction materials can provide stable high friction for very-high-duty applications. Ceramic friction materials are expensive and require specialist rotor materials. They can present NVH and wear disadvantages.
- Carbon. Usually 'carbon' friction materials are carbon fibre composites that operate against a rotor of a similar material. They are expensive and are sensitive to temperature and humidity, but are capable of withstanding very-high-duty levels for prolonged times, and are used in F1 racing cars and aircraft.

The formulations of commercial friction materials are regarded as highly confidential so they are proprietary to the manufacturers; a typical formulation is given by Barton (2013) in Table 2.2.

The function of each constituent may be summarised as follows:

- The resin provides the matrix or binder and has a strong influence on the mechanical and tribological properties of the friction material.
- Steel fibres and Aramid fibres (previously known by its DuPont® name of 'Kevlar') provide mechanical strength to the material, and also strongly influence the baseline tribological characteristics, particularly the friction level.
- Mineral constituents such as vermiculite, which is a form of expanded mica, contribute to the mechanical strength and heat resistance properties of the friction material. Chalk is used as a filler to add bulk at low cost; it has the advantage that it improves the wear resistance without adversely affecting frictional performance.
- Abrasives and lubricants such as sand and graphite can be tuned in the formulation to provide a good balance between friction and wear. Abrasive friction is characteristically stable under most operational and environmental conditions, but if overdone tends to promote high wear of both parts of the friction pair. Lubricants can mitigate the undesirable effects of abrasives by improving the tribological conditions at the friction interface. Typically, friction material formulators will use abrasives and lubricants together to achieve the desired friction characteristics and wear rates. Lubricants tend to promote coatings or deposits on the contacting surfaces, which protect them from abrasive action, while the abrasives help to control the build-up and make-up of such coatings, scouring the friction surface to create a dynamic stability on the friction surface coatings.
- 'Friction dust' is the common name of a naturally occurring thermoset organic material; the traditional source of friction dust is cashew nut shells or cashew nut shell liquid (CNSL). CNSL resins have been used for many years in the manufacture of friction materials, contributing to the binding of the constituents as well as friction enhancers in the form of dust obtained from the cured resin. It appears to promote adhesion even at higher operating temperatures. Rubber, usually in the form of crumb from recycled waste such as tyres, also promotes adhesion, but its effect is more limited in terms of temperature. One particular advantage of rubber as a constituent of resin-bonded composite friction materials is its damping enhancement properties.
- Alloys or metals such as bronze and copper are used in powder form to enhance the tribological behaviour of the friction material; such metals are very good bearing materials against iron, and also tune the heat transfer properties of the friction material. Although the transfer of frictionally generated heat through the friction material must be kept low to avoid damage to the hydraulic actuation mechanism in passenger car and light van disc brakes, there is some advantage to conducting more heat away from the

friction interface through the pad of a disc brake if this can be tolerated. However, with the significant presence of iron in a friction material, the heat transfer benefits of other metals such as bronze and copper are less important today. NAO disc brake pad formulations in particular have included metals such as copper because they have been found to influence performance, wear and noise characteristics beneficially through the formation of a transfer layer. There is, however, a new demand for copper-free materials due to environmental concerns, which is a significant challenge to the friction material formulators (Morbach et al., 2012).

The manufacturing process starts with the preparation of a mix of the constituents. This is usually done in batches with a mechanical mixer into which pre-weighed amounts of each constituent are added in a prescribed sequence. The type of mixer used depends on the class of friction material. Sufficient mixing is required to ensure good distribution of each constituent, but care must be taken not to destroy the structure of the individual components by mechanical over-working. The curing agents for the polymeric resin are usually added towards the end of the mixing cycle because the mechanical work input is sufficient to raise the mix temperature and start the polymerisation reaction. Once the mixing is completed, the mixture is transferred to the next stage, taking care not to allow settling, e.g. of the higher density components, to affect the distribution of the constituents in the mixture.

- NAO materials are usually mechanically mixed and dried in mixers that are temperature controlled (heating and cooling), and have 'choppers' to disperse the fibres. Because the constituent materials do not flow easily (the mix is 'dry'), the fibres are added during the mixing operation. Once the mixing is complete, the batch is emptied into containers for transport to 'press cure' processing machinery.
- Low steel and semi-metallic friction materials tend to be mixed 'wet' because the formulations may often contain liquid resins that are added towards the end of the mixing cycle. A different type of mixer is required to disperse the resin evenly throughout the material; failure to do so could result in local concentrations of resin in the finished brake pads. After mixing (wet), the mixes may be dried for transport to the next stage of the manufacturing process.

The next stage of the manufacturing process for disc brake pads (or drum brake linings) after mixing is based upon the polymerisation process or 'cure' of the phenolic resin. This is a batch process that should follow the mixing process with a minimum of delay. Wet mixes may have a shelf life of 2 or 3 days, while 'dry' NAO mixes may last up to 1 week. Generally a 'press-cure' process is used, where the mix is cured under high temperature and pressure. Commonly used press-cure processes include:

- Die moulding. Disc brake pads are manufactured by the compression moulding of individual pads, usually straight on to backplates.

- Slab moulding. Larger disc pads and drum brake linings can be manufactured by pressing large slabs of friction material, which are then cut to size and attached to shoes or backplates in a separate operation.

A third process is 'roll processing', which is often used to manufacture smaller drum brake linings. In this process a wet mix is squeezed between two rollers in a 'calendering' type continuous process to form a 'pre-form' that is subsequently cut into arc lengths, formed into the required curvature, and cured under light clamp pressure in an oven.

Disc brake pads are usually press-cured straight on to the metal (usually mild steel) backplate, which must be prepared by a process of degreasing, shot blasting and applying an adhesive layer by roller coat or spraying. The strength of the adhesive bond between the friction material and the mild steel is very important to avoid debonding in use. Moisture ingress and corrosion in any debonded area between pad and backplate can cause the bond integrity to deteriorate rapidly. An interlayer is often incorporated to improve the adhesive bond; the backplate to interlayer bond is designed to be better than the friction material to backplate bond, and the interlayer to friction material bond is also designed to be extremely good. The interlayer may have different properties from the friction material, which can be used to advantage by insulating the friction material from the backplate to reduce heat transfer through the pad and hence the likelihood of fluid vaporisation in hydraulically actuated systems in heavy-duty use. However, the frictional properties of the interlayer are usually inferior to those of the friction material, so a heavily worn brake pad is likely to be associated with reduced brake performance. However, where wear indicators are required to be fitted these should prevent such extreme conditions of wear in practice.

The chemical process of polymerisation or cure of the resin is a condensation reaction that creates water that must be allowed to escape as vapour (steam); otherwise manufacturing flaws and associated poor integrity may result. Die and slab moulding processes therefore release the steam from the die cavity at regular intervals using a moulding cycle of the type:

$$\text{Press cure time} = (w + x) \cdot y + z \tag{2.4}$$

where

w = time at set pressure (s);
x = time at zero pressure − designed to release the steam (s);
y = number of cycles for $(w + x)$;
z = Time at set pressure (which may be different from the set pressure for w cycles above) (s).

The rate of polymerisation or cure is exponentially related to temperature. This means that for a rapid cure, the processing temperature must be high (e.g. 250°C or above). However, there is a limit to the temperature that can be used, because friction material is a poor conductor of heat, and too high a processing temperature can degrade the outer regions of the material while leaving the inner regions incompletely cured. Manufacturers therefore seek a compromise between minimising processing time and maximising the degree of cure, and it is important to understand that a new piece of friction material may not be fully cured in the sense that polymerisation of the binder resin may not be complete. Processing time can be reduced by the use of a number of automated stages; e.g. one way of manufacturing disc brake pads is to cold press the mix into a pre-form (some tackiness is necessary to provide adequate mechanical strength in the pre-form), and then to transfer the pre-forms to a curing die. Another method is to part-cure the friction material in a high-pressure die, and then to finish the process under low pressure in an oven.

The precise manufacturing process used depends on the class of friction material. For press curing NAO materials, each die is heated to 140–200°C before adding the underlayer mix and then the friction material mix to the die cavity. The resin is then cured under temperature and pressure; the pressure and time can be adjusted to give the required pad density and compression. After completion of the press-cure cycle the pad assembly is ejected and subjected to a 'bake', i.e. further cure at temperature under atmospheric pressure, before it is finished by surface grinding to remove the surface layer of material, which may have different friction and wear properties. Sometimes a post-processing treatment such as 'scorch' or 'heat searing' is applied to stabilise the friction properties of the newly machined surface, and sometimes a protective coating is applied to prevent surface contamination. Low steel friction materials that are based on wet mixes are usually press formed, i.e. the mix is added to the die cavity and pressed at ambient temperature. The pressed pads can either be 'stack cured', i.e. they are clamped in a stack at much lower pressure than that in the press-cure process, and cured under temperature or baked. Again, after baking, the pads are finished by grinding where required and then usually scorched. The density of the finished pads is largely determined by the clamp force in a stack. Different customers may require additional finishing operations, including the machining of chamfers and slots, and the fitment of shims, piston clips and wear indicators.

Friction Material Specification

Friction material manufacturers usually provide specification sheets for their friction materials in which mechanical properties, friction and wear characteristics, and an indication of the effects of temperature and the recommended operating duty conditions (usually in the form of peak working temperatures), are described. Examples are shown in Figure 2.1 and Table 2.3.

Such data are only a starting point for the brake or vehicle engineer. Dynamometer tests are used to evaluate friction levels, stability/fade performance, noise performance and wear

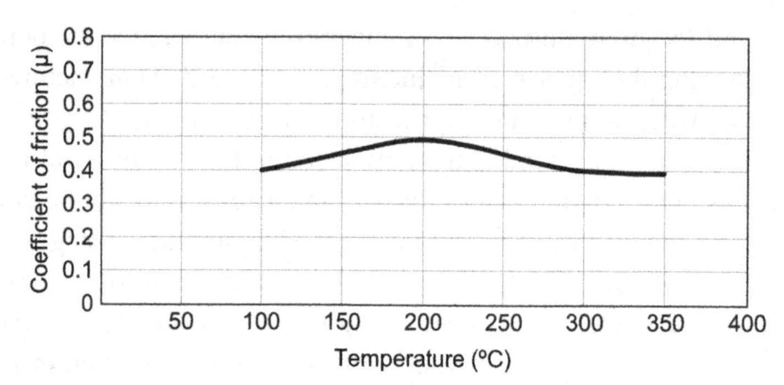

Figure 2.1: Example of a Resin-Bonded Composite Friction Material μ Specification.

performance (disc and pad). The frictional performance is usually quoted on the basis of results from an industry standard test (see Chapter 9), in the form of a nominal or design μ value with an associated indication of variation with temperature as measured by a standard sensor arrangement, e.g. an embedded thermocouple in the friction material or the rotor, or a rubbing thermocouple on the rotor (see Chapter 9). The mechanical properties quoted are also usually quoted as values based on results obtained under specified conditions from industry standard tests, e.g. ultimate tensile strength, ultimate shear strength, hardness and density. There is no industry standard test for 'recommended peak working temperatures', and this information is usually empirically derived with an emphasis on wear life. The operating temperature of a brake on a specific vehicle under any defined brake application varies with time (see Chapter 7) and peak working temperature is often exceeded. What is important is the amount of time spent at high temperatures, since this can permanently damage the friction material. Vehicle designers use power density to size their brakes (see Chapter 6) and this is a much more useful design parameter.

Table 2.3: Friction Material Design Data Specification Example

Design data (average values)	
Ultimate tensile strength	15 MN/m^2
Ultimate shear strength	25 MN/m^2
Rockwell hardness	HRL 80
Density	1950 kg/m^3
Recommended peak working temperatures	
Continuous	250°C
Intermittent	350°C

Other thermophysical properties of the friction material and/or brake pad assembly that may be usefully measured include:

- Ambient compression
- Hot compression
- Thermal swell
- Thermal conductivity
- Hardness
- pH
- Ambient shear strength
- Hot shear strength
- Pad density/porosity
- Surface flatness
- Parallelism.

Calculations, design predictions and computer-aided engineering (CAE) techniques are used extensively in brake and brake system design to predict system performance ranging from vehicle deceleration to brake pedal feel (see Chapter 6). To make accurate predictions, a brake or vehicle engineer requires much more information about the thermophysical and mechanical properties of friction materials than is readily available from friction material manufacturers' data sheets. In this case the vehicle manufacturer has the option of asking the friction material manufacturer to provide detailed data, which often means asking them to undertake experimental work to obtain the required properties, or conducting experimental tests themselves. Some examples of friction material thermophysical properties are shown in Table 2.4 to illustrate the level of detail required for accurate CAE modelling and prediction. Some of the challenges of modelling and prediction that arise from the friction material thermophysical data include:

- Time and rate dependence. Friction material incorporates constituents such as the resin binder (thermoset) polymer and friction modifiers that are visco-elastic in nature. 'Swell' (relaxation expansion) and 'set' (permanent strain, usually compression) are characteristic effects and tests should always be based on load and unload cycles at controlled and different rates.
- Temperature dependence. Not only are the properties temperature dependent, but they change with time and are not always completely reversible so there is often a strong hysteresis effect associated with temperature. It is therefore essential to control temperature precisely and repeatably, noting that because of the low thermal conductivity of friction materials the precise measurement of temperature depends strongly on the position of the sensor.

Table 2.4: Examples of Friction Material Thermophysical Properties for CAE Analysis (Day, 1983)

Material	Transition temperatures		Property	Value of property at temperature
Disc brake resin-bonded composite friction material	Lower (formation)	Higher (degradation)		50 ... 200 250 ... 400 ... 700°C
New	-	200°C	E (MPa)	300 280 260 240
			ρ (kg/m^3)	2250 ----------> (to 250)
			γ (K^{-1})	$14E^{-6}$ $23E^{-6}$ $32E^{-6}$ $78E^{-6}$
			k (W/m K)	0.9 ----------> (to 250)
			C_p (J/kg K)	1200 ----------> (to 250)
Reaction zone	200°C	400°C	E (MPa)	300 280 260 240 220 ------> (to 400)
			ρ (kg/m^3)	2250 ----------> (to 400)
			γ (K^{-1})	$14E^{-6}$ $23E^{-6}$ $32E^{-6}$ $57E^{-6}$ $81E^{-6}$ $85E^{-6}$
			k (W/m K)	0.9 ----------> (to 400)
			C_p (J/kg K)	1000 ----------> (to 400)
Char layer	400°C	1000°C	E (MPa)	-
			ρ (kg/m^3)	1500 ---------->
			γ (K^{-1})	-
			k (W/m K)	0.2 ---------->
			C_p (J/kg K)	700 ---------->
Wear debris	-	-	E (MPa)	Negligible ---------->
			ρ (kg/m^3)	Negligible ---------->
			γ (K^{-1})	Negligible ---------->
			k (W/m K)	0.07 ---------->
			C_p (J/kg K)	1000 ---------->

- Shape dependence. Many thermophysical property tests are carried out on samples of friction material. Because of the heterogeneous nature of friction materials, with different physical shapes and sizes of component materials randomly distributed in the matrix, test samples must be physically many times larger than the largest constituent, and repeat tests must be carried out to determine the effects of variation.
- Environmental dependence. As well as temperature, friction materials are susceptible to humidity and contamination, so such factors must be carefully controlled.

An example of friction material thermophysical properties for different 'phases' of the material measured at different temperatures is shown in Table 2.4. The 'phases' relate to

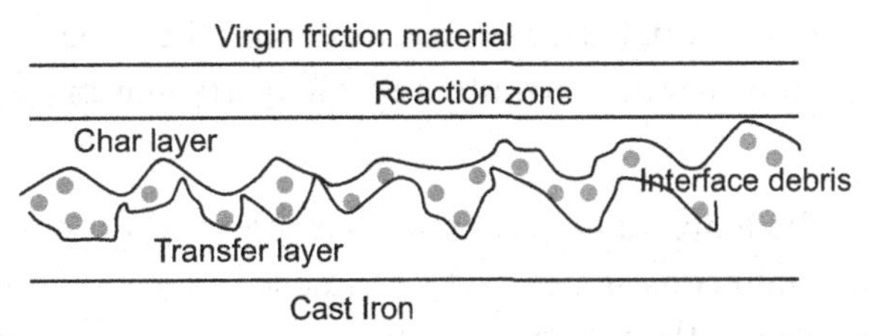

Figure 2.2: The 'Five-Phase' Model of a Brake Friction Pair Comprising a Resin-Bonded Composite Friction Material and a Cast Iron Rotor (Day, 1983).

the 'five-phase' model of a brake friction pair (resin-bonded composite friction material and cast iron; Day, 1983) – see Figure 2.2.

Operational Effects

In principle, Amontons' laws of friction apply for friction materials; however, the coefficient of friction of a resin-bonded composite/cast-iron friction pair does not stay constant and vehicle and brake designers must therefore be prepared to design for variation. It is useful to understand the physical reasons why variation in friction coefficient occurs. The main cause of variation is temperature; as brakes work, they get hot, and the effect of heat is to raise the temperature of the friction material, and very high temperatures can be generated at the friction interface even under relatively low-duty operation because of the low thermal diffusivity of the friction material. The thermoset resin binder thermophysical properties are temperature dependent and those of many of the other constituents will also be changed by temperature. Chemical reactions may occur, and in particular the thermal degradation of the friction material at the interface is known to be an ablation process. The net result is that friction coefficient changes with temperature; typically μ increases slightly up to disc or drum temperatures of about 200–250°C and subsequently decreases as illustrated in Figure 2.1. The precise variation with temperature depends on the friction material.

In brake terms, operating temperature can be usefully defined in terms of the temperature of the brake rotor. There is some debate as to the best way of measuring this; for conventional resin-bonded composite/cast-iron friction pairs rubbing thermocouples can be used, but embedded thermocouples are often preferred, especially for legislative testing, but whichever method is used, consistency is important (see Chapter 9). Friction material manufacturers may prefer to use their own temperature measurement techniques, which are consistent within the company but may not be directly comparable with other methods used elsewhere. More recently, infrared pyrometry has become popular, and provided that the problems of varying surface emissivity can be overcome, this is a good method for identifying surface temperature variations. No method gives an exact measure of the temperatures generated at the actual friction interface, but all can be reliable as a good measure of the temperature generally prevailing for the particular brake operating conditions.

As the brake is applied, the temperature increases and the friction coefficient changes as explained above. To maintain consistency and equivalence in testing, the 'start-of-stop' temperature is usually taken as the reference temperature. Thus, in comparing different applications, the rotor temperature on initial application of the brake is taken as the defining parameter. A typical example of resin-bonded composite friction material performance at different 'start' temperatures, as measured against a cast iron rotor on a small sample friction test rig, is shown in Figure 2.3. These data indicate how the coefficient of friction varies during a test sequence and between test sequences. The test utilised a 10 mm diameter friction material specimen sliding on a cast iron disc rotating at a constant speed equivalent to 7.15 m/s. A constant normal load was applied for 20 s, removed and repeated for 20 applications on a 1-minute cycle. The first application of the 20 was made when the disc had reached the required initial temperature of 80, 100 or 120°C. Natural convection cooling was provided.

The first test (80°C start temperature) indicated μ increasing from about 0.46 to 0.49. The second test (100°C start temperature) showed a fairly stable μ of about 0.48. The third test (120°C start temperature) showed a fairly stable μ reduced to about 0.46. The fourth test returned to a start temperature of 80°C and showed an increase from the 0.46 of the 120°C test to the level indicated in the first 80°C test, but rather surprisingly it then fell back towards the 120°C level. These results show fairly good friction material behaviour for example purposes only; the test was not particularly demanding or long, and the friction pair demonstrated quite high μ.

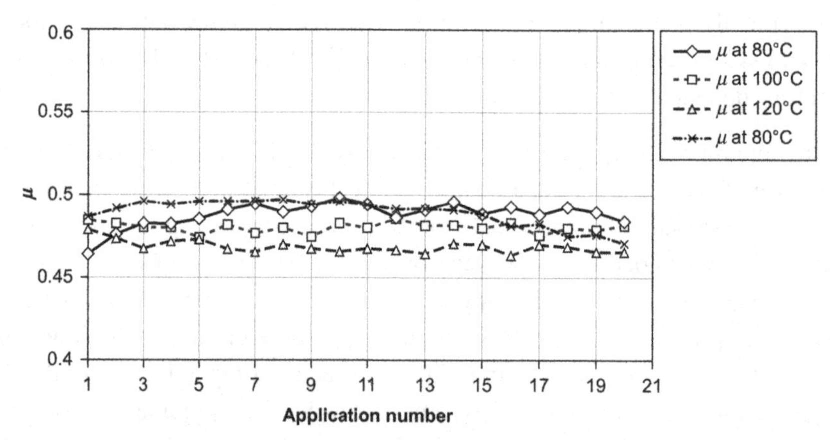

Figure 2.3: Friction Coefficient Measurements from a Small Sample Test Rig.
Disc initial temperature 80°C, 20 s drag application, linear slip speed 7.15 m/s.

The reduction of friction coefficient with temperature is commonly referred to as 'fade'. One physical explanation of fade is that volatile organic components from the resin and other constituents generate regions of pressurised vapour or gas at the interface, separating the sliding surfaces and essentially creating pseudo-hydrodynamic sliding conditions. Because such volatile components are in much greater supply in partly cured friction materials, new or 'green' material frictional performance is likely to be noticeably different from that of used friction material, often showing more variation with temperature. For this reason new brake linings should be treated carefully and not exposed to high-duty, high-temperature operation until they have had a chance to bed-in and burnish. In the USA the terms 'burnishing' and 'bedding in' are used interchangeably, with burnishing being the more commonly used. As explained in Chapter 9, bedding-in can be regarded as a process to achieve geometric conformity between the stator and rotor at the friction interface, and burnishing as a process to achieve a steady condition of sliding or tribological contact at the friction interface, which includes exposing new friction material to temperature to fully cure it and release volatiles from the reaction zone (Figure 2.2).

If a friction material is exposed to high-temperature operation sufficient to cause fade, then it would be expected that when the temperature is allowed to return to a lower value, μ will return to its original value as indicated in Figure 2.3. Although this temperature effect is largely reversible, there is often an effect known as 'delayed fade' that can occur and catch out the unwary. In its extreme manifestation, the brakes of a vehicle can be allowed to cool down, but when they are next applied, a low value of μ is generated (see Chapter 9). For resin-bonded composite friction materials paired with a typical cast iron rotor, prolonged sliding at temperatures in excess of around 300°C (depending on the material and the operating conditions) will result in changes in the surface friction material and possibly through the thickness of the pad or lining. The organic components that are there to control the friction and wear characteristics start to degrade thermally, the friction material's performance is significantly affected, and the mechanical strength of the material is reduced. In the extreme, the friction material surface becomes 'denatured' as all the organic constituents are burnt away and only the temperature-resistant components are left (see Figure 2.4). The friction and wear performance deteriorates irreversibly.

Speed can also affect frictional performance. There is a definite transition zone between the static coefficient of friction μ_s and the sliding coefficient of friction. The former is usually higher than the latter, so at very low speeds brakes can overperform, producing vibration effects such as 'creep-groan'. With resin-bonded composite friction materials, speed effects are almost entirely related to temperature distributions and thermal conditions. Higher vehicle speed means higher sliding speed at the friction interface, and a higher rate of energy dissipation. Higher interface temperatures are generated and μ

Figure 2.4: Example of a 'Denatured' Disc Brake Pad Caused by Excessive Duty and High Temperature.

decreases accordingly. This is a phenomenon known as 'speed sensitivity', and is particularly noticeable in heavy commercial vehicles (Day, 1988). The effect of speed and temperature for a typical resin-bonded composite friction material operating against a cast iron on the same small sample test rig as before is shown in Figure 2.5. Note that the speed axis extends from 1000 to 2500 rev/min, and then returns to 1500 rev/min to

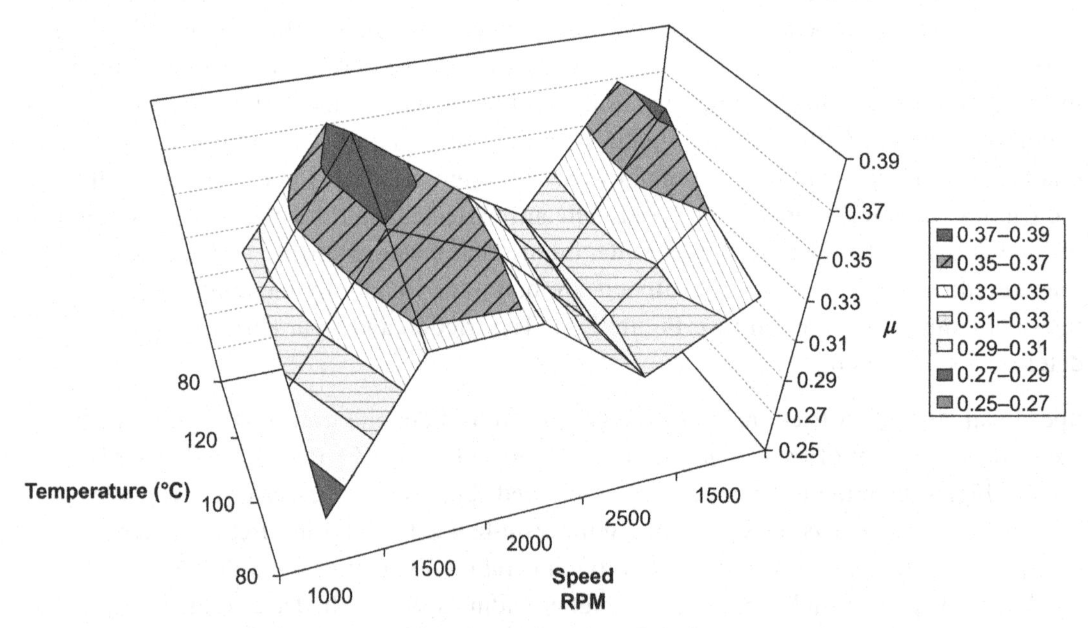

Figure 2.5: Surface Plots of μ, Speed and Temperature.

indicate the repeatability of the frictional performance. It is standard practice to complete a friction material test sequence by repeating a test at the start conditions to check the 'recovery' (see Chapter 9). Data from tests like these can be used to define friction models for use in computational analyses.

There are many other operational and environmental conditions that can influence frictional performance. Water can have two opposing effects: high humidity can raise μ, so that a vehicle's brakes can appear to be very sharp (and noisy) on cold damp mornings, but a few applications can raise the temperature, dry off the water, and bring μ down to the normal operating level. Soaking or immersion in water can reduce frictional performance because of the presence of a lubricating film (liquid or vapour) between the friction surfaces. (It is interesting to note that the controlled introduction of water to a highly thermally loaded friction surface has been used in truck racing to improve brake performance by increasing the heat dissipation through the latent heat of evaporation of the water.)

Most of the μ variation so far considered has been related to high-duty usage. As mentioned above, μ can also be affected by a usage regime of low-duty brake operation, e.g. when the vehicle is driven on short journeys at relatively low speeds with infrequent light braking and resultant low temperatures. This type of usage can result in films being generated on the surface of the friction material and on the mating surface that are associated with low frictional performance (low μ), and is often referred to (in Europe) as 'glazing'. The surface films would need to be removed or replaced before a return to the characteristic steady-state frictional performance can be achieved. The traditional way of dealing with glazing is to apply some high-duty usage, but this does not always work with modern friction materials where the coatings may be particularly tenacious. The term 'glazing' should not be confused with the use of the same term in the USA to describe the result of overheating the friction material, e.g. in high-duty usage or fade and recovery testing.

When a conventional resin-bonded composite disc brake pad or a drum brake lining is newly applied to a cast-iron mating surface (often referred to as 'green' conditions), the tribological conditions at the interface are very different from those 'steady-state' conditions that exist between used and worn brake friction pairs. The process by which steady-state tribological operating conditions are established is termed 'bedding-in' as previously discussed, but it is often called 'burnishing' particularly in the USA, where burnishing is primarily considered to be exposing the friction material to heat cycles to fully cure them and disperse the volatile compounds while bedding-in is a result of the burnishing process. To explain this in more detail, there can be considered to be two aspects of preparing a new brake friction pair for operation:

1. Through the process of wear, geometric conformity between the two surfaces will be generated so that the whole of the apparent area of the friction surfaces of the stator

and rotor are in full contact. This is regarded as 'bedding-in', and if the brake is subjected to heavy-duty usage before bedding-in is complete, thermal damage to the stator and rotor is likely to result because the frictional work is done over a smaller area than either the rotor or stator has been designed for and the work rate or duty level is too high as a result. During this bedding-in process, the friction material (because it has the smaller area of the two components of the friction pair, and is also the less wear-resistant) wears to accommodate the geometric constraints of the brake. Typically a brake lining or pad will not initially make full contact with the brake drum or disc, as evidenced by an unworn region on the rubbing surface, and if this is found on inspection of the friction surfaces, common practice is to estimate the amount of contact and refer to it as 'percentage bedding'. So if an inspection of a disc brake pad indicates that three-quarters of the friction surface is in contact with the disc, this would be recorded as '75% bedded'. Subsequent usage and wear would be expected to bring all the rubbing surfaces into contact to achieve '100% bedded'.

2. The process of sliding between the friction material and the rotor will cause the friction surfaces to transform by the thermal, mechanical and chemical processes involved in friction until a quasi-steady-state of tribological contact is established at the interface. Transfer films will be generated on the stator and rotor surfaces, which may be polymeric films arising from the binder resin and its components, filler, friction modifiers, etc., or the 'packing' of third body wear debris at the interface, or the modification of surface topography and metallurgy or microstructure. This is regarded as 'burnishing'.

An example of bedding/burnishing is illustrated in Figure 2.6, which shows the friction surface of a passenger car front disc brake pad at three conditions at the start, interim and final stages of the bedding cycle on an inertia dynamometer test (see Chapter 9). It is actually quite difficult to capture the state of bedding in a photograph; the bedded area in the interim condition (the centre photograph of Figure 2.6) is highlighted by light reflection from the shiny contact region, which would be described as burnished. In the 95% bedded state (bottom photograph) the pad friction surface is burnished but is a matt rather than shiny surface, which is more difficult to distinguish. Representative steady-state brake performance is unlikely to be obtained until the rubbing surfaces are both bedded-in and burnished. Studies of contact effects on local heat frictional generation at a brake friction interface, e.g. by Eriksson et al. (2002) and Qi et al. (2004), provide insight into the science of burnishing as well as friction variation in terms of local contact zones, thermal expansion and wear.

As explained earlier, the prediction of the friction and wear characteristics of friction materials from first principles by analysis and calculation is not possible, so development and testing are essential (see Chapter 9). Variations in the μ of disc brake pads and drum brake linings must be anticipated, and good brake and system design can help to minimise

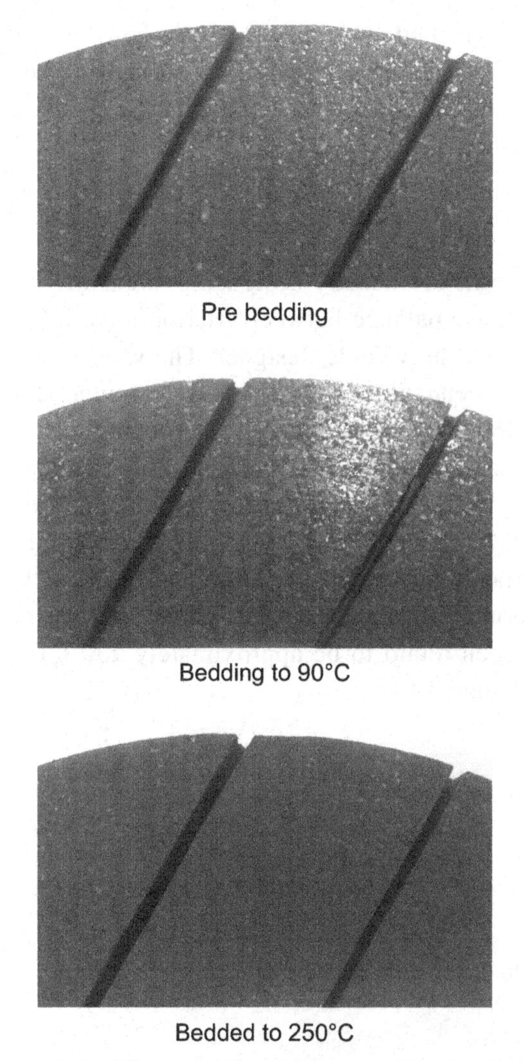

Figure 2.6: Bedding and Burnishing on Disc Brake Pads.
Top: unbedded new condition 0% bedded; centre: estimated 25% bedded; bottom: estimated 95% bedded.

the effects of such variations.The value of μ and any associated variation with operating environment or conditions fundamentally defines the 'performance' of the brake, and achieving the required level and consistency of μ forms an essential part of the friction material design and verification. As a general rule, the coefficient of friction μ of a modern friction material can be expected to vary by $\pm 10\%$ from the nominal; thus when a value of μ is used in this book for brake and system design purposes, the performance of the designed system should always be evaluated at these upper and lower limits. As an example, a disc brake pad specified at a μ of 0.4 should be considered to have a friction coefficient of $0.36 \leq \mu \leq 0.44$. Particular usage or environmental conditions may cause the friction material to exhibit performance that might appear to be outside even this $\pm 10\%$ range.

Wear

Automotive friction materials are usually designed to wear in preference to the rotor material, although the precise balance between friction material life and disc or drum life is a matter of choice for the vehicle designer. The wear rate of organic friction materials is temperature dependent, and since interface temperature is proportional to μpv (the product of friction coefficient, local pressure and local sliding speed), it is also speed and pressure dependent. The temperature effect is exponential, so the material will wear many times faster at an operating temperature of, for example, 500°C than it will at 100°C. It has been shown (Liu and Rhee, 1976) that resin-bonded composite friction material wear follows an Arrhenius rate law that governs the energy activation process of thermal decomposition of the phenolic resin. Below a 'threshold' temperature (which has been found to be approximately 230°C) the wear (weight loss per unit area) can be calculated from:

$$\Delta w' = \beta' p^{a'} v^{b'} t \tag{2.5}$$

while above the threshold temperature:

$$\Delta w = \beta p^a v^b t \exp\left[-Q_A / \overline{R}\theta_K\right] \tag{2.6}$$

where:

Δw = wear (weight loss per unit area kg/m^2);
β, a, b are constants;
p = pressure (Pa);
v = linear sliding speed (m/s);
t = time (s);
Q_A = Arrhenius activation energy (J);
$\overline{R}$ = universal gas constant (8.314462 J/mol·K);
θ_K = absolute temperature (K).

This demonstrates the exponential dependence of resin-bonded composite friction material wear and also explains why temperatures in the region of 200–250°C are significant in μ: temperature dependence. Wear rate and the coefficient of friction tend to be related, i.e. higher μ materials tend to wear more quickly even when they are used under the same conditions. When vehicles are driven in such a way that they require frequent braking, the 'mean journey temperature' of the brake is a useful indicator of friction material wear life, but for intermittent braking, e.g. motorway driving, average application temperature, rather than average journey temperature, is much more relevant. This is because the long periods between braking allow the brakes to cool down and keep mean journey temperatures low while the wear occurs under braking conditions that individually can generate quite high temperatures because of the speeds involved. The relationship between the wear life of the friction material and that of the rotor depends largely upon the preference of the vehicle manufacturer. Normal practice is to specify the brake type and size and then the friction material to work within the manufacturer's recommended duty and/or temperature ranges. Wear life of the friction material and rotor can then be experimentally evaluated in field trials (see Chapter 9).

No discussion of resin-bonded composite friction materials would be complete without mention of two characteristics that, although they do not directly affect performance, can be of major concern to vehicle (especially car) manufacturers, representing 'error states' in the brake's ideal function. Brake noise and judder constitute the first characteristic, and the part that the friction material has to play in their generation (and suppression) is covered in Chapter 10. The second is 'brake dust', which is a more recent phenomenon in the EU and the USA that causes customer dissatisfaction, although it does not affect braking performance. This brake dust appears to be generated as the brake friction pair wears away and is frequently seen to be deposited on the road wheels, causing them to appear black and dirty. More importantly the particles potentially present a health hazard in the environment. The black brake dust is essentially comprised of wear particles from both the friction material and the rotor with a measured particle size distribution around 350 nm. The dust seems to be attracted to the vehicle's road wheels and is remarkably adherent to the wheel surfaces, for which the physical reason is unclear. One explanation is that the organic components of brake dust combine with other environmental components and moisture to create an adhesive paste on the wheel surface, while another is that iron oxide in the form of magnetite (Fe_3O_4), which is present in some resin-bonded composite formulations and is a product of cast iron disc wear, is electrostatically attracted to the wheel. The black dust is certainly ferromagnetic, as it can be collected by a magnet, so this may also be a contributing factor. It is suspected that high-μ disc brake pad friction materials increase brake dust generation and complaints, most likely because the higher abrasive content scours iron particles from the brake disc, which are then partially oxidised to Fe_3O_4, although this material is used in some friction material formulations, so it may also be released as wear product from the disc brake pads. The problem of discoloration of road wheels by brake dust has been less well known in markets outside Europe, and since these markets tend to use different formulations of friction materials the root cause is almost certainly the friction material formulation. Since brake noise propensity is also known to be increased with higher μ materials (see Chapter 10), some car manufacturers have started to give preference to brake system designs based on different formulations of lower friction materials, in particular the NAO formulations described earlier.

Chapter Summary

- Resin-bonded composite friction materials operating against a cast iron rotor continue to be the most widely used type of friction material for road vehicles. Other friction materials and friction pairs (e.g. ceramic, carbon) do find application on road vehicles, but they tend to have specialist uses and can be expensive, and their use is limited, e.g. to specialist high-performance cars.
- The basic laws of friction (Amontons and Coulomb) represent a good practical basis for brake friction pairs. The physical explanation of friction mechanisms is complicated, but the three basic processes that underpin brake friction generation are adhesion, abrasion and deformation. The formulations of modern friction materials aim to achieve a durable and effective mixture of these three processes to meet customer requirements, which include consistency, durability and reliability, coupled with environmental acceptability and cost-effectiveness. An overview of the role of individual constituents in friction material formulations is given.
- Friction materials are a vitally important part of automotive brakes, and vehicle and brake designers need to understand the characteristics of their friction and wear behaviour and the causes of variation in their properties. The friction material is only one-half of the friction pair in a brake so the tribological (friction and wear) performance of a brake depends on the stator and the rotor material combination and may change if one of that pair is substituted. The design and composition of the mating body material is therefore as important as the design and composition of the friction material.
- Specifications for commercially available friction materials include mechanical properties, friction and wear characteristics, and an indication of the effects of temperature and the recommended operating duty conditions. The frictional performance of these materials depends upon their formulation, and vehicle manufacturers can select friction materials to suit their requirements; e.g. for passenger cars they may use friction materials with a nominal μ value in the range $0.38 < \mu < 0.45$ while commercial vehicles often emphasise durability over high μ and use a lower nominal μ range, e.g. $0.35 < \mu < 0.40$. These represent a starting point for brake system design.
- The manufacture of resin-bonded composite friction materials is summarised, and the press-cure process for die and slab moulding is briefly explained.
- Once in operation, frictional performance is primarily affected by heat and temperature. A friction pair basically follows Amontons' laws of friction, but when braking duty increases and the temperatures generated at the friction interface increase significantly,

the coefficient of friction (μ) will change. Most changes in μ are related to temperature, and how much μ changes depends on the specification of the friction material. The friction coefficient (μ) of any friction material can only be assessed accurately by testing (see Chapter 9). Because μ is a fundamental parameter in brake and brake system design, its characteristic behaviour is very important.

- The performance of resin-bonded composite friction materials changes during the bedding-in stage, when at the same time the process of burnishing is also important in establishing a stable friction and wear performance. Bedding-in and burnishing are important to avoid thermal damage to the friction pair early in the life of a new or newly refurbished brake.
- Brake temperatures generated in operation affect the frictional performance. If the brake usage is sufficiently severe to generate high temperatures and cause a significant reduction in μ termed fade, then usually when the temperature is allowed to return to a lower value, μ will return to its original value as well. However, where the brake usage is extremely severe, there may be other effects such as delayed fade while prolonged operation at high temperatures will cause thermal degradation of the friction material when the organic components that are there to control the friction and wear characteristics start to degrade irreversibly, known as denaturing.
- The wear of resin-bonded composite friction materials is exponentially dependent on temperature at high operating temperatures, typically above 200–250°C, and follows an Arrhenius rate law. The wear properties of friction materials can be tuned to suit the vehicle manufacturer's requirements, and experience has indicated that for resin-bonded composite friction materials the higher the friction level, the higher the wear rate. Abrasives in friction materials are useful to raise friction levels and control disc (or drum) coatings, including corrosion, but there is a trade-off in terms of rotor wear and life.
- Brake noise and judder, and brake dust are two important 'error states' associated with friction materials that may be experienced in use and are mentioned here. A short explanation of the generation of brake dust is included in this chapter while brake noise and judder are covered in Chapter 10.

References

Barton, D.C., 2013. Materials design for disc brakes, Braking of Road Vehicles 2013. University of Bradford. ISBN 978185143 269 1.

Chapman, B.J., Hatch, D., 1976. Cast Iron rotor metallurgy. Proc. IMechE. Conference on Braking of Road Vehicles, 1976, paper C35/76, pp. 143–152.

Day, A.J., 1983. Energy transformation at the friction interface of a brake. Loughborough University. PhD thesis.

Day, A.J., 1988. An analysis of speed, temperature, and performance characteristics of automotive drum brakes. Trans. ASME. J. Tribol. 110, 298–305.

Dowson, D., 1979. History of Tribology. Longman. ISBN 0582 44766-4.

Eriksson, M., Bergman, F., Jacobson, S., 2002. On the nature of tribological contact in automotive brakes. Wear 252 (1–2), 26–36.

Liu, T., Rhee, S.K., 1976. High temperature wear of asbestos-reinforced friction materials. Wear 37, 291–297.

Morbach, M., Paul, H.-G., Severit, P., 2012. Systematic approach for structured product development of copper free friction materials. Eurobrake Conference 2012.

Newcomb, T.P., Spurr, R.T., 1989. A Technical History of the Motor Car. Adam Hilger. ISBN 0852 74074-3.

Qi, H.S., Day, A.J., Kuan, K.H., Rosala, G.F., 2004. A contribution towards understanding brake interface temperatures, PEP, Braking 2004, Vehicle Braking and Chassis Control.

Spurr, R.T., 1976. The equation for the friction of metals. Wear, 40, 389–393.

CHAPTER 3

Braking System Design for Passenger Cars and Light Vans

Introduction

This chapter introduces the basic theory for the high-level design of the braking system for any road vehicles of a two-axle rigid configuration. This is the most straightforward (in terms of basic braking theory) road vehicle configuration and includes passenger cars, light vans and solo commercial vehicles (buses and trucks without trailers). Some rigid commercial vehicles can have more than two axles, in which case the analyses presented here still apply, with the force systems assumed to be concentrated on one 'composite' axle. The analysis presented aligns with the UN Braking Regulations 13 and 13H (UN, Mar 2014 and UN, Feb 2014) (Regulation 13 applies to commercial vehicles- including light vans-, and Regulation 13H applies to passenger cars and light vans; see Chapter 8), which are adopted in many world regions. The categories of vehicle in these Regulations to which the theory presented here applies include M_1 (referred to in this book as passenger cars), M_2 and M_3 (buses and coaches), N_1 (goods vehicles that are similar to passenger cars in general layout, having a maximum mass not exceeding 3.5 tonnes, which are referred to in this book as light vans), and rigid commercial vehicles in categories N_2 and N_3. Trailers (category O) are studied in Chapter 4, which focuses on commercial vehicles and vehicle/trailer combinations. Other categories of vehicle, L_1-L_5 (two- and three-wheel vehicles), L_6 and L_7 (quadricycles), and T/C and R/S (agricultural tractors and trailers), are not referred to explicitly, although the basic theory is applicable.

The fundamental theory and analysis relating to road vehicle braking system high-level design is based on the 'bicycle' model, which is a two-dimensional (2D) representation of the vehicle that assumes that there is no lateral variation in the wheel and brake forces that act on the vehicle during braking. The analysis presented here does not include advanced braking technologies such as Antilock Braking (ABS) and Electronic Brakeforce Distribution (EBD), which are covered in Chapter 11. These technologies are intended to offer safety benefits in terms of the control of the vehicle under conditions that are at the extremes of operation and usage, not to compensate for sub-optimal basic braking system design. Hence this chapter focuses on the theory, knowledge and understanding needed to design a good basic braking system, which can then be further enhanced by the incorporation of brake-related safety technologies.

Weight Transfer During Braking

The primary function of a road vehicle braking system is to generate deceleration of the vehicle. Associated with this deceleration are dynamic effects relating to the vehicle and its components; perhaps the most obvious is the 'dive' often observed while braking, where the nose of the vehicle dips because of 'weight transfer' between the vehicle's axles. The analysis of the dynamic effects is based on the ISO configuration global axis set (Figure 3.1), and the symbols and notation used in this chapter are based on the nomenclature adopted in UN Regulations 13 and 13H (UN, Mar 2014 and UN Feb 2014). The vehicle configuration shown in Figure 3.1 defines X as the vehicle longitudinal axis, Y as the lateral axis, and Z as the vertical axis. To avoid unnecessary complication in the equations presented in this book, subscripts defining the direction of forces, deflections, etc., are only used where necessary, e.g. in the combined effect of longitudinal and lateral forces at the tyre/road interface.

A 'free-body' diagram of a two-axle, four-wheel rigid vehicle (a passenger car) in dynamic equilibrium is shown in Figure 3.2. This is referred to as a bicycle model where the forces on the wheels at each end of an axle are the same, i.e. it is a 2D representation in which there is no side-to-side difference in forces. A 3D vehicle model would be necessary for a full dynamic simulation of a road vehicle during braking, e.g. a multiple degree of freedom (DoF) model including longitudinal, lateral, pitch, roll, and yaw forces and motions etc., but this is beyond the scope of this book and the reader is referred to readily available texts on Vehicle Dynamics.

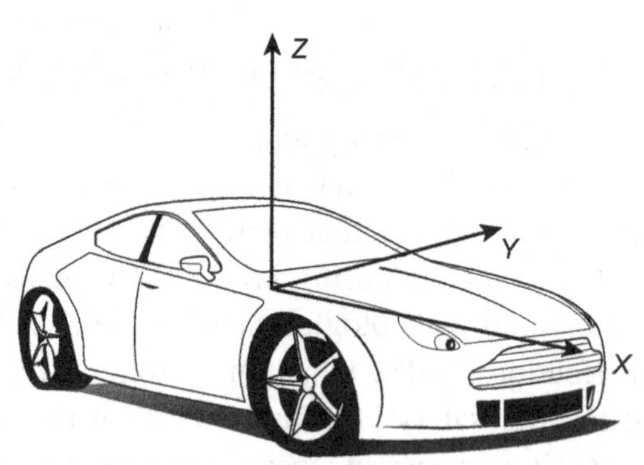

Figure 3.1: Vehicle Coordinate System (ISO).

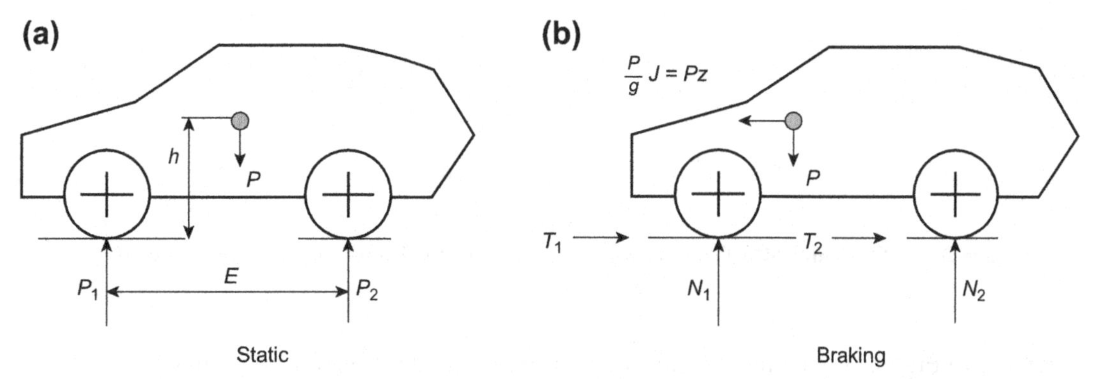

Figure 3.2: 2D Road Vehicle Force System Under Static (a) and Dynamic (b) Braking Conditions.

Nomenclature

M	Mass of the vehicle (kg)
P	Vehicle weight (N); this differs from the UN Regulations 13 and 13H, where P is specified as the 'mass of the vehicle'. In this book the mass of the vehicle is denoted as 'M' (kg)
P_i	Normal reaction at the road surface of axle 'i' under static conditions (N). Note that subscript '$_i$' refers to the axle, so for the vehicle illustrated here '$_1$' indicates the front axle and '$_2$' the rear axle
N_i	Dynamic normal reaction at the road surface of axle 'i' under braking conditions (N)
T_i	Brake force exerted by the brakes on axle 'i' under normal braking conditions on the road (N)
f_i	T_i/N_i = Adhesion used by axle i
J	Deceleration of the vehicle (m/s^2)
g	Acceleration due to gravity (= 9.81 m/s^2)
z	J/g = Rate of braking of the vehicle. In road vehicle braking parlance 'z' is usually quoted as a decimal, e.g. '0.5', or as a percentage. Thus, the statements '$z = 0.5$' and '$z = 50\%$' both mean that the deceleration of the vehicle is 0.5×9.81 m/s^2
h	Height of the centre of gravity of the vehicle above the road surface (m)
E	Wheelbase (m) = $(L_1 + L_2)$
L_1	Horizontal distance from the front axle to the vehicle's centre of gravity (m)
L_2	Horizontal distance from the rear axle to the vehicle's centre of gravity (m)
μ	Coefficient of sliding friction associated with the friction material in the brake (also sometimes written as μ_D for clarity when compared with the static coefficient of friction μ_S)
k	Theoretical coefficient of adhesion between the tyre and the road
μ_t	Coefficient of sliding friction associated with a tyre on a locked wheel sliding on the road surface
F	Friction drag force (N)
X_i	Proportion of the total vehicle braking force generated at axle 'i'

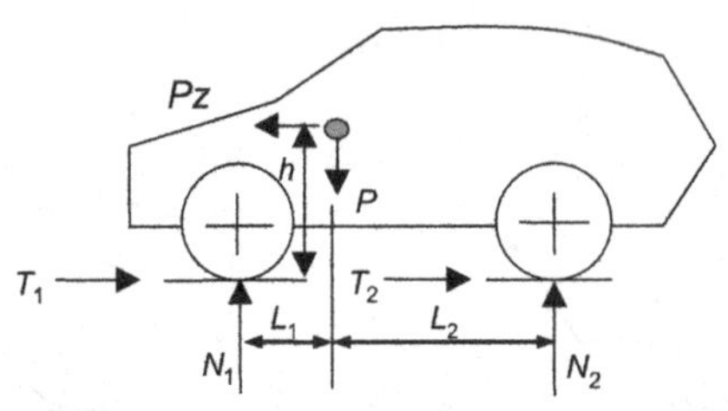

Figure 3.3: 2D Model of a Car Driving on a Horizontal Road with all Wheels Braking but not Locked.

Referring to Figure 3.3, taking moments about the instantaneous point of contact of the rear wheels with the road gives:

$$N_1(L_1 + L_2) - PL_2 - Pzh = 0 \tag{3.1}$$

At rest:

$$J = 0; \quad P_1 = \frac{PL_2}{L_1 + L_2}$$

Decelerating at J m/s^2 where $z = J/g$:

$$N_1 = P\left[\frac{L_2}{L_1 + L_2} + \frac{zh}{(L_1 + L_2)}\right] = P_1 + \frac{Pzh}{E} \tag{3.2}$$

Note that:

$$N_1 + N_2 = P \quad \text{and} \quad L_1 + L_2 = E$$

The term P_1 is the static normal reaction at the front wheels, and Pzh/E is called the 'weight transfer' term, indicating the extra normal reaction transferred to the front axle (from the rear axle in this case) during braking.

Similarly, at the rear wheels:

$$N_2 = P\left[\frac{L_1}{L_1 + L_2} - \frac{zh}{(L_1 + L_2)}\right] = P_2 - \frac{Pzh}{E} \tag{3.3}$$

In this case P_2 is the static normal reaction at the rear wheels, and the weight transfer term Pzh/E is negative, indicating that weight is transferred from the rear axle (to the front axle). This basic analysis ignores other retarding forces on the vehicle such as aerodynamic drag and other sources of downward force, e.g. aerodynamic down thrust.

The magnitude of the weight transfer depends upon the vehicle's deceleration, so when the vehicle is not decelerating (rate of braking $z = 0$), the weight transfer term is zero. It also depends upon two vehicle design parameters, the height of the centre of gravity above the road surface (h) and the wheelbase (E), so a tall short vehicle will experience

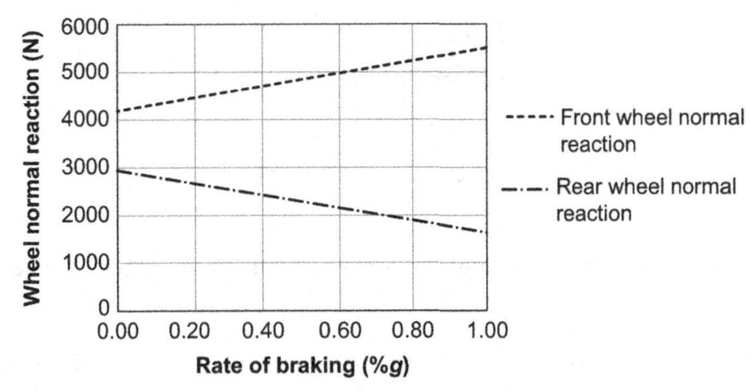

Figure 3.4: Effect of Braking Weight Transfer on Normal Reactions at the Front and Rear Wheels of a Passenger Car (Per Wheel).

more weight transfer to the front wheels while braking than a long low vehicle. The vehicle occupants might notice weight transfer by the extra compression deflection of the front road springs, i.e. the dip of the vehicle bonnet during braking. Suspension design can incorporate anti-dive geometry to reduce the dip effect, but the weight transfer is still there. Figure 3.4 shows an example of the calculated weight transfer during braking in a passenger car of the design specification summarised in Table 3.1, in the unladen condition. The 'unladen' condition usually includes the driver, and is often referred to as 'driver-only weight', DoW, while the 'laden' condition usually refers to the vehicle's 'gross vehicle weight', GVW.

Under normal conditions of vehicle braking, a retarding torque is generated at each rotating wheel by the brake at that wheel. The retarding torque from the brake is reacted by the product of the braking force at the tyre/road interface and the wheel and tyre rolling radius (see Chapter 6, Equation (6.1a)). As explained in Chapter 6, the amount of vehicle retardation and thus the braking force at each wheel is determined by the driver effort

Table 3.1: Example Passenger Car Design Specification

Design Parameter	Specification (DoW)	Specification (GVW)
Wheelbase, $E = L_1 + L_2$ (mm)	2750	2750
Centre of gravity height, h (mm)	500	575
Position of centre of gravity behind the front axle, L_1 (mm)	1130	1300
Vehicle mass, M (kg)	1450	1950
Vehicle weight, P (N) $= Mg = 1450 \times 9.81 =$	14,225	19,130
Front/rear braking ratio (X_1/X_2)	70/30	70/30

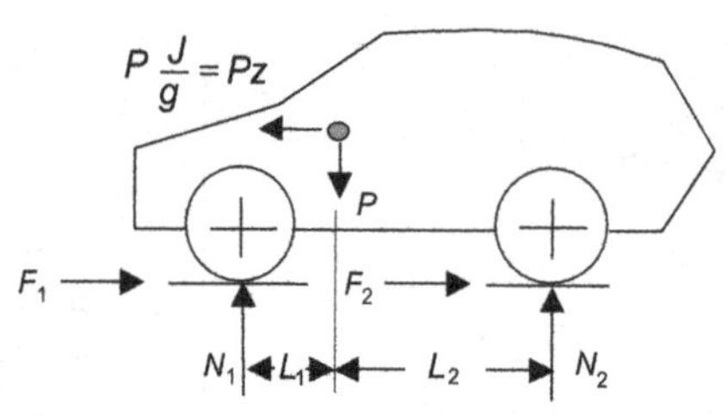

Figure 3.5: Vehicle Skidding (Sliding) on a Horizontal Road with all Wheels Locked.

(i.e. the force he/she presses with on the brake pedal). If the driver effort is large enough so that the retarding torque exceeds the tyre/road reaction torque, the wheel will lock (stop rotating) and if all the road wheels on the vehicle were locked and sliding, as illustrated in Figure 3.5, the vehicle would be in a skid condition and its deceleration would depend only on the sliding coefficient of friction (μ_t) between the tyres and the road. This is obviously an undesirable situation for two reasons: (a) the vehicle is uncontrollable, and (b) the kinetic energy is being dissipated through the tyre/road sliding friction and not rotor/stator sliding friction in the brakes. Tyres are not designed to dissipate kinetic energy by sliding friction and will quickly overheat and suffer serious damage if they skid. However, the basic weight transfer analysis still applies, as shown below.

Referring to Figure 3.5, where the vehicle is in dynamic equilibrium and all wheels are locked and sliding:

Resolving forces vertically:

$$P = N_1 + N_2 = Mg \tag{3.4a}$$

Newton's second law:

$$F_1 + F_2 = MJ \tag{3.4b}$$

Tyre/road sliding friction:

$$(F_1 + F_2) = \mu_t(N_1 + N_2) \tag{3.4c}$$

Hence, for all wheels sliding:

$$\frac{(F_1 + F_2)}{(N_1 + N_2)} = \mu_t = \frac{J}{g} = z \tag{3.4d}$$

Note that μ_t is the coefficient of sliding friction between the tyre and the road.

Weight transfer from the rear axle to the front axle still occurs in this case, i.e. Equations (3.1)–(3.3) still apply even though the wheels are locked and sliding. The difference is that the frictional work to decelerate the vehicle is being done by the tyres on the road, not by the brake friction material on the rotor as is intended. A decelerating vehicle should never be in the situation of having its wheels locked and sliding because the driver would then have no directional control over it so braking systems must be designed to avoid wheel lock as much as possible. As previously mentioned, modern road vehicles in Europe and many other countries are fitted with ABS and ESC to ensure that wheels do not lock under braking. However, in order to understand road vehicle braking system design, it is necessary to consider what would happen when the wheels on one or more of its axles were on the point of locking.

Referring to Figure 3.3, N_1 and N_2 define the normal (i.e. perpendicular to the road surface) dynamic contact forces between the tyre and the road. These forces ultimately determine the total vehicle braking force that can be transmitted through the tyres to the road, and hence the deceleration of the vehicle, and to achieve this the brake system must be designed to generate the appropriate brake forces at each wheel. This is termed the 'braking distribution' on the vehicle. Maximum vehicle deceleration can only be achieved when the tyre/road adhesion at all road wheels is utilised to the full, and this must be achieved within the constraints of legislative or manufacturers' standards (e.g. deceleration and wheel lock sequence), physical constraints (e.g. brake size vs. wheel size and space envelope), and operational requirements (e.g. acceptable wear in the brakes on all axles).

In operation, all vehicles experience lateral forces while driving. Wind, road camber and cornering introduce lateral forces that affect the contact forces between the tyre and the road at the wheels on each end of an axle. Thus, the normal reactions (contact forces) between the tyre and the road are not only affected by longitudinal weight transfer (along the X-axis) during braking but also by lateral weight transfer during cornering generated by 'roll' of the vehicle about the X-axis. The car illustrated in Figure 3.6 is driving towards the reader along a right-hand corner, where:

P is the vehicle weight acting through its centre of gravity (CG);
P_i and P_o are the normal reactions at the inner and outer wheels respectively (front wheels only shown) when the car is stationary;
N_i and N_o are the normal loads at each wheel (front wheels only shown) when the car is cornering.

Taking moments about the outside tyre contact patch:

Stationary car:

$$N_i t - P\frac{t}{2} = 0 \tag{3.5a}$$

$$\text{i.e. } N_i = N_o = \frac{P}{2} \tag{3.5b}$$

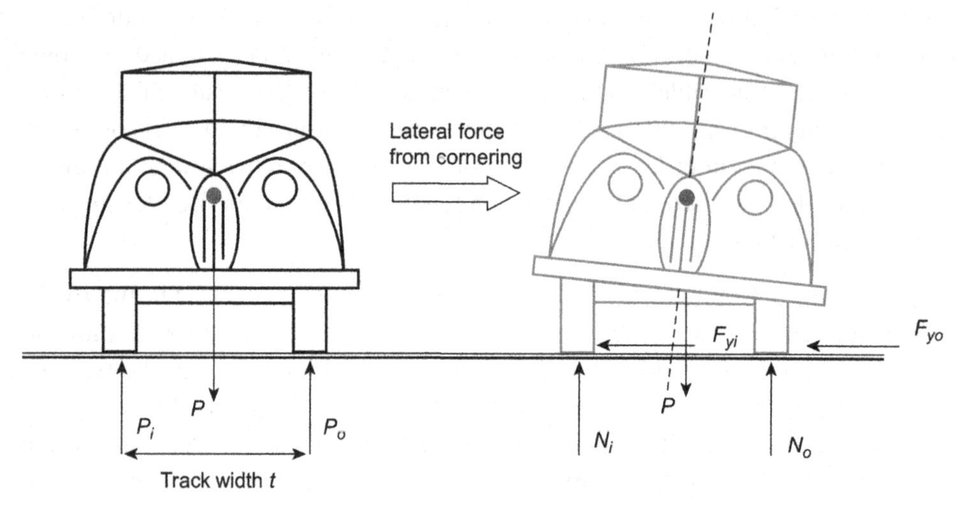

Figure 3.6: Car Cornering on a Horizontal Road.
(a) Stationary. (b) Cornering.

Cornering car:

The cornering force is $F_c = mv^2/R$, where v is the speed of the car (m s^{-1}) and R is the radius of the corner (m). Taking moments about the outside tyre contact patch:

$$N_i t - P\frac{t}{2} + hF = 0 \tag{3.6a}$$

$$\text{i.e. } N_i = \frac{P}{2} - \frac{hF}{t} \tag{3.6b}$$

$$\text{and } N_o = \frac{P}{2} + \frac{hF}{t} \tag{3.6c}$$

By extension from the previous analysis, the term $P/2$ is the static load on the wheels, and hF/t is called the 'cornering weight transfer' term, indicating the extra normal force transferred to the outer wheel (from the inner wheel) during cornering. When the vehicle is not cornering or subject to any other lateral force, this term is zero. The analysis ignores any correction for the roll angle (γ) of the car; the roll angle is determined by the vehicle's 'roll stiffness', which is a function of the suspension geometry and spring stiffness at each axle.

Figure 3.7 shows the lateral weight transfer (wheel normal reactions) calculated for the DoW car (DoW) specified in Table 3.1 with a track width of 1550 mm while cornering. Lateral weight transfer is one of the reasons why braking during cornering can be a potentially hazardous manoeuvre. During cornering, the outer wheel dynamic normal reaction is increased, while the inner is decreased. As a result, the outer wheel could be

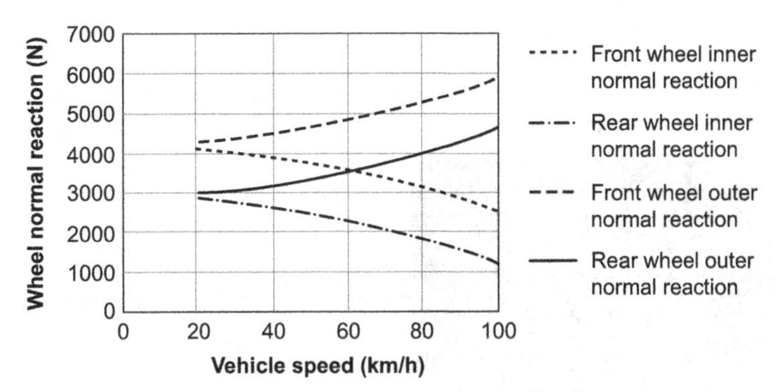

Figure 3.7: Effect of Lateral Weight Transfer on the Front and Rear Wheel Normal Reactions of a Passenger Car While Cornering.

operating below its adhesion limit, while the inner wheel could be operating above its adhesion limit, which could lead to wheel lock. The ability to adjust braking forces not only between the front and rear wheels but also from side to side of a vehicle through Electronic Brakeforce Distribution (EBD), has introduced safer braking in everyday vehicle usage and further advantages of vehicle stability control, as explained in Chapter 11.

Tyre/Road Adhesion

Tyres are fundamentally important to vehicle braking in that they provide the 'grip' between the vehicle and the road that enables the dynamic forces on the vehicle to be reacted by the road. If the tyres do not work properly, e.g. when the tyre is damaged or worn, or the road surface is contaminated, braking safety is compromised (as is the whole vehicle safety). So a basic understanding of tyre characteristics in terms of their grip on the road is a necessary part of understanding road vehicle braking. Wheel lock during braking, wheel spin during acceleration, and skidding off the road during enthusiastic cornering are regular features of excitement in motorsport events but are completely undesirable in everyday safe motoring. All of these are situations when the limit of the tyre's functionality has been exceeded and as a result it is unable to do its job properly, so in order to achieve safe driving on the road under all reasonable operating conditions, the vehicle designers make every effort to ensure that the limit of the tyre's functionality is not reached. However, many operational and environmental factors are not under the vehicle designer's control, e.g. slippery roads due to rain, snow or ice, and in these circumstances although there are active safety enhancements associated with the braking system on modern road vehicles, the skill and judgement of the driver must ultimately be relied upon to ensure that the vehicle is driven safely.

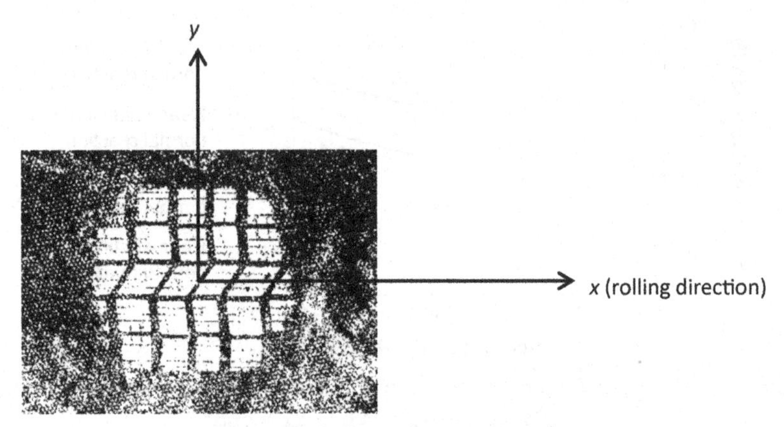

Figure 3.8: Tyre Contact Patch.

The tyre contacts the road over an area known as the 'contact patch'or the 'footprint' (Figure 3.8), which plays a major role in defining the tyre/road grip or adhesion of the tyre in use. Under radial and sideways loading the tyre acts as a spring, and the characteristics of the radial stiffness affect the size of the contact patch and the ride comfort of the vehicle. Under cornering, acceleration and braking forces, the contact patch will change its shape and size, but in radial-ply tyres under 'normal' driving conditions the magnitude of any changes is designed to be small, and this is one reason why the modern radial-ply tyre has superior grip compared to the old design of cross-ply tyre. Tyre load capacity is determined by tyre size and inflation pressure; a correctly inflated tyre will provide an even distribution of load across the footprint, thereby minimising tyre wear and rolling resistance, maximising durability, and most importantly maximising tyre/road grip. Low-profile tyres have been developed to allow larger wheel rim diameters to be used, which provide room for larger brakes in high-performance cars. These do not necessarily increase tyre/road grip, but, because they have higher radial and lateral stiffness, handling can be improved in terms of sideways compliance, although this is often associated with reduction in ride comfort because of the increased radial stiffness. (The 'profile' of a tyre is the ratio of the tyre radial width to the rim diameter.)

Rolling resistance, i.e. the resistance to motion when a tyre rolls on a surface, provides vehicle retardation without using the brakes, and thus should be minimised to avoid wasting energy (Passmore and Le Good, 1994). It is caused by the hysteresis in the deformation of the tyre and the road surface, the contribution from each depending on their physical characteristics. Additional factors that affect tyre rolling resistance include wheel radius, speed, tyre mass, type of construction, carcass and tread compound, road surface roughness, tyre temperature, inflation pressure, drive/braking torque, steer angle, camber, and toe of the vehicle. More compliant or 'loose' road surfaces, e.g. sand, gravel or snow, can make a large difference to tyre rolling resistance.

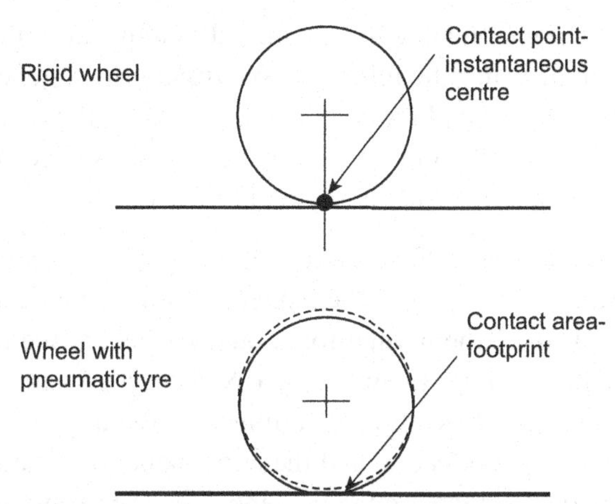

Figure 3.9: Instantaneous Centre and Contact Patch of a Rolling Wheel (The Dashed Circle Shows the Undeformed or Rigid Wheel Circumference).

Tyre/road grip or adhesion refers to the tangential force between the tyre and the road. When the wheel is rolling, the contact patch is an 'instantaneous centre' where conditions of static friction apply (Figure 3.9), i.e. there is no relative motion or sliding between the tyre and the road. It is therefore more accurate to use the term 'adhesion' rather than 'friction' to describe the tyre/road grip, since the overriding condition is that of static friction at an interface with no relative (sliding) motion. Effectively, at that instant, the contact patch on the tyre is adhering to the road surface, only to be peeled off as the wheel rotates. This is the condition that enables motor vehicle tyres to provide reliable and consistent tyre/road grip, because the contact patch can be peeled off only in the direction of rotation. It is also the condition that provides lateral stability in a rolling wheel; any lateral force is reacted by the adhesion force until that force reaches the limit of adhesion and then lateral skid would ensue under conditions of sliding friction. If the available adhesion force at the tyre/road interface is exceeded in the rolling direction, sliding friction occurs and the wheel would tend towards the 'locked' condition, i.e. not be rotating.

Detailed analyses of road vehicle tyre mechanics and dynamics can be found in an appropriate textbook (e.g. Pacejka, 2012), and only a simple overview of the physics of tyre/road adhesion is given here. Among the many physical mechanisms governing the grip between the tyre and the road the two most important are considered to be:

- Adhesion forces generated by intermolecular bonding between the rubber and the aggregate in the road surface.
- Mechanical interlocking as the tyre rubber compound deforms locally to accommodate the topography of the road surface. The deformation associated with this interlocking is associated with hysteresis in the tyre constituent materials and construction, and is the source of consequent energy loss.

On dry roads the adhesion component is the larger, but is reduced in the presence of road surface contaminants such as water; therefore the tyre/road grip is generally lower on wet roads. Adhesion between the tyre and the road is influenced by the thermophysical properties of the tyre tread material and the physical characteristics of the road surface type and condition, an example of which is shown in Figure 3.10.

Tyre/road adhesion is generally specified in terms of an 'adhesion coefficient', which is the maximum or limiting value of the ratio of the tangential force generated by a road wheel that is rolling (i.e. not locked) to the normal force on the wheel at the tyre/road interface. European (UN) Braking Regulations 13 and 13H (UN, Mar 2014 and UN, Feb 2014) state that 'the coefficient of adhesion (k) shall be determined as the quotient of the maximum braking forces without locking the wheels and the corresponding dynamic load on the axle being brake (d).' Any attempt to increase the tangential force beyond the maximum or limiting value of adhesion coefficient will result in wheel lock, as explained below. Modern tyre/road combinations can give an adhesion coefficient of 0.85 or higher on a clean dry road surface, but this may reduce to 0.3 or lower on a worn, wet or contaminated road surface, while an icy road can have an adhesion coefficient of 0.2 or lower.

Braking Force and Wheel Slip

The brake on a road wheel creates retarding torque by the generation of friction force between the rotor and the stator. This retarding torque is then reacted by the tangential

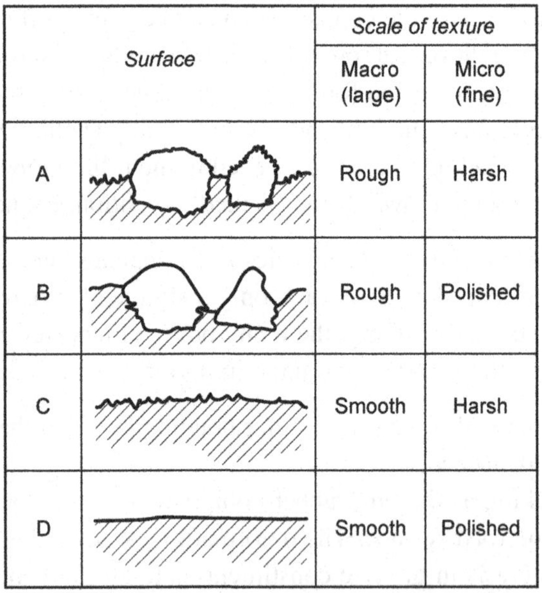

Surface		*Scale of texture*	
		Macro (large)	Micro (fine)
A		Rough	Harsh
B		Rough	Polished
C		Smooth	Harsh
D		Smooth	Polished

Figure 3.10: Road Surface Texture.

force at the tyre/road interface. The nature of rubber friction is that the tangential force is generated by the successive formation and release of adhesion over macroscopic regions of the tyre's rubber compound in the tread. Thus, tangential force in a tyre is always associated with the microscopic movement of the tyre rubber compound relative to the road surface, a phenomenon known as 'wheel slip', as illustrated in Figure 3.11.

Wheel slip is defined in terms of the ratio of the actual speed of rotation of the road wheel to the free-rolling speed of rotation, as in Equation (3.7):

$$\text{Wheel slip} = (V - \omega r_r)/V \tag{3.7}$$

where:

V = vehicle forward speed (m/s)
ω = tyre rotational speed (rad/s)
r_r = rolling radius of the tyre (m).

If a road vehicle is braked in the 'straight-ahead' condition, wheel slip increases as the deceleration increases under increased driver effort. This is not a linear relationship, and ultimately a limit to the available braking force is reached, where the slip is 100%, which is the condition of wheel lock and skid. The term 'braking force coefficient' (BFC) can be used to represent the instantaneous ratio between the retarding force and the normal reaction force at the tyre/road interface:

$$\text{BFC} = \text{Braking force on the wheel/Dynamic wheel load} \tag{3.8}$$

The limit of the BFC is the tyre/road 'adhesion coefficient', which is illustrated in the example of BFC vs. wheel slip in Figure 3.12, showing how the BFC increases to a peak value (point C) and then decreases to a lower value at wheel lock (point D, 100% slip).

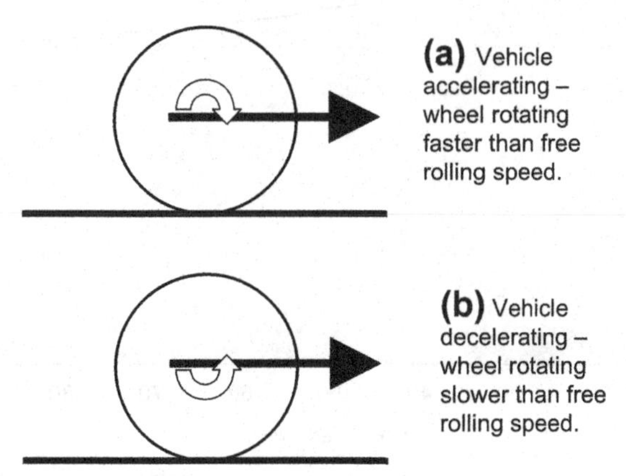

Figure 3.11: Direction of Wheel Slip in Acceleration (a) and Braking (b).

The BFC vs. wheel slip characteristic shown in Figure 3.12 is an idealised example, and different tyres, loading conditions and road surfaces will all show different characteristics, although the points A–D are usually clearly identifiable. These points correspond to:

A. Free rolling (no braking)
B. Stable slip zone
C. Peak adhesion
D. Wheel lock.

This diagram indicates that at any value of slip, the corresponding maximum BFC value represents the limiting value of tyre/road adhesion at that point. The maximum value of BFC at point C is known as the peak adhesion and occurs at only one value of slip. This value is often quoted (and is used as such in this book) as the tyre/road adhesion coefficient, for which the symbol 'k' is used, and therefore it represents the limit at which it is no longer possible to increase the braking force transmitted to the road by the tyre. The maximum braking force at this one wheel would therefore be determined by the relationship:

$$T_i = kN_i \tag{3.9}$$

At point D the wheel is completely locked, the wheel slip is 100% and the braking force coefficient is represented by the symbol μ_t, which represents the dynamic (sliding) coefficient of friction between the tyre and the road and is significantly lower than k.

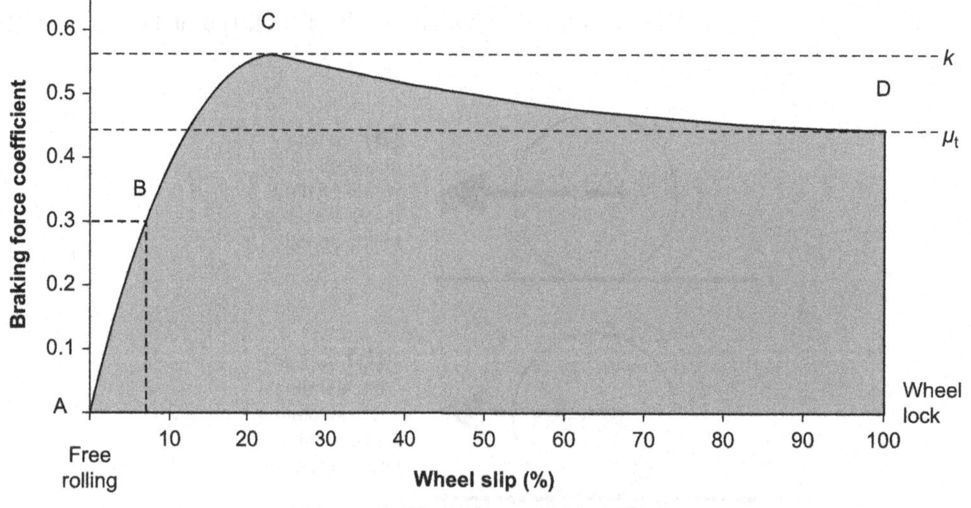

Figure 3.12: BFC vs. Wheel Slip.

The maximum dynamic braking force at this one wheel would therefore be determined by the relationship:

$$F_i = \mu_t N_i \tag{3.10}$$

Clearly $F_i \leq T_i$ since $\mu_t \leq k$.

The region A–B–C is termed the 'stable zone' where it is possible to transmit the developed braking force to the road, and the region C–D is termed the 'unstable zone' where it is no longer possible to transmit the developed braking force to the road. Once a braking force has been generated that brings the wheel to peak adhesion (point C), any attempt to increase braking force further will result in increased slip and the wheel will enter the unstable section of the slip curve (C–D). Because it is not possible to transmit any increased braking force to the road surface (because the BFC is decreasing) the wheel will decelerate rapidly to point D. The same effect would be produced if the dynamic normal load on the wheel reduced. It is therefore not possible for a wheel to remain at a constant slip level in the section C–D. Thus, to prevent a wheel from locking or remaining locked it is necessary to reduce the braking force, which will enable the wheel to accelerate and reduce the amount of slip generated to once again maintain the wheel within the stable slip zone (A–C). Achieving and maintaining a wheel at the peak adhesion (point C) is effectively impossible because of:

- Changes in the BFC vs. wheel slip curve due to variations in the road surface characteristics
- Brake hysteresis where changes in brake actuation force do not result in a corresponding reduction in braking force
- Dynamic changes in wheel normal forces due to dynamic road force inputs.

These considerations have been taken into account in the development of ABS technology (see Chapter 11).

Braking and Lateral Forces in Tyres

Road vehicle tyres have to operate under combined conditions of longitudinal and lateral forces at the tyre/road interface. A tyre can only develop a lateral (or sideways) force, i.e. a force parallel to the wheel axle, by a lateral deformation of the contact patch, which in turn is associated with 'lateral slip'. Thus, for the wheel to roll straight ahead while at the same time reacting to a lateral force, the tyre must run at a 'slip angle', as illustrated in Figure 3.13.

Any change in the magnitude of the lateral force would change the slip angle, but if the lateral force were to continue to increase, eventually the lateral reaction force would reach

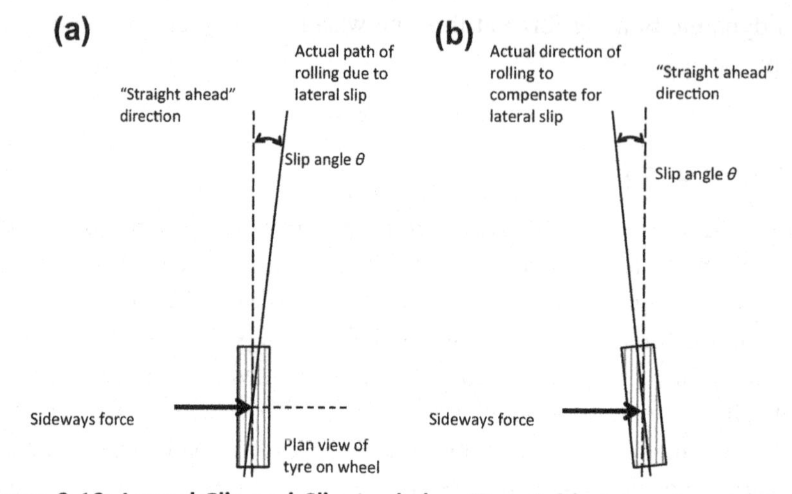

Figure 3.13: Lateral Slip and Slip Angle in a Tyre Subject to a Lateral Force.
(a) A lateral force causes the wheel to roll at an angle to the required ('straight ahead') direction. (b) The wheel is steered at the slip angle to compensate for the lateral slip caused by the lateral force.

the limit of adhesion between the tyre and the road at the contact patch. At this limit, no further increase in sideways force can be developed and thus where the required cornering force exceeds the available lateral force, a skid would result. The term 'sideways force coefficient' (SFC, often termed 'cornering coefficient' in the USA) is often used to represent the (empirically derived) relationship between the lateral force and the normal load on the tyre:

$$\text{SFC} = \text{Lateral force/Dynamic wheel load} \qquad (3.11)$$

The SFC vs. slip angle graph shown in Figure 3.14 represents the limit of the available tyre/road adhesion under lateral force loading. The way this graph should be interpreted is that at any value of slip angle, the corresponding SFC value represents the limiting value of tyre/road adhesion at that point. The maximum value of SFC again represents the peak adhesion that occurs at only one value of slip angle. This may be the same as the tyre/road adhesion coefficient (k) previously quoted for longitudinal braking, but is not necessarily the same, and needs to be evaluated for individual tyre/road combinations. It represents the point at which it is no longer possible to increase the lateral force transmitted to the road by the tyre, and the maximum sideways force (T_{yi}) at this wheel would therefore be determined by the relationship:

$$T_{yi} = kN_i \qquad (3.12)$$

Beyond point C the wheel is skidding sideways, and the SFC again represents the dynamic (sliding) coefficient of friction between the tyre and the road. The maximum lateral

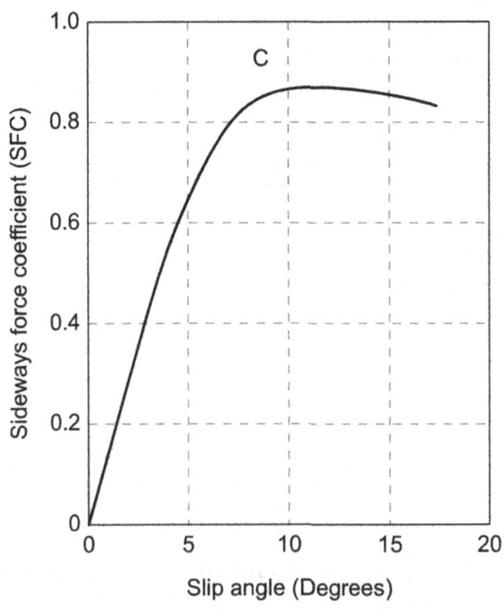

Figure 3.14: SFC vs. Slip Angle.

dynamic friction force at this one wheel would therefore be determined by the relationship:

$$F_{yi} = \mu_t N_i \tag{3.13}$$

$F_{yi} \leq T_{yi}$ since $\mu_t \leq k$.

Under combined conditions of braking and lateral forces, e.g. cornering, the resultant force generated at a tyre/road interface is a combination of the longitudinal and lateral forces. A useful method to estimate the maximum values of the braking (longitudinal) and the lateral forces is based on the assumption that since the limit of adhesion is essentially the transition from static to dynamic friction and the latter is independent of the apparent contact area, the value of μ_t (sliding coefficient of friction between the tyre and the road) is the same in any direction. Hence the vector addition of lateral and braking forces on any road wheel (T_{xi} and T_{yi}) would give the resultant force, which can only be reacted by the product of the dynamic normal force and the coefficient of adhesion k:

$$\mu_t N_i \ll \sqrt{T_{xi}^2 + T_{yi}^2} \tag{3.14}$$

If the resultant of the longitudinal (braking) force and the lateral force demands are too high, a skid will ensue. Figure 3.15 illustrates an idealised example of the relationship between braking and cornering, showing how the limit of the available BFC reduces as

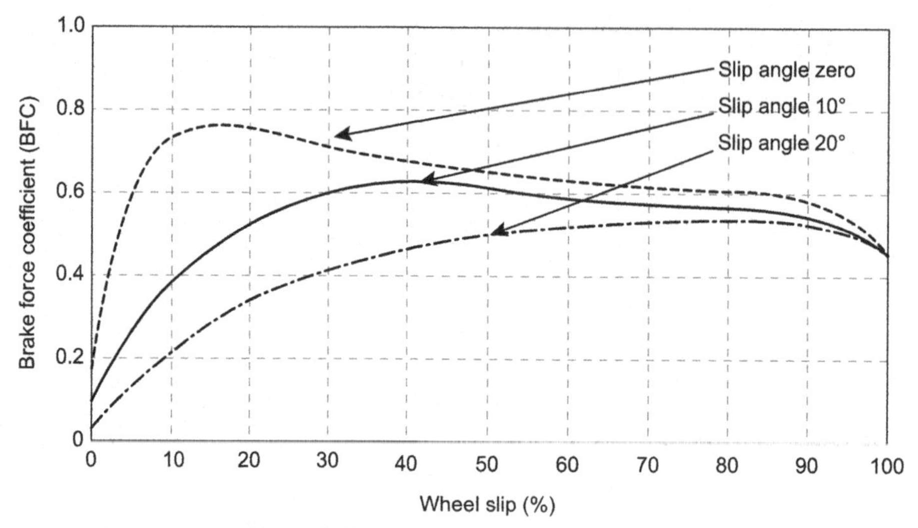

Figure 3.15: Effect of Slip Angle on BFC—Wheel Slip Relationship.

lateral force and thus the slip angle increase. Mathematical models (formulae) are available (Pacejka, 2012) for use in CAE simulation and prediction of vehicle braking and cornering.

Brake Force Distribution

As explained earlier, the normal forces between the tyre and the road can be calculated from the static weight distribution of the vehicle, any weight transfer due to braking or acceleration, and any lateral weight transfer due to cornering. It is assumed that the road surface is horizontal unless stated otherwise, so that normal forces are perpendicular to the road surface. There are also dynamic forces input to the vehicle from the road surface but these are ignored for the purposes of the analyses presented here. The braking weight transfer has been determined (Equations (3.2) and (3.3)) from the total mass of the vehicle, its rate of braking, wheelbase and centre of gravity height above the road surface. In real life a road vehicle is moving and the normal forces at each wheel are always dynamic, but the basic theory of braking starts from the quasi-static approach, which is continued here. The analysis assumes no ABS or ESC operation.

If wheel lock does not occur, the braking force generated at each tyre/road interface (in the longitudinal direction) is determined by the specification of the brake on each wheel, as set out in Chapters 5 and 6. As discussed above, wheel lock will occur if the magnitude of the longitudinal brake force on any wheel exceeds the maximum that the tyre/road interface can sustain. Maximum brake force is developed just at the critical point (C in Figure 3.12), so it follows that for the maximum vehicle deceleration or rate of braking to be achieved on any given road surface, all wheels must simultaneously approach the point of locking. Referring again to Figure 3.2, which shows a four-wheeled (two-axle rigid) passenger car driving on a horizontal road with all wheels braking but not locked, then according to Newton's Second Law:

$$\text{Total braking force} = \text{Mass} \times \text{Deceleration} = \frac{PJ}{g} = Pz \tag{3.15}$$

The vehicle total braking force is the sum of the braking torque at each individual road wheel and is reacted by the tyre/road adhesion at each tyre/road interface, so that the total retarding force on the vehicle is:

$$Pz = T_1 + T_2 \tag{3.16}$$

The maximum braking force on the vehicle depends upon the tyre/road coefficient of adhesion k:

$$Pz_{max} = kP \tag{3.17}$$

So $z_{max} = k$ when all wheels are simultaneously at the point of locking.

Since this represents the case of maximum vehicle deceleration for a given set of road/tyre conditions, the maximum rate of braking z in those conditions therefore cannot exceed the adhesion coefficient k unless the downforce from the vehicle weight P is augmented, e.g. by aerodynamic effects. The following analysis (based on the bicycle model with symmetrical brake forces and dynamic tyre loadings about the vehicle centreline) assumes that the only downward force on the vehicle is its weight.

For all wheels to be on the point of locking simultaneously, the braking force at each wheel must be in proportion to the dynamic normal reaction at that wheel, so:

$$T_1/T_2 = N_1/N_2 \tag{3.18}$$

The ratio of the braking force generated by the front wheels to the braking force generated by the rear wheels of a two-axle rigid road vehicle such as a passenger car can be specified as the ratio X_1/X_2, where X_1 and X_2 are the proportion of the vehicle's total braking force generated at the front and rear axles respectively. By definition:

$$T_1/T_2 = X_1/X_2 \text{ and } X_1 + X_2 = 1 \tag{3.19a}$$

because:

$$T_1 + T_2 = PzX_1 + PzX_2 = Pz(X_1 + X_2) = Pz \tag{3.19b}$$

So the total braking force on this vehicle where all wheels are on the point of locking simultaneously is made up of $T_1 = X_1Pz$, and $T_2 = X_2Pz$, and by extension for a vehicle with 'n' axles: $T_n = X_nPz$.

As previously explained there is dynamic weight transfer between the axles during braking, so the dynamic normal forces on each road wheel vary with the rate of braking. Thus, a variable brake distribution ratio (X_1/X_2) would be required to give ideal braking under all driving conditions of the vehicle. Equations (3.2) and (3.3) previously showed that the dynamic normal reactions at the front and rear wheels during braking are:

$$N_1 = P_1 + \frac{Pzh}{E} \quad \text{and} \quad N_2 = P_2 - \frac{Pzh}{E} \quad \text{respectively}$$

Taking the case of a four-wheel, two-axle rigid vehicle with a fixed brake distribution:

$$X_1/X_2 = \text{constant}$$

which is designed so that the front wheels will lock first on a surface of tyre/road coefficient of adhesion k, then the braking force at the front axle when the front wheels are about to lock is given by:

$$T_1 = kN_1 = k\left\{P_1 + \frac{Pzh}{E}\right\} \tag{3.20}$$

The corresponding braking force at the rear axle is less than the limiting adhesion value because the braking system has been designed to lock the front wheels first. Since $T_1/T_2 = X_1/X_2$, therefore $T_2 = T_1(X_2/X_1)$ and

$$T_2 = k\left\{P_1 + \frac{Pzh}{E}\right\}\frac{X_2}{X_1} \tag{3.21}$$

Then the total braking force on the vehicle is:

$$T_1 + T_2 = Pz = k\left\{P_1 + \frac{Pzh}{E}\right\} + k\left\{P_1 + \frac{Pzh}{E}\right\}\frac{X_2}{X_1} = k\left(1 + \frac{X_2}{X_1}\right)\left\{P_1 + \frac{Pzh}{E}\right\} \tag{3.22}$$

i.e.

$$Pz = k\left(\frac{X_1 + X_2}{X_1}\right)\left\{P_1 + \frac{Pzh}{E}\right\} = k\left(\frac{1}{X_1}\right)\left\{P_1 + \frac{Pzh}{E}\right\} \tag{3.23}$$

since $X_1 + X_2 = 1$.

Rearranging this equation gives the rate of braking z when the front wheels are about to lock:

$$z = \frac{k}{P}\left(\frac{1}{X_1}\right)\left\{P_1 + \frac{Pzh}{E}\right\} = \frac{k}{P}\left(\frac{P_1}{X_1} + \frac{Pzh}{EX_1}\right) \tag{3.24}$$

Hence:

$$z\left(1 - \frac{kh}{EX_1}\right) = \frac{kP_1}{PX_1} \tag{3.25}$$

And therefore:

$$z = \frac{kP_1}{PX_1}\left(\frac{EX_1}{EX_1 - kh}\right) = \frac{kEP_1}{P(EX_1 - kh)} \tag{3.26}$$

A similar analysis for the rear wheels about to lock gives:

$$z = \frac{kEP_2}{P(EX_2 + kh)} \tag{3.27}$$

As previously explained, the rate of braking z of the vehicle cannot exceed the value of the adhesion coefficient k. Equation (3.20) (i.e. front wheels locking) therefore holds up to the point where $z = k$, depending upon the braking ratio X_1/X_2. This is the value at which both front and rear wheels of the vehicle are on the point of locking simultaneously. For values of k greater than this, the calculation of z from Equation (3.20) will yield z being greater than k, which is not possible. Therefore, braking will then be limited by the rear wheels locking and Equation (3.21) will be applicable. This is illustrated in Figure 3.16 for the passenger car example previously used (Table 3.1) with the front/rear braking ratio specified as $X_1/X_2 = 70/30$.

Figure 3.16 illustrates the lowest value of adhesion coefficient that is theoretically required for the vehicle specified in Table 3.1 to achieve a particular rate of braking using Equations (3.26) and (3.27). Up to the point of simultaneous front and rear wheel lock, i.e. where the front and rear lines cross over, the front wheels require higher adhesion coefficient and therefore will lock before the rear wheels. Above the crossover point, the

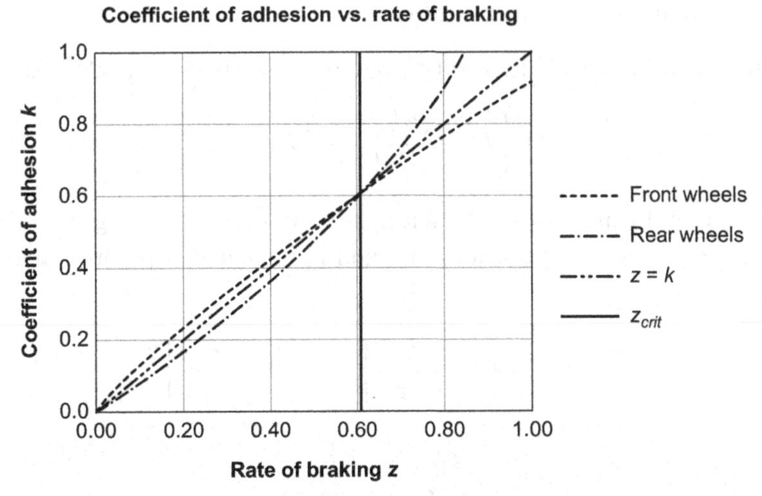

Figure 3.16: Limiting Rate of Braking (z) for Specified Adhesion Coefficient (k) for Vehicle Specified in Table 3.1.

rear wheels require a higher adhesion coefficient and therefore will lock before the front wheels. The rate of braking at simultaneous front and rear wheel lock is defined as z_{crit}, where $z_{crit} = k$, and therefore k can be substituted for z in Equations (3.20) and (3.21) to give Equations (3.28) and (3.29). The parameter z_{crit} is a useful brake system design parameter, as explained and used in Chapter 6.

$$z_{crit} = k = \left(X_1 - \frac{P_1}{P}\right)\frac{E}{h} \tag{3.28}$$

$$z_{crit} = k = \left(\frac{P_2}{P} - X_2\right)\frac{E}{h} \tag{3.29}$$

Note that for simultaneous wheel locking to be possible, X_1 must be greater than the proportion of static normal reaction at the front wheels (P_1/P). Similarly X_2 must be less than the proportion of static normal reaction at the rear wheels (P_2/P). On small passenger cars and light vans, the front and rear axle dynamic normal reactions can change significantly over the range of permissible loading conditions from DoW to GVW, so simultaneous locking may be possible at some loading conditions but not others if X_1/X_2 is fixed.

When one or more wheels on a vehicle have passed the point of locking, the tyre(s) will be skidding on the road and the deceleration of the vehicle will decrease because the brake force at the locked wheel(s) is reduced by the difference between μ_t and k. In practice, if it is the front wheels that lock first the driver may not notice the small contribution from the rear wheels but if the driver continues to increase pedal effort, the unlocked wheels will deliver increased brake force until those wheels also lock, and then the driver would notice a further reduction in the vehicle deceleration.

The equation for the front wheels locked (μ_t) but the rear wheels at peak adhesion k is:

$$Pz = \mu_t\left\{P_1 + \frac{Pzh}{E}\right\} + k\left\{P_2 - \frac{Pzh}{E}\right\} \tag{3.30}$$

To solve this equation it can be assumed that μ_t is proportional to k, e.g. $\mu_t = 0.7k$. This assumption is based on the relative values of k and μ_t indicated at points C and D in Figure 3.12.

$$Pz = 0.7k\left\{P_1 + \frac{Pzh}{E}\right\} + k\left\{P_2 - \frac{Pzh}{E}\right\} \tag{3.31}$$

This yields:

$$z = k\left\{\frac{0.7P_1 + P_2}{1 + 0.3k\frac{h}{E}}\right\} \tag{3.32}$$

Based on the previous example, the effect of front wheel lock at $k = 0.50$ would be to reduce the vehicle deceleration from $z = 0.49$ with the front wheels on the point of lock (i.e. they are still rotating) to $z = 0.40$ with the front tyres skidding. The alternative situation where the rear tyres are skidding (μ_t) while the front wheels are at peak adhesion (k) is too unstable to be of practical interest, as discussed next.

Wheel Lock and Vehicle Stability During Braking

So far, the principles of vehicle braking have been explained based on the basic design of a braking system for a four-wheel, two-axle rigid vehicle such as a passenger car or a light van. It has been shown how and why wheel lock can occur (in the absence of ABS and EBD), and although ABS is now a normal fitment to road vehicles in Europe and many other countries for safety reasons, there are still potential conditions under which wheel lock could occur, e.g. ABS failure. As previously explained, when one wheel on an axle locks, that wheel loses directional stability. However, if the wheels on an axle or on a set of close-coupled axles (for reasons of load capacity as with heavy commercial vehicles; see Chapter 4) are not all locked, a wheel that is still rolling will provide some directional stability to that axle. But when all the wheels on one axle or a close-coupled set of axles are locked, the vehicle's stability is compromised by the loss of directional stability from those wheels.

If both wheels on the front axle of a two-axled rigid vehicle are locked, the vehicle cannot be steered because there is no preferential direction of motion for a sliding tyre; it simply moves in the direction it is pushed, i.e. the instantaneous direction of vehicle travel. The driver is therefore faced with the choice of continuing in that direction, or releasing the pedal force sufficiently to allow the front wheels to rotate and restore directional control from steering input. The practice of 'pumping the brakes' or 'cadence braking' in earlier generations of non-ABS road vehicles involved the driver's cyclically releasing and reapplying the brake pedal. Given that wheel lock is usually associated with dangerous driving conditions, this required confidence and great presence of mind by a highly capable driver. But in the condition of front wheel lock, even though the driver has no steering control, the vehicle is at least still moving in the forward direction.

If both wheels on the rear axle are locked, the vehicle becomes unstable, and a 180° yaw about the vertical Z-axis occurs. This is particularly disorientating for the driver because it happens very quickly with the result that the vehicle continues in its original direction but facing backwards. At this stage the driver has little opportunity to recover from a very dangerous 'out-of-control' situation, and this is the reason why braking legislation in Europe prior to the general adoption of ABS required vehicle braking system designs to demonstrate preferential front wheel lock within a specified range of deceleration and adhesion (z and k).

The design of braking systems for other types of road vehicle, e.g. commercial vehicles with trailers, is covered in Chapter 4. The analysis of vehicle dynamics under wheel lock conditions is a very interesting field of modern computer modelling and simulation that is beyond the scope of this book.

Braking Efficiency

Braking efficiency is a measure of the use which a vehicle's braking system is able to make of the available coefficient of adhesion. Since it has already been shown that z cannot exceed k, braking efficiency can be defined as:

$$\eta = \frac{z}{k} = \frac{J}{gk} \tag{3.33}$$

At conditions where simultaneous locking of all the road wheels on the vehicle occurs:

$$z = z_{crit} = k, \quad \text{so} \quad \eta = 1.0 \tag{3.34}$$

For the case of a two-axle rigid vehicle, at the condition where the front wheels are on the point of lock, and the rear wheels are braked but still rotating, Equation (3.20) gives:

$$\eta = \frac{z}{k} = \frac{EP_1}{P(EX_1 - kh)} = \frac{P_1}{P\left(X_1 - \frac{kh}{E}\right)} \tag{3.35}$$

Similarly, for the condition where the rear wheels are on the point of lock, and the front wheels are braked but still rotating:

$$\eta = \frac{P_2}{P\left(X_2 + \frac{kh}{E}\right)} \tag{3.36}$$

These two equations yield different values of braking efficiency so it should be understood that the braking efficiency can never exceed unity, i.e. $\eta \leq 1.0$. So if one equation yields a value for braking efficiency which is >1, then the other equation should be used. This is illustrated below for the example vehicle specified in Table 3.1.

Figure 3.17 shows a graph of braking efficiency against adhesion coefficient k for the example vehicle. Assuming no lateral differences in tyre/road normal reaction and braking force:

- In the DoW condition maximum braking efficiency occurs at $k = 0.61$ when $z = z_{crit} = k$, and the front and rear wheels are on the point of locking simultaneously as indicated in Figure 3.16. Below $k = 0.61$, the front wheels will lock first and above $k = 0.61$ the rear wheels will lock first.
- In the GVW condition maximum braking efficiency occurs at $k = 0.83$ when $z = z_{crit} = k$, and the front and rear wheels are on the point of locking simultaneously. Below $k = 0.83$, the front wheels will lock first and above $k = 0.83$ the rear wheels will lock first.

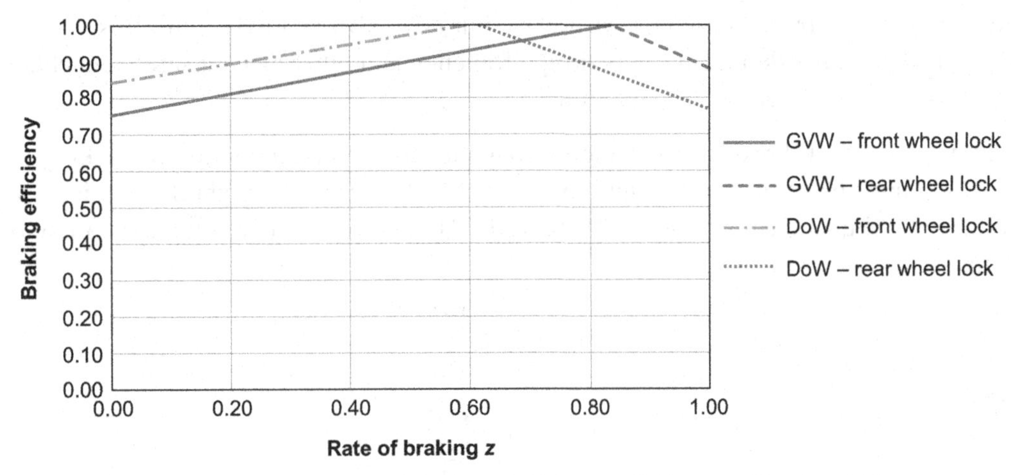

Figure 3.17: Comparison of Braking Efficiency, Vehicle at GVW and DoW.

The design of the braking system on this example vehicle in terms of the braking distribution X_1/X_2 is reasonably satisfactory because the value of z_{crit} across the range of loading conditions (DoW to GVW) is above 0.6, which many car manufacturers view as a sensible system design target in the basic braking system design, i.e. without ABS or EBD (see Chapter 6). Before the implementation of ABS on passenger cars and light vans, the UN Legislation required that for all states of load of the vehicle, the adhesion utilisation curve of the rear axle should not be above that for the front axle for rates of braking between 0.15 and 0.8, and this still applies to non-ABS vehicles, (see Chapter 8). In the event of an electrical functional failure that only affects the ABS operation, a mean fully developed deceleration (MFDD; see Chapter 8) > 5.15 m/s^2 is required (equivalent to a rate of braking of $z = 0.525$).

The braking distribution specified as $X_1/X_2 = 70/30$ gives reasonably good braking efficiency between $0.40 \leq z \leq 0.80$ for the DoW vehicle, and between $0.50 \leq z \leq 0.95$ for the GVW vehicle. Since the maximum rate of braking z cannot exceed the coefficient of adhesion k, this means that this vehicle would have reasonably good braking efficiency under higher deceleration conditions on good road surfaces. But on lower adhesion surfaces the braking efficiency is lower, which means that the driver effort (i.e. the pedal force) would need to be higher to achieve the same deceleration than if the braking efficiency were higher. Road vehicles generally cannot achieve an acceptable compromise between stability and efficiency with a fixed braking distribution ($X_1/X_2 =$ constant), so in order to improve the braking efficiency over a greater range of vehicle loading and usage conditions, the braking ratio X_1/X_2 needs to be designed to be variable. In the past this was partially achieved by the incorporation of valves to adjust the hydraulic actuation pressure to the rear brakes, but now ABS and EBD fulfil this function and can provide maximum braking efficiency and other driving safety benefits over the full range of vehicle deceleration.

To achieve the 'ideal' braking distribution through a variable braking ratio, the braking distribution must be continuously adjustable to match the dynamic weight distribution for all values of z. Equations (3.2) and (3.3) showed that for a vehicle decelerating at a rate of braking z:

$$N_1 = P_1 + \frac{Pzh}{E} \tag{3.2}$$

$$N_2 = P_2 - \frac{Pzh}{E} \tag{3.3}$$

The variable front to rear braking ratio for a two-axle, four-wheel rigid vehicle such as a passenger car is defined as X_1^{var}/X_2^{var}, and this would ensure that the front and rear wheels of the vehicle lock simultaneously under all conditions of loading and usage. This would represent the best case of 100% vehicle braking efficiency where $z = k$ and the maximum deceleration of the vehicle is limited only by the tyre/road adhesion coefficient k:

$$\frac{X_1^{var}}{X_2^{var}} = \frac{P_1 + \frac{Pzh}{E}}{P_2 - \frac{Pzh}{E}} \quad \text{for all values of } z \tag{3.37}$$

or

$$X_1^{var} = \frac{N_1}{P} = \frac{P_1}{P} + \frac{zh}{E} \tag{3.38}$$

and

$$X_2^{var} = \frac{N_2}{P} = \frac{P_2}{P} - \frac{zh}{E} \tag{3.39}$$

Figure 3.18 shows the'ideal'front/rear braking ratio X_1^{var}/X_2^{var} for the example vehicle, which indicates that at low rates of braking the ideal braking ratio is close to the static weight distribution between the front and rear axles (ignoring lateral variations in wheel loading). The fixed braking ratio specified in the example is 70/30 ($X_1/X_2 = 2.33$), so at low rates of braking the fixed braking ratio is lower than ideal, and at higher rates of braking it is higher than ideal. A complication is that the adhesion coefficient k for the different wheels may not be the same because of different types or makes of tyre, wear, normal forces, and inflation pressure, etc. Ideal braking distribution is implemented on modern cars via some form of Electronic Braking Distribution (EBD) with ABS to sense wheel or axle loadings to follow a 'map' of ideal braking ratio for the vehicle. The ABS would respond to local differences around the limit of adhesion for individual wheels.

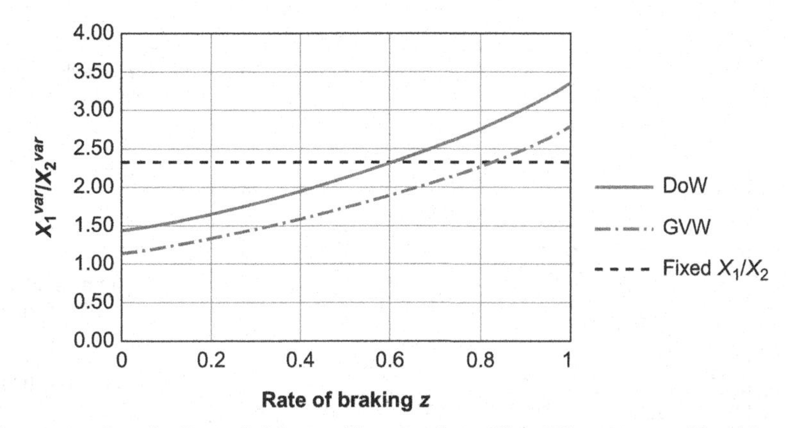

Figure 3.18: 'Ideal' Variable Braking Ratio X_1^{var}/X_2^{var} vs. rate of braking z.

Adhesion Utilisation

UN Regulations 13 and 13H (UN, Mar 2014, UN, Feb 2014) define adhesion utilisation (f_i) as the ratio of the braking force on the axle to the dynamic normal reaction on the axle, i.e.

$$f_i = T_i/N_i \tag{3.40}$$

The adhesion utilisation is therefore the minimum coefficient of adhesion required by a particular axle to achieve a given rate of braking. The theoretical braking performance of a road vehicle can be illustrated by a graph of the same form as Figure 3.16, where the y-axis represents the adhesion utilisation instead of the adhesion coefficient. This is illustrated in Figure 3.19, which shows the adhesion utilisation at each axle (ignoring any lateral variation) versus the rate of braking z.

For a fixed braking ratio X_1/X_2:

$$f_1 = \frac{T_1}{N_1} = \frac{X_1 zP}{P_1 + \frac{Pzh}{E}} \tag{3.41}$$

$$f_2 = \frac{T_2}{N_2} = \frac{X_2 zP}{P_2 - \frac{Pzh}{E}} \tag{3.42}$$

The lines in Figure 3.19 indicate which axle will lock first for any particular road surface and operating conditions as defined by the prevailing adhesion coefficient k. The adhesion utilisation f_1 or f_2 cannot exceed k so the axle with the higher adhesion utilisation will be the one that locks first. So, if the value of k is 0.5, the front wheels will lock first for both the DoW and GVW conditions, with peak rates of braking z being

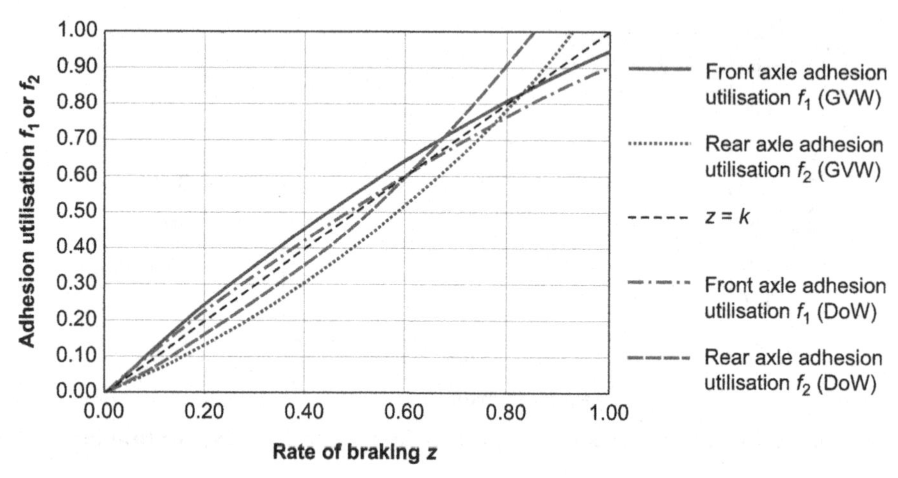

Figure 3.19: Adhesion Utilisation for Example Vehicle.

approximately 0.48 and 0.44 respectively. This indicates that the braking efficiency is less than 100% at these two conditions; the actual value of braking efficiency can be calculated as indicated by Equation (3.33) and Figure 3.17, namely 0.97 and 0.89 respectively. The two curves f_1 and f_2:z for any given loading condition intersect on the line $k = z$, which as previously explained is the only point at which the braking efficiency reaches 100%. Large separation of the adhesion curves from the ideal $k = z$ indicates poor braking efficiency.

Where ABS (and EBD) are standard fitment on passenger cars and light vans in Europe, there are several control options for the brake system designer to consider to maximise the braking efficiency over the full range of loading and operational conditions, of which two are considered here. If some form of load sensing is provided on the vehicle to indicate the front and rear wheel dynamic normal forces, often termed 'axle loadings' (the associated technology is not covered here), this information can be used as follows.

- Option 1. The front/rear braking ratio X_1/X_2 could be adjusted by EBD to take account of the static loading conditions for each journey, and then maintained constant until the journey is complete. Wheel lock would be managed by the ABS. Referring again to the example vehicle, it can be seen that the braking efficiency in the GVW condition is lower than in the DoW condition. So if the GVW condition were detected in the example car, X_1/X_2 could be adjusted from the fixed value of 70/30 to, for example, 66/34. At $k = 0.5$ the peak rates of braking z would be 0.47 increased from 0.44 and the braking efficiency would be 0.95 increased from 0.89. Below z_{crit} the EBD would not be required because the efficiency is already quite

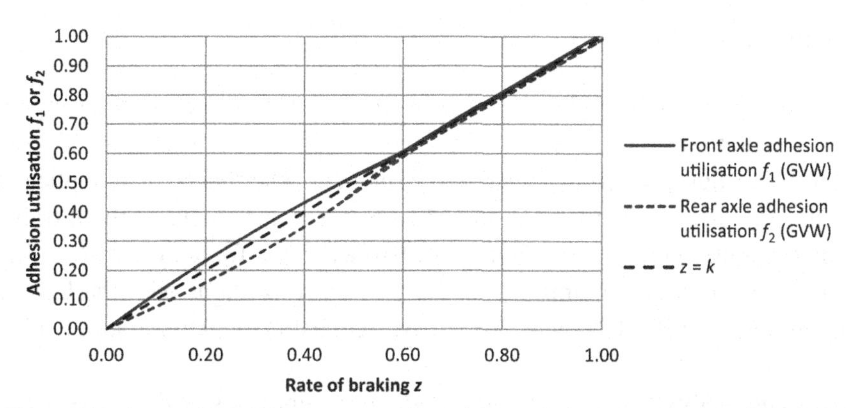

Figure 3.20: Adhesion Utilisation for Example Vehicle with ABS and EBD Control Option 1.

high. Above z_{crit} the ABS would intervene to ensure rear wheel lock does not occur. The EBD and ABS intervention to prevent premature rear wheel lock above z_{crit} is illustrated in Figure 3.20 assuming no lateral variation in dynamic wheel load or braking force.

- Option 2. The front/rear braking ratio X_1/X_2 could be continually adjusted by EBD to take account of the dynamic loading and operating conditions. The hydraulic pressure to each brake would be modulated and if incipient wheel lock were detected, the ratio X_1/X_2 could be adjusted to increase the amount of braking on the wheels on the other axle. The result would be that the adhesion utilisation curves f_1 and f_2 closely follow the $z = k$ line of maximum braking efficiency.

Option 1 is relatively unsophisticated in terms of the chassis control technology deployed on modern passenger cars in Europe. The ratio X_1/X_2 would be evaluated and set as part of the vehicle ignition start cycle, and not adjusted again until the next start cycle. The benefits would be lower complexity and cost, and improved braking efficiency compared with a fixed ratio braking system. Option 2 offers the most sophistication, including its extension to ESC, where individual brake control is needed. Both options would tend to reduce ABS interventions arising from premature rear wheel lock, which is advantageous in terms of ABS durability and driver comfort.

ABS and EBD are explained in more detail in Chapter 11 of this book. The underlying purpose of EBD is to reduce the percentage slip difference between the front wheels and the rear wheels while retaining a small slip bias towards the front for stability. By avoiding excessive rear slip at high rates of braking it permits a greater proportion of braking at the rear at low rates of braking. This improves vehicle handling during simultaneous braking and cornering and helps to equalise wear rates between the front and rear brakes and tyres.

Chapter Summary

- The basic theory for the high-level design of the braking system for the simplest road vehicle configuration of a four-wheeled two-axle rigid vehicle (category M_1) such as a passenger car or light van is presented and explained. The theory also applies to solo commercial vehicles (categories M_2 and M_3, i.e. buses and trucks without trailers) and where these (rigid) vehicles have more than two axles the force systems can be assumed to be concentrated on one 'composite' axle. The analysis aligns with the UN Braking Regulations 13 and 13H (UN, Mar 2014, UN, Feb 2014), which are adopted in many world regions.
- The theory and analysis of dynamic longitudinal weight transfer during braking, and lateral weight transfer during cornering, are developed and explained in terms of an example vehicle.
- The basic principles of tyre/road adhesion are explained including the grip developed by adhesion at the footprint or contact patch. The influence of different types of road surface and different environmental operating conditions is briefly discussed. Braking force and wheel slip are explained, and a typical brake force coefficient (BFC) versus wheel slip characteristic is illustrated and discussed. The two regions of stable and unstable slip zone are explained together with their importance for ABS design. The generation of lateral forces at the tyre/road contact patch is discussed in the context of sideways force coefficient (SFC) and slip angle. This is extended to the combination of longitudinal braking forces and lateral forces at the contact patch.
- The distribution between the two axles of the generated brake forces at each wheel is analysed, and the front/rear braking ratio (X_1/X_2) parameter is introduced. Methods of calculating the maximum rate of braking assuming that the front wheels lock first, and alternatively assuming that the rear wheels lock first, are developed and discussed. The rate of braking at simultaneous front and rear wheel lock is defined as z_{crit}, where $z_{crit} = k$; this is a useful brake system design parameter and is used in Chapter 6.
- Wheel lock and vehicle stability during braking is discussed. If the vehicle's brakes are applied and lock the wheels on the rear axle, the vehicle becomes unstable and rotates (yaws) about the vertical Z-axis, which is dangerous and disorientating. If both wheels on the front axle are locked, the vehicle cannot be steered because there is no preferential direction of motion for a sliding tyre.
- Braking efficiencyis a measure of the use which a vehicle's braking system is able to make of the available adhesion coefficient between the tyre and the road. It can never exceed unity. An example is given in which the braking efficiencies in two loading cases of a passenger car (GVW and DoW) are compared, and shown to be reasonably satisfactory. 100% braking efficiency can only be achieved if the braking ratio (X_1/X_2) is continuously adjustable to match the dynamic weight distribution for all values of z.

- Adhesion utilisation is defined as the ratio of the braking force on the axle to the dynamic normal reaction on the axle and represents the the minimum adhesion coefficient required by a particular axle to achieve a given rate of braking. Graphs of adhesion utilisation versus rate of braking are widely used to demonstrate braking system performance and compliance with legal requirements such as the UN Regulations 13 and 13H, and an example is presented and discussed. Two control options to maintain good adhesion utilisation and braking efficiency for all vehicle loading conditions are described and discussed, and the operation of ABS and EBD is briefly discussed in this context as a precursor to the more detailed coverage in Chapter 11.

References

UN, Feb 2013. UNECE Regulation 13H. Uniform provisions concerning the approval of passenger cars with regard to braking, E/ECE/324/Rev.2/Add.12H/Rev.3, February 2014.

UN, Feb 2014. UNECE Regulation 13. Uniform provisions concerning the approval of vehicles of categories M, N and O with regard to braking, E/ECE/324/Rev.1/Add.12/Rev.8, March 2014.

Pacejka, H., 2012. Tire and Vehicle Dynamics, 3rd ed. Butterworth Heinemann, ISBN 9780080970165.

Passmore, M.A., Le Good, G.M., 1994. A detailed drag study using the coastdown method, SAE 940420.

CHAPTER 4

Braking System Design for Vehicle and Trailer Combinations

Introduction

The theory for the high-level design of the braking system for a two-axle rigid road vehicle was introduced in Chapter 3 and is extended here to cover road vehicle/trailer combinations. Since these combinations are mostly commercial vehicles with more than two axles for load capacity purposes, the simplifying assumption is made that a set of close-coupled axles can be treated as one axle subject to combined loading and braking forces from all the axles in the set. The analysis is again based on the 'bicycle' model and does not include advanced technology such as ABS and EBD. The types of vehicle studied are predominantly commercial vehicles (categories M_3, N_2 and N_3) and trailers (category O), but also include small trailers and caravans that may be towed by passenger cars and light vans (categories M_1 and N_1) (UN, Mar 2014). The coupling of a trailer to a towing vehicle, whether it is a heavy goods vehicle (HGV) combination such as an articulated lorry comprising a 'tractor' unit and a semi-trailer or a passenger car towing a caravan, significantly affects the braking performance of the towing vehicle irrespective of the type of braking system fitted to the trailer. Road vehicle combinations with ill-matched braking can exhibit problems of vehicle instability, extended stopping distances and component durability, all of which can be minimised by good basic design of the braking system. The braking distribution and adhesion utilisation for different types of vehicle combinations are examined, and the legal requirements for braking compatibility between the vehicles are discussed. The vehicle combinations considered are:

- Passenger cars (or light vans) and light trailers or caravans
- Commercial vehicles (trucks) and full trailers
- Articulated commercial vehicles comprising 'tractors' and semi-trailers.

Calculations of adhesion utilisation are carried out for the different vehicles in various states of loading so that the results can graphically illustrate the effects of the braking distribution on braking performance and stability. Such calculations and graphs are not required by European legislation (UN Regulation 13) where towing vehicles and their trailers are always considered separately, but are presented here to illustrate the effects.

Braking of Road Vehicles.

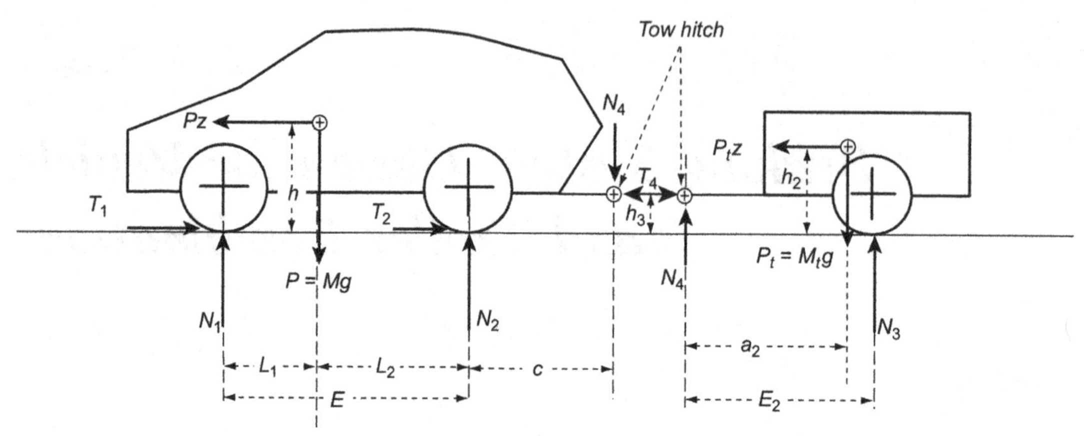

Figure 4.1: 2D road Vehicle and Light Trailer Combination Force System Under Dynamic Braking Conditions.

Car and Light Trailer

A category O_1 trailer (maximum mass not exceeding 0.75 tonnes) is not required by the European legislation to be fitted with wheel brakes. This type of trailer may be towed by a passenger car or light van, which is therefore required to provide the required braking capability for the combination from the vehicle's brakes alone. Most passenger car and light van manufacturers therefore include the possibility of the extra duty from towing such a trailer in the design and specification of their braking systems.

The force system on a car and light trailer combination in dynamic equilibrium is shown in Figure 4.1 in the form of a 'free-body' diagram. The two parts of the combination are connected at the tow hitch, but they are analysed separately as indicated in Figure 4.1. The coupling forces transmitted between the car and the trailer at the tow hitch are:

- The tow hitch vertical load N_4
- The horizontal drawbar force T_4.

These are equal and opposite on the car and the trailer. When designing the braking systems for a road vehicle towing a trailer, the towing vehicle and the trailer are considered separately and the two parts (car and trailer) in Figure 4.1 represent separate free-body diagrams of the forces on the towing vehicle and the trailer. Because the trailer is braked by the car in this configuration, the drawbar will be in compression so the horizontal drawbar force will be negative on the car and positive on the trailer as illustrated in Figures 4.2 and 4.3.

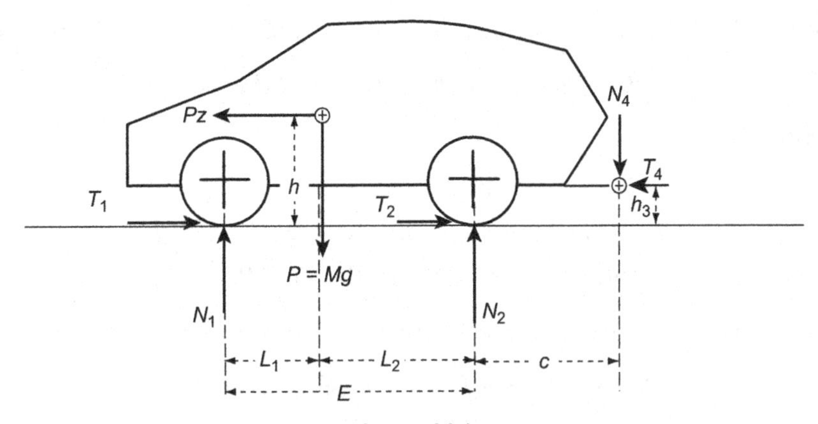

Figure 4.2: Towing Vehicle Force System.

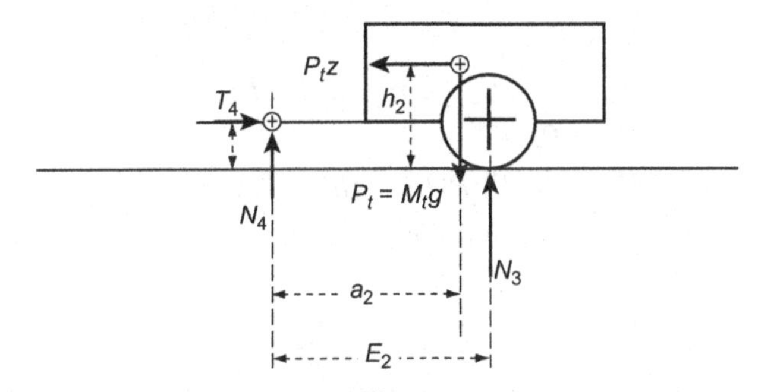

Figure 4.3: Trailer Force System.

The nomenclature is the same as in Chapter 3, with additional dimensions and forces as follows:

Subscript '$_i$' refers to the axle, so '$_3$' indicates the trailer axle.

- M_t Mass of the trailer (kg).
- P_t Trailer weight (N).
- h_2 Height of the centre of gravity of the trailer above the road surface (m).
- h_3 Height of the tow hitch (car and trailer) above the road surface (m).
- E_2 Horizontal distance from the trailer hitch to the trailer axle centre line (m).
- a_2 Horizontal distance of the trailer centre of gravity behind the trailer tow hitch (m).
- c Horizontal distance from the car rear axle to the car tow hitch (m).

Note that because there are no brakes on the trailer axles, $T_3 = 0$.

$$\text{Resolving forces vertically: } N_1 - P + N_2 - N_4 = 0 \tag{4.1}$$

$$\text{Resolving forces horizontally: } T_1 + T_2 - Pz - T_4 = 0 \tag{4.2}$$

Taking moments about the point of contact between the rear wheels and the road surface:

$$N_1E - PL_2 - Pzh + N_4c - T_4h_3 = 0 \quad (4.3)$$

Braking distribution:

$$T_1/T_2 = X_2/X_1 \quad (4.4)$$

Resolving forces vertically: $N_4 + N_3 - P_t = 0$ (4.5)

Resolving forces horizontally: $T_4 - P_tz = 0$ (4.6)

Taking moments about the point of contact between the trailer tyre and the road surface:

$$N_4E_2 + T_4h_3 - P_tzh_2 - P_t(E_2 - a_2) = 0 \quad (4.7)$$

From these equations:

$$T_4 = P_tz$$
$$N_4 = (P_t/E_2)[z(h_2 - h_3) + E_2 - a_2]$$
$$N_3 = P_t - N_4$$
$$N_1 = (1/E)[P(L_2 + zh) - N_4c + T_4h_3]$$
$$N_2 = P - N_1 + N_4$$
$$T_2 = T_1(X_2/X_1)$$
$$T_1 = (Pz + T_4)/(1 + (X_2/X_1))$$

Based on the example car from Chapter 3 at DoW towing a 750 kg unbraked trailer of the dimensions shown in Table 4.1, the effect of the trailer on the weight transfer and the wheel normal reactions at each axle are shown in Figure 4.4. The front wheel normal reactions are slightly increased while the rears are slightly reduced. Because the car is braking the 750 kg trailer, the duty on its brakes at DoW is increased by 52% (750/1450). In the GVW (laden) condition the trailer would increase the braking duty by 39%.

The braking efficiency for this combination can be calculated based on the theory presented in Chapter 3, assuming that the towing vehicle's front wheels are on the point of

Table 4.1: Trailer Data

M_t	750	kg
P_t	7358	N
E_2	1750	mm
h_2	600	mm
a_2	1715	mm
h_3	450	mm
c	1250	mm

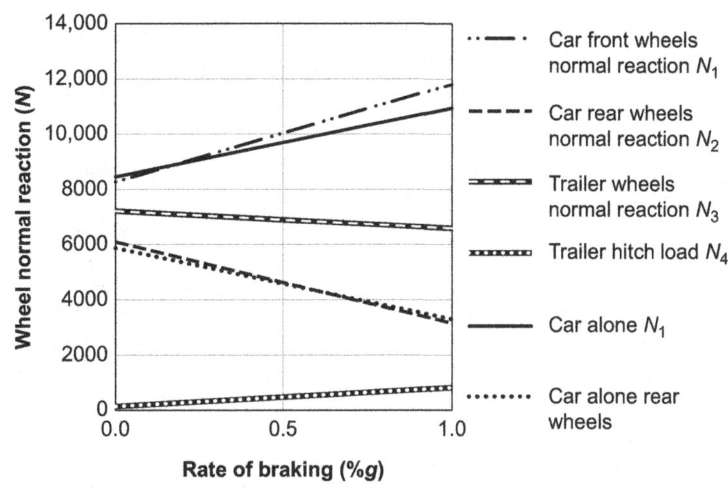

Figure 4.4: Effect of Weight Transfer on the Front and Rear Wheel Normal Reactions of an Unladen Car (DoW) Towing a Laden (750 kg) Unbraked Trailer.

locking (the rear wheels and the trailer wheels are still rotating). (The alternative scenario where the rear wheels of the towing vehicle lock first is not considered.) Equation (3.33) states that $\eta = z/k$, therefore:

$$T_1 = kN_1 \tag{4.8}$$

$$T_2 = T_1(X_2/X_1) \tag{4.9}$$

As shown in Figure 4.5(a), the braking efficiency of the combination is lower than that of the car alone (cf. Figure 3.17), and is especially low for the unladen car where the braking efficiency is 0.71 at $z = 0.5$ (cf. approximately 0.85 for the laden car towing). This indicates that the wheels on the front axle of the car are 'overbraked', i.e. they are providing too much of the braking force, and the rears are

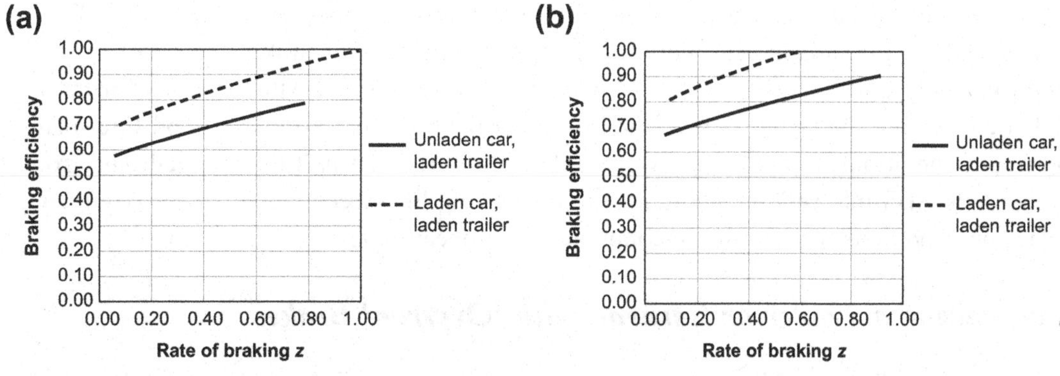

Figure 4.5: Braking Efficiency of a Car Towing a Laden (750 kg) Unbraked Trailer.
(a) $X_1/X_2 = 70/30$. (b) $X_1/X_2 = 60/40$.

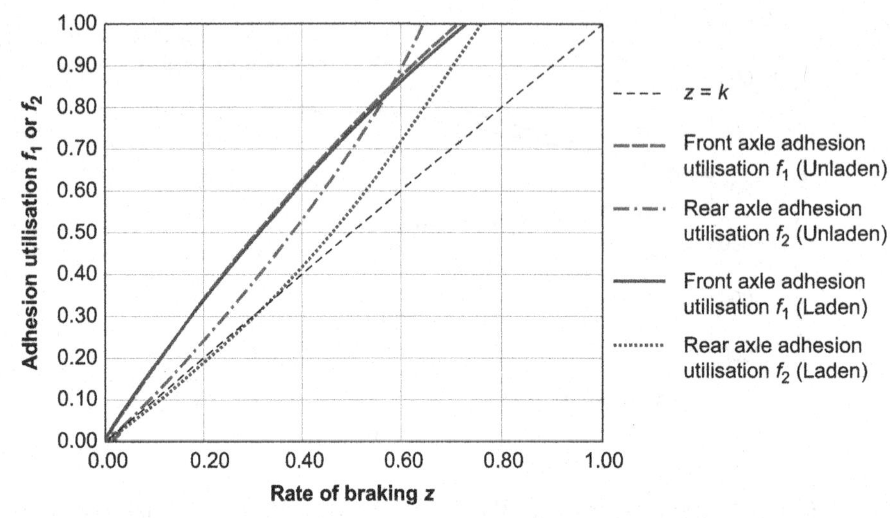

Figure 4.6: Adhesion Utilisation for the Example Vehicle Towing a 750 kg Unbraked Trailer.

'underbraked', i.e. they are providing too little of the braking force, with the set braking ratio X_1/X_2 of 70/30. Premature front wheel locking (or ABS operation) may thus occur on any type of road surface. This emphasises the need for a variable braking ratio; the driver would not only notice the effect of the trailer by the need for increased pedal effort to brake the trailer mass, but also early ABS intervention on the front wheels and poor efficiency (approximately 75% at $k = 0.2$ and 90% at $k = 0.65$). Changing the braking ratio X_1/X_2 from 70/30 to 60/40 would increase the braking efficiency to approximately 86% at $k = 0.2$ and 100% at $k = 0.6$, as illustrated in Figure 4.5(b).

The adhesion utilisation (see Chapter 3) for the car (as the trailer is unbraked, $f_3 = 0$) is shown in Figure 4.6, which illustrates the serious effect this unbraked trailer has on the braking efficiency of the combination. The front and rear axle adhesion utilisation curves are displaced from the $z = k$ line because the trailer wheels are unbraked. The rear wheels of the unladen car would lock first above a rate of braking (z_{crit}) of approximately 0.55 for which k would need to be at least 0.82. Although z_{crit} for the laden car towing the trailer is increased to about 0.74 (under extremely good tyre/road adhesion conditions, $k > 1$), front wheel adhesion utilisation is little changed, and front wheel lock/ABS intervention would occur at relatively low decelerations under tyre/road adhesion conditions that are not uncommon on European roads.

Car Towing a Trailer or Caravan with 'Overrun' Brakes

Category O_2 trailers including 'touring' caravans having a maximum mass exceeding 0.75 tonnes are required to have brakes that are actuated when the towing vehicle's brakes are applied. These may be inertia, 'overrun' or 'surge' braking systems, in which the compressive force in the coupling developed by the trailer as the towing vehicle starts to decelerate is used to mechanically apply brakes on the trailer wheels. An alternative to the overrun braking system on these types of trailer that is used in some countries, e.g. the USA, is the 'electric over hydraulic' system, where the trailer brakes are separately actuated by hydraulic pressure generated by an electric hydraulic pump in the trailer braking system. The driver can adjust the brake gain to match the trailer to the towing vehicle, and the trailer brake controller may also adjust the trailer brakeforce to match the deceleration. An important advantage of the overrun braking system for trailers is that it can be designed to be inherently stable; if the brake gain is not adjusted correctly, compatibility problems of uneven duty and wear, and possible wheel lock and vehicle instability, may result as with HGV combinations, which are explained later in this chapter. The overrun braking actuation system is usually a hydraulic master cylinder mounted between the tow hitch and the trailer frame; the tow hitch horizontal force provides an input force to the master cylinder that creates hydraulic pressure to actuate the wheel brakes (the system must be disabled when reversing). The main advantage of this system is that it reduces the additional braking load of the trailer on the towing vehicle's foundation braking system without requiring coupling to its braking system.The amount of braking force developed at the wheels of the trailer is defined by the 'brake gain' (G) of the overrun braking system, which is defined as the ratio of the braking force at the wheels of the trailer to the force in the coupling. Referring to Figure 4.7:

$$G = T_3/T_4 \tag{4.10}$$

and a typical value of G is about 7.5.

The force system on a vehicle combination of a car and caravan with overrun brakes in dynamic equilibrium is shown in Figure 4.7 in the form of a 'free-body' diagram. As

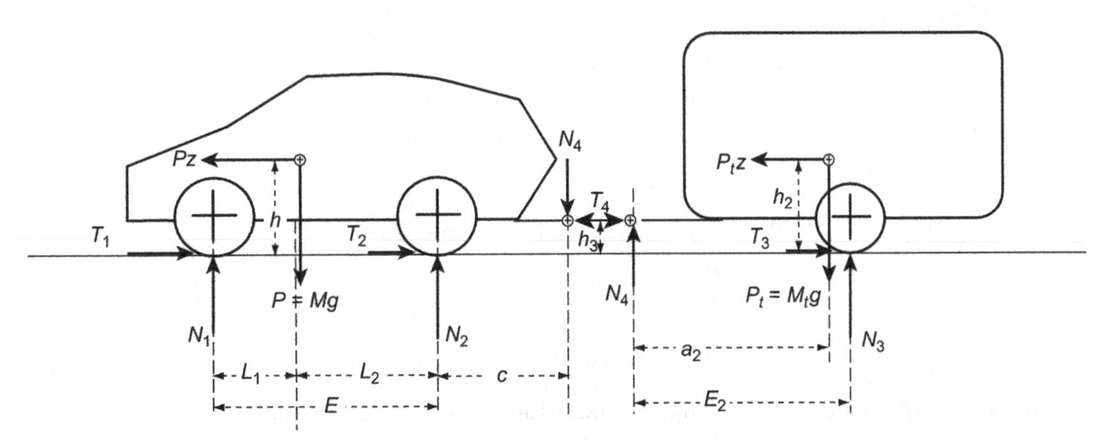

Figure 4.7: 2D Force System on a Towing Vehicle and Trailer with Overrun Brakes Under Dynamic Braking Conditions.

previously explained the two parts of the combination are connected at the tow hitch, but they are analysed separately. The compressive force in the caravan drawbar will always be in compression, i.e. negative on the car and positive on the trailer. The nomenclature is the same as previously with the addition of:

T_3 = Braking force at the caravan wheels (N)

Considering the car (towing vehicle) part of the combination:

$$\text{Resolving forces vertically: } N_1 - P + N_2 - N_4 = 0 \tag{4.11}$$

$$\text{Resolving forces horizontally: } T_1 + T_2 - Pz - T_4 = 0 \tag{4.12}$$

Taking moments about the point of contact between the rear wheels and the road surface:

$$N_1 E - PL_2 - Pzh + N_4 c - T_4 h_3 = 0 \tag{4.13}$$

Considering the caravan part of the combination:

$$\text{Resolving forces vertically: } N_4 + N_3 - P_t = 0 \tag{4.14}$$

$$\text{Resolving forces horizontally: } \mathrm{T}_4 + T_3 - P_t z = 0 \tag{4.15}$$

Taking moments about the point of contact between the trailer wheels and the road surface:

$$N_4 E_2 + T_4 h_3 - P_t z h_2 - P_t (E_2 - a_2) = 0 \tag{4.16}$$

The overrun braking gives:

$$T_3 = G T_4 \tag{4.17}$$

and the car has a fixed braking ratio:

$$T_1/T_2 = X_1/X_2 \tag{4.18}$$

From these equations:

$$\begin{aligned}
T_4 &= P_t z/(1+G) \\
T_3 &= G T_4 \\
N_4 &= (1/E_2)[P_t(zh_2 + (E_2 - a_2)) - T_4 h_3] \\
N_3 &= P_t - N_4 \\
N_1 &= (1/E)[P(L_2 + zh) - N_4 c + T_4 h_3] \\
N_2 &= P - N_1 + N_4 \\
T_1 &= T_2(X_1/X_2) \\
T_2 &= (Pz + T_4)/(1 + (X_1/X_2))
\end{aligned}$$

Based on the example car from Chapter 3 (see Table 3.1) towing a caravan of the dimensions shown in Table 4.2, the effect of the caravan on the weight transfer and the wheel normal reactions at each axle are shown in Figure 4.8.

Table 4.2: Caravan Data

M_t	1000	kg
P_t	9810	N
E_2	3500	mm
h_2	900	mm
a_2	3307	mm
h_3	380	mm
c	1250	mm
Brake gain G	7.5	

Figure 4.8(a) and (b) indicates that the car rear wheels are significantly under-braked in both unladen and laden conditions, and as with the previous example of the unbraked light trailer, the front wheels are over-braked such that premature front wheel locking (or ABS operation) might occur on any type of road surface. Should this in fact occur, the

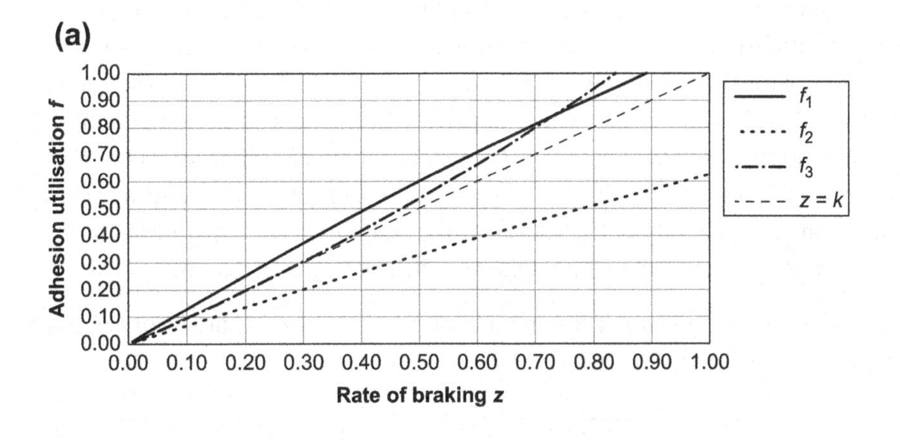

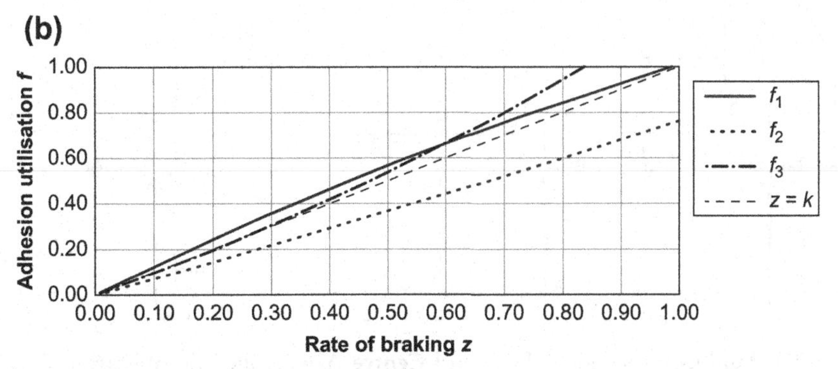

Figure 4.8: Adhesion Utilisation for a Car Towing a Caravan.
(a) Unladen car. (b) Laden car.

combination will remain stable, but to minimise the stopping distance the driver should continue to increase the brake pedal effort after the onset of wheel lock or ABS operation, in order to continue to increase the contribution of the braking at the rear wheels.

Rigid Truck Towing a Centre-Axle Trailer

Category O_3 trailers with a centre axle or a set of close-coupled axles may be towed by a rigid truck such as the combination illustrated in Figure 4.9, for which brakes are required to be actuated as part of the towing vehicle's braking system. Since most commercial road vehicles utilise compressed air (pneumatic) actuation, this means that the trailer must be fitted with a compatible braking system that is connected to that of the towing vehicle. An explanation of the types of braking systems used on commercial road vehicles is given in Chapter 6. The theory for this type of truck/trailer combination is similar to the car/caravan combination previously studied, but with the brake gain parameter G replaced by a braking ratio for the combination of the towing vehicle and the trailer. The truck (towing vehicle) and trailer are connected by the drawbar, which transmits horizontal and vertical forces between the vehicles. The drawbar is in tension when the truck is pulling the trailer, but may be in either tension or compression when the truck and trailer are being braked. For the analysis shown here, it is assumed that it is in compression, i.e. the trailer is under-braked relative to the truck, which is generally considered to be the better arrangement. The truck is also designed to operate without a trailer attached, and prior to the mandatory adoption of ABS had to meet the adhesion utilisation requirements for solo rigid vehicles in UN Regulation 13 (ECE, 2010), and although the trailer cannot operate on its own, it was subject to the same adhesion utilisation requirements.

The force system for a rigid truck towing a centre-axle trailer is shown in Figure 4.9. The nomenclature is the same as previously used.

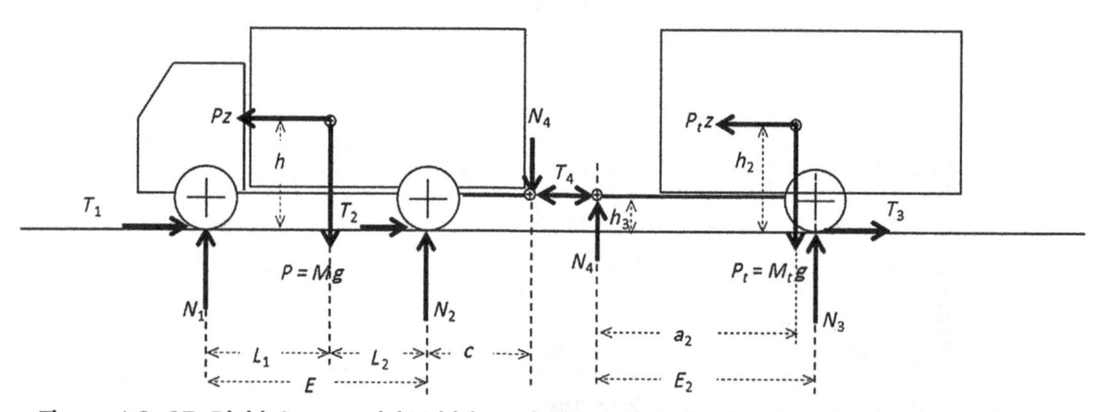

Figure 4.9: 2D Rigid Commercial Vehicle and Centre-Axle Trailer Combination Force System Under Dynamic Braking Conditions.

Considering the truck (towing vehicle) part of the combination:

$$\text{Resolving forces vertically: } N_1 - P + N_2 - N_4 = 0 \tag{4.19}$$

$$\text{Resolving forces horizontally: } T_1 + T_2 - Pz - T_4 = 0 \tag{4.20}$$

Taking moments about the point of contact between the front wheels and the road surface:

$$N_4(E + c) + PL_1 - N_2E - Pzh - T_4h_3 = 0 \tag{4.21}$$

Considering the trailer part of the combination:

$$\text{Resolving forces vertically: } N_4 + N_3 - P_t = 0 \tag{4.22}$$

$$\text{Resolving forces horizontally: } T_4 + T_3 - P_tz = 0 \tag{4.23}$$

Taking moments about the tow hitch:

$$P_ta_2 - N_3E_2 - T_3h_3 - P_tz(h_2 - h_3) = 0 \tag{4.24}$$

With commercial vehicle truck and trailer combinations, it is important to design the braking system to be as efficient as possible, which as previously explained means that all wheels should be on the point of lock at the required rate of braking. However, the towing vehicle (truck) may also be designed to operate on its own without the trailer, in which case the analysis could assume that the truck has a fixed braking ratio, and the proportion of braking on the trailer is set to maintain the required (compressive) drawbar force. However, because vertical as well as horizontal forces are transmitted to the truck by the trailer during braking, this would not be a very satisfactory solution. As with the car towing the caravan, the overall braking efficiency of the combination would be reduced considerably by the trailer, and some form of variable braking distribution between the axles would be essential to avoid this and frequent ABS intervention.

A better approach is to base the analysis on an optimum braking distribution, i.e. when the braking on each axle is proportional to the normal load carried by the axle and the braking efficiency is therefore 1. For a fixed braking ratio system, this can only occur at one rate of braking, which is specified as $z = z_1$:

$$T_1/N_1 = T_2/N_2 = T_3/N_3 = z_1 \tag{4.25}$$

If the braking distribution between the axles is specified as $X_1/X_2/X_3$, where again $X_1 + X_2 + X_3 = 1$, then (by Newton's second law):

$$T_1 = X_1(P + P_t)z_1 = N_1z_1 \tag{4.26}$$

$$T_2 = X_2(P + P_t)z_1 = N_2z_1 \tag{4.27}$$

$$T_3 = X_3(P + P_t)z_1 = N_3z_1 \tag{4.28}$$

From Equations (4.24) and (4.28):

$$N_3 = (P_t a_2 - N_3 z_1 h_3 - P_t z(h_2 - h_3))/E_2$$

Thus:

$$N_3 = P_t(a_2 - z(h_2 - h_3))/(E_2 + (z_1 h_3)) \quad (4.29)$$

At the selected rate of braking for $\eta = 1$, $z = z_1$:

$$T_3 = N_3 z_1 \quad \text{(from 4.28)}$$
$$N_4 = P_t - N_3 \quad \text{(from 4.22)}$$
$$T_4 = P_t z_1 - T_3 \quad \text{(from 4.23)}$$
$$N_2 = (P(L_1 - z_1 \text{h}) + N_4(E + c) - T_4 h_3)/E \quad \text{(from 4.21)}$$
$$N_1 = P - N_2 + N_4 \quad \text{(from 4.19)}$$

T_1 and T_2 can be calculated from Equations (4.26) and (4.27).

Since $X_1 + X_2 + X_3 = 1$, the braking ratio at the selected rate of braking ($z = z_1$) can be calculated from Equations (4.26)–(4.28), and used to calculate the normal reactions and braking forces at each axle over the required range of vehicle decelerations.

With the fixed braking ratio $X_1/X_2/X_3$, Equation (4.24) gives:

$$N_3 = (P_t a_2 - P_t z(h_2 - h_3) - X_3(P + P_t)zh_3)/E_2 \quad (4.30)$$

Table 4.3: Truck and Centre-Axle Trailer Data

	Towing Vehicle		Trailer	
Parameter	**Unladen**	**Laden**	**Unladen**	**Laden**
Wheelbase, $E = L_1 + L_2$ (mm)	5000	5000	–	–
Truck centre of gravity height, h (mm)	950	1600	–	–
Position of truck centre of gravity behind the front axle, L_1 (mm)	2357	3397	–	–
Position of centre of gravity in front of the rear axle, L_2 (mm)	3143	2103	–	–
Tow hitch position, c (mm)	2250	2250	–	–
Vehicle weight, P (N)	70,000	170,000	25,000	90,000
Vehicle mass, M (kg)	7136	17,329	2548	9174
Distance of trailer cg behind tow hitch, a_2 (mm)	–	–	3920	3978
Height of trailer centre of gravity, h_2 (mm)	–	–	900	1500
Static vertical load on coupling (N)	–	–	500	500
Length from tow hitch to trailer axle centre, E_2 (mm)	–	–	4000	4000
Height of tow hitch above road, h_3 (mm)	–	–	600	600

An example is given below for the vehicle combination specified in Table 4.3.

The adhesion utilisation graphs for this example truck and centre-axle trailer combination with a fixed braking ratio for each of four loading conditions designed to provide maximum braking efficiency at $z = 0.6$ are shown in Figure 4.10(a)–(d). These demonstrate that the braking ratio $X_1/X_2/X_3$ needs to change significantly to accommodate different loading conditions; the extremes are the unladen truck towing a laden trailer (0.27/0.24/0.49) and the laden truck towing the unladen trailer (0.48/0.43/0.10). In the first case the trailer would provide almost half the braking force required for the vehicle combination, while in the latter case it would only be providing 10%. It is clear from these graphs how effective a variable braking distribution control system based on axle loading information (ignoring lateral effects) would be in providing a very efficient braking system under all loading conditions. In comparison, the braking ratio for the truck operating solo would be $X_1/X_2 = 0.64/0.36$, and laden 0.48/0.52.

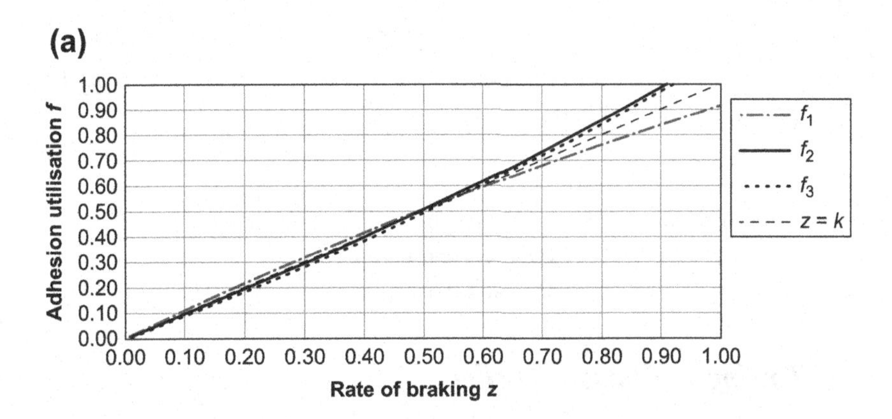

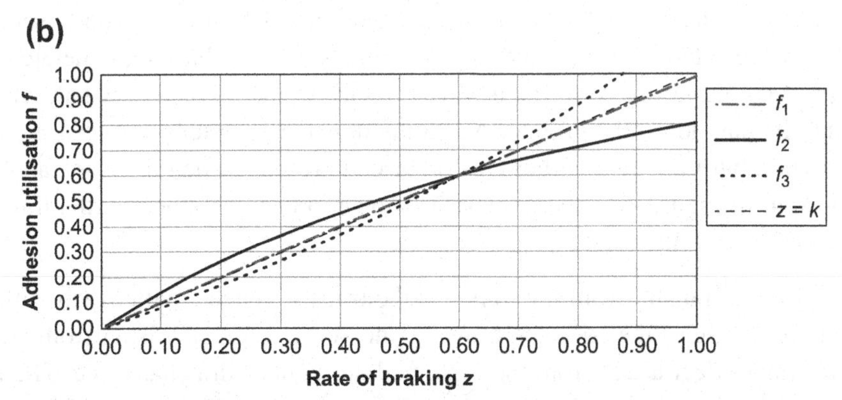

Figure 4.10: Adhesion Utilisation for a Two-Axle Rigid Truck Towing a Centre-Axle Trailer. (a) Truck and trailer unladen, $X_1/X_2/X_3 = 0.48/0.32/0.2$. (b) Truck and trailer laden, $X_1/X_2/X_3 = 0.35/0.35/0.3$. (c) Truck laden and trailer unladen, $X_1/X_2/X_3 = 0.48/0.43/0.1$. (d) Truck unladen and trailer laden, $X_1/X_2/X_3 = 0.27/0.24/0.49$.

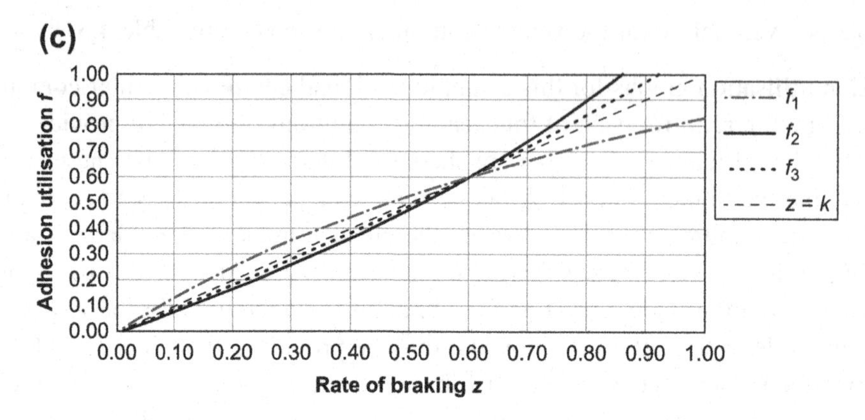

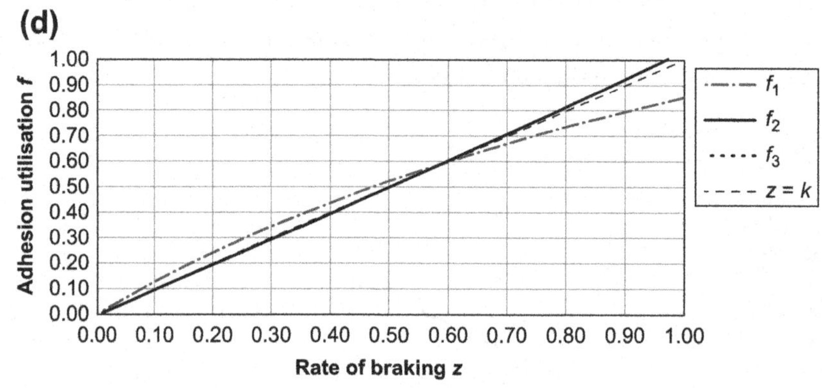

Figure 4.10 cont'd

Rigid Truck Towing a Chassis Trailer

Category O_3 and O_4 chassis trailers as illustrated in Figure 4.11 are essentially separate vehicles that are free-standing and are designed to transmit only braking-generated forces (not any part of the trailer mass) to the towing vehicle via the pivoted drawbar, which can be assumed to be parallel to the ground. Again the drawbar is in tension when the truck is pulling the trailer, but may be in either tension or compression when the truck and trailer are being braked. In Figure 4.11 the drawbar is shown in compression, i.e. the trailer is under-braked relative to the truck.

The effect of weight transfer from the trailer to the truck is much smaller than with a centre-axle trailer because the coupling transmits no vertical forces arising from the trailer mass) and the only effect arises from the (assumed horizontal) drawbar force. The force system for a rigid truck towing a centre-axle trailer is shown in Figure 4.11. The nomenclature is the same as previously used with some adjustments to account for the differences in trailer design. The coupling force between the truck and trailer is defined as the 'drawbar force' *D*.

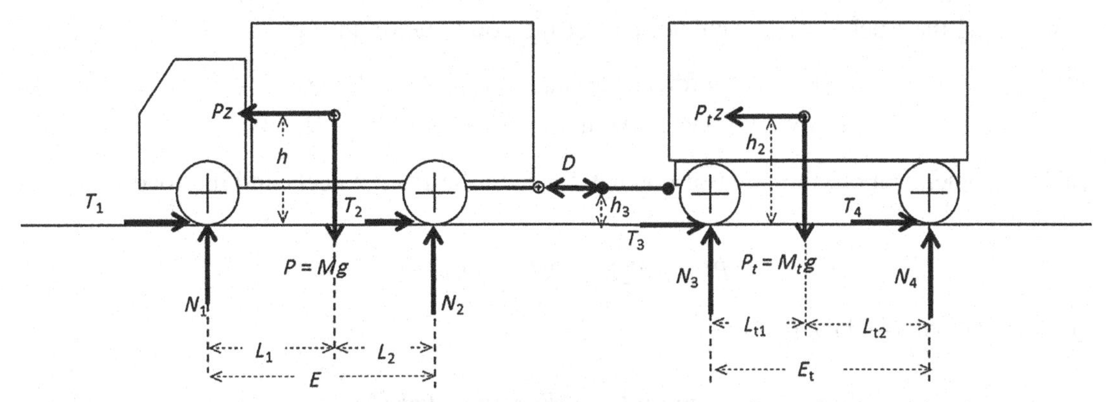

Figure 4.11: 2D Rigid Commercial Vehicle and Two-Axle Chassis Trailer Combination Force System Under Dynamic Braking Conditions.

The braking distribution of the truck (X_1/X_2) can be optimised for solo operation, but when it is coupled to a trailer the drawbar force will affect the adhesion utilisation of each part of the combination. Similarly the braking distribution of the trailer (X_3/X_4) can also be optimised for solo operation, but again when it is coupled to a truck the drawbar force will affect the adhesion utilisation of each part of the combination. This means that ideally when braking, the drawbar force should be zero or as close to it as possible, which is also desirable from a user point of view; a non-zero drawbar force means that the truck is contributing to the braking of the trailer or vice versa, with a consequent effect on the wear and durability of the brake components. It is also an operating requirement that trucks and trailers should be safely interchangeable.

In order to calculate the adhesion utilisation of the different axles of this type of coupled combination, it is necessary to consider the towing vehicle and trailer as two separate vehicles connected by the force D in the drawbar. Assuming the combination is fitted with a pneumatic actuation system, the air pressure supplied to the trailer at the coupling head can be adjusted in proportion to the truck actuation air pressure to vary the braking force generated by the trailer brakes. This would control the proportion of braking provided by the truck and the trailer to decelerate the vehicle combination, which can be defined as Y_1/Y_2 where:

$$Y_1 = (T_1 + T_2)/z(P + P_t) \tag{4.31}$$

and

$$Y_2 = (T_3 + T_4)/z(P + P_t) \tag{4.32}$$

where

$$Y_1 + Y_2 = 1 \tag{4.33}$$

Considering the truck (towing vehicle) part of the combination in Figure 4.11:

$$\text{Resolving forces vertically: } N_1 + N_2 - P = 0 \tag{4.34}$$
$$\text{Resolving forces horizontally: } T_1 + T_2 - Pz - D = 0 \tag{4.35}$$

Taking moments about the point of contact between the front wheels and the road surface:

$$PL_1 - N_2E - Pzh - Dh_3 = 0 \tag{4.36}$$

From Equations (4.31) and (4.33):

$$N_1 = (P(L_2 + zh) + Dh_3)/E = P_1 + Pzh/E + Dh_3/E \tag{4.37}$$
$$N_2 = (P(L_1 - zh) - Dh_3)/E = P_2 - Pzh/E - Dh_3/E \tag{4.38}$$

P_1 and P_2 are the static normal reactions on the truck front and rear wheels respectively. These equations are the same as those for a solo rigid vehicle (static normal load plus the weight transfer term) with an additional term for the drawbar force, which reduces the rear axle normal reaction and increases the front axle normal reaction when the drawbar is in compression.

Similarly, for the trailer part of the combination in Figure 4.11:

$$N_3 + N_4 - P_t = 0 \tag{4.39}$$
$$T_4 + T_4 - P_tz + D = 0 \tag{4.40}$$
$$P_tL_{t1} - N_4E_t - P_tzh_2 + Dh_3 = 0 \tag{4.41}$$

From Equations (4.39) and (4.41):

$$N_3 = (P_t(L_{t2} + zh_2) - Dh_3)/E_t = P_3 + P_tzh_2/E_t - Dh_3/E_t \tag{4.42}$$
$$N_4 = (P_t(L_{t1} - zh_2) + Dh_3)/E_t = P_4 - P_tzh_2/E_t + Dh_3/E_t \tag{4.43}$$

P_3 and P_4 are the static normal reactions on the trailer front and rear wheels respectively, and again these equations are the same as those for a solo rigid vehicle (static normal load plus the weight transfer term) with an additional term for the drawbar force, which increases the rear axle normal reaction and decreases the front axle normal reaction when the drawbar is in compression.

From Equations (4.31) and (4.35):

$$D = T_1 + T_2 - Pz = z(Y_1(P + P_t) - P) \tag{4.44a}$$

and from Equations (4.32) and (4.40):

$$D = P_tz - T_3 - T_4 = z(P_t - (Y_2(P + P_t)) \tag{4.44b}$$

From Equation (4.31), the braking force on the truck with a non-zero drawbar force is:

$$T = T_1 + T_2 = Y_1 z(P + P_t) \tag{4.45}$$

Since $T_1 = X_1 T$ and $T_2 = X_2 T$:

$$T_1 = X_1 Y_1 z(P + P_t) \tag{4.46}$$

and

$$T_2 = X_2 Y_1 z(P + P_t) \tag{4.47}$$

Similarly from Equation (4.32), the braking force on the trailer with a non-zero drawbar force is:

$$T_t = T_3 + T_4 = Y_2 z(P + P_t) \tag{4.48}$$

Since $T_3 = X_3 T$ and $T_4 = X_4 T$:

$$T_3 = X_3 Y_2 z(P + P_t) \tag{4.49}$$

and

$$T_4 = X_4 Y_2 z(P + P_t) \tag{4.50}$$

An example is given below for the vehicle combination specified in Table 4.4.

The adhesion utilisation graphs for this example truck and chassis trailer combination, each with a fixed braking ratio set for each of four loading conditions to provide maximum braking efficiency at $z_{crit} = 0.6$ and zero drawbar force, are shown in Figure 4.12(a)–(d). These demonstrate that the braking ratios X_1/X_2 and X_3/X_4 for the truck and trailer respectively are slightly affected by the loading conditions (under zero drawbar force they are effectively two separate vehicles ideally matched as a combination) with no weight transfer or braking duty transfer between the two, while the

Table 4.4: Truck and Chassis Trailer Data

	Towing Vehicle		Trailer	
Parameter	Unladen	Laden	Unladen	Laden
Wheelbase, $E = L_1 + L_2$ (mm)	5000	5000	4500	4500
Centre of gravity height, h (mm)	950	1600	900	1500
Position of centre of gravity behind the front axle, L_1 (mm)	2357	3397	2250	2250
Position of centre of gravity in front of the rear axle, L_2 (mm)	3143	2103	2250	2250
Vehicle weight, P (N)	70,000	170,000	40,000	160,000
Vehicle mass, M (kg)	7136	17,329	4077	16,310
Height of tow hitch above road, h_3 (mm)	600	600	600	600

proportion of braking between the truck and the trailer Y_1/Y_2 needs to change significantly to accommodate different loading conditions. The extremes are the unladen truck towing the laden trailer ($Y_1/Y_2 = 0.30/0.70$) and the laden truck towing the unladen trailer ($Y_1/Y_2 = 0.81/0.19$).

If the towing vehicle and the trailer are not ideally matched, the drawbar force will not be zero. An example of the effect of this is shown in Figure 4.13(a)–(d), where the proportion of braking between the truck and the trailer Y_1/Y_2 has been set for zero drawbar force for the fully laden truck towing the fully laden trailer. As before, the truck and the trailer braking ratios are separately set to provide maximum braking efficiency at $z_{crit} = 0.6$. The drawbar forces are shown in Table 4.5 for $z = 0.5$. These are very high forces that demonstrate the point but are unrealistic; the amount of braking duty taken from the trailer on to the truck or vice versa is excessive.

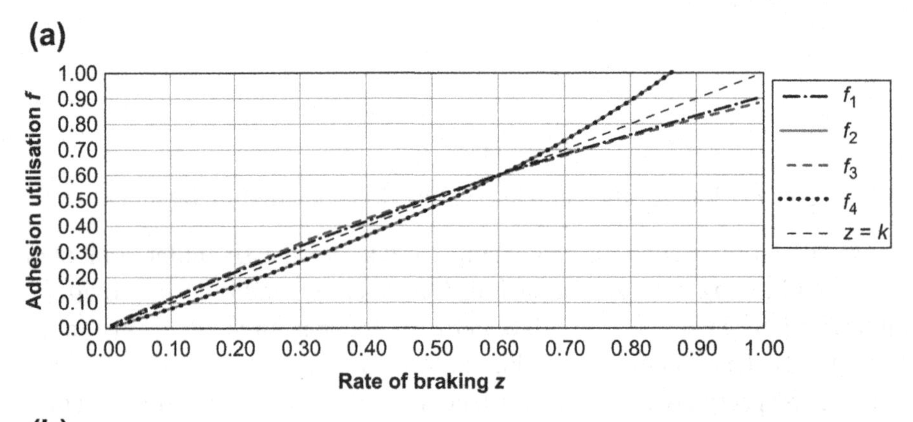

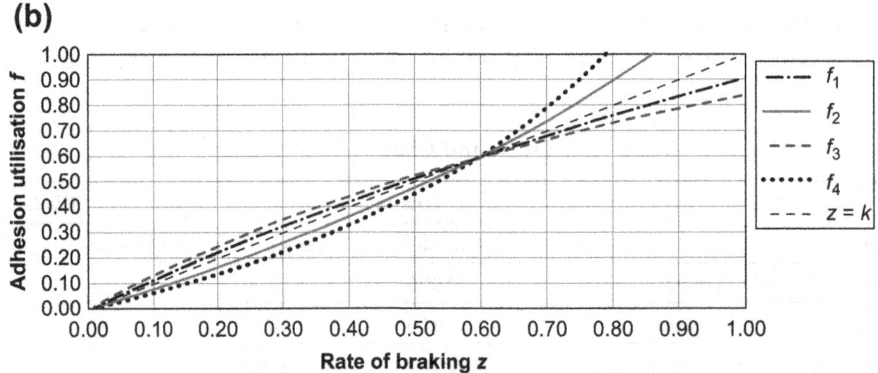

Figure 4.12: Adhesion Utilisation for a Two-Axle Rigid Truck Towing a Chassis Trailer, with a fixed braking ratio set for each of four loading conditions to provide maximum braking efficiency at $z_{crit} = 0.6$ and zero drawbar force.

(a) Truck and trailer unladen, $X_1/X_2 = 0.68/0.32$, $X_3/X_4 = 0.62/0.38$, $Y_1/Y_2 = 0.64/0.36$.
(b) Truck unladen and trailer laden, $X_1/X_2 = 0.68/0.32$, $X_3/X_4 = 0.70/0.30$, $Y_1/Y_2 = 0.30/0.70$.
(c) Truck laden and trailer unladen, $X_1/X_2 = 0.56/0.44$, $X_3/X_4 = 0.62/0.38$, $Y_1/Y_2 = 0.81/0.19$.
(d) Truck and trailer laden, $X_1/X_2 = 0.56/0.44$, $X_3/X_4 = 0.70/0.30$, $Y_1/Y_2 = 0.52/0.48$.

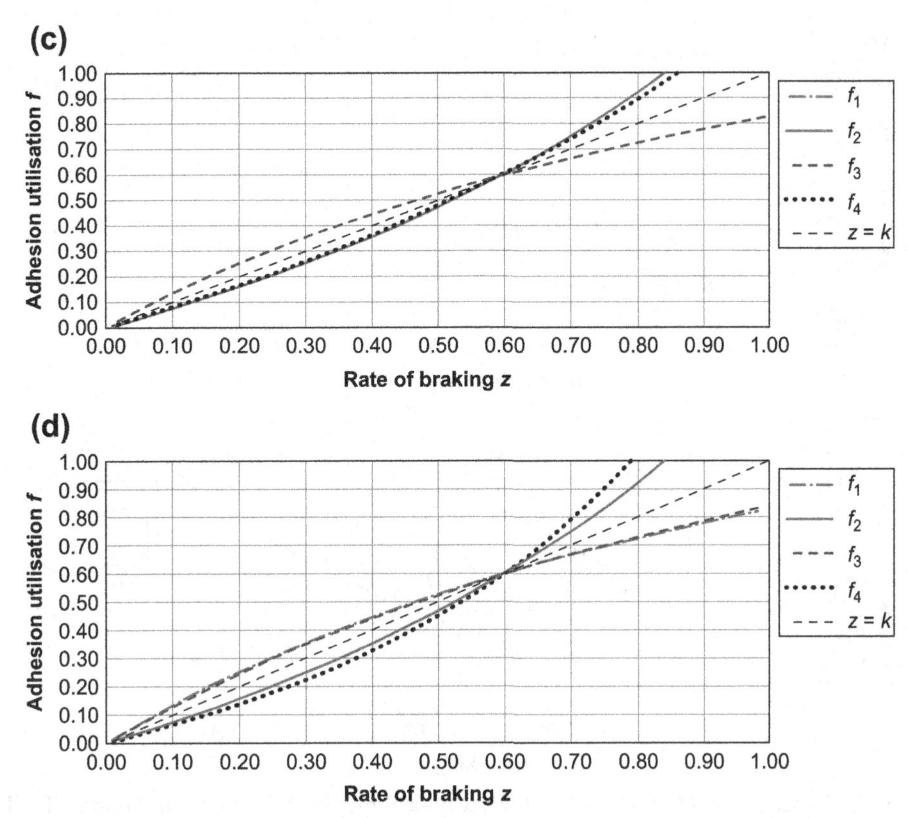

Figure 4.12 cont'd

The graphs shown in Figure 4.13 clearly demonstrate the characteristics of the braking problems that would arise if the braking system on this type of truck/trailer combination were set up in this way. The unladen truck towing the laden trailer would create a compressive drawbar force, and comparing Figure 4.12(b) with Figure 4.13(b) shows the effects of this compressive drawbar force. Whereas on the correctly matched combination a tyre/road adhesion of 0.5 would enable a rate of braking (z) of 0.49 to be achieved, with a mismatched combination where the braking is biased towards the truck, a tyre/road adhesion of 0.5 would only enable a rate of braking of about 0.28 before the front wheels reached the point of lock. The truck is required to contribute to the braking of the trailer, whilst also reducing the dynamic load on the truck rear axle.

It is not unusual for a laden truck to be required to pull an unladen trailer, and Figure 4.12(c) shows the adhesion utilisation for such a combination with ideal braking set up (the trailer could be fitted with a load-sensing device to enable it to adjust the control pressure to the trailer brakes to achieve this in practice). In Figure 4.13(c) the drawbar is in tension, and the over-braked trailer is trying to hold back the truck, whilst possessing

(a)

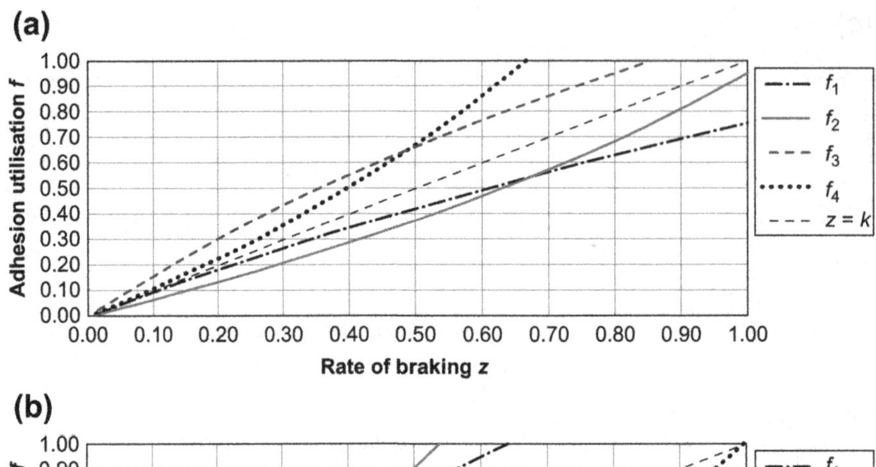

(b)

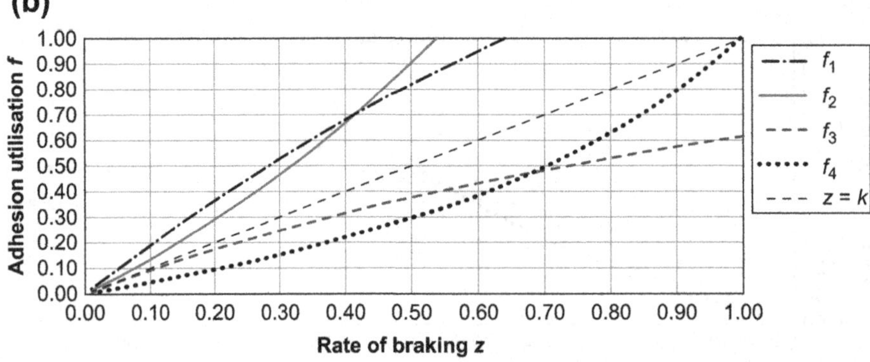

Figure 4.13: Adhesion Utilisation for a Two-Axle Rigid Truck Towing a Chassis Trailer, $Y_1/Y_2 = 0.52/0.48$ for all Loadings.

(a) Truck and trailer unladen, $X_1/X_2 = 0.68/0.32$, $X_3/X_4 = 0.62/0.38$. (b) Truck unladen and trailer laden, $X_1/X_2 = 0.68/0.32$, $X_3/X_4 = 0.70/0.30$. (c) Truck laden and trailer unladen, $X_1/X_2 = 0.56/0.44$, $X_3/X_4 = 0.62/0.38$. (d) Truck and trailer laden, $X_1/X_2 = 0.56/0.44$, $X_3/X_4 = 0.70/0.30$.

very low normal reaction forces at the tyre/road interfaces and thus only low braking forces can be generated by the trailer brakes before wheel lock occurs. The trailer wheels will lock before the truck brakes, as indicated in Figure 4.13(c), which is an extremely bad arrangement.

Articulated Commercial Vehicles — Tractors and Semi-Trailers

The 'articulated lorry' or 'tractor and semi-trailer' combination represents the largest group of heavy commercial vehicles on modern roads. Generally they comprise category O_4 trailers drawn by purpose designed tractor units, and although the tractors may run solo, it is unusual as they have no load-carrying capability on their own (they must still meet braking regulations when solo). The tractor and trailer are connected through the

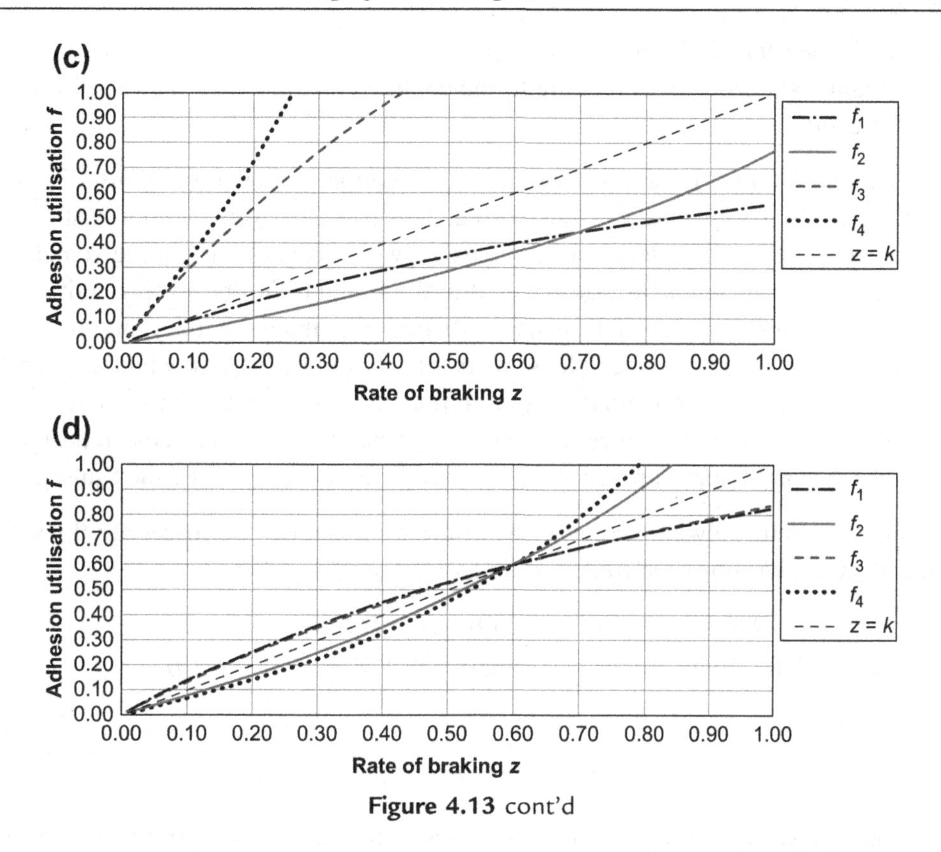

Figure 4.13 cont'd

coupling ('kingpin' or the 'fifth wheel'), which transmits horizontal and vertical forces between the two parts of the vehicle. The tractor is designed to carry a substantial proportion of the weight of the trailer via the coupling, and the retardation forces provided by its brakes must be capable of decelerating both the tractor mass and this proportion of the trailer mass. The analysis of the forces acting on the tractor and semi-trailer

Table 4.5: Drawbar Force for Proportion of Braking Between the Truck and the Trailer set as Ideal for Fully Laden Condition

Vehicle Combination Loading Condition	Drawbar Force (N) at $z = 0.5$	Condition
Truck unladen, trailer unladen	−6667	Drawbar in tension — trailer over-braked, truck under-braked
Truck unladen, trailer laden	+24,242	Drawbar in compression — trailer under-braked, truck over-braked
Truck laden, trailer unladen	−30,909	Drawbar in tension — trailer over-braked, truck under-braked
Truck laden, trailer laden	0	Ideal

combination is the same as the truck and centre-axle combination previously studied, but with some significant differences relating to the configuration of tractor and semi-trailer that are examined here.

The force system for a tractor and semi-trailer combination is shown in Figure 4.14, and the nomenclature is the same as previously used. This particular vehicle combination includes a semi-trailer with a 'bogie' of three axles, which is necessary in the vehicle design to achieve acceptable axle loadings that in the UK and Europe are typically a maximum of 10 tonnes per axle. The analysis assumes that the three axles are 'close-coupled' and can be treated as one axle with six wheels and six brakes, which acts at the transverse centre line of the three-axle bogie. In practice the tyre/road normal reactions on each axle may be different, but since this depends on the bogie and suspension design, for simplicity the tyre/road normal reactions on each axle are assumed to be the same.

Considering the tractor (towing vehicle) part of the combination, the forces are the same as for the rigid truck/centre-axle trailer combination, i.e.:

$$\text{Resolving forces vertically: } N_1 - P + N_2 - N_4 = 0 \tag{4.19}$$
$$\text{Resolving forces horizontally: } T_1 + T_2 - Pz - T_4 = 0 \tag{4.20}$$

Taking moments about the point of contact between the front wheels and the road surface:

$$N_4(E + c) + PL_1 - N_2E - Pzh - T_4h_3 = 0 \tag{4.21}$$

Note that 'c' is negative in this case, i.e. the coupling between the tractor and the semi-trailer is in front of the tractor rear wheels.

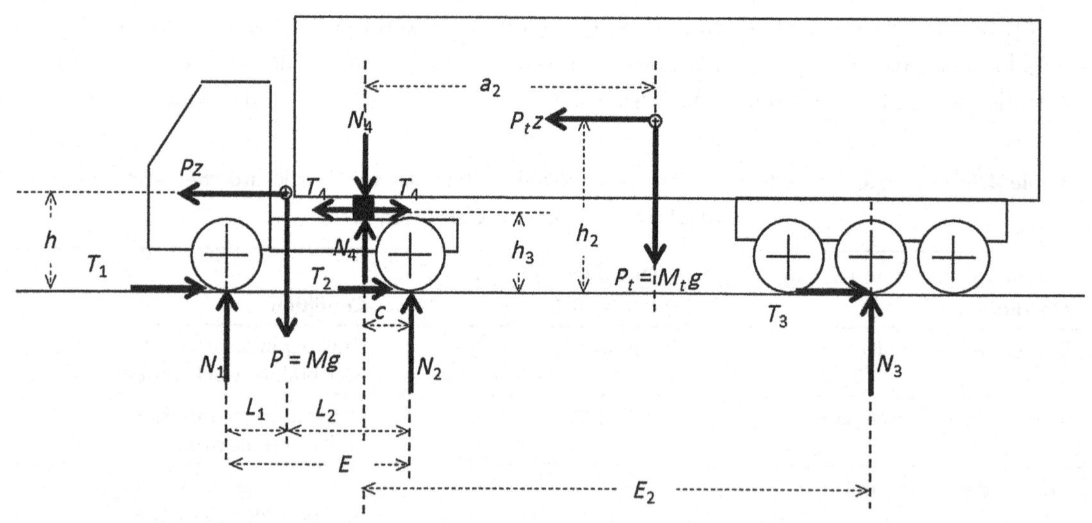

Figure 4.14: 2D 'Articulated Lorry' or Tractor and Semi-Trailer Combination Force System Under Dynamic Braking Conditions.

Considering the semi-trailer part of the combination:

$$\text{Resolving forces vertically}: N_4 + N_3 - P_t = 0 \tag{4.22}$$

$$\text{Resolving forces horizontally}: T_4 + T_3 - P_t z = 0 \tag{4.23}$$

Taking moments about the coupling (fifth wheel):

$$P_t a_2 - N_3 E_2 - T_3 h_3 - P_t z(h_2 - h_3) = 0 \tag{4.24}$$

As previously explained for the two-axle rigid truck and centre-axle trailer combination, the best approach might be to base the analysis on an optimum braking distribution, i.e. when the braking on each axle is proportional to the normal load carried by the axle and the braking efficiency is therefore 1. For a fixed braking ratio system this can only occur at one rate of braking, which is again specified as $z = z_1$,

i.e.

$$T_1/N_1 = T_2/N_2 = T_3/N_3 \tag{4.25}$$

and as before (from Equations (4.24) and (4.28)):

$$N_3 = P_t(a_2 - z(h_2 - h_3))/(E_2 + z_1 h_3) \tag{4.29}$$

and the braking distribution between the axles is specified as $X_1/X_2/X_3$, where again $X_1 + X_2 + X_3 = 1$, having determined T_1, T_2, T_3 and N_1, N_2, N_3, the values of $f_i = T_i/N_i$ can be calculated for all values of z for the axles.

An example is now given for the vehicle combination specified in Table 4.6.

The adhesion utilisation graphs for this example covering articulated combination with a fixed braking ratio set for the tractor with the laden semi-trailer to provide maximum braking efficiency at $z = 0.5$, is shown in Figure 4.15. The braking ratio $X_1/X_2/X_3$ is 0.254/0.263/0.483, and would need to change to 0.457/0.182/0.361 for the combination with the unladen semi-trailer. In comparison, the braking ratio for the truck operating solo would be $X_1/X_2 = 0.80/0.20$.

Figure 4.15 shows that on a road surface with a coefficient of adhesion of 0.5, the maximum rate of braking is $z = 0.5$, i.e. the braking efficiency is 100%. On lower adhesion surfaces the wheels on the tractor front axle are the first to reach the limit of adhesion, and on a very low adhesion road where $k = 0.15$, a rate of braking of just over 0.1 can be achieved, equivalent to an efficiency of around 72%. This assumes no lateral variation in tyre/road normal reactions or braking forces. As previously explained, the semi-trailer in this example has three axles in a close-coupled bogie, which is modelled as one axle where the tyre/road normal reactions and the braking forces are assumed to be the same for all six wheels and the sums of these forces act at the geometric transverse

Table 4.6: Articulated Lorry (Tractor and Semi-Trailer) Data

Parameter	Tractor (Towing Vehicle) Unladen	Trailer Unladen	Trailer Laden
Wheelbase, $E = L_1 + L_2$ (mm)	3800		
Truck centre of gravity height, h (mm)	750		
Position of truck centre of gravity behind the front axle, L_1 (mm)	1150		
Position of centre of gravity in front of the rear axle, L_2 (mm)	2650		
Coupling position (negative because in front of tractor rear axle), c (mm)	−660		
Vehicle weight, P (N)	66,200	60,000	330,000
Vehicle mass, M (kg)	6748	6116	33,639
Distance of trailer cg behind tow hitch, a_2 (mm)		5760	4800
Height of trailer centre of gravity, h_2 (mm)		900	1800
Length from coupling to trailer axle centre, E_2 (mm)		7200	7200
Height of coupling above road, h_3 (mm)		1300	1300

centre line. When the required rate of braking is greater than 0.5, the trailer wheels will be the first to lock, almost immediately followed by the tractor rear wheels.

Commercial vehicles are operated over a much wider loading range than passenger cars and a fixed braking ratio between the axles of any commercial vehicle and trailer combination is incapable of providing stable braking without some form of control of the braking ratio depending on the loading condition. For example, if this articulated combination were to be operated unladen with the braking distribution for the laden

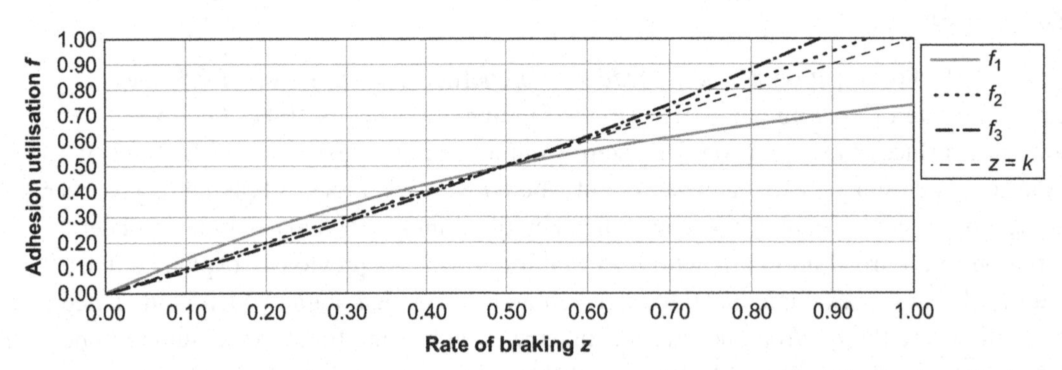

Figure 4.15: Adhesion Utilisation for a Two-Axle Tractor Towing a Semi-Trailer in a Typical Articulated Lorry Combination, Laden.

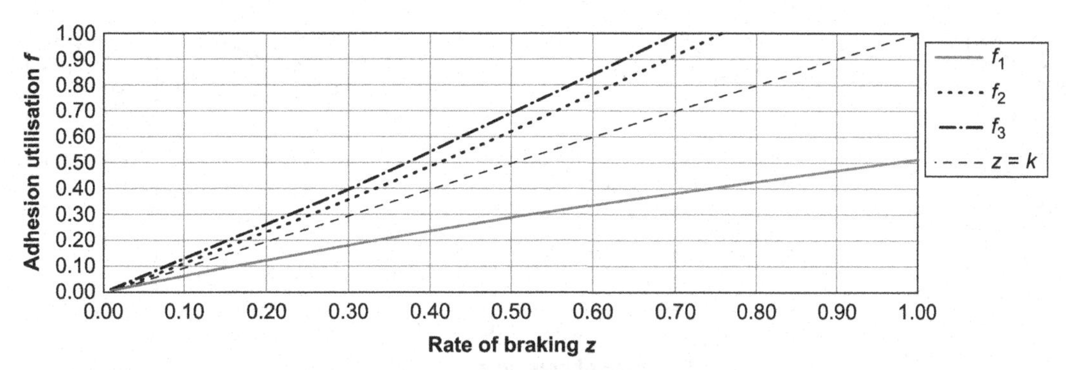

Figure 4.16: Adhesion Utilisation for the Tractor/Semi-Trailer Unladen Combination with the Laden Braking Ratios.

combination of 0.254/0.263/0.483, Figure 4.16 shows clearly that premature locking of the trailer axle closely followed by the tractor rear axle will occur, and that the tractor front axle is significantly under-braked.

This is unacceptable, being a very unstable and dangerous condition that illustrates clearly the need for some method of adjusting the braking distribution according to the state of loading of the vehicle combination. Recalculating the braking distribution for the unladen combination to give 100% braking efficiency at $z = 0.5$ requires the braking ratios to be 0.457/0.182/0.361. If load-sensing technology were applied in an air-actuated braking system to control the amount of braking on the tractor rear and semi-trailer axles but not on the tractor front axle, then the actuation pressure supplied to the tractor rear and semi-trailer axles could be adjusted by means of pressure-reducing valves. The valve ratios R_2 and R_3 (defined as the input pressure vs. the output pressure for the tractor rear axle and the semi-trailer axles respectively) would be calculated as follows:

$$X_{2(unladen)} / \left(X_{2(laden)} / R_2\right) = X_{1(unladen)} / X_{1(laden)} \tag{4.51}$$

Hence,

$$R_2 = \left(X_{1(unladen)} X_{2(laden)}\right) \Big/ \left(X_{1(laden)} X_{2(unladen)}\right) \tag{4.52}$$

and

$$R_3 = \left(X_{1(unladen)} X_{3(laden)}\right) \Big/ \left(X_{1(laden)} X_{3(unladen)}\right) \tag{4.53}$$

For this case:

$$R_2 = (0.457 \times 0.264)/(0.254 \times 0.182) = 2.61$$

and similarly:

$$R_3 = (0.457 \times 0.483)/(0.254 \times 0.361) = 2.41$$

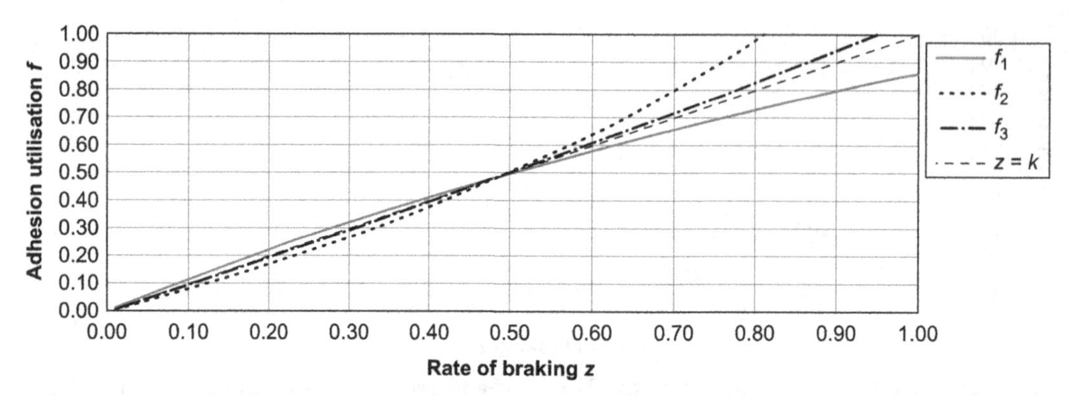

Figure 4.17: Adhesion Utilisation for the Tractor/Semi-Trailer Unladen Combination with Adjusted Braking Ratios.

Figure 4.17 shows the adhesion utilisation of the unladen combination adjusted in this way.

As previously mentioned, most commercial vehicles will in practice often operate partially laden, and be required to carry many different kinds of load. In the case of tractor/semi-trailer combinations, the provision of load-sensing valves on the tractor rear and semi-trailer axles can be very effective for correcting the braking distribution based on the static axle loads, but are less effective in compensating for unusual or large amounts of dynamic weight transfer under braking, e.g. when the loading on the semi-trailer creates a very high centre of gravity. The height of the centre of gravity of the typical semi-trailer considered here, assumed to be carrying a normal full load, was 1.8 m. The same vehicle, fully loaded with a high-density payload, e.g. steel sheets, might have a lower centre of gravity height of about 1.35 m, but if carrying a low-density payload of the same total weight the centre of gravity height could be 2.3 m, an increase in height and therefore of dynamic weight transfer of 70%. In comparison, the addition of a roof rack carrying 75 kg to a passenger car might raise the centre of gravity height by about 10%.

The effects of these two loading conditions on the adhesion utilisation are shown in Figures 4.18 and 4.19. In Figure 4.18 the semi-trailer wheels would lock first above a rate of braking of about 0.44, which would initiate an instability known as 'trailer swing' as the locked trailer wheels provide no directional stability to the trailer. In practice of course, all six wheels on this semi-trailer would rarely all reach the point of wheel lock simultaneously.

In Figure 4.19, the tractor rear wheels would lock first above a rate of braking of about 0.44, which would initiate an instability known as 'jack-knife' as the locked wheels provide no directional stability to the tractor, which would attempt to yaw rapidly about the vertical (Z) axis under what is effectively rear wheel lock, but would reach an abrupt limit as the tractor cab met the trailer. This happens quickly even at low road speeds and is an extremely dangerous and damaging instability. Modern commercial vehicle braking systems with ABS have reduced considerably the occurrence of jack-knife instabilities on

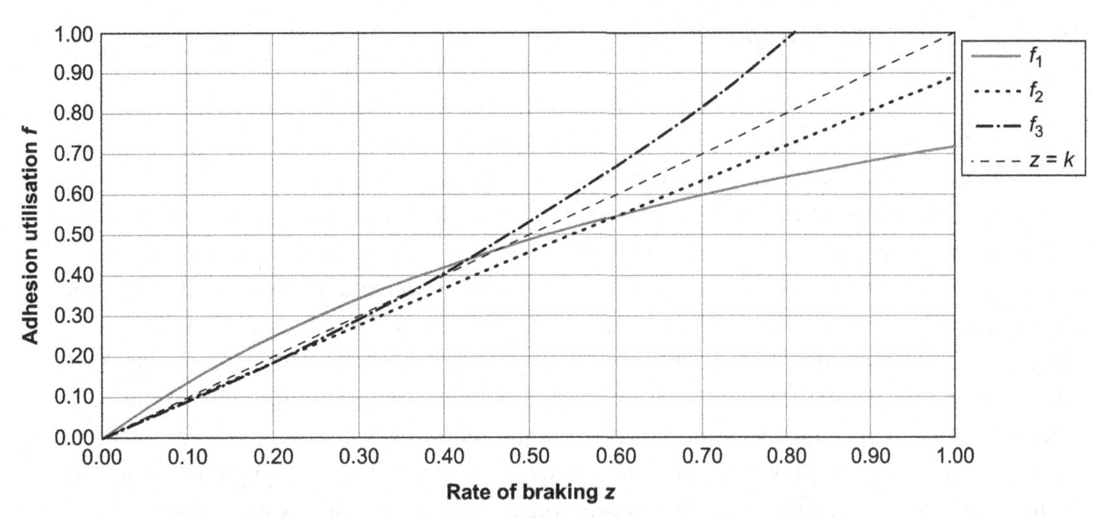

Figure 4.18: Adhesion Utilisation for the Tractor/Semi-Trailer Combination (Braking Ratios Specified for Standard Laden Conditions) with Low-Density Payload, i.e. High Centre of Gravity 2.3 m.

articulated vehicles, but the braking system designers still have to consider the worst case, e.g. in the event of partial system failure. These two figures illustrate how the behaviour of a vehicle under emergency braking can change from trailer-swing to jack-knifing depending solely on the type of load being carried, and emphasise the importance of ABS in preserving vehicle stability under the wide variety of operating conditions experienced by commercial vehicles.

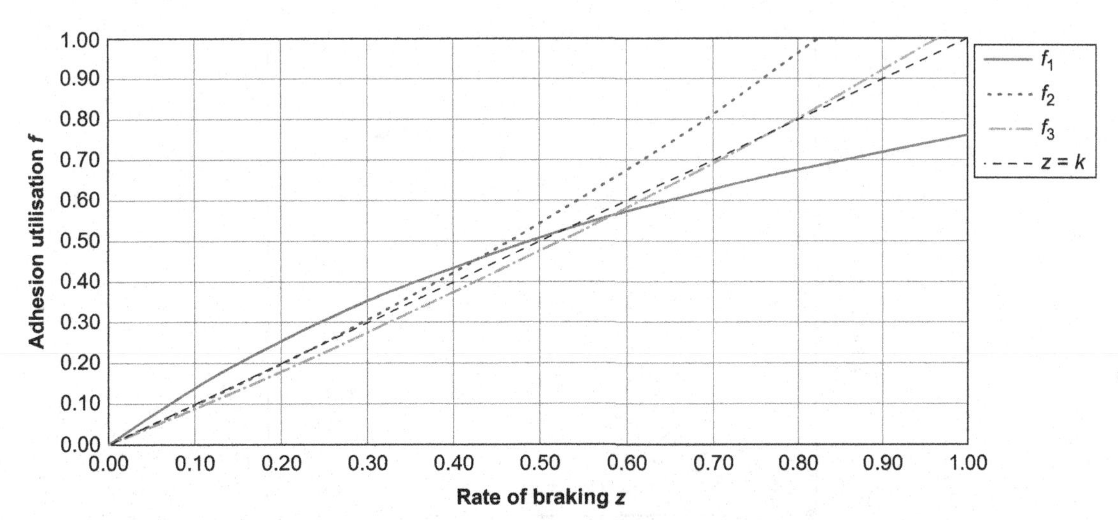

Figure 4.19: Adhesion Utilisation for the Tractor/Semi-Trailer Combination (Braking Ratios Specified for Standard Laden Conditions) with High-Density Payload, i.e. Low Centre of Gravity 1.35 m.

Load Sensing and Compatibility

Load-sensing valves were first used on articulated vehicle tractors in the 1960s, and their use was extended to rigid trucks, trailers and semi-trailers in the 1970s. They were generally fitted to those axles that were subject to the largest variations in static weight, namely the rear axles of trucks and tractors, and were seldom fitted to front axles. On semi-trailers, only one load-sensing valve was installed, which controlled the brake actuation pressure to all axles. As chassis trailers have independent axles both are subject to significant load, so valves were installed on both front and rear axles but as an alternative it was found to be possible to utilise only one load-sensing valve on the front axle (Shilton, 1969–70). The purpose of load-sensing valves is to adjust the brake actuation pressure to control the brake force generated at each axle to achieve a vehicle braking distribution that takes account of the wheel normal reactions, which depend on the loading condition of the vehicle. As previously mentioned, this is particularly important for commercial vehicles because of the large differences between loading states and the effect these have on braking performance and efficiency. Such valves function by reducing the actuation pressure delivered to the brakes on one or more axles from the sensed axle relative to that supplied by the footbrake valve in the case of the towing vehicle, or the coupling head pressure in the case of trailers, in accordance with the load carried by the sensed axle. Typical characteristics are shown in Figure 4.20, the pressure ratio being variable between 1:1 and typically 6:1. The 'inshot' pressure of 0.5 bar is to prevent the pressure reduction from taking effect before the brake threshold pressures have been overcome. On a vehicle with steel suspension springs the load would

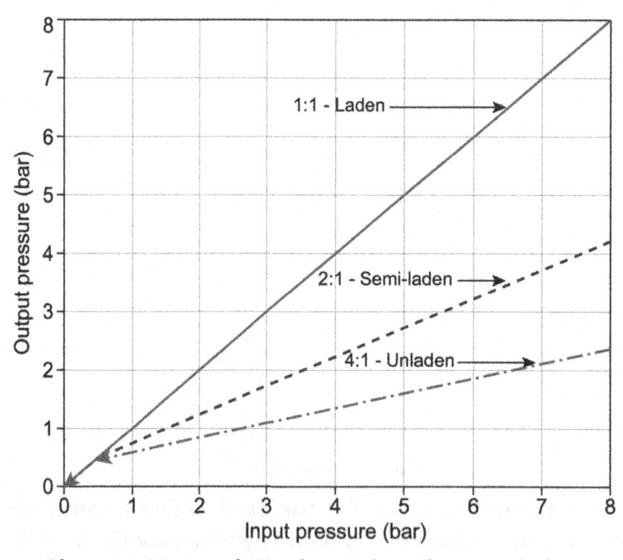

Figure 4.20: Load-Sensing Valve Characteristics.

be sensed by the axle to chassis deflection, while vehicles with air suspension monitor the air spring pressure, the load-sensing ratio being determined by the slope of the line from the inshot pressure to the delivery pressure for a given input pressure. More modern systems are capable of reacting to dynamic weight transfer as well as static axle load variations.

Commercial vehicle braking systems are carefully designed to avoid as far as possible the effects on vehicle stability of locking of the wheels of a particular axle, e.g. loss of steering control on front wheel lock-up and 'jack-knifing' or 'trailer swing' if the tractor rear or semi-trailer wheels lock first. With an ABS system fitted (as has been required for commercial vehicles in Europe since 1991), wheel lock and the associated instabilities can in principle be avoided irrespective of the basic braking system design. UN Regulation 13, which is explained in Chapter 8, exempts towing vehicles and full trailers fitted with ABS from having to meet the adhesion utilisation requirements, and only the laden compatibility requirements must be met by towing and towed vehicles with pneumatically actuated brakes. (The compatibility requirements require that the relationship between the individual vehicle deceleration and the control line air pressure lies within certain boundaries; this is explained in Chapter 8.) Despite this, there is no substitute for good basic braking system design and whilst it might not initiate drastic instability, the premature operation of ABS on one axle due to inappropriate braking distribution can adversely affect a road vehicle's braking behaviour, and is carefully avoided in modern vehicle braking system designs. The vehicle and braking system design should try to achieve the best compromises for adhesion utilisation and compatibility, and only rely upon ABS for emergency situations. Because of this, many truck and trailer manufacturers continue to fit load-sensing valves and other devices to improve distribution and compatibility in addition to ABS. Without load sensing, tractor/trailer compatibility can be very poor because of the uneven braking distribution and under normal usage conditions (i.e. non-ABS) can result in excessive brake wear on over-braked axles or 'glazing' of the brake linings as a result of under-braking. Both of these conditions have an effect on the efficiency of the brakes and hence the stopping capability of the vehicle or combination as explained in Chapter 3. Electronic Braking Distribution (EBD; see Chapter 11) is now fitted to many air-braked commercial vehicles, and has the advantage of being able to sense and control the amount of braking applied to an axle more accurately in proportion to the dynamic axle load.

Chapter Summary

- The theory for braking system design is extended to cover road vehicle and trailer combinations, which are predominantly commercial vehicles (categories M_3, N_2 and N_3) and trailers (category O), but may also include small trailers and caravans, which may be towed by passenger cars and light vans (category M_1 and N_1). The types of vehicle studied are:

- Passenger cars (or light vans) towing light trailers or caravans without and with overrun brakes
- Commercial vehicles (trucks) towing full trailers (centre-axle and chassis)
- Articulated commercial vehicles comprising tractors and semi-trailers.

- The force system equations are derived for each type of vehicle combination, and examples are provided to explain and discuss the features affecting braking system design for these types of vehicle, based on the calculation of adhesion utilisation and braking efficiency.
- It is important to design the braking system of any commercial vehicle truck and trailer combinations to be as efficient as possible. Because of the wide range of loading and operating conditions under which the combination has to operate, plus the fact that the towing vehicle may have to operate solo, it is explained why some form of variable braking distribution between the axles is essential.
- Strategies to achieve a good basic braking system design are explained, and examples are presented and discussed. The effects of compatibility between the towing vehicle and the trailer are explained in terms of braking performance, efficiency and durability.
- Strategies for achieving and maintaining compatibility between the towing vehicle and trailer are described and discussed, focusing on load sensing. The purpose of load-sensing valves is to adjust the brake actuation pressure to control the brake force generated at each axle to achieve a vehicle braking distribution that takes account of the wheel normal reactions, which depend on the loading condition of the vehicle.
- A short consideration of the role of ABS and EBD with load sensing in commercial vehicle combination braking system design is provided in preparation for Chapter 11.

References

UN, Mar 2014. UNECE Regulation 13. Uniform provisions concerning the approval of vehicles of categories M, N and O with regard to braking, E/ECE/324/Rev.1/Add.12/Rev.8, March 2014.

Shilton, B.R., 1969–70. Braking system load sensing valves. Proc. IMechE. 184 (12), 30–37.

CHAPTER 5

Brake Design Analysis

Introduction

The function of an automotive friction brake is to generate a braking torque to retard the road wheel and thus the vehicle to which it is fitted. The working principle of a friction brake is that friction material stators are pressed against a rotor, generating frictional forces that slow down the rotor and anything connected to it such as a road wheel. The torque reaction is transmitted from the stators to the vehicle structure, usually via the suspension components. There are two main types of friction brake that have been in common use on road vehicles for many years: the disc brake works by pressing'pads' of friction material against each side of a rotating disc, while in a drum brake the brake linings mounted on 'shoes' are expanded outwards against the inner surface of a rotating brake drum. The kinetic energy of the vehicle's motion is converted into heat by the process of friction, and a key part of the brake design process is that this heat must be effectively and efficiently dissipated by the brake components to avoid problems of overheating. Consistent brake friction torque generation under all conditions of use, hot and cold, dry and wet, is a major design objective.

Disc brakes have been fitted to the front axles of passenger cars and light vans in Europe for over 30 years because of their better heat dissipation ability and fade resistance compared with drum brakes. Drum brakes continue to be popular for the rear brakes of many small cars and vans where the braking duty is light. There are two main reasons for this: the first is that a disc brake is usually more expensive and can have a greater mass than the equivalent drum brake (in terms of torque capacity); the second is that provision of a parking brake function within a disc brake can be complicated and expensive. It requires either a dual actuation caliper that incorporates a mechanical or electromechanical actuation mechanism in combination with its normal hydraulic actuation system or an alternative brake such as a miniature drum brake installed inside the cylindrical section of the rotor (which connects the friction ring to the hub) (often termed the 'top hat' region) or a separate transmission brake. Disc brakes have replaced drum brakes on many European commercial motor vehicle axles, while for commercial vehicle trailers and semi-trailers (see Chapter 4) many operators still use drum brakes because of the ease of maintenance and the very long lining replacement intervals that can be achieved.

Braking of Road Vehicles.

On most road vehicles the foundation brakes are fitted immediately inside the road wheels with the associated advantage that the brake can easily be accessed for maintenance by removing the wheel. The major disadvantage of this arrangement is that the brakes contribute to the 'unsprung' mass of the vehicle, and since friction brake components tend to be large and heavy in order to provide thermal mass for good heat transfer away from the friction interface, and the strength to accommodate very high forces and temperatures, their contribution to a vehicle's unsprung mass can be considerable. An alternative is to mount the foundation brakes 'inboard', i.e. attached to the chassis, subframe or differential casing. This arrangement has the advantage of reducing the unsprung mass, but also complicates maintenance access and can introduce problems of brake cooling, and torsional vibration in the axle or drive shaft, which can lead to durability problems. Also, because the axle or driveshaft effectively becomes part of the braking system, greater safety considerations are required in their design.

Apart from disc and drum brakes, there are other types and designs of friction brakes that may be found on road vehicles. The most common design of automotive disc brake is often described as an 'open', 'spot' or 'caliper' type of disc brake to differentiate it from the 'annular' or 'plate' type, which comprises a 'stack' of multiple annular rotor and stator plates. Usually annular disc brakes are fully enclosed in a rigid housing so that they can operate 'wet' in an oil-immersed environment to maintain cleanliness, reduce wear and promote cooling, but 'dry' annular disc brakes are also available. Enclosed annular types of brake are usually only used in heavy-duty off-road applications such as agricultural tractors, construction vehicles and military vehicles, where severe deterioration of braking performance by contamination of the friction surfaces in an 'open' brake would be expected. The rotor plates, usually with friction material on both faces, can move axially on a splined shaft connected to the hub and thus rotate with the wheel. In between the rotor plates are plain stator plates that can move axially in a rigidly mounted splined housing. Actuation force is applied to the braking surfaces by an annular actuator device that is concentric with the splined shaft. This type of brake is expensive and heavy, and usually demonstrates a significant 'off-brake' drag torque, but can generate large torques within compact dimensions, making it better suited to off-road vehicles. The oil in which the rotors and stators operate provides cooling but heat dissipation is usually the critical limitation and such brakes are limited to use on low-speed vehicles, typically less than 40 km/h. The design and analysis of fully enclosed multi-plate annular disc brakes is not considered further here.

The brakes on most modern cars and light commercial vehicles are hydraulically actuated using 'muscular' energy (i.e. from the force applied to the brake pedal by the driver's leg) to actuate the brake, and a servo (or booster) to amplify the driver's effort on the brake pedal is usually incorporated. Some smaller and/or specialist commercial vehicles can also be hydraulically actuated, but usually as a 'power hydraulic' system, where actuation pressure is generated by a hydraulic pump and pressure control system actuated by the driver. Most commercial vehicles use an air-actuated system that is lower cost, more robust and easier to maintain than hydraulic systems. Actuation systems are covered in Chapter 6.

Disc Brakes

Basic Principles

The modern automotive disc brake is an 'open' type of 'spot' disc brake, i.e. the friction surfaces are not enclosed for protection (a 'dust' shield may be fitted to prevent the ingress of road debris and water spray, but does not enclose the brake). The rotor is a disc that is attached to the wheel hub and rotates with it while the stator, which is attached to the axle or suspension (e.g. the steering knuckle), consists of two opposing brake pads that are held in a 'caliper' and clamped against the disc by the actuation forces. The friction surface of the brake pad only covers a portion (typically no more than 15%) of the rotor friction surface area. As the pads are clamped against the disc by the actuation force, the friction force generated opposes the motion of the disc and slows it down. Because the disc is attached to the road wheel of the vehicle, the vehicle is decelerated. The principle of the disc brake was first patented by Lanchester in 1902 but it was not until the 1950s (Jaguar cars at Le Mans) that their advantages were fully demonstrated. Their resistance to temperature effects, especially fade, compared with even the most advanced drum brakes of the time, enabled racing drivers to brake harder and later than their competitors. This resistance to temperature effects (see Chapter 7) provides greater consistency of performance, which is the main reason why disc brakes have superseded drum brakes for higher duty automotive applications, especially on vehicle front axles.

Disc brake pads have no significant self-servo effect (as explained later), and the friction force generated is usually considered to be directly proportional to the actuation force applied. This means that greater actuation force is required to deliver a particular brake torque output from a disc brake compared with a drum brake, which has a significant self-servo effect. Since the brake pads of a disc brake contact only the side faces of the disc, radial thermal expansion during operation does not affect the torque generated while in the axial direction thermal expansion is small and of little consequence (except in parking brake mode). This permits small running clearances and large area actuators with a high mechanical advantage. Most modern passenger cars and light vans with hydraulically actuated brakes are also fitted with brake boosters to multiply the actuation force that originates from the driver effort (force on the brake pedal). Although the rotors of open disc brakes may be exposed to dirt, dust and water contamination, they have the advantage of being largely self-cleaning via the centrifuge action whereas braking debris tends to accumulate inside a drum brake. Disc rotation also helps dust and gases released during braking to escape. The inherent self-cleaning characteristic can be further assisted by slots in the pads, and grooves or holes in the rotors, although this also reduces their thermal mass (see Chapter 7).

In a hydraulically actuated disc brake (as fitted to a passenger car or light van) slave pistons in the caliper are forced against the pad backplates by hydraulic pressure, generating a normal force at each pad/disc friction interface. The hydraulic seals in the system are designed to provide a small amount of pad retraction via the mechanism of seal 'rollback', so that springs or other devices to move the pads clear of the disc when not being used are not required. This means that threshold pressures, i.e. the hydraulic actuation pressure required to overcome the forces of the springs or retraction mechanisms, are reduced and braking performance becomes more linear. Despite seal 'rollback', disc brake pads do often touch the disc surface while rotating, so residual brake drag losses are not completely eliminated, and some manufacturers are considering positive retraction of the brake pads in their quest for CO_2 emission reduction. For mechanically (or electromechanically) actuated disc brakes, e.g. on larger commercial vehicles that use compressed air for brake actuation, positive pad retraction is required, usually by a spring in the actuator chamber, and therefore threshold pressures are more significant. Threshold pressures and forces should always be included in brake performance calculations. It is possible that hydraulically and/or air-actuated disc brakes will eventually be superseded by electromechanical actuation. Potentially these could provide faster response and better control of running clearances as well as eliminating concerns about the risk of fluid vaporisation and end-of-life recovery of brake fluid in the case of hydraulically actuated systems. Concerns about safety and compliance with legislation coupled with the technological challenges of the high electric current demands have so far been a major inhibitor to the implementation of electromechanical actuation of road vehicle braking systems.

There are two designs of automotive disc brake caliper predominantly in use today. The original design of caliper had a separate actuator for each brake pad either side of the disc in the 'fixed' or 'opposed piston' hydraulic caliper arrangement, illustrated in Figure 5.1.

For each pad there may be one, two or more pistons to ensure that actuation force is uniformly distributed over the pad/disc friction interface. This is especially important for high aspect ratio pads, i.e. where the pad circumferential length is more than about twice its radial width. This design has very few moving parts and the actuation pistons on each side of the rotor give good actuation force equalisation on both sides, but do require brake fluid to be transferred from one side of the caliper to the other, across the 'bridge' part close to the disc, which may be hot. As a result, this design is susceptible to brake fluid vaporisation. The disadvantages include weight and size; the outboard piston requires space inside the road wheel and constrains the position of the disc with respect to the wheel, which is not compatible with the preferred design of many modern front suspensions and steering geometry (Figure 5.2). However, where very wide section tyres

Figure 5.1: 'Fixed' or 'Opposed Piston' Disc Brake Caliper (Courtesy Meritor).

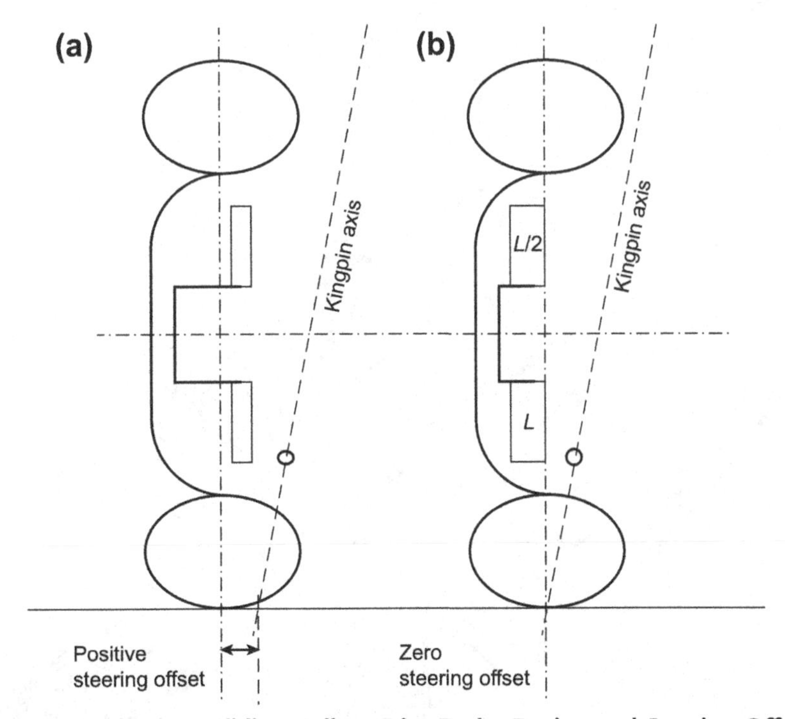

Figure 5.2: Fixed vs. Sliding Caliper Disc Brake Design and Steering Offset.
(a) Fixed caliper. (b) Sliding caliper.

permit positioning the brakes deep into large offset wheels, e.g. in high-performance luxury cars, fixed calipers with two or more pistons each side actuating large area high aspect ratio pads to accommodate the power dissipation requirements are popular.

The other, more recent, design of disc brake caliper is the 'sliding' (or 'fist') type caliper (Figure 5.3), which has a fixed carrier attached to the axle casing or mounting, e.g. the steering knuckle, and this carrier is fitted with two rods or pins on which the body of the caliper slides. The caliper body has an actuator on one side only (the inner side) because the force it applies to the inner pad backplate is reacted by the opposing force generated

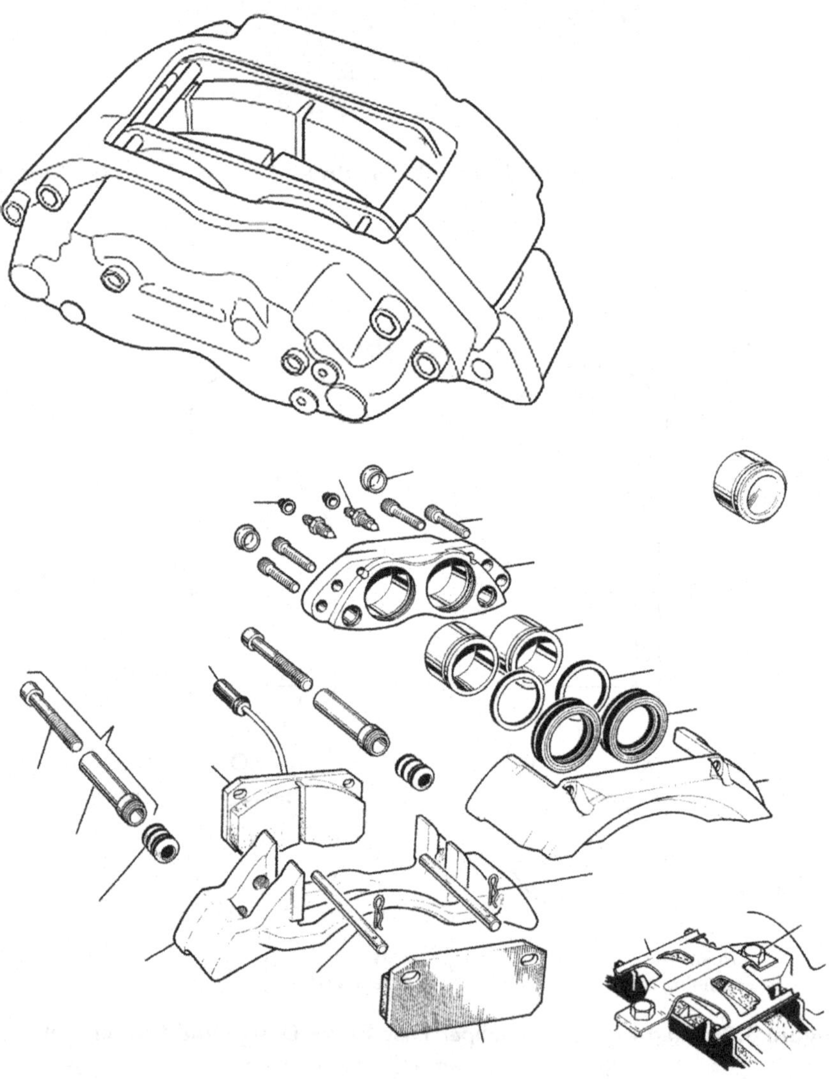

Figure 5.3: Sliding Type Disc Brake Caliper (Courtesy Meritor).

on the pad on the outer side of the disc as the caliper body slides. The applied force and the reaction force are almost equal and opposite; there is a small difference because of friction in the slide pins. Although this is low when new, it can adversely affect force equalisation when the slide pins are worn or damaged, e.g. by corrosion or water ingress and road debris contamination, so effective sealing of the slides is essential. The absence of an outboard actuator allows the rotor to be positioned further outboard, which in turn allows optimum positioning of the suspension lower ball joint to achieve the desired steering geometry. For hydraulic actuation, the brake fluid vaporisation risk is much reduced because there is no fluid path across the caliper.

Mounting the calipers ahead of the centre of the wheel can help caliper (and pad) cooling, but can also deflect cooling air away from the rotor. It can also increase the load carried by the wheel bearings because of the torque reaction developed at the calipers. For these reasons calipers are preferably mounted to the rear of the brake but for the front axles of front wheel drive (FWD) passenger car types of vehicles, packaging requirements of the steering linkage often dictate mounting the calipers forward of the wheel centre.

Torque Generated by a Disc Brake

A disc brake is illustrated in Figure 5.4, which shows an idealised sector-shaped brake pad in contact with one side of the friction ring of a brake disc superimposed on a

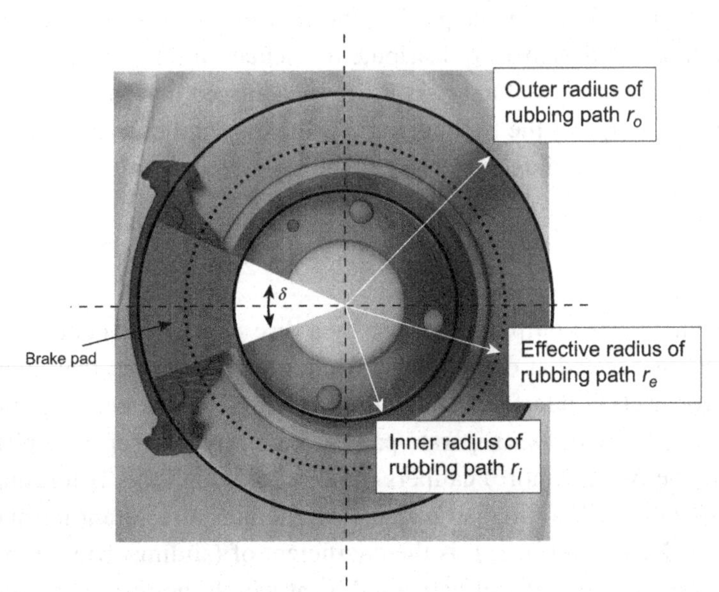

Figure 5.4: Disc Brake Showing the Pad as a Sector on the Rubbing Path.

photograph of an actual brake disc and pad. As the disc rotates its surface sweeps under the stationary disc brake pad and the 'swept area', 'rubbing path' or 'friction surface area' of the disc is calculated from the outer and inner radii (r_o and r_i respectively), which bound the swept area. This part of the disc is often termed the 'friction ring', and the part of the disc that connects the friction ring to the wheel hub is often termed the 'hat' or 'top hat' section.

$$\text{Friction surface area of the disc: } A_s = \pi\left(r_o^2 - r_i^2\right) \tag{5.1}$$

The friction surface area of the pad A_p depends upon the design of the pad; it is the area of the pad that is in contact with the disc. In Figure 5.4, the idealised pad is shown as a sector with inner and outer radii that match those of the disc rubbing path, while the actual pad on which it is superimposed is different in shape. Although disc brake pads come in many different shapes and sizes, usually the inner and outer edges are circular arcs equivalent to the inner and outer radii of the disc rubbing path, as illustrated. The friction surface areas of the pad and the disc are important in the thermal design of the disc brake (see Chapter 7).

The actuation force (P_a) applied to the brake pad backplate by the actuation mechanism is perpendicular (normal) to the disc plane and generates a pressure distribution over the pad friction surface area where it is in contact with the disc. This pressure distribution can be represented by a resultant normal force referred to as the 'clamp force' N_c, which acts at the 'centre of pressure' (CoP) of the pad friction contact area perpendicular to the disc plane. In practice the actuation force is applied by hydraulic pistons, 'fingers' on the outside pad of a sliding caliper, or 'tappets' in the case of large pneumatically actuated disc brakes, and these are designed to distribute the actuation force as required about the pad centre. However, for an initial analysis it can be assumed that the centre of pressure acts at the mean radius (r_m) of the rubbing path of the disc and thus the friction drag force also acts on the disc at this radius, which is termed the 'effective radius' (r_e), to generate the braking torque:

$$r_e = r_m = (r_o + r_i)/2 \tag{5.2}$$

When calculating the torque output generated by a disc brake, it is good practice to consider the torque generated at one pad/disc friction interface, and then multiply by the number of pads (or friction interfaces) acting on the disc. This is because some disc brakes may be designed with two separate pads actuated by two separate pistons each side, or there may be two (or more) calipers per disc, although this is unusual. If the actuation force (P_a) exerted on the pad backplate generates a resultant normal clamp force N_{c1}, at the pad/disc interface, μ is the coefficient of (sliding) friction between the pad friction material, and r_e is the effective radius at which the resultant friction drag force acts (for one pad Amontons' law states: $F_1 = \mu N_{c1}$), the braking torque τ_1 generated by one brake pad is:

$$\tau_1 = F_1 r_e \tag{5.3a}$$

Therefore:

$$\tau_1 = \mu N_{c1} r_e \tag{5.3b}$$

and for a standard design of automotive disc brake comprising one caliper and two pads the torque τ_w generated by the brake is:

$$\tau_w = \mu(N_{c1} + N_{c2}) r_e \tag{5.4a}$$

where N_{c1} and N_{c2} are the inner and outer pad clamp forces respectively. Assuming that $N_{c1} = N_{c2} = N_c$:

$$\tau_w = \mu \cdot 2N_c r_e \tag{5.4b}$$

It makes no difference whether the caliper is fixed or sliding; both types apply the same force to each pad, neglecting frictional losses in the sliding mechanism of a sliding caliper.

The classical definition of brake factor (BF) is the ratio of the total friction drag force generated by the brake stators to the total actuation force applied to the brake stators, which for the disc brake would be:

$$\text{Brake factor} = \mu \cdot 2N_c / 2N_c = \mu \tag{5.4c}$$

Or:

$$\text{Brake factor} = \tau_w (r_e \cdot 2N_c) = \mu \tag{5.4d}$$

From Equations (5.4c,d) it can be seen that the brake factor of a disc brake is equal to the coefficient of friction between the pad and the disc. However, for ease of comparison between different types of brake and actuation associated with the definition of parameters in common usage such as internal and external ratio and ηC^*, which are explained later, disc brake factor is now usually quoted as BF $= 2\mu$:

$$\text{BF} = \tau_w / (r_e N_c) = 2\mu \tag{5.4e}$$

$$\tau_w = \text{BF}(r_e N_c) \tag{5.4f}$$

For a hydraulically actuated disc brake, the actuation force P_a applied to a single pad backplate is:

$$P_a = (p - p_t) A_a \eta \tag{5.5}$$

where:

A_a is the total actuation area of the hydraulic piston(s) actuating the pad (m^2);

p is the actuation hydraulic line pressure (MPa);
p_t is the threshold pressure (MPa);
η is the efficiency of the hydraulic actuation system (discussed in Chapter 6).

So assuming for now (discussed later) that the pad actuation force (P_{ai}) is the same as the clamp force at the pad/disc interface (N_{ci}), combining Equations (5.4b) and (5.5) gives the wheel brake torque (τ_w):

$$\tau_w = 2\mu(p - p_t)A_a\eta r_e \tag{5.6a}$$

$$\tau_w = \mathrm{BF}(p - p_t)A_a\eta r_e \tag{5.6b}$$

If the axle is fitted with two brakes, each having one caliper with two pads, the axle has a total of four pads contributing to the braking torque, so the axle braking torque is:

$$\tau_{axle} = 4\mu(p - p_t)A_a\eta r_e \tag{5.7}$$

Example. A disc brake for a small passenger car has the following dimensions:

Disc: $r_o = 110$ mm, $r_i = 75$ mm, giving $r_m = 92.5$ mm.
Caliper: Piston diameter $D_p = 48$ mm, actuator area $A_a = 1810$ mm^2.

Assuming an actuation pressure of 40 bar (= 4 MPa), a threshold pressure of 0.5 bar (= 0.05 MPa) and a hydraulic actuation efficiency of $\eta = 0.95$:

$$P_a = 68.7 \text{ kN} \quad \text{(from Equation (5.5))}$$

$$\tau_w = 503 \text{ Nm} \quad \text{(from Equation (5.6))}$$

Friction Interface Contact and Pressure Distribution in Disc Brakes

The previous analysis assumed firstly that the clamp force is a point force at the friction interface that is the same as the brake pad actuation force (P_a) and secondly that this clamp force acts at an effective radius on the disc, which is the same as the mean radius of the rubbing path. In practice neither of these assumptions may be completely correct. The actuation force applied to the disc pad backplate is reacted by the distribution of clamp force over the pad/disc contact interface, which is usually termed the 'interface pressure distribution', and depends upon the flexural characteristics of the pad assembly: friction material thickness and compressive modulus; backplate thickness and Young's modulus; shape and size; and the way the actuation force is applied to the backplate, e.g. the number, size, and position of pistons in a hydraulically actuated brake (Tirovic and Day, 1991). Also important is the initial contact between the pad friction surface and the disc, which is governed by factors including the geometric conformity between the pad and the disc, determined by pad wear, disc distortion and actuation compliance. Figure 5.5

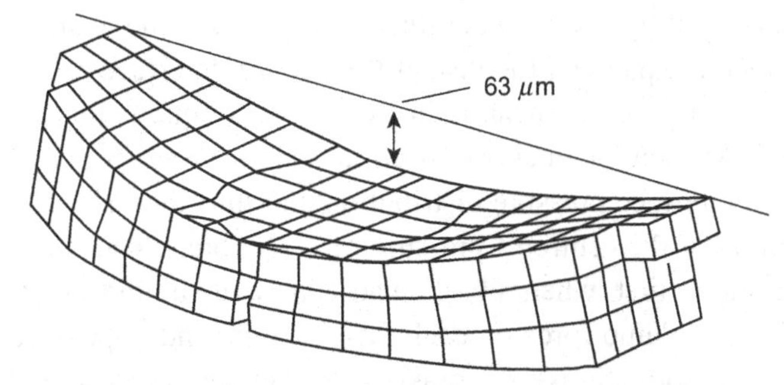

Figure 5.5: Disc Brake Pad Showing Deformation Under Actuation Force (Tirovic and Day, 1991).

(Tirovic and Day, 1991) illustrates how the pad compresses and deforms; clearly, if the backplate were infinitely stiff the friction material would compress uniformly and the friction interface pressure distribution would be uniform. The friction interface pressure distribution can be predicted by finite element (FE) analysis as illustrated in Figure 5.6; this indicates that the peak pressure is displaced towards the leading end of the pad, and also that the edges of the pad experience low interface pressure, possibly zero, indicating loss of interface contact. This is examined later in the context of the 'critical length' parameter. When the clamp force is distributed over the friction interface as illustrated in Figure 5.6, the pressure distribution can be considered to be represented by a single resultant normal force that acts at the centre of pressure (CoP) and is equivalent to the resultant normal force N_a previously defined.

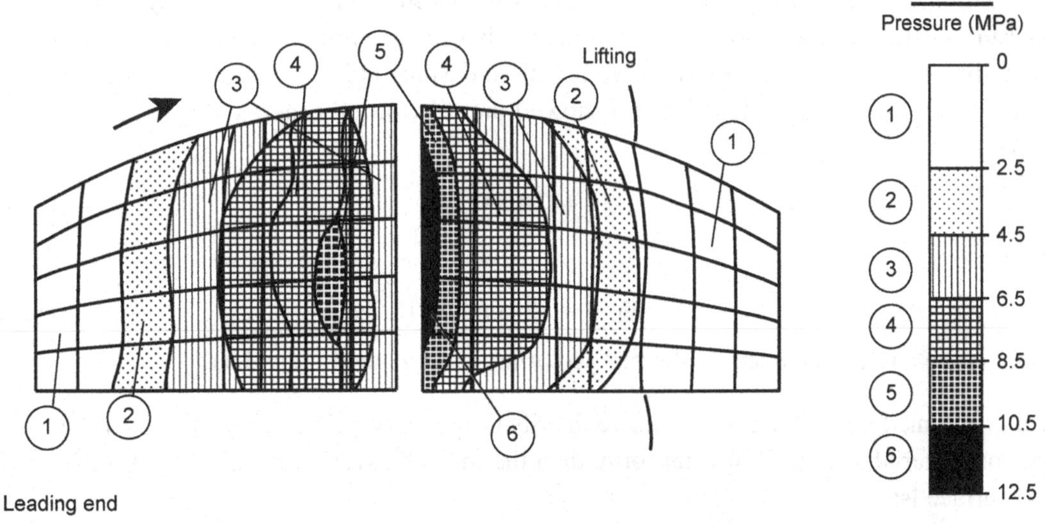

Figure 5.6: Disc Brake Pad Friction Interface Pressure Distribution Predicted by FE Analysis (Tirovic and Day, 1991).

The brake torque generated by a disc brake thus depends upon the position of the centre of pressure (CoP) between the pad and the disc at the friction interface relative to the disc rotational axis; if it sits outside the mean radius the brake torque is increased, and if it sits inside it is decreased. As mentioned above, friction material wear is one of the factors that influences the friction interface pressure distribution and thus the radial and circumferential position of the centre of pressure of a disc brake pad. Considering the radial pressure distribution first, when a brake pad is new (assuming the disc does not deflect or deform, i.e. the clamp force on each face is equal and opposite and there are no mechanical or thermal deformations), the brake actuation may be designed so that the centre of pressure sits at the mean radius. The linear or sliding speed (ωr) increases as the radius r increases, and since friction material wear is proportional to linear sliding speed and pressure, and temperature above a certain 'threshold' temperature (see Chapter 2), the work rate ($= \mu p v$) is higher towards the outer edge of the brake pad so the interface temperature and the wear rate will be higher than at the inner edge. Opposing this is the effect of relative surface area of the friction material outside and inside the mean radius. If the pad is designed to be in the form of a sector, the area outside the mean radius is greater than that inside the mean radius and since this means that there is more friction material to wear in the region where the wear rate is higher, the result may be that the pad does not wear tapered in the radially outward direction. In this condition a steady state of operation might have been established at the friction interface where the pressure distribution is uniform radially.

No brake is ever in a truly 'steady-state' operational condition all the time and despite the best efforts of brake designers, some radial wear variation of the friction material at the friction interface always occurs. The effect is that the pad will tilt slightly and the pressure distribution will no longer be uniform, but under normal usage this will not be excessive. In this state of operation, Newcomb and Spurr (1967) related the effective radius (r_e) to the mean radius (r_m) and showed that the torque generated by a disc brake is:

$$\tau_w = 2\mu N K_1 r_m \tag{5.4c}$$

where

$$K_1 = \delta/(2\sin(\delta/2)) \tag{5.8a}$$

and δ is the angle subtended by the pad sector arc length (see Figure 5.4).

If it is assumed instead that the pressure distribution over the contact area between the friction material and the disc is uniform, then the mean or average pressure (p_m) over the pad surface is:

$$p_m = N_c/A_p = (p - p_t)A\eta/A_p \tag{5.9}$$

where p_m is the mean pressure over the contact area between the friction material and the disc. This must not be confused with the pneumatic or hydraulic pressure in the actuation system. According to Newcomb and Spurr (1967) for the case of uniform pressure distribution the factor K_1 also depends on r_i and r_o:

$$K_{1(uniform\ pressure\ distribution)} = 2\delta\left(1 - \left((r_i r_o)/(r_i + r_o)^2\right)/3\ \sin(\delta/2)\right) \tag{5.8b}$$

For the same small passenger car brake disc as above, where $r_o = 110$ mm and $r_i = 75$ mm, giving $r_m = 92.5$ mm (e.g. for a small passenger car), and $\delta = 60°$, and the pressure distribution over the pad surface is uniform, $K_1 = 1.06$ (it would be 1.05 for the non-uniform pressure distribution case, Equations (5.8b) and 5.8(a) respectively). The brake torque generated would thus be 526 Nm for $K_1 = 1.05$.

In this design, $r_o/r_i = 1.47$. If the inner radius r_i were reduced to 55 mm, i.e. $r_o = 2r_i$, and $\delta = 60°$, K_1 would increase to 1.09 (it would remain at 1.05 for the non-uniform pressure distribution case). Newcomb and Spurr (1967) suggested that good practice is for $r_o/r_i \leq 1.5$ to minimise the effect of brake pad wear on the effective radius. Trends in modern automotive disc brakes have been towards larger outer diameters accommodated within larger diameter wheel rims with lower profile tyres. In these designs, the ratio r_o/r_i may be 1.4 or lower, which may require a longer (higher aspect ratio) pad design to provide enough pad friction area to meet power density criteria (see Chapter 6), but even so δ is unlikely in practice to exceed 60°, for which the value of K_1 is 1.05, while for $\delta = 45°$, K_1 is 1.03. The effective radius (r_e) can therefore be assumed to be the mean radius (r_m; Equation (5.2)) with less than 5% change from new to 'bedded' provided that the 'good practice' rule of $r_o/r_i \leq 1.5$ is maintained.

Disc brake designers can thus define the disc inner and outer radii to minimise the effects of radial pad wear and can also define the geometric shape of the pad, starting with the angle subtended by the pad sector arc length; effectively these two parameters define the 'aspect ratio' of the pad. Additionally designers may adjust the geometry of the caliper to bias the centre of pressure radial offset from the mean radius, and to counteract the effect of wear. The shape of the pad friction surface can be adjusted, e.g. to give more (or less) friction surface area outside the mean radius. Examples of disc brake pad shapes are shown in Figure 5.7 to illustrate this.

The pressure distribution at the friction interface of the brake pad also varies in the circumferential direction as illustrated in Figure 5.6, which shows the peak pressure to be displaced towards the leading end of the pad. This is because of the previously mentioned 'self-servo' action generated by the friction couple between the pad/disc interface and the pad backplate abutment, in this case compressive. Assuming the contact between the pad

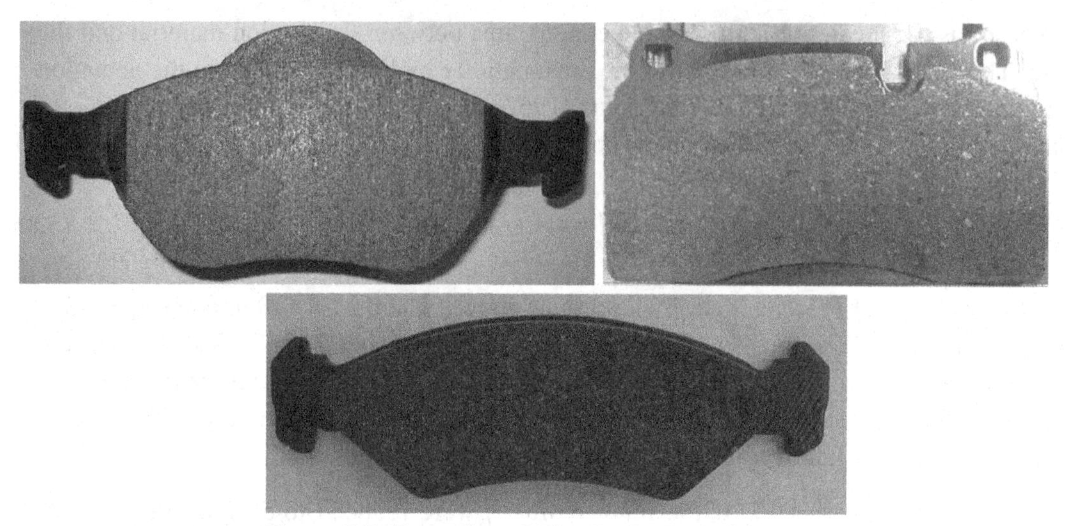

Figure 5.7: Examples of Disc Brake Pad Shapes.

and the backplate at the caliper abutment to be sliding friction, the force system is illustrated in a 2D model in Figure 5.8, where:

L = pad length (m);
x = distance from the leading end of the pad to the centre of pressure (m);
h_{ab} = height of the pad abutment reaction above the disc friction surface (m);
P_a = applied actuation force, assumed to be at the pad centre (N);
N = normal reaction between the pad and the disc (N);
μ = coefficient of friction between the pad and the disc;
μ_{ab} = coefficient of friction between the pad and the abutment.

Resolving forces vertically and horizontally and taking moments about the contact point of the pad backplate on the abutment:

$$x = ((L/2)(1 - \mu\mu_{ab})) - (\mu h_{ab}) \tag{5.10}$$

$$N_a = P_a/(1 + \mu\mu_{ab}) \tag{5.11}$$

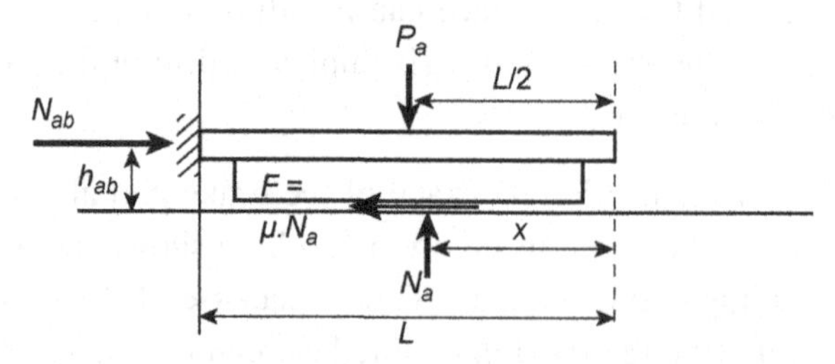

Figure 5.8: 2D Friction Force System on a Disc Brake Pad.

For the same small passenger car brake disc as above the brake pad has a length (L) of approximately 100 mm, thickness (h_{ab}) 15 mm, $\mu = 0.4$, $\mu_{ab} = 0.1$ and $x = 42$ mm, i.e. the centre of pressure is displaced by 8 mm towards the leading end of the pad. The pad normal reaction (N) would be 4% lower than the applied force (P_a) with a consequent effect on the generated torque; the 'servo factor' would be 0.96 and the brake torque based on r_e would be 505 Nm, very close to the 503 Nm originally calculated. The net result is that for this disc brake the assumption that $r_e = r_m$, is reasonable, and the servo factor may be regarded as insignificant and ignored. So the prediction of the torque generated by a disc brake according to Equations (5.5) and (5.6) is basically correct.

This analysis assumes that the backplate/caliper abutment contact is in compression, and is modelled as sliding friction contact. The coefficient of friction between the backplate and caliper abutment thus affects the force system, such that a well-lubricated contact (e.g. $\mu = 0.05$) would cause the centre of pressure to be displaced towards the leading end by 7 mm, and the pad normal reaction would be 2% lower than the applied force. In this condition the servo factor is insignificant and is usually ignored. In some high-performance brakes with multiple cylinders, this effect can be compensated for (only in the forward direction of rotation) by providing slightly smaller diameter cylinders at the leading end of the pad. If there is only one cylinder per pad, the same can be achieved by designing an offset between the centre of the cylinder and the centre of the pad. Both of these design features mean that the calipers would be 'handed' left and right.

If the backplate/abutment contact were completely seized so that no sliding could occur, it could be modelled as a pivoted joint, in which case the position of the centre of pressure would be determined not by force equilibrium but by the geometry of the mechanism, especially the wear profile of the pad friction material. A badly corroded or contaminated abutment can lead to reduced and erratic performance of the disc brake, usually accompanied by noise.

Some pad/caliper abutments are designed to react the friction drag force on the pad in tension; in this design the pad backplate has 'lugs' at each end, which engage with slots in the caliper abutments. The analysis of the force system for this design, assuming that the backplate/caliper abutment reaction is carried in tension at the leading end, gives the equation:

$$x = ((L/2)(1 + \mu\mu_{ab})) - (\mu h_{ab}) \tag{5.12}$$

where x is again the distance from the leading end of the pad to the centre of pressure. The result is that the displacement of the centre of pressure towards the leading end of the pad is reduced; in the example given the centre of pressure would be 46 mm from the pad leading end for the tension abutment compared with 42 mm for the compression abutment. This implies a slightly more uniform pressure distribution and consequently reduced tendency for taper wear in the tangential direction.

Manufacturers claim advantages for both designs of abutment but based on their usage on road vehicles there appears to be no strong preference for one design over the other. However, the clearance between the abutment edges of the pad, and the caliper abutment features, must be carefully controlled to allow axial movement of the pad (normal to the disc surface) under all conditions of use. Because the brake has to work in the forward and reverse directions, the pad backplate must present a clearance fit between the caliper abutment features at all times. If at any stage this clearance fit becomes tight, the pad assembly will stick in the caliper, usually exerting a significant normal force on the disc even when the actuation force is released. The result is a dragging brake that continues to heat up the backplate and jam it between the caliper abutments even more firmly. For a trailing abutment (compression) design this means that the thermal expansion of the backplate and the caliper, the potential for contamination or corrosion at the abutments, and the manufacturing tolerances on the pad backplates and the caliper abutments must all be examined carefully. For a leading abutment (tension) design, these parameters are equally important with the additional complication that they can combine to have the effect of changing the leading abutment design to a combination of leading and trailing, or even a trailing abutment with consequent changes in the brake characteristics. Because the brake is required to work backwards as well as forwards, too much abutment clearance may result in impact noise of the backplate on the abutment under friction force reversal, and this is unacceptable to many users.

So far it has been explained how the pressure distribution at the pad/disc friction interface is not uniform even for the simplest case where the pad, caliper and disc are assumed to be rigid bodies, i.e. flexural effects are ignored. This is because the interface pressure distribution and the position of the centre of pressure are influenced by the actuation and friction forces acting in the system and by the characteristic wear of the friction material. In practice, the pad, caliper and disc are not rigid bodies and therefore the effects of flexure, deformation and displacement in the brake assembly must be examined. Pressure and contact at the brake friction interface can be considered at three levels (Tirovic and Day, 1991):

1. Large-scale pressure variation over the full rubbing surface, which is induced by bulk deformation (usually flexural and thermal) effects in the application of actuation forces.
2. 'Macroscopic' interface pressure variation that arises from localised deformation or distortion of the rubbing surfaces; in resin-bonded composite friction materials this can be caused by localised thermal expansion, wear and surface coatings or films.
3. Frictional contact on the microscopic scale: asperity contact and interaction.

Large-scale pressure variations over the brake friction interface in the form of flexural effects in disc brake pads were studied by Harding and Wintle (1978), who introduced the concept of 'critical length' associated with the actuation of a disc brake pad. Based on 'beam on elastic foundation' theory they derived a formula to estimate the critical length of a brake pad under central point loading:

$$L_c = \pi^4\sqrt{((td^3E)/3E_c)} \tag{5.13}$$

where

L_c = critical length (m);
t = pad friction material thickness (m);
d = backplate thickness (m);
E = Young's modulus of the backplate (MPa);
E_c = compressive modulus of the friction material (MPa).

For the small pad previously taken as an example (length 100 mm, t = 12 mm, d = 4.5 mm, E = 210 GPa, E_c = 500 MPa) the critical length L_c = 62 mm. This indicates that if this pad were actuated by a point force at the centre of the pad backplate, the pad assembly would deflect to create a contact/pressure distribution that extends only over a length of 62 mm, i.e. 31 mm either side of the pad centre line. Since the pad is 100 mm long, this means that 38 mm of the friction material is not initially in contact with the disc and thus the frictional work is not done by this pad over its full surface area as designed (until it has bedded-in, i.e. worn to bring the non-contacting regions into contact). Since the working area is reduced in the ratio of 38%, the power dissipation is correspondingly increased, which may overload the friction material thermally and cause thermal cracking in the disc because of the localised high heat flux (see Chapter 7). The friction material will wear excessively at the centre of the pad, and when the actuation force is released the friction surface becomes concave, which can lead to more operational problems. It has also been shown that the pad flexural deformation introduces tensile stresses at the bond line between the friction material and the backplate, which can lead to bond failure and eventual separation of the friction material from the backplate. The process of failure is accelerated by the ingress of moisture and contamination at the bond line cracks.

In practice, actuation forces are not applied as a point force acting on the pad backplate, but are distributed over the area under the actuation piston or pistons in the case of a hydraulically actuated brake, the fingers of the outer (reaction) side of a sliding caliper, or the tappet or load carrier in an air or electromechanically actuated brake. The result is that the critical length is increased, but the concept is still valuable in assessing the need for distributed actuation forces, e.g. with high aspect ratio pads. As previously mentioned, FE analysis can be used to predict the contact/pressure distribution for different designs of pad and actuation arrangements that enable the caliper actuation design to be optimised. As yet there is no reliable or accurate way of measuring the pressure distribution at the friction interface of a disc brake pad dynamically, i.e. while it is operating; measurements can only be made under static conditions, e.g. using pressure-sensitive paper.

The deformation, displacement and distortion of the disc and caliper, as well as the pad assembly, must also be considered in disc brake design. The following list indicates the major causes and effects:

- Caliper deflection under clamp force. The force applied by the actuation mechanism on the pads is reacted by the body of the caliper and because the caliper bridge (fixed design) or pins (sliding design) is offset from the actuation forces, the caliper tends to 'open up' under clamp load. The effect is to increase the friction interface pressure towards the outer edge of each pad, increasing radial taper wear at high braking duty. 'Frame' type sliding calipers aim to minimise this by aligning the clamp load and actuation forces (see Figure 5.3).
- Caliper twisting under friction drag loading. The caliper mounting (to the axle or suspension) is offset from the disc plane so there is a moment arm that causes the caliper to twist in the direction of wheel rotation. With a fixed caliper design, the effect is to increase the pressure at the trailing end of the outer pad, and the leading end of the inner pad, but for the sliding caliper there is the additional effect of twisting the caliper on the guide pins or sliding elements. This may lead to high slide friction or even seizure of the caliper on the slides, especially when badly worn.
- Disc coning. The disc temperature rises during braking (see Chapter 7) and thermal stresses are generated that cause the disc to 'cone', i.e. the axial deflection of the disc at the outer radius is different to that at the inner radius. The result is that one pad sees the centre of pressure displaced towards the outer radius, and the other pad sees it displaced towards the inner radius. This thermomechanical deflection is analysed in Chapter 7; the effect can be to concentrate the braking work over a smaller area of the friction surfaces than designed, with increased potential for thermal damage, increased wear and failure.

These effects can easily be investigated using FE analysis of disc brake individual components with appropriate thermal and mechanical boundary conditions and load cases being applied. The interactions between individual components, especially the effect of frictional contact and pressure distribution on the magnitude and distribution of mechanical and thermal loads, are more complicated to investigate and there are many examples in the published literature to guide advanced brake modelling for design analysis.

Drum Brakes

Basic Principles

A modern automotive drum brake is the 'internal expanding' design, in which friction material 'linings' on stators known as 'shoes' are forced radially outwards against the internal surface of the rotor, which takes the form of a brake 'drum' (see Figure 5.9). The friction surface of the brake linings covers most of the drum's friction surface area, typically around 70% compared with 15% for the disc brake. The brake shoes usually consist of a 'platform', which is a cylindrical section to which the brake lining(s) is attached, and a 'web' (also termed 'rib') to provide flexural stiffness to the assembly. Small drum brakes use fabricated mild steel shoes, while commercial vehicle brakes may use cast iron shoes. The brake shoes are mounted on a torque plate (also known as a backplate or an anchor bracket, especially in commercial vehicle drum brakes), which is attached to the axle or suspension of the vehicle. The brake drum is attached to the wheel hub and rotates with it, so as the shoes are forced out against the drum inner surface, the brake drum and the vehicle are retarded. Most automotive drum brakes have two brake shoes, which can be configured in several different ways, but the simplest design is the 'leading/trailing' (simplex or L&T) drum brake because it has one leading shoe and one trailing shoe (as illustrated in Figure 5.9). The brake shoe whose actuation causes it to rotate in the same direction as the brake drum is called a 'leading' shoe, while the brake shoe that is actuated to rotate in the opposite direction from the drum is called a 'trailing' shoe. A summary of different drum brake configurations and types is given in Table 5.1.

In small drum brakes as fitted to many modern mass-produced passenger cars, the actuation is by a hydraulic slave cylinder, which acts directly on the shoe tip to apply the actuation force. For larger brakes on commercial vehicles (e.g. heavy trucks) a

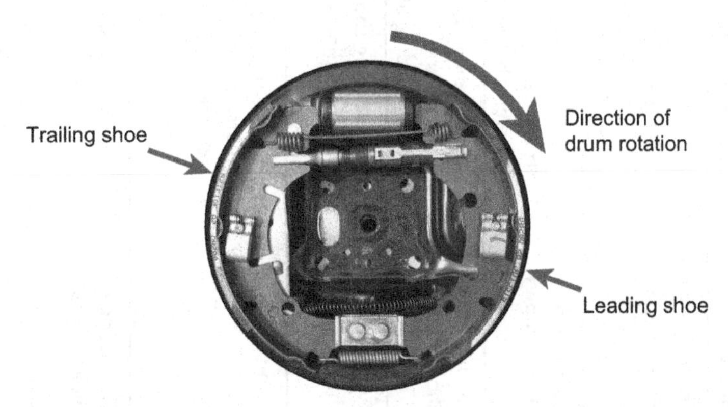

Figure 5.9: Simplex Drum Brake Showing Leading and Trailing Shoes.

Table 5.1: Types of Drum Brake Configuration and Actuation

Type of Drum Brake	Configuration	Actuation	Applications	Comments
Leading/trailing shoe (simplex), Figure 5.14	One leading and one trailing shoes Parallel or angled sliding abutments	Hydraulic (mechanical or electromechanical for parking brake actuation)	Passenger car and van rear axles Light commercial vehicles	Pivoted abutments now only used on very light duty vehicles, e.g. motorcycles
Leading/trailing shoe (simplex)	One leading and one trailing shoes Pivoted abutments	Air S-cam	Commercial vehicles and semi-trailers	Also with 'Z-cam' actuation Pivoted abutments, also termed 'anchors'
Leading/trailing shoe (simplex)	One leading and one trailing shoe Parallel sliding abutments	Air wedge	Commercial vehicles	
Twin leading shoe (duplex)	Two leading shoes Parallel sliding abutments	Air wedge	Used on commercial vehicles with packaging restraints	Reverse operation, much lower torque than forward operation
Duo-servo	Two leading shoes, abutment of primary shoe actuates secondary shoe	Hydraulic	No significant current usage as service brake Used as secondary brake in 'drum-in-hat' disc brake designs	Very high brake factor, easily leading to unstable performance
Twin leading shoe (duplex)	Two leading shoes	Hydraulic	No significant current usage on passenger cars	Reverse operation, much lower torque than forward operation
Leading/trailing shoe (simplex)	One leading and one trailing shoe Parallel or angled sliding abutments	Hydraulic	Passenger car and van rear axles Light commercial vehicles	Pivoted abutments now only used on very light duty vehicles, e.g. motorcycles

pneumatically powered mechanical actuator linked to a cam or a wedge mechanism is commonly used, although electromechanical actuation is also possible. The friction drag force generated on each brake lining is reacted through the brake shoe at an abutment on the torque plate. These abutments can take the form of a pivot, where the shoe is mechanically constrained to rotate about a fixed axis parallel to the vehicle's axle, or a sliding contact that permits radial movement as well as axial rotation. The torque plate also holds the springs or devices that locate the shoes correctly in the wheel plane, and mechanisms for adjusting and maintaining the correct running clearances between the friction material and the drum surface. Drum brakes are convenient for actuation as a parking brake and light vehicle drum brakes with hydraulic actuation for service braking often incorporate separate mechanical actuation for parking. 'Drum in hat' parking brakes are sometimes used on passenger cars with disc brakes on all four wheels; these comprise a small drum brake inside the 'hat' region of a brake disc. Where a parking function is required on a heavy vehicle drum brake, a pneumatic 'spring brake' actuator can be mounted to operate through the same cam or wedge mechanism. A spring brake actuator uses the air actuation to hold the brake off, and is a failsafe device; if control line air pressure is lost, the brake is automatically applied by the action of the spring actuator.

Torque Generated by a Drum Brake

Calculating the torque generated by a drum brake starts by considering the total friction drag generated by the brake lining (friction material) on each brake shoe as it is forced against the rotating brake drum. At any angular position on the brake lining the friction drag force is the product of the radial force at that point and the coefficient of friction μ, and since it acts at a constant radius (r_e, the inner radius of the brake drum) its direction varies; it is always tangential to the brake drum. In order to calculate the torque generated by a drum brake, it is therefore necessary to know what the normal force distribution (more commonly known as the pressure distribution) is around the lining. Classical brake analysis methods have now largely been replaced by computer-based methods including finite element analysis (Day et al., 1979), but the basic mechanics of the drum brake are illustrated here using a simplified drum brake, as shown in Figure 5.10. This brake comprises a leading shoe and a trailing shoe, pivoted about points marked H_1 and H_2 and lined with a small pad of friction material that is modelled as being in point contact with the brake drum. The actuation forces P_a on the leading and trailing shoes are assumed to be the same. When analysing the mechanics of a drum brake, it is good practice to analyse each brake shoe individually and then combine the two to give the total brake torque.

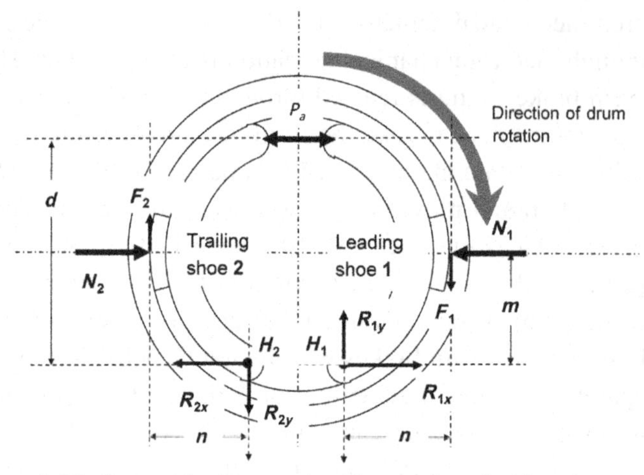

Figure 5.10: Forces Acting on the Shoes of a Simple Drum Brake.

Leading shoe:

$$\text{Resolving forces horizontally: } P_a + R_{1x} - N_1 = 0 \tag{5.14}$$

$$\text{Resolving forces vertically: } R_{1y} - F_1 = 0 \tag{5.15}$$

$$\text{Moments about } \mathrm{H}_1 : P_a d + F_1 n - N_1 m = 0 \tag{5.16}$$

The radial force generated on the pad of friction material is:

$$N_1 = P_a d/(m - \mu n) \tag{5.17}$$

Amontons' law states $F_1 = \mu N_1$; therefore, the friction drag force generated is:

$$F_1 = \mu N_1 = \mu P_a d/(m - \mu n) \tag{5.18}$$

Hence the ratio of the friction drag force generated to the actuation force applied is:

$$F_1/P_a = \mu d/(m - \mu n) \tag{5.19a}$$

Similarly for the trailing shoe:

$$F_2/P_a = \mu d/(m + \mu n) \tag{5.19b}$$

It can be seen that there is a servo factor associated with each brake shoe and the relationship between the applied force (P_a) and the friction force (F_1 or F_2) is a non-linear function of μ as defined by Equations (5.19). The ratio of the frictional force generated by each brake shoe at the drum radius to the applied force is called the 'shoe factor' (S_1 for the leading shoe and S_2 for the trailing shoe). This means that the total friction drag generated by this drum brake is:

$$F_1 + F_2 = P_a(S_1 + S_2) \tag{5.20}$$

The sum of the two shoe factors can be termed the 'combined shoe factor' (CSF) and this is often called the 'brake factor' of a drum brake. As previously stated in the context of the disc brake, the classical definition of brake factor (BF) is the ratio of the total friction drag force generated by the brake stators to the total actuation force applied to the brake stators. This would be the 'mean shoe factor' (= CSF/2) but again for ease of comparison between different types of brake, which is explained later, drum brake factor is usually quoted as the combined shoe factor (BF = CSF) nowadays.

The torque output generated by a drum brake is calculated from the total friction drag generated by all the brake shoes in the brake multiplied by the friction (or inner) radius r_e of the brake drum (neglecting any losses in the actuation system), which is a general rule for drum brakes. The brake torque developed by this brake (same as the disc brake) is given by:

$$\tau = \mu(N_1 + N_2)r_e \tag{5.4a}$$

which can be written specifically for the drum brake as:

$$\tau = (F_1 + F_2)r_e = P_a(S_1 + S_2)r_e \tag{5.21}$$

where r_e (the effective radius) is the inner radius of the drum that, unlike the disc brake, is constant.

The abutment reactions are:

Leading shoe:

$$R_{1x} = P_a(d - m + \mu n)/(m - \mu n) \tag{5.22a}$$

$$R_{1y} = \mu P_a d/(m - \mu(n - b)) \tag{5.22b}$$

Trailing shoe:

$$R_{2x} = P_a(d - m - \mu n)/(m + \mu n) \tag{5.22c}$$

$$R_{2y} = \mu P_a d/(m + \mu n) \tag{5.22d}$$

In this simple design, the friction drag force is reacted entirely by the pivoted abutment force R_{1y}, whereas all practical designs of drum brake have friction material linings extending over longer arcs so the brake drum itself provides some of the reaction force. This enables sliding abutments to be employed; the shoes have radiused ends that react against flat abutments to allow the brake shoe to 'float' on the torque plate, self-centring to allow a more even distribution of lining wear over the arc length and accommodate any lack of concentricity between the drum and lining. This design adopts an equilibrium position more quickly than a pivoted shoe, which being mechanically constrained by the pivot has to rely upon wear of the brake lining to reach a steady state of operation. For this reason, sliding abutment designs are used extensively on brakes that are required to reach a steady-state performance as quickly as possible, especially where manufacturing costs are important, since the alternative is to machine the linings to match the drum inner radius (termed 'radius grinding'), as is often necessary with large commercial vehicle drum brakes. As will be discussed later, the contact and pressure distribution between the brake linings and the drum can significantly affect the shoe factor and hence the brake torque performance, so where pivoted shoe brakes are used, e.g. with fixed mechanical actuators such as the 'S-cam', great care has to be taken to ensure good geometric conformity of the brake lining and drum friction surfaces in the new condition. Sliding abutments may be parallel, where the abutment provides no contribution to reacting the vertical component of the friction drag force (R_{1y} in Figure 5.10), or angled, where the abutment provides a small contribution. In practice the angled abutment (typically no more than 30°) seems to be preferred as it enables designers to optimise lining arc length and disposition on the shoe with respect to performance and wear. Friction at the abutment interfaces of a floating shoe does not appear to cause any appreciable difference in shoe factor provided that seizure is avoided by good maintenance.

The distribution of contact, pressure and friction drag forces over friction material linings that extend over longer arc lengths is the main reason why the design analysis of drum brakes is complicated. However, the basic principles demonstrated above still apply, in particular the relationship between the applied actuation force and the friction drag force generated (the shoe factors), which generally takes the form:

$$S_1 = \frac{\mu\alpha}{(\beta - \mu\chi)} \tag{5.23a}$$

and

$$S_2 = \frac{\mu\alpha}{(\beta + \mu\chi)} \tag{5.23b}$$

where α, β and χ are positive value constants determined by the geometrical dimensions of the brake, including the position and arc length of the friction material on each shoe, and may be different for the two shoes in the brake. These relationships for the leading and trailing shoe factors indicate that S_1 is greater than unity while S_2 is less than unity; thus, a leading shoe factor is always greater than a trailing shoe factor. Both leading and trailing shoe factors depend upon the coefficient of friction μ of the friction material lining. The trailing shoe factor tends to a constant value as μ increases, while the leading shoe factor tends to infinity as μ increases to approach a value of β/χ. This condition is known as 'spragging' and must be avoided as it indicates a situation where the shoe is completely self-energising, would lock into engagement with the drum, and cannot be released simply by removing the actuation force. In drum brake design it is essential to evaluate the level of lining friction coefficient for sprag to occur at the design stage (see Chapter 12).

Classical methods (i.e. by analysis and not by computer modelling) of calculating the torque generated by a drum brake start by considering the total friction drag ($F_1 + F_2$), for which it is necessary to know what the pressure distribution is around the lining. It has generally been assumed that the brake drum and shoe are rigid and the friction material lining behaves elastically in compression, so that the friction force at any point on the lining surface is proportional to the normal force, i.e. the radial component of the lining compression force at that point. Thus, the radial force at any individual small element of contact along the lining arc is directly proportional to the virtual displacement of that element. The pressure (p), defined as the force per unit area over a small element located at an angle β from the lining centre line, is thus (Newcomb and Spurr, 1967):

$$p(\beta) = Q\cos(\psi - \beta) \tag{5.24}$$

where:

β is the angular position of the point on the lining surface measured from the centre of the lining arc length, where
$-\delta \leq \beta \leq +\delta$, and 2δ is the lining arc length;
ψ is the angle between the horizontal centre line of the brake and a line perpendicular to the line joining the shoe abutment point to the central axis of the brake (see Newcomb and Spurr (1967) for more information);
Q is a numerical factor that depends upon the actuation force and the total lining surface area.

The result for the case of the pivoted shoe is the classical 'sinusoidal' pressure distribution, as shown for a leading shoe in Figure 5.11. The peak pressure is displaced towards the actuator end, which is the leading end for the leading shoe (and the trailing end for the trailing shoe). For design calculations it has been found to be more representative of steady-state operating conditions to assume a uniform pressure over the lining arc length. This has been shown to give results that are in better agreement with experimental results and finite element analysis predictions, which include flexural effects of the shoe and thermal expansion effects of the drum (Day et al., 1979; Day, 1987; Day et al., 1991). The pressure distribution between the lining and the drum is also affected by the geometric conformity between the drum and the lining on the brake shoe. Mechanical and thermal distortion of the brake components can have a significant effect on this contact and thus the pressure distribution between the brake shoes and the brake drum. If the drum inner radius is greater than the radius of the lining

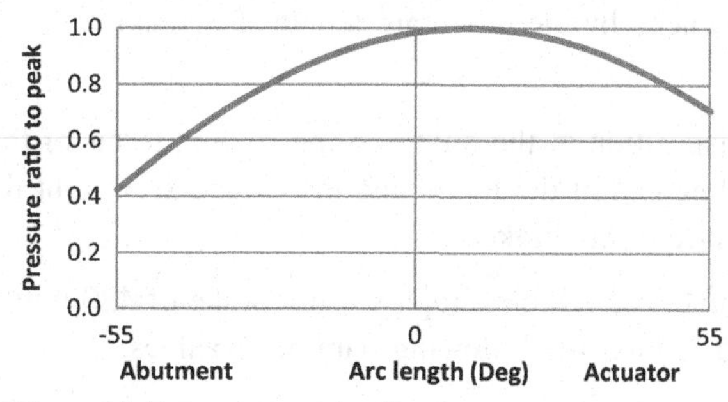

Figure 5.11: Classical 'Sinusoidal' Pressure Distribution Between the Brake Lining and Drum of a Leading Shoe.

surface arc, which may occur for example with a worn drum and new brake shoes, the pressure can be concentrated at the maximum pressure point, giving the condition known as 'crown contact'. Conversely, if the drum inner radius is less than the radius of the lining surface arc, which may occur for example with a new drum and worn brake shoes, the pressure can be concentrated at the ends of the lining, giving the condition known as 'heel-and-toe contact'. In the former case the shoe factor is reduced, while in the latter case the shoe factor will be considerably increased. Drum brake performance is influenced by lining wear, which tends to stabilise the brake performance at particular pressure/temperature conditions, while at the same time accentuating the effect of variations from these conditions. As with disc brake pads, wear of the friction material will always cause the friction interface pressure to tend towards uniform distribution since regions of high pressure experience higher rates of wear than lower pressure regions. The effect of lining contact and pressure distribution is discussed later in this chapter.

Limpert (2011) presents detailed methods of design analysis for different types of drum brakes based on the calculation of shoe factors and brake factors. For a pivoted brake shoe the shoe factors are calculated from:

$$S_1 = \frac{\mu \cdot \frac{d}{r_e}}{\frac{a'}{r_e}\left\{\frac{\alpha_0 - \sin\alpha_0 \cos\alpha_3}{4\sin\left(\frac{\alpha_0}{2}\right)\sin\left(\frac{\alpha_3}{2}\right)}\right\} - \mu\left\{1 + \left(\frac{a'}{r_e}\right)\cos\left(\frac{\alpha_0}{2}\right)\cos\left(\frac{\alpha_3}{2}\right)\right\}} \tag{5.25a}$$

$$S_2 = \frac{\mu \cdot \frac{d}{r_e}}{\frac{a'}{r_e}\left\{\frac{\alpha_0 - \sin\alpha_0 \cos\alpha_3}{4\sin\left(\frac{\alpha_0}{2}\right)\sin\left(\frac{\alpha_3}{2}\right)}\right\} + \mu\left\{1 + \left(\frac{a'}{r_e}\right)\cos\left(\frac{\alpha_0}{2}\right)\cos\left(\frac{\alpha_3}{2}\right)\right\}} \tag{5.25b}$$

where:

d is the distance between the actuation point and the pivot point of the shoe (parallel to the central axis of the brake, which is defined as the axis running through the mid points of the actuator(s), the pivot(s) and the brake centre; see Figure 5.12) (m);
R_e is the drum inner radius (m);
a' is the distance from the centre of the brake to the pivot centre (m);
α_0 is the angle subtended by the lining arc length (Degrees);
$\alpha_3 = \alpha_0 + 2\alpha_1$.

where α_1 defines the disposition of the lining on the shoe in terms of the angle between the position of the leading end of the lining and the extended straight line between the pivot centre and the centre of the brake.

Example. A large commercial vehicle simplex drum brake of 420 mm diameter with pivoted shoes (Figure 5.13) has the following parameter values:

$d = 324$ mm;
$R_e = 210$ mm;

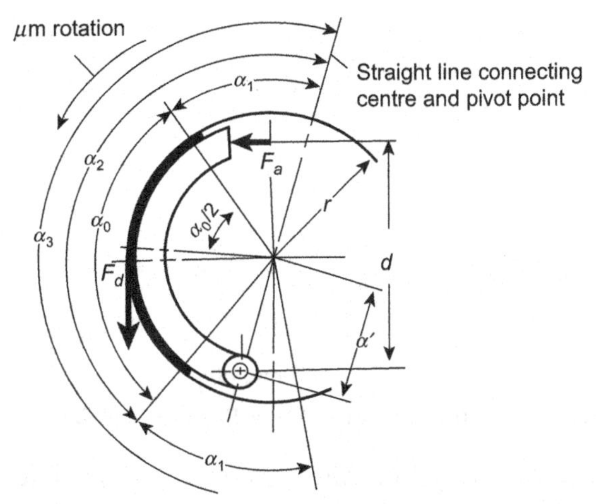

Figure 5.12: Pivoted Shoe Drum Brake Analysis (Limpert, 2011).

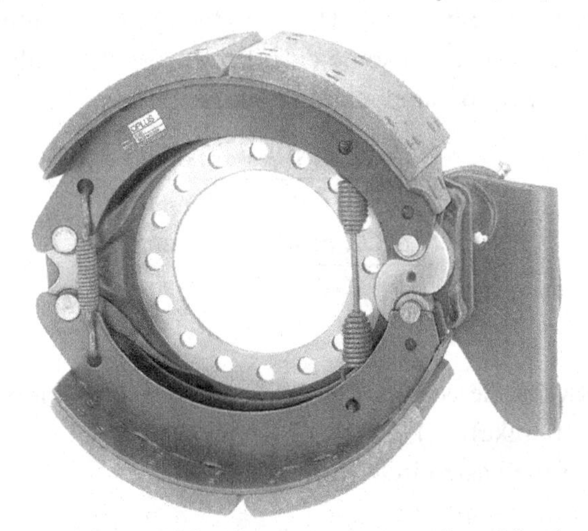

Figure 5.13: Large Commercial Vehicle Simplex Drum Brake with Pivoted Shoes and S-Cam Actuation (Courtesy Meritor).

$a' = 159$ mm;
$\alpha_0 = 110°$;
$\alpha_1 = 57.6°$;
$\alpha_3 = 225°$.

The calculated shoe factors from Equations (5.25) for a range of μ values are shown in Figure 5.14. At a typical μ of 0.38, the leading shoe factor S_1 is 1.78, while the trailing shoe factor S_2 is 0.61. These results are compared with those from other methods of brake factor calculation later in this chapter.

For a sliding abutment brake shoe, the position of the centre of pressure (or drag) is not specified by the geometry of the brake (because the shoe is floating), but this is

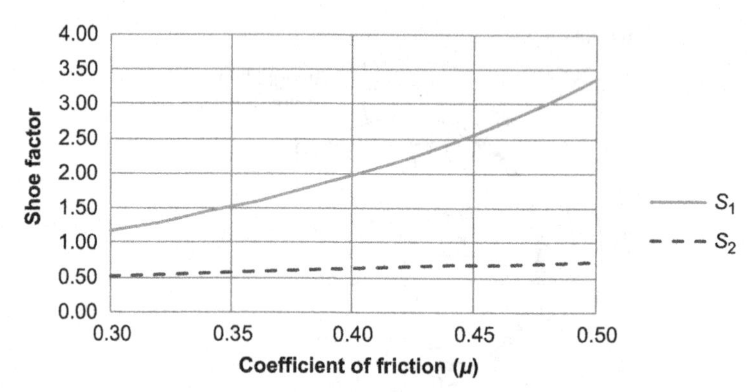

Figure 5.14: Shoe Factor vs. μ for a Pivoted Shoe Brake (Calculated from Limpert, 2011).

compensated for by the knowledge of the direction of the abutment reaction force between the shoe tip and the abutment face (the direction is perpendicular if the shoe/abutment friction coefficient is assumed to be zero). The shoe factors for a sliding abutment brake are given by Limpert (2011) but can also be calculated from the analysis given by Newcomb and Spurr (1967) for parallel sliding abutments:

$$S_1 = \frac{d(OZ\cos(\phi - \chi)\sin\phi)}{r_e\{b - OZ\cos(\phi - \chi)\sin\phi\}} \tag{5.26a}$$

$$S_2 = \frac{d(OZ\cos(\phi - \chi)\sin\phi)}{r_e\{b + OZ\cos(\phi - \chi)\sin\phi\}} \tag{5.26b}$$

where:

d is the distance between the actuation point and the pivot point of the shoe (parallel to the central axis of the brake) (m);
OZ is the diameter of the basic circle (m) $= 2kr_e$, where
 $k = 2\sin(\alpha_0/2)/(\alpha_0 + \sin(\alpha_0))$;
 α_0 is the angle subtended by the lining arc length;
ϕ is $\tan^{-1}\mu$;
χ is the angle between the centre of the lining arc and the line perpendicular to the central axis of the brake (Degrees);
b is the distance between the brake centre and the shoe abutment (parallel to the central axis of the brake) (m).

Example. The hydraulically actuated simplex brake with equal diameter pistons actuating the leading and trailing shoes shown in Figure 5.9 is fitted to the rear axle of a medium-sized passenger car. The geometry of this brake for classical analysis is as follows:

Leading shoe:

$\chi = 10°$
$\alpha_0 = 110°$, $\delta = 55°$

$OZ = 131$ mm
$b = 88.8$ mm
$d = 176$ mm
$d/r_e = 1.54$
$\phi = \tan^{-1}\mu$

Trailing shoe:

$\chi = -7°$ (above horizontal),
$\alpha_0 = 110°$, $\delta = 55°$
$OZ = 131$ mm
$b = 88.8$ mm
$d = 176$ mm
$d/r_e = 1.54$
$\phi = \tan^{-1}\mu$

$$OZ = 2kr, \quad \text{where } k = \frac{2 \sin \delta}{2\delta + \sin 2\,\delta} = \frac{1.6383}{1.92 + 0.94} = 0.573$$

$$\therefore OZ = 2 \times 0.573 \times \frac{228.6}{2} = 131 \text{ mm}$$

From Equations (5.26), S_1, S_2 and the combined shoe factor can be calculated for a range of values of lining friction coefficient μ, as shown in Table 5.2 and in Figure 5.15. At a typical μ of 0.38, the leading shoe factor S_1 is 1.64, while the trailing shoe factor S_2 is 0.52. These results are compared with those from other methods of brake factor calculation later in this chapter.

Millner (1972) showed that based on the classical theories the effect of the various design parameters of drum brakes was as follows:

- The leading shoe factor (S_1) increases with d, the distance between the actuation force and the abutment reaction force.
- The leading shoe factor (S_1) increases as the distance (b) between the abutment and the brake centre decreases.
- Increasing the sliding abutment angle increases the leading shoe factor (S_1) and decreases the trailing shoe factor (S_2).

Table 5.2: Shoe Factors and Brake Factor for a Parallel Abutment Sliding Shoe Passenger Car Drum Brake (Equations 5.26)

μ	S_1	S_2	$S_1 + S_2$
0.2	0.63	0.33	0.96
0.3	1.12	0.43	1.55
0.4	1.78	0.50	2.28
0.5	2.65	0.55	3.20

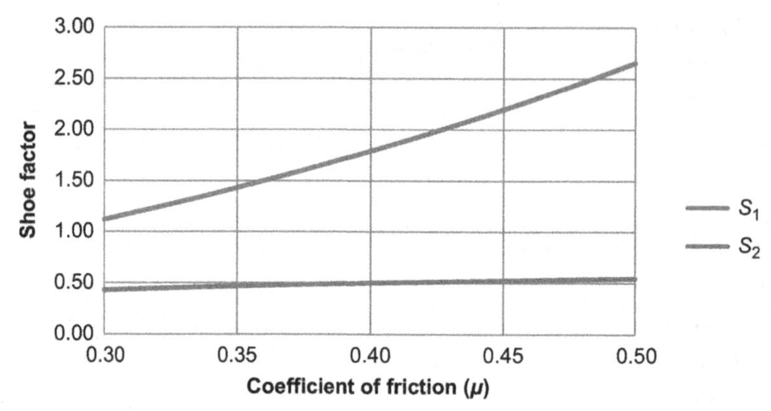

Figure 5.15: Shoe Factor vs. μ for a Parallel Sliding Abutment Shoe Brake (Equations 5.26).

- The leading shoe factor (S_1) increases as the lining position is moved towards the trailing end of the shoe.
- The leading shoe factor (S_1) increases as the lining arc length increases.

These included the assumption that the brake lining was in contact with the drum over the full arc length with a defined pressure distribution. Compared with the leading shoe, the trailing shoe is relatively insensitive to geometry changes, but care must always be taken to avoid unintended consequences from apparently minor design changes to the trailing shoe as well as the leading shoe of a drum brake. In one design of drum brake fitted to the rear axle of a small passenger car the disposition of the friction material lining on the trailing shoe lining was found to initiate a 'sprag' effect in the forward direction when the parking brake was applied. Another 'sprag' type instability, this time in the reverse direction, was also identified and investigated by the same authors and found to be associated with the leading shoe, which was acting as the trailing shoe in reverse rotation. This is explained in more detail in Chapter 12 (Harding and Day, 1981).

The same hydraulically actuated simplex brake has been analysed using the FE method to predict the shoe factors S_1 and S_2 (see Table 5.3). The shoe factor$-\mu$ relationship is compared with the results from classical analysis in Figure 5.16. The agreement between the two methods can be seen: both sets of results demonstrate how the shoe factor is dependent on the coefficient of friction of the brake lining (μ); for a typical value of $\mu =$ 0.36, the FE analysis predicts the leading shoe factor (S_1) is about 1.67 and the trailing shoe factor (S_2) is about 0.51. The combined shoe factor (CSF) is 2.18. These values are slightly higher than those calculated by the classical method, which is attributed to the effect of the friction interface contact/pressure distribution. It has been demonstrated (Day, 1987) that the use of computer modelling methods for drum brake analysis is better able to predict the effects of design changes and usage (wear, temperature), as discussed later.

Table 5.3: Shoe Factors and Brake Factor for a Parallel Abutment Sliding Shoe Passenger Car Drum Brake – FE Analysis

μ	S_1	S_2	$S_1 + S_2$
0.2	0.70	0.35	1.05
0.3	1.25	0.45	1.70
0.4	2.00	0.55	2.55
0.5	3.25	0.60	3.85

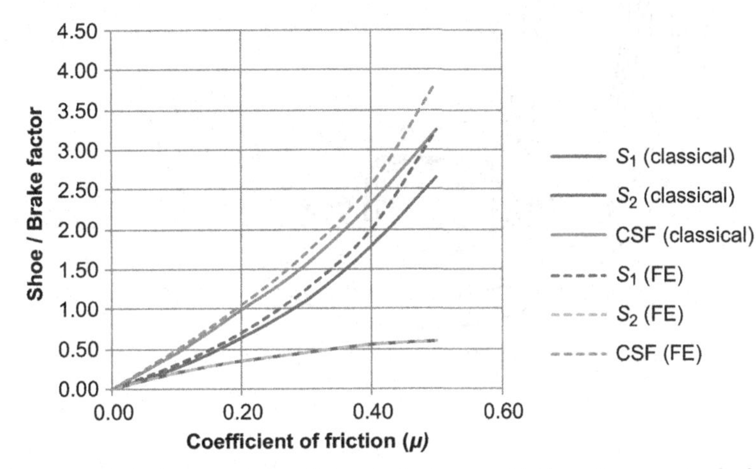

Figure 5.16: Comparison of Analytical and FE Shoe Factor Analysis.

Types of Drum Brake

As shown in Table 5.1, there are three types of drum brake that are used in modern road vehicle applications:

- Leading/trailing shoe (simplex)
- Twin leading shoe (duplex)
- Duo-servo.

Other types of automotive drum brake do exist, but these are mainly of historical interest. The simplex (leading/trailing shoe; see Figures 5.9 and 5.13) design is by far the most commonly used drum brake on all types of road vehicles; the duplex (Figure 5.17) finds application on a small number of commercial vehicles, and the duo-servo (Figure 5.18) is found in a small number of parking brakes in Europe, but has been more popular in the USA. The torque characteristics of the duplex and duo-servo designs of drum brakes are derived here from the individual shoe factor characteristics for illustrative purposes, and for more detailed design analysis the reader is referred to Limpert (2011).

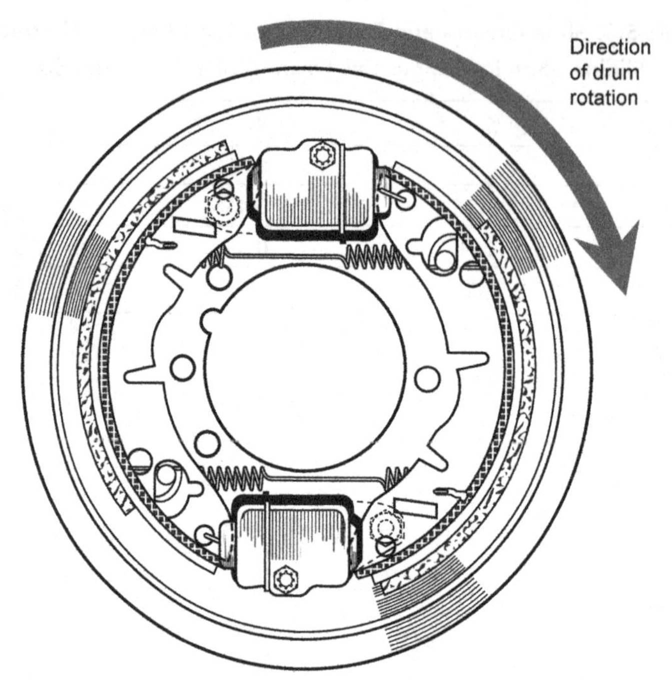

Figure 5.17: Duplex Drum Brake (Internet source).

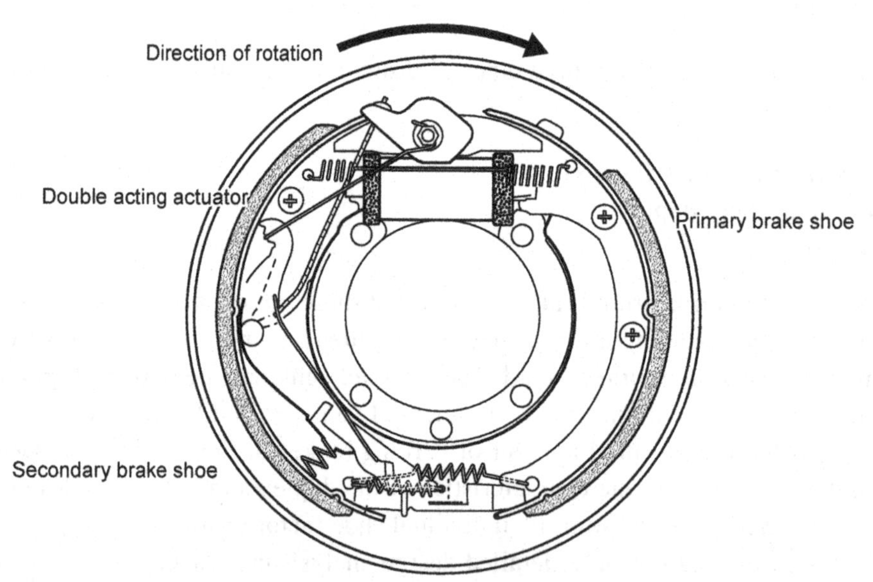

Figure 5.18: Duo-Servo Drum Brake (Internet source).

In the simplex drum brake analysed above (Figure 5.9) the actuation force on each shoe is the same, so the brake factor (BF) = combined shoe factor ($S_1 + S_2$) = 2.18 for a typical coefficient of friction $\mu = 0.36$, and therefore the torque output can be calculated from Equation (5.21), remembering that P_a is the actuation force on each brake shoe (see Equation (5.5)). For the specified hydraulic slave pistons in the example drum brake analysed above, the brake torque is:

$$\tau = (p - p_t)A_a\eta(S_1 + S_2)r_e \tag{5.27}$$

where:

A_a is the actuation area of the hydraulic slave pistons actuating the brake shoe (m) (this piston is 20.64 mm diameter, 334.6 mm^2);
p is the actuation hydraulic line pressure (MPa);
p_t is the threshold pressure (0.3 MPa);
η is the efficiency of the hydraulic actuation system (0.95 – discussed in Chapter 6);
r_e is the drum inner radius (m) (for this brake r_e = 114.5 mm).

Because leading shoes in particular have a high self-servo effect where the friction force helps to apply the brake shoe, it is important to avoid any unintended contact between lining and drum when no braking is required, i.e. when the brake is 'off'. For this reason, drum brakes have pre-loaded 'pull-off' springs to return the shoes positively to their adjuster stops. When the brakes are applied, the initial increment of actuator force is absorbed in overcoming pull-off spring pre-load before the lining can contact the drum and start to develop braking torque. This reduction in the effective force applied to the shoes must be allowed for in calculations. For hydraulic and air-actuated brakes the 'threshold' force is usually expressed as 'threshold pressure', i.e. in terms of the equivalent pressure in the hydraulic cylinder or air chamber.

If the axle is fitted with two drum brakes, the axle braking torque is:

$$\tau_{axle} = 2(p - p_t)A_a\eta(S_1 + S_2)r_e \tag{5.28}$$

If the brake lining μ is 0.36, for an input pressure of 40 bar (4MPa), the combined shoe factor is 2.18 and the torque output of the brake is therefore 293 Nm (in comparison, the previous example of the small passenger car disc brake showed a brake torque of 503 Nm for 40 bar actuation hydraulic pressure; the two brakes have approximately the same rotor outer diameter and would be packaged within the same size road wheel). As previously mentioned, care is required when faced with a specified 'brake factor' for the simplex brake as sometimes it is quoted as the sum of the individual shoe factors ($S_1 + S_2$), i.e. the 'combined shoe factor' as above, and sometimes as the 'mean shoe factor' ($S_1 + S_2$)/2. This is further discussed later; the rationale behind the use of the mean shoe factor is that the total actuation force is twice the force generated by one piston since there are two pistons actuating the two shoes.

For a hydraulically actuated duplex brake (Figure 5.17) with the same actuation force on each shoe the brake factor is still the sum of the two shoe factors but since there are two leading shoes, this equates to $2S_1$ in the forward direction and $2S_2$ in the reverse direction. For $\mu = 0.36$ and the same input pressure of 40 bar (4MPa), the torque generated by a duplex brake of the same geometry as analysed above would be:

Forward direction: brake torque is 449 Nm, axle braking torque is 898 Nm.
Reverse direction: brake torque is 137 Nm, axle braking torque is 274 Nm.

This means that since the ratio of the leading shoe factor to the trailing shoe factor is 3.27 for this design, the torque generated while the vehicle is travelling backwards is 31% of the torque generated while the vehicle is travelling forwards for the same line pressure. This is a major reason why duplex brakes are seldom used in modern road vehicles.

In a duo-servo brake (Figure 5.18) the actuation force is applied to the primary shoe and the abutment reaction from this shoe is used to actuate the secondary shoe. Modern duo-servo brakes are designed to operate equally effectively in the forward and reverse directions, and this means that both brake shoes operate as if they have floating abutments. In analysing the duo-servo brake the primary and secondary shoe factors must be calculated, and then the reaction force at the abutment end of the primary shoe is calculated and applied as the actuation force to the secondary shoe. If two leading shoes of the same design as in the example shown above for the simplex brake were connected in the duo-servo configuration, then a simple estimate of the brake factor can be made assuming that the abutment reaction force on the primary shoe reacts the friction drag entirely, and for $\mu = 0.36$ it would therefore be $1.65P_a$, which is assumed to be the actuation force for the secondary shoe. The friction drag generated by the secondary shoe (assuming it were identical to the primary shoe) would then be $1.67 \times 1.67P_a$, and the total friction drag generated by the two shoes would be $(1.67 + 1.67^2)P_a$, which is equivalent to a brake factor of approximately 4.4. For a full analysis of the duo-servo brake, the reader is again referred to Limpert (2011).

A comparison of the brake factors for these three types of drum brake and a disc brake is shown in Figure 5.19, which demonstrates the magnitude of the servo factor associated with each design, and the sensitivity of these different brakes to the coefficient of friction (μ) of the friction material. Note that the brake factor for a disc brake has been shown to be equal to the coefficient of friction μ, but for comparison with the use of the combined shoe factor for the brake factor of a drum brake, only one actuation force is considered and the disc brake factor as shown in Figure 5.19 is 2μ.

The sensitivity of brakes to variation in the coefficient of friction can be evaluated from the brake factor. Taking the $\pm 10\%$ tolerance on the design value of μ suggested in Chapter

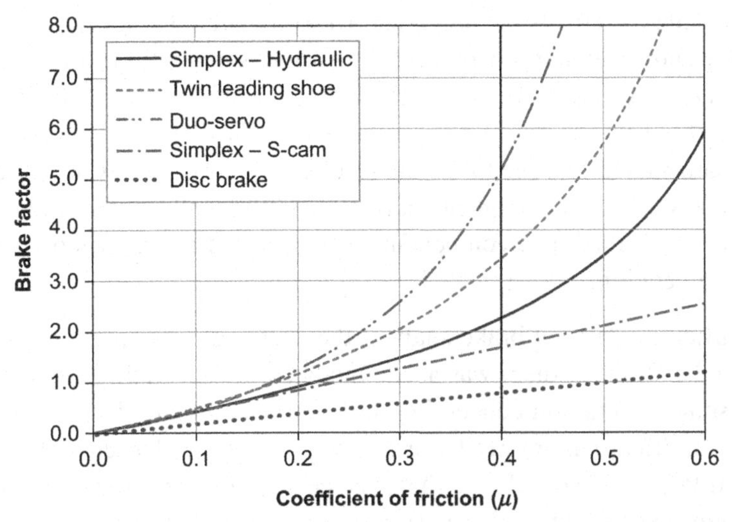

Figure 5.19: Comparison of the Brake Factors for Four Types of Drum Brake and a Disc Brake.

2, at a nominal value of $\mu = 0.36$ for a drum brake lining, the variation in brake factor would be as follows:

Simplex: −14%/+16%
Twin leading shoe Forward: −17%/+20%
Reverse: −6%/+ 5%
Duo-servo −26%/+34%
Disc brake: −10%/+10%

It is clear from Figure 5.19 that the brake factor : μ characteristics of the high-factor drum brakes, especially the duo-servo, are very sensitive to variation in the coefficient of friction, which is the main reason that high factor drum brakes are not used for service braking on modern road vehicles. The twin leading shoe (duplex) brake is sensitive to μ variation in the forward direction (but significantly less sensitive than the duo-servo) and is very stable in reverse rotation, being a twin trailing shoe brake in this configuration, which would require very high driver pedal effort. The disc brake factor is linearly related to the coefficient of friction and provides the best consistency of performance with modern friction materials.

Friction Interface Contact and Pressure Distribution in Drum Brakes

The finite element method (FE method) has been used for over 40 years in the form of commercial software packages and procedures that were used originally for the prediction of stresses, strains and temperature distributions in brake components under operational conditions. FE procedures have been developed for drum and disc brake performance prediction that enable the effects of design and friction material characteristics to be evaluated quickly and accurately for brake design optimisation. Typical examples of 'routine' FE analyses include brake pad backplate or brake shoe stress and deflection calculations, and disc or drum thermal stress and strain calculations. However, when brake components are assembled into the brake the challenge in brake analysis is the relationship and interactions between the components during operation with respect to the boundary conditions and loads on the brake components, which stem from the friction forces generated during operation.

As explained earlier, in classical brake analysis the frictional contact between the brake lining and the brake drum (or disc) was accommodated by calculating a resultant normal force based on some assumption concerning the interface pressure distribution, and then applying a resultant frictional or drag force perpendicular to it. Subsequent computational analyses (e.g. Millner, 1972) developed the basic concept of Amontons' laws of friction to elements or regions over the brake lining friction surface so that over a small element 'i' the friction drag force and the normal reaction were related:

$$F_i = \mu N_i \tag{5.29}$$

and these were summed over the whole of the lining contact area with the brake drum.

In FE analysis, forces are transmitted between elements at the nodes, and local nodal forces and friction forces that are normal and tangential to the friction interface respectively related by Equation (5.29) can be used to simulate frictional contact between a brake lining and a brake drum (or disc). Because the friction forces depend upon the normal forces, which in turn depend upon the deflection of the brake stator and rotor components resulting from the total force loading on these components, the computation of the distribution of the local normal force distribution (the pressure distribution) and the friction force distribution over the brake friction interface must be iterative. This type of procedure is now automated within modern FE packages, and in principle allows a sliding friction interface boundary condition to be applied to the brake shoe, brake pad, disc or drum components.

The technique was first developed (Day et al., 1979) for the calculation of shoe factor in a duplex brake with angled sliding abutments. The brake drum was assumed to be rigid, and the friction material was assumed to be linear elastic in tension and compression. A simple 2D FE mesh (Figure 5.20), based on an 'equivalent section' of the brake shoe web and platform, demonstrated the forces and enabled the analysis to be made economically to the required accuracy. It was subsequently shown that for brake torque prediction for basic design purposes, the increased cost and complexity of 3D FE modelling was unnecessary, although modern FE analysis packages work in 3D almost by default.

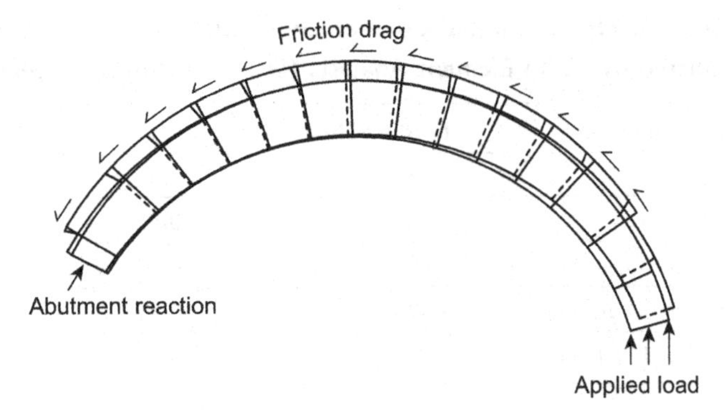

Figure 5.20: Drum Brake Shoe 2D FE Mesh for Shoe Factor Prediction (Day et al., 1979).

The predicted shoe factor based on a lining coefficient of friction μ of 0.44 was 2.18, compared with the measured value on an inertia dynamometer of 2.19. In comparison, the calculated shoe factor using the classical analysis was 10–20% lower. One reason for this was the interface pressure distribution predicted for this brake lining, which is illustrated in Figure 5.21 for the initial condition of 'perfect initial contact' between the brake lining friction surface and the drum inner surface. This condition represents full geometric

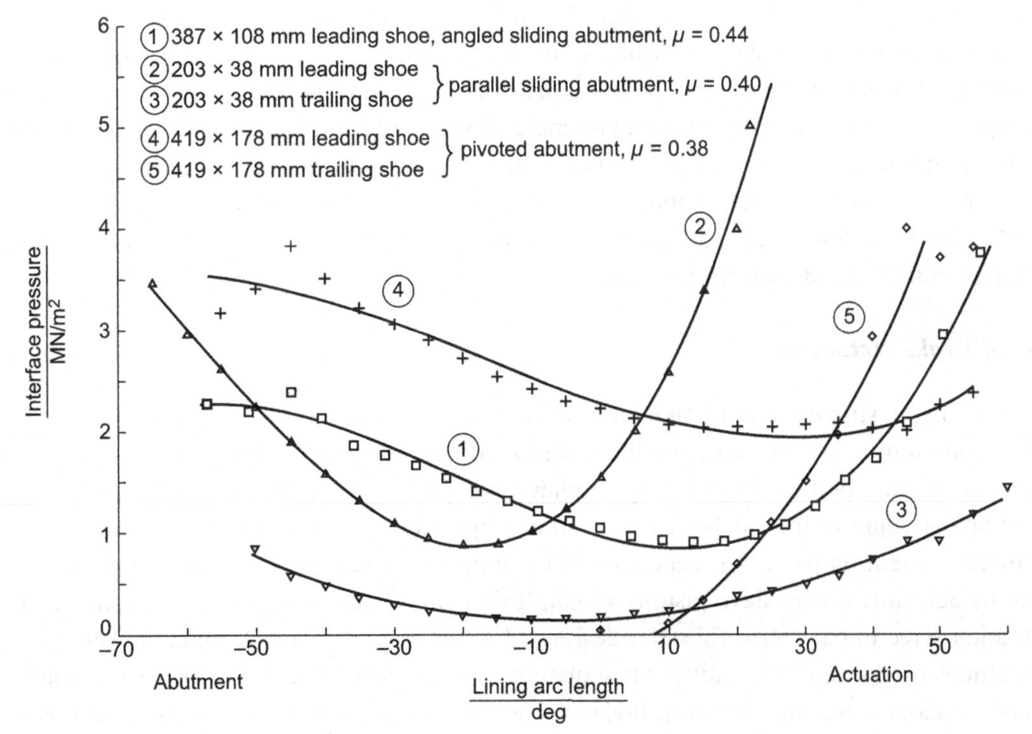

Figure 5.21: Drum Brake Lining Pressure Distribution Predicted by FEA (Day et al., 1991).

Table 5.4: Effect of Initial Contact Conditions on Shoe Factors Predicted by Finite Element Analysis (Day and Harding, 1983)

Contact Condition	Leading Shoe Factor	Trailing Shoe Factor
Perfect	1.46	0.45
0.2 mm crown	1.37	0.45
0.2 mm heel-and-toe	1.65	0.48
Cam block 0.1 mm thin	1.46	0.46
Anchor block 0.1 mm thin	1.42	0.45

conformity between the lining surface and the drum inner surface just as they come into contact, and in practice would represent a perfectly 'bedded' system. The form of the pressure distribution was substantially different from the sinusoidal distribution that had previously been the accepted form for this type of brake.

Results from FE analysis (Day et al., 1979) predicted the effect of different shoe and lining stiffnesses on the interface pressure distribution. The most uniform pressure distribution was achieved with a relatively soft (compressible) lining and a stiff shoe, a result that was also confirmed for disc brake pads (Tirovic and Day, 1991). Further work (Day and Harding, 1983) confirmed that the uniform pressure distribution defines the nominal level of drum brake performance for design torque prediction, and investigated the effect of different contact and pressure distributions. Using the FE method, the shoe factors of a large pivoted shoe S-cam commercial vehicle brake were predicted for a range of different initial contact conditions representing crown, heel and toe, and the effect of thick or thin blocks (the brake linings were in two parts, the 'anchor' block and the 'cam' block); see Table 5.4. These indicated the amount of variation that could be expected from different conditions of manufacture and use.

Drum Brake Actuation

In the example simplex drum brake analysed above, the actuation forces applied to each brake shoe were equal. In principle there are two different types of actuation system for automotive drum brakes, the 'floating' actuator and the 'fixed' actuator, and the analysis must take account of their individual design and operating features. The hydraulic actuation system on passenger cars and light commercial vehicles is an example of the floating actuator; it uses slave pistons which, if they are of equal diameter, apply an equal actuation force to each shoe (in older designs of drum brake 'stepped' cylinders were sometimes used to provide a different actuation force on each shoe, but these are seldom found on modern automotive drum brakes). The force is independent of displacement as it depends only on the hydraulic line pressure. The 'cam' or 'wedge' type of actuation used on larger brakes for commercial vehicles represents the fixed actuator since it applies a displacement to each shoe and therefore the actuation force depends on the system's response to that actuation displacement, which may be different, e.g. for a leading and trailing shoe. This is discussed in more detail later in this chapter.

The S-cam drum brake illustrated in Figure 5.13 is an example of a brake that incorporates 'fixed' actuation. It is widely used in HGV applications, being generally considered to be a robust, reliable, brake design with two shoes in a simplex arrangement (leading/trailing shoe). Actuation is provided by the S-cam (Figure 5.22), which is rotated by a linear actuator attached to a lever (see Chapter 6) and creates a 'lift' or displacement to each brake shoe so that both shoes are in principle constrained to move by equal amounts if the cam is truly 'fixed'. The force applied to each shoe will then depend upon the system's response to that actuation displacement. Because the leading shoe factor (S_1) is greater than the trailing shoe factor (S_2), what happens in practice is that the leading shoe lining initially wears more than the trailing shoe lining because it is doing more work. The S-cam actuation displacement then becomes more biased towards the trailing shoe because the leading shoe displacement has to accommodate the extra clearance arising from the wear of the lining. Eventually an equilibrium is reached where the actuation displacement apportions the force to each brake shoe so that the wear on the linings is nominally the

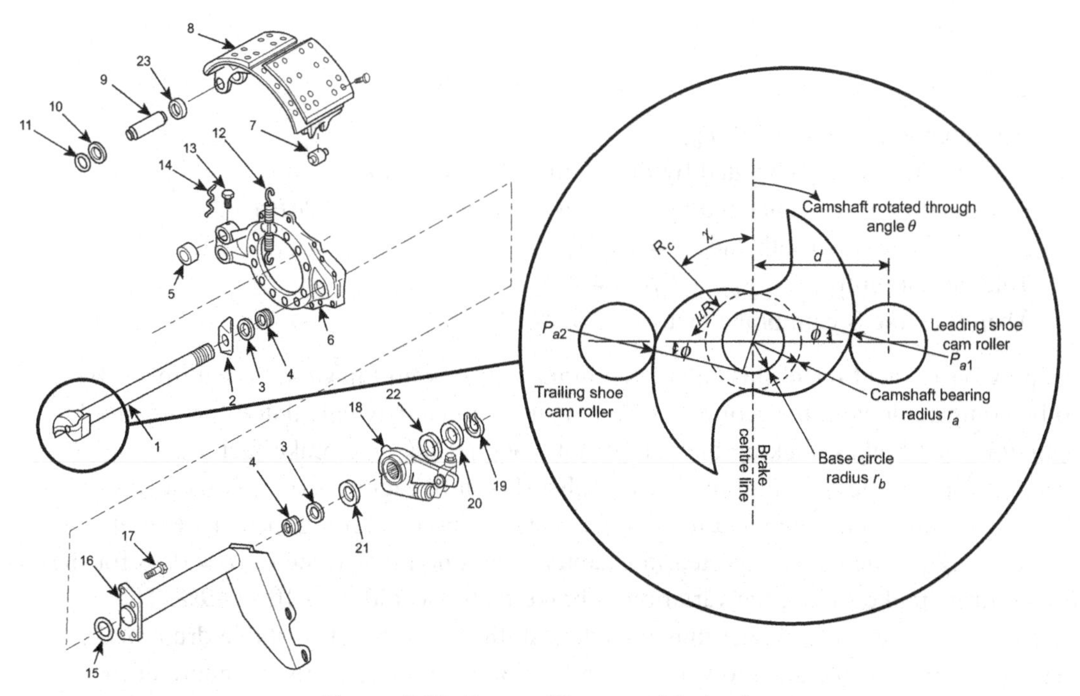

Figure 5.22: S-cam (Courtesy Meritor).

same and in this case the applied shoe tip forces must be inversely proportional to the shoe factors. This is known as the 'equal work' mode, which is usually assumed for purposes of design analysis.

In the 'equal work' mode the brake factor of an S-cam brake cannot be calculated from the sum of the two shoe factors because the actuation forces are not equal, so a different approach is required. Following the principle that the shoe factor is the ratio of the frictional force generated by each brake shoe at the drum radius (F_{s1} or F_{s2}) to the actuation force (P_{a1} or P_{a2}) on the shoe, the brake factor for a drum brake can be defined as the ratio of the total frictional force generated by the brake shoes at the drum radius to the total actuation force. This would have to be multiplied by 2 to be comparable with the 'combined shoe factor' previously calculated for 'floating' actuator drum brakes. The calculation is illustrated below, using a coefficient of friction $\mu = 0.38$, which is common for this type of commercial vehicle brake (Day and Harding, 1983). It should be noted that the actuation forces (P_{a1} or P_{a2}) on the brake shoes are not the same as defined in Equation (5.20) because in the S-cam brake the force generated by the fixed actuator is amplified mechanically, in this case by the S-cam. This is considered further later in this chapter in the form of the parameter ηC^*.

Example S-cam drum brake:

$r_e = 210$ mm
$S_1 = 1.46$
$S_2 = 0.45$
$S_1/S_2 = 3.24$
For equal work, $P_{a2} = 3.24P_{a1}$
Friction drag force generated by the leading shoe $= 1.46P_{a1}$
Friction drag force generated by the trailing shoe $= 0.45 \times 3.24P_{a1}$
Total friction drag (both shoes) $= 2.92P_a$
Total actuation force $= P_{a1} + P_{a2} = 4.24P_{a1}$
Therefore, the S-cam brake factor $= 2.92/4.24 = 0.69$.

As previously mentioned, this calculated brake factor would have to be multiplied by 2 to be comparable with the 'combined shoe factor' previously calculated for 'floating' actuator drum brakes, making the comparable factor 1.38. The brake factor : μ characteristic of the S-cam drum brake is also shown in Figure 5.19, and indicates that the S-cam brake is a stable and relatively low factor brake. Taking the $\pm 10\%$ tolerance on the design value of μ suggested in Chapter 2, at a nominal value of $\mu = 0.38$ for the S-cam drum brake lining, the variation in brake factor would be $-10\%/+9\%$. The advantages of the S-cam brake thus include stability; it is the most stable drum brake design in terms of tolerance to variation in friction coefficient. This is important in heavy-duty commercial vehicle brakes where large vehicle weights coupled with high road speeds (the power to weight ratio of modern commercial vehicles enables mean journey speeds to approach legal speed limits on many long distance journeys) places considerable thermal loads on the brake components with the consequent expectation of significant variation in μ.

The brake factor that has been calculated above, in order to illustrate the relationship with μ in comparison with other drum brake designs, is not the definition of brake factor that is used in the heavy commercial vehicle brake industry. Instead, 'brake factor' is defined as 'specific torque'. Commercial vehicle brake actuation systems are explained in detail in Chapter 6, but as previously mentioned, most commercial vehicles use compressed air (pneumatic) brake actuation systems. In these systems an air actuator, which is essentially a piston inside a cylinder, creates a rotary motion by attaching the piston pushrod to a lever (sometimes called a 'slack adjuster' because on a drum brake it combines the function of a lever with an automatic 'ratchet' type of adjustment for the effect of wear), as illustrated in Figure 5.23. In the case of an S-cam brake the torque generated rotates the S-cam to generate the displacement of the brake shoes.

The 'brake factor' is then conveniently specified in terms of 'specific torque' (τ_s), where:

$$\tau_s = \text{Brake torque}/\text{Actuation torque} = \tau/\tau_a \tag{5.30}$$

Referring to Figure 5.23, the actuation torque is:

$$\tau_a = \text{Actor force} \times \text{Lever moment arm} \tag{5.31}$$

Since the force generated by the actuator is:

$$P_a = (p - p_t)A_a\eta \qquad \text{(see Equation (5.5))}$$

Then:

$$\tau = P_a l = (p - p_t)A_a\eta l \tag{5.32}$$

The detail of the pneumatic actuation system design is covered in Chapter 6, but it is necessary to consider the S-cam theory here. As explained earlier, the S-cam essentially converts camshaft torque to actuation forces that are applied to the tips of the brake shoes via rollers (to minimise friction losses in the actuation), as shown in Figure 5.22. Ideally the centres of the cam rollers should lie on a straight line that passes through the centre of the cam. Each shoe tip force (P_1 and P_2) passes through the centre of one roller and the point of contact between the cam and the roller, being tangential to the 'base circle' of the cam, which is defined as having a radius of r_b. Thus:

$$P_a = \tfrac{1}{2}\tau_{input}/r_b \tag{5.33}$$

The shoe tips and the rollers actually move in an arc about the shoe pivots rather than in a straight line, but the analysis presented above is sufficiently accurate for design purposes.

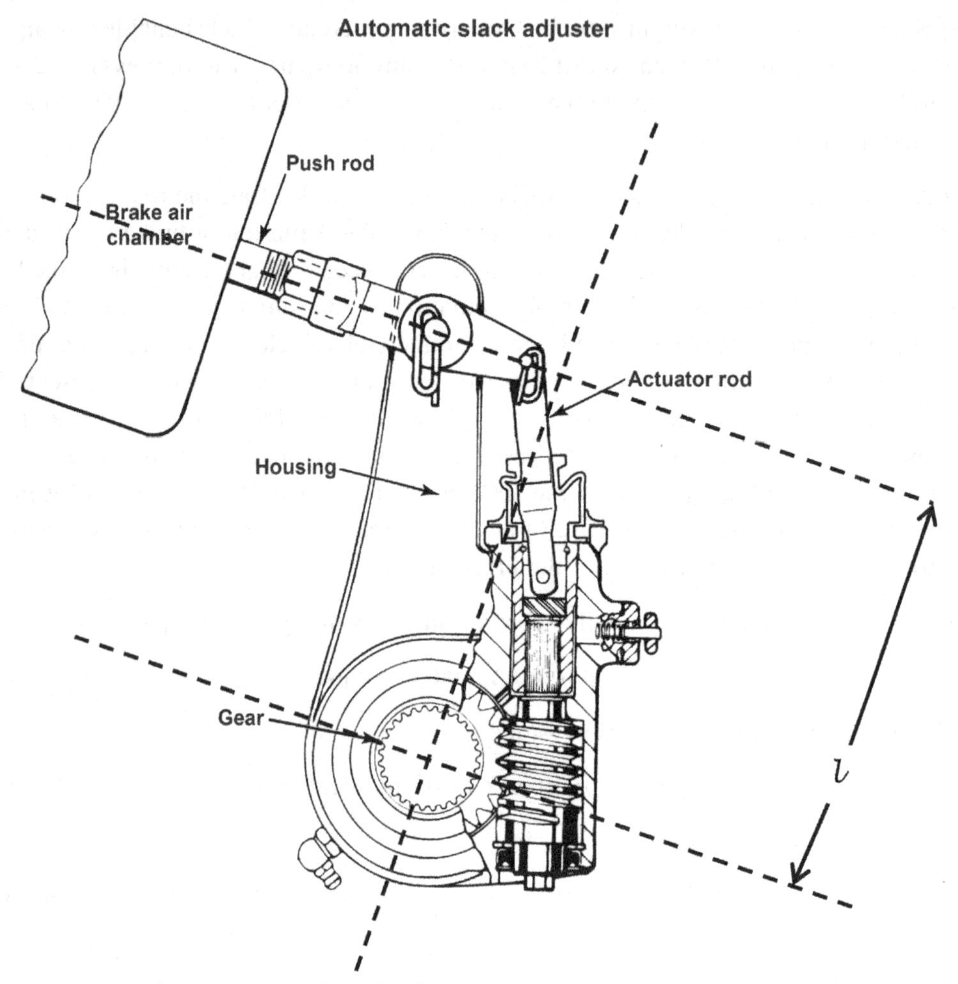

Figure 5.23: Air Actuator Acting Via the Piston Pushrod to a 'Slack Adjuster' Lever to Generate a Rotary Motion to Actuate the Brake (Courtesy Meritor).

The direction of the applied shoe tip force depends on the angular position of the cam as defined by the angle ϕ, where:

$$\phi = \arcsin(r_b/d) \tag{5.34}$$

and d is the distance between the centre of the roller and the centre of the cam, as illustrated in Figure 5.22.

For a large commercial vehicle S-cam drum brake of 420 mm diameter, the value of the angle ϕ varies from 21° to 15° over a 60° cam rotation, and the effect that the actuation force vector has on the deformation of the brake shoe is one reason why an S-cam brake must always be mounted in the orientation where in order to actuate the brake, the S-cam must rotate in the same direction as the wheel is rotating, in the forward direction of vehicle travel. It has previously been explained that the actuation forces P_{a1} and P_{a2} on the leading and trailing shoes respectively are unequal, and therefore there must be a resultant force (R_c) on the camshaft bearing, which can be calculated as follows:

$$R_c = (P_{a1} + P_{a2})/\sqrt{1+\mu_c^2} \tag{5.35}$$

where μ_c is the coefficient of friction between the camshaft and the bearing in which it is located.

The resultant force vector (R_c) is at angle ψ to the line of symmetry of the brake (Figure 5.22), where:

$$\tan\psi = (\mu_c \sin\phi + \cos\phi)/(\mu_c \cos\phi + \sin\phi) \tag{5.36}$$

and the torque ($\tau_{input} = \tau_c$) necessary to rotate the camshaft while supplying the specified shoe tip actuation forces P_{a1} and P_{a2} is therefore:

$$\tau_{input} = \tau_c = \mu_c R r_a + (P_1 + P_2) r_b \tag{5.37}$$

and substituting for R:

$$\tau_c = \left[\frac{\mu_c r_a (P_1 - P_2)}{\pm\sqrt{(1+\mu_c^2)}}\right] + (P_1 + P_2) r_b \tag{5.38}$$

For this S-cam brake (420 mm diameter) the specific torque in the 'floating cam' mode was calculated (Day and Harding, 1983) to be 15.3, using the design parameters:

$r_a = 20$ mm
$r_b = 13.1$ mm
$P_{a1} = P_{a2} = 40$ kN.

In comparison, the calculation for the 'equal work' mode, where:

$P_{a1} = 17$ kN
$P_{a2} = 56$ kN,

gave the results shown in Table 5.5.

Although the S-cam brake is designed as a 'fixed' actuator brake as explained earlier, in practice there is always a certain amount of compliance and clearance in the brake, and this can permit the cam to float to some extent. If it could float freely enough, the brake would operate as a simplex brake with equal actuation forces on both shoes, in which case the camshaft/bearing reaction force would be zero and there would be no friction losses

Table 5.5: Predicted S-Cam Specific Torque in 'Equal Work' Mode with Camshaft Friction (Day and Harding, 1983).

Camshaft Friction μ_c	Brake Torque τ (Nm)	Camshaft Actuation Torque τ_c (Nm)	Specific Torque $\tau_s = \tau/\tau_c$
0	10,475	956	11.0
0.1	10,475	1034	10.1
0.2	10,475	1109	9.4
0.3	10,475	1180	8.9

(arising from μ_c) in the actuation system, as illustrated in the example above. But even if there were some initial clearance and compliance in the system, the brake would tend towards the equal work mode as previously explained, bearing reaction would be generated and the effect of friction in the camshaft bearings (usually there are two bearings: one at the cam end and one at the actuation end, modelled here as one bearing) would have to be taken into account, as indicated in Table 5.5. This is the reason why it is important to design a camshaft bearing system that minimises camshaft friction and is able to be maintained in this condition in between services, often in very arduous conditions. Based on the example shown here, S-cam brake performance could range from a specific torque of around 15 to 11 during bedding-in (from 'floating cam' to 'equal work'), and friction between the camshaft and its bearings can reduce this still further to as low as 8.9.

Because brakes are used over a range of different actuation forces and operating temperatures, the wear of the friction material, although it always tends towards a condition of uniform pressure over the friction surface, creates contact conditions that can affect the brake performance. The effect of different lining/drum friction interface contact and pressure distributions on the torque generated by an S-cam brake were investigated (Day and Harding, 1983) taking into account different levels of camshaft bearing friction with the following conclusions:

- A change in S-cam brake performance during the early part of the working life of a set of brake linings can (and generally will) occur as the brake operates in some intermediate stage between floating cam and equal work modes depending on the clearances and compliances in the actuation system.
- The brake torque in the floating cam mode is higher than in the equal work mode, and initial lining/drum crown contact is recommended to minimise over-performance when new.
- The equal work mode will predominate during the working life of the S-cam brake and is therefore correctly used for design purposes, but the time taken to reach this mode of operation depends on the wear properties of the friction material lining.

The conditions of contact between the lining and the drum in a drum brake are important factors in the performance of any drum brake, including the S-cam brake. Perfect initial contact occurs when the rubbing radius of the lining is exactly the same as the rubbing radius of the drum as the two engage. Crown contact occurs when the rubbing radius of the lining is marginally less than that of the drum, and heel-and-toe contact occurs when the lining rubbing radius is marginally greater than that of the drum. In the case of the S-cam brake, the brake linings are often fitted as two half-blocks for manufacturing reasons, so contact can also be altered significantly by incorrect tolerance thick or thin half-blocks. Close attention should always be paid to the conformity of the brake linings to the brake drum during the manufacture, assembly and servicing of an S-cam brake to ensure that the brake linings are initially in good contact with the brake drum.

When a brake drum becomes hot it expands away from the lining. Since it then has a greater radius, contact pressure will be concentrated at the centre of the lining with a consequent change in shoe factor; if the lining becomes bedded at this temperature then subsequent braking at lower temperatures will result in contact at the ends of the lining, once more affecting the shoe factor. This is an additional cause of instability with drum brakes that does not affect disc brakes. Brake drum expansion is restrained by the hub flange and therefore causes the drum to 'bell', i.e. become conical, which is similar to the effect of coning in disc brakes. In addition to the effect of the contact and pressure distribution on the shoe factors and consequently the torque generated, incomplete lining contact may cause thermal damage to the linings and to the drum (see Chapter 7).

Brake Factor and ηC^* for Air-Actuated Commercial Vehicle Brakes

In the analyses presented above, three interpretations of 'brake factor' for drum brakes have been discussed all based on the ratio of the total frictional force generated by the brake at the effective radius to the actuation force. These are:

- Combined shoe factor $= (S_1 + S_2)$
- Mean shoe factor $= (S_1 + S_2)/2$
  (Both used for floating actuator (hydraulic actuation) drum brakes)
- Specific torque $\tau_s = \tau/\tau_a$.

As previously explained, in the case of hydraulically actuated brakes as used on passenger cars and light commercial vehicles, the actuation force is directly calculated from the hydraulic actuation pressure (Equation (5.5)) because the piston acts directly on the pad backplate (disc brake) or the shoe tip (drum brake). In most commercial vehicle brake systems a pneumatic actuator creates a force that is amplified through a cam or similar mechanism to provide the actuation force on the brake shoes (drum brake) or the brake pads (disc brake). The parameter C^* defines the ratio between the change in output force (i.e. the change in the total friction drag force generated by the brake) and the corresponding change in the input force (applied by the brake actuation mechanism, e.g. the S-cam described above on one brake shoe or pad) for any type of brake and the friction material for design purposes. C^* refers to the 'internal' brake ratio (theoretical ratio of the total frictional force generated by the brake to the actuation force), and is linked to the actual (i.e. measured) ratio of the total frictional force to the actuation force by the parameter η, which represents the mechanical efficiency of the actuation system. The combined parameter ηC^* is usually quoted by commercial vehicle brake manufacturers to include the effect of actuator efficiency. The definitions of C^* and ηC^* are:

$$\text{Theoretical (calculated) } C^* = \Delta F / \Delta P_{a(on\ 1\ shoe\ or\ pad)} \tag{5.39a}$$

$$\text{Measured (experimental) } \eta C^* = \Delta F / \Delta P_{a(on\ 1\ shoe\ or\ pad)} \tag{5.39b}$$

where ΔF is the (theoretical or measured) increment in friction drag generated by the brake for an input force increment of ΔP_a (on one shoe or pad). The 'increment' is used to account for the effects of threshold pressure.

For actuation systems that use a cam, helix, wedge, lever or similar device to amplify the actuator force, the force generated by, for example, the pneumatic actuator is not the same as the force applied to the pad backplate or the shoe tip. In the case of a hydraulically actuated brake, where the actuating piston acts directly on the pad backplate or the shoe tip, the force generated by the actuator is the same as the force applied to the pad backplate or the shoe tip and C^* would be the combined shoe factor for a simplex or duplex drum brake, or 2μ for a disc brake, multiplied by the actuation efficiency η.

For a commercial vehicle brake actuated by a pneumatic actuator via a lever (see Figure 5.23), the brake factor (BF) is generally defined as the specific torque described above for the S-cam brake, i.e.:

$$\text{BF} = \tau_s = \Delta \tau / \Delta \tau_a \tag{5.40}$$

(where $\Delta\tau$ is the measured or theoretical increment in torque generated by the brake for an input torque increment of $\Delta\tau_a$).

This can be very confusing because for a hydraulically actuated brake the 'brake factor' is defined in terms of the ratio of the total friction drag force generated to the actuation force (see Equation (5.4c) for disc brake factor), while for an air-actuated brake the 'brake factor' is defined in terms of the ratio of the brake torque generated to the actuation torque applied. Then there is the additional terminology of the combined parameter ηC^* that is used to include the effect of actuator efficiency.

Since:

$$\Delta \tau = \Delta F r_e \tag{5.41a}$$

the relationship between the measured torque output increment ($\Delta\tau$) and the applied actuator force increment (ΔP_a) is given by:

$$\Delta\tau = \eta C^* \Delta P_a r_e \tag{5.41b}$$

where P_a is the actuation force on one shoe or pad.

The parameters on which C^* and the brake factor of an air-actuated commercial vehicle brake depend are:

- The cam design, e.g. the 'base circle' of an S-cam (r_b).
- The effective radius of the friction pair, e.g. the brake drum inner radius or the mean radius of a brake disc (r_e).
- The type of brake, e.g. simplex, duplex, duo-servo, S-cam (drum brakes) or disc brake.
- The type of friction material, in particular the friction coefficient μ.
- The mechanical efficiency of the brake actuation η.

These are summarised in Table 5.6.

For the same large commercial vehicle S-cam drum brake used earlier as an example, from Equation (5.39a):

$$C^*_{(drum)} = \Delta F / \Delta P_{a(on\ 1\ shoe)} \tag{5.42}$$

From Equation (5.33):

$$\Delta P_a = ½ \Delta \tau_a / r_b \tag{5.43a}$$

Table 5.6: Nomenclature for Air-Actuated CV Brake Factor Calculations Based on ηC^*

Symbol	Parameter
r_b	Cam radius to calculate ΔP_a on one brake shoe or pad (mechanical actuator)
i_A	External ratio (see Figure 5.23)
l	Actuation lever length (see Figure 5.23)
r_e	Effective friction radius
ΔF	Increment of friction force at the effective friction radius
ΔP_a	Increment of input force to one brake shoe
ΔP_a	Increment of input force to one brake pad
ΔP_{ac}	Increment in brake actuation chamber force
$\Delta\tau_a$	Increment in input (actuation) torque to the brake
$\Delta\tau$	Increment in output torque of the brake
η	Mechanical efficiency of the brake actuation

The factor of ½ in Equation (5.43a) arises because the S-cam has two working surfaces so there are one on each shoe, which are assumed to be equal. These two actuation forces multiplied by the cam moment arm (the base circle radius r_b) thus are theoretically equal to the applied actuation torque. Therefore:

$$\Delta \tau_a = 2r_b \Delta P_a \quad (5.43b)$$

(It should be noted that in the 'equal work' mode, as previously explained, the actuation forces to each shoe are not the same. The losses that arise from, for example, the camshaft bearing friction contribute to the mechanical efficiency η.)

From Equation (5.40), the brake factor:

$$\mathrm{BF} = \Delta \tau / \Delta \tau_a = C^*_{(drum)} r_e / 2r_b \quad (5.44)$$

giving the relationship between C^* and the brake factor for this brake (defined in terms of specific torque τ_s):

$$C^*_{(drum)} = 2\mathrm{BF} \cdot r_b / r_e \quad (5.45)$$

For the example S-cam brake studied above:

Theoretical brake factor (specific torque T_s) = $\Delta T / \Delta T_{input}$ =11.0 for the equal work mode, zero camshaft friction (Table 5.5)
$r_e = 210$ mm
$r_b = 13.1$ mm.

Hence, from Equation (5.45):

$$C^*_{(drum)} = 1.37$$

Assuming a mechanical efficiency η of 0.85, $\eta C^*_{(drum)} = 1.17$.

Example: Mechanically actuated disc brake. This disc brake is a sliding caliper actuated mechanically by a cam that provides a single actuation force (P_a) on the backplate of one pad, as indicated in Figure 5.24. P_a passes through the point of contact between the cam and the backplate (or tappet), and is tangential to the 'base circle' of the cam of radius r_b. The reaction force acting through the cam centre generates the actuation force for the other pad.

From Equation (5.39a):

$$C^*_{(disc)} = \Delta F / \Delta P_{a(on\ 1\ pad)} \quad (5.46a)$$

The friction drag force generated by the brake is $\Delta F = 2\mu P_a$. Therefore:

$$\eta C^*_{(disc)} = 2\mu \text{ (assuming 100\% actuation efficiency)} \quad (5.46b)$$

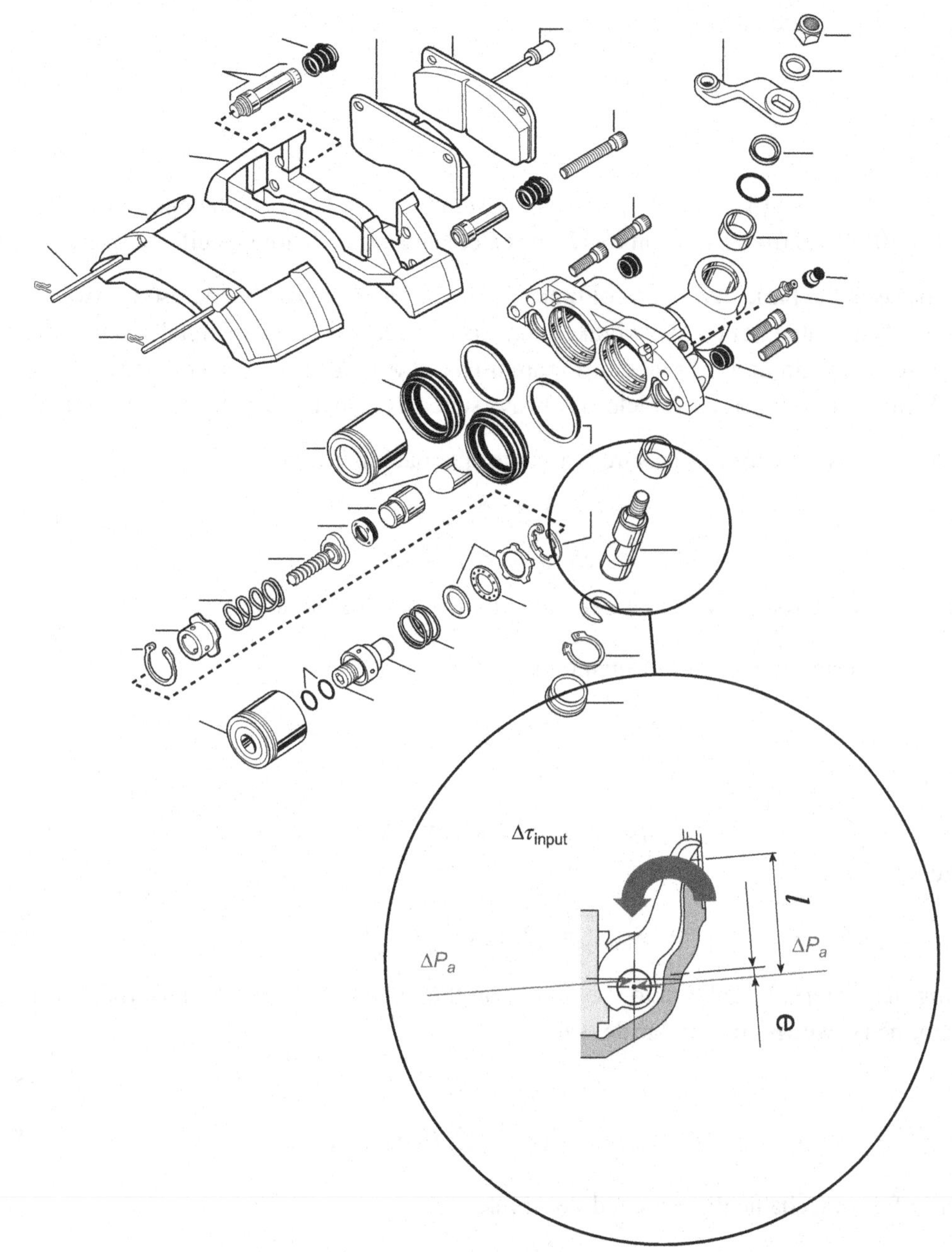

Figure 5.24: Cam Actuated Disc Brake Mechanism.

Because BF= $\Delta\tau/\Delta\tau_a$, the brake torque $\Delta\tau = 2\mu r_e$, and the actuation force $\Delta\tau_{input} = \Delta P_a r_b$, and therefore:

$$BF = \tau_s = 2\mu r_e / r_b \tag{5.47}$$

For an example air-actuated disc brake:

$r_e = 173$ mm
$r_b = 5.3$ mm
$\mu = 0.4.$

Hence, from Equation (5.46b), $C^*_{(disc)} = 0.8$. Assuming a mechanical efficiency η of 0.95, $\eta C^*_{(disc)} = 0.76$ and from Equation (5.47) the calculated brake factor (specific torque τ_s) = 26.

As can be seen, the basic definitions of brake factor and ηC^* are easily confused between hydraulically actuated light vehicle brakes and air-actuated heavy commercial vehicle brakes. Because several definitions of brake factor are used in the industry, it is recommended that 'brake factor' for commercial vehicles is always stated as 'specific torque' to avoid confusion.

ηC^* can be calculated from measured brake torque data as follows:

$$\Delta F / \Delta \tau / r_e \tag{5.48}$$

$$\Delta \tau_{input} = \Delta P_{ac} l \tag{5.49}$$

Where ΔP_{ac} is the force increment generated by the pneumatic actuator.

For an S-cam drum brake, from Equation (5.39b):

$$\text{BF} = \Delta \tau / \Delta \tau_a = \eta C^*_{(drum)} r_e / 2 \cdot r_b \tag{5.50}$$

which leads to:

$$\eta C^*_{(drum)} = 2 r_b \Delta \tau / \Delta \tau_a r_e$$

and thus:

$$\eta C^*_{(drum)} = (2 r_b \Delta \tau) / (\Delta P_{ac} l r_e) \tag{5.51}$$

Defining the 'external' brake ratio (i.e. based on the pneumatic actuator, pushrod, lever and cam, helix, wedge, or similar device):

$$i_{A(drum)} = l / 2 r_b \tag{5.52}$$

$$\eta C^*_{(drum)} = \Delta \tau / \left(\left(\Delta P_{ac} i_{A(drum)} r_e \right) \right) \tag{5.53}$$

Similarly for a mechanically actuated disc brake:

$$\Delta F = \Delta \tau / r_e \tag{5.54}$$

$$\Delta \tau_a = \Delta P_{ac} l \tag{5.55}$$

For an air-actuated commercial vehicle disc brake, from Equation (5.47):

$$\text{BF} = \Delta \tau / \Delta \tau_a = 2 \mu r_e / r_b \tag{5.56}$$

which leads to:

$$\eta C^*{}_{(disc)} = r_b \Delta \tau / \Delta \tau_{input} r_e \quad (5.57)$$

and thus:

$$\eta C^*{}_{(disc)} = (r_b \Delta \tau)/(\Delta P_{ac} l r_e) \quad (5.58)$$

Again defining the 'external' brake ratio:

$$i_{A(disc)} = l/r_b \quad (5.59)$$

$$\eta C^*{}_{(disc)} = \Delta \tau / ((\Delta P_{ac} i_{A(disc)} r_e)) \quad (5.60)$$

Note that Equations (5.53) and (5.60) are the same.

The brake torque ($\Delta\tau$), actuation force (ΔP_{ac}), external brake ratio ($i_{A(drum)}$ or $i_{A(disc)}$) and the effective radius of the drum or disc (r_e) are known or can be measured, therefore ηC^* can be evaluated experimentally and compared with the theoretical value, with any difference being attributed to the efficiency of the internal actuation system (η). Typical values of η range from 0.to 0.9 for drum brakes, and from 0.9 to 0.95 for disc brakes.

Chapter Summary

- An introduction to brake types and designs for road vehicles is given and the basic design and operation principles of disc and drum brakes are introduced and explained; this covers 'spot' type disc brakes, and simplex, duplex (twin leading shoe), duo-servo and S-cam types of drum brakes. The performance of these different types of brakes is compared, and their advantages and disadvantages for light and heavy vehicles are discussed.
- Basic analysis methods are explained for the calculation of brake torque based on the brake geometrical design parameters and the coefficientof friction. The importance of the contact and pressure distribution between the friction material and the rotor at the friction interface of drum and disc brakes is explained. Some guidelines are provided for good brake design.
- Definitions of brake factor, specific torque and ηC^* with the associated internal and external brake ratios are explained.
- Mechanisms for the actuation of brakes are described and their analysis explained. This includes 'S-cam' drum brakes. Internal losses, e.g. arising from friction in actuator mechanisms, hysteresis and retraction, must be taken into account in brake design calculations. Drum brakes require positive retraction of the brake shoes away from the drum to prevent off-brake drag and the effect of 'pull-off' springs on threshold

actuation pressure and force/torque must be included. Disc brake pads do not have positive retraction and threshold pressures/forces are lower, but still need to be included.
- Effects on brake performance that can be experienced in use, including sensitivity to the coefficient of friction and wear of the friction material, are described and discussed. Friction variation can have a significant effect upon brake performance; disc brakes are least sensitive, and duplex and duo-servo drum brakes are most sensitive. Brakes should be designed to operate equally effectively in the forward and reverse directions, but some types of drum brake, e.g. duplex, are inherently direction dependent unless the actuation mechanism can accommodate reverse rotation as in modern duo-servo designs. However, because of the wear of the friction material over the drum brake lining or disc brake pad length, which establishes an interface contact and pressure distribution in the forward direction, drivers may notice performance differences in reverse operation.
- Some worked examples for disc and drum brake analysis are explained and discussed, and an overview of computational methods for brake analysis is included.

References

Day, A.J., 1987. An analysis of speed, temperature, and performance characteristics of automotive drum brakes. Trans. ASME. 87-Trib-11.

Day, A.J., Harding, P.R.J., 1983. Performance variation of cam operated drum brakes, IMechE Conference on the Braking of Road Vehicles, C10/83, 69–77.

Day, A.J., Harding, P.R.J., Newcomb, T.P., 1979. A finite element approach to drum brake analysis. Proc. IMechE. vol. 193 (no. 37), 401–406.

Day, A.J., Tirovic, M., Newcomb, T.P., 1991. Thermal effects and pressure distributions in brakes. Proc. IMechE. vol. 205, 199–205.

Harding, P.R.J., Day, A.J., 1981. Instability in the handbrake performance of cars and vans. Proc. IMechE. vol. 195 (no. 27), 315–323.

Harding, P.R.J., Wintle, J.B., 1978. Flexural effects in disc brake pads. Proc. IMechE. vol. 192, 1–7.

Limpert, R., 2011. Brake Design and Safety, 3rd edn. SAE. ISBN 978 0 7680 3438 7.

Millner, N., 1972. A study of some torque characteristics of drum brakes. University of Leeds. PhD thesis.

Newcomb, T.P., Spurr, R.T., 1967. Braking of Road Vehicles. Chapman & Hall. ISBN 0412092700.

Tirovic, M., Day, A.J., 1991. Disc brake interface pressure distributions. Proc. IMechE. vol. 205, 137–146.

CHAPTER 6

Brake System Layout Design

Introduction

Previous chapters have shown how the braking forces generated at the road wheels of a vehicle must be designed to maximise the vehicle's ability to decelerate with acceptable driver effort and maximum stability, and how foundation brakes are designed to generate the required braking torque. In this chapter these aspects are brought together in the design or layout of the braking system. Two of the most commonly used actuation systems are studied: the hydraulic system used on most passenger cars and light commercial vehicles (vans), and the pneumatic system used on most heavy commercial vehicles. Other actuation systems are fitted to road vehicles, including power hydraulic and electromechanical ('brake-by-wire' - BBW), and whilst these are mentioned, the focus is on the principles and practice associated with the hydraulic and pneumatic systems.

A correctly designed braking system operates as the driver and passengers expect under all driving conditions at minimum weight and cost. The hydraulic braking system was invented over 100 years ago (Newcomb and Spurr, 1989) and has been universally used in passenger cars for over 60 years. Recent developments in electronic stability systems (see Chapter 11) have been associated with significant technological advances over the basic hydraulic braking system. These advances may, in time, lead to the introduction of brake-by-wire systems on passenger cars, but as yet no commercial system has reached the market. The hydraulic braking system relies upon the 'muscular' energy of the driver, which may be amplified by a suitable 'booster'. The basic principle is that incompressible 'brake fluid' is pressurised by a 'master cylinder' piston connected to the brake pedal, and the pressure generated creates an actuation force from a 'slave cylinder' piston, which actuates the brake shoe or pad.

The pneumatic braking system fitted to most commercial vehicles was introduced on to commercial road vehicles also around 60 years ago, and quickly became the standard braking system for such vehicles. It is lower cost, more robust and easier to maintain than power hydraulic systems, which might otherwise be used on heavy commercial vehicles, and can easily accommodate electronic control. The actuation energy is provided by a separate energy source, in this case a pneumatic compressor. Pressurised air is stored in reservoirs, which are an essential element to fulfil braking system response time and energy consumption requirements that are carried out with the engine stopped. The driver actuates the brakes by means of a footbrake valve that controls the air pressure supplied to actuate the foundation

Braking of Road Vehicles.

brakes relative to the driver demand. Air pressure is then supplied to an air actuator, which converts pressure to actuation force to apply the brakes. Because of the compressible nature of the air actuation medium, with large commercial vehicles such as articulated tractor and semi-trailer combinations there can be a delay in the actuation pressure generation at an axle. This delay can be reduced by the use of relay valves where full pressure is available at an axle, where it is controlled by either a pneumatic or an electric signal from the control valve.

The vehicle braking system layout design process can proceed as summarised in the following four-step brake system design procedure:

Step 1. Design the basic braking system parameters based on the vehicle configuration:
- 1.1. Calculate the basic braking ratio for the worst case vehicle
- 1.2. Calculate the maximum brake torque required at each brake
- 1.3. Calculate the maximum power dissipation at each brake.

Step 2. Specify the foundation brakes:
- 2.1. Determine the rotor type and size
- 2.2. Determine the thermal mass of the rotor
- 2.3. Confirm the brake factor or C^* (ηC^*)
- 2.4. Confirm the friction material size in terms of thermal and mechanical loading.

Step 3. Design the actuation system:
- 3.1. Specify the actuation mechanism, e.g. hydraulic or pneumatic
- 3.2. Design the brake actuators (hydraulic cylinder sizes, air actuator sizes, slack adjuster lever lengths)
- 3.3. Design the 'master actuator' (brake pedal) system:
  - 3.3.1. Hydraulic systems: pedal configuration, master cylinder piston size and stroke, servo boost ratio and knee point, fluid 'consumption'
  - 3.3.2. Air systems: maximum pressure, response time, air reserve.

Step 4. Verify the design and check compliance with legislative requirements (e.g. UN Regulations 13 and/or 13H):
- 4.1. Intact system
- 4.2. Partial system failure
- 4.3. Parking.

Each step is explained below, and two examples are fully worked through for a passenger car (a two-axle rigid vehicle – Chapter 3) and a commercial vehicle (an articulated commercial vehicle comprising a tractor and semi-trailer – Chapter 4). After completing the design of the vehicle braking system, there are four more steps that will be covered in later chapters:

Step 5. Evaluate operational effects: loading and usage; heat and temperature; wear and durability; environmental; consistency and stability.

Step 6. Refine and optimise: system response and pedal feel; noise and vibration; cost and weight; manufacturability.
Step 7. Integrate with safety systems: ABS; ESC; traction control; automated emergency braking; etc.
Step 8. Brake performance verification by testing.

Overview of the Vehicle Braking System Layout Design Process

Step 1

Step 1.1 Design the Basic Braking System Parameters Based on the Vehicle Configuration – Calculate the Basic Braking Ratio for the Worst Case Vehicle

The analyses derived in Chapter 3 (two-axle rigid vehicles, e.g. passenger cars and light commercial vehicles) and Chapter 4 (multi-axle commercial vehicle combinations including towing vehicles and full trailers or semi-trailers) are used in this step. For passenger cars and light commercial vehicles the 'worst case vehicle' is usually defined in terms of wheel lock sequence in the unladen condition (driver-only weight, DoW), because this is when the adhesion utilisation on the rear wheels is the highest because of the static axle loading, which is biased towards the front especially with front wheel drive (FWD) designs, and longitudinal weight transfer during braking. The starting point for the brake system design on this type of vehicle is to set a value of z_{crit}, the rate of braking at which the front and rear wheels would lock simultaneously (see Equations (3.18) and (3.19)) under the applied hydraulic pressure in the actuation system, ignoring any possible intervention from any additional systems such as EBD or ABS, or any other mechanism such as load-sensing valves. This parameter z_{crit} is set according to the vehicle manufacturer's in-house standards, and may range between 0.4 and 0.9, with a typical value in the range 0.6–0.7 for the DoW case.

Two aspects of the UN Regulations 13 and 13H (ECE, 2010) (UN, Mar 2014, UN Feb 2014) that are used as the legislative basis for braking system design in this book (see Chapter 8) may influence the choice of value of z_{crit}. Vehicles fitted with ABS do not have a requirement for wheel lock sequence as it is implicitly assumed that the ABS will prevent wheel locking as part of its intended function. In the event of failure of the ABS system, the wheel lock sequence will be determined by the basic design of the system including any mechanisms such as valves. UN Regulations 13 and 13H require at least 80% of the prescribed (Type 0) performance in the event of a single electrical failure affecting the antilock function, so the vehicle must be able to achieve $z = 0.525$ (mean fully developed deceleration (MFDD) $= 5.15\ \text{m/s}^2$) safely. This implies that in all load states z_{crit} should be above this value. The other consideration is that for vehicles without ABS in all load states, the wheels on the rear axle must not lock before the brakes on the front axle over the range $0.15 \leq z \leq 0.8$. This would imply $z_{crit} \geq 0.8$, but since ABS is required in UN Regulation 13 (ECE, 2010) for heavy vehicles and is indirectly mandated in UN Regulation 13H (UN, Feb 2014) for passenger cars, this consideration is no longer particularly relevant for most modern passenger cars and light vans in countries that subscribe to the harmonised regulations (see Chapter 8).

For multi-axle commercial vehicles including combinations with full and semi-trailers, the difference between the unladen and fully laden axle loadings is much greater than for passenger cars and the approach described above based on setting a value of z_{crit} in the unladen (worst case) condition is not used. Instead a better approach is to base the analysis on an optimum braking distribution, i.e. when the braking on each axle is proportional to the normal load carried by the axle and the braking efficiency is therefore 1 at the selected rate of braking (see Equation (4.25)). It is also often assumed for heavy commercial vehicles that the brakes on each axle should be designed to provide sufficient braking force to provide the required maximum deceleration for the maximum weight being carried on that axle. Most commercial vehicle axles have a specified maximum axle loading, and in some countries this is a legal specification, e.g. in the UK a drive axle maximum mass may be 11.5 tonnes while a non-drive axle may be limited to a maximum of 10 tonnes (actual limits depend on the vehicle configuration). Total vehicle mass is also specified for different configurations so that a range of vehicle loadings (in terms of the position of the centre of gravity when laden) can be accommodated without exceeding the individual permitted static axle loading.

Step 1.2 Design the Basic Braking System Parameters Based on the Vehicle Configuration – Calculate the Maximum Brake Torque Required at Each Wheel

Brakes are usually specified in terms of the maximum braking torque that they are designed to deliver in use and this is related to the braking force T_i at each axle of the vehicle using Equation (3.16) (for two-axle rigid vehicles such as cars and light vans), and the braking ratios established in Step 1.1. For multi-axle commercial vehicles the calculation of the braking force T_i at each axle is presented in Chapter 4, e.g. Equations (4.20) and (4.23) for the towing vehicle and a centre-axle trailer respectively, using the optimum braking distribution Equation (4.25).

Assuming no lateral variation in brake torque, the axle braking force T_i at the tyre/road interface (see Chapters 3 and 4) is generated by two brakes, so for each wheel on the axle the brake torque (τ_{wi}) is calculated from the braking force (T_{wi}) as illustrated in Figure 6.1 and shown in Equation (6.1a):

$$\tau_{wi} = T_{wi} r_r = T_i r_r / 2 \tag{6.1a}$$

where r_r is the rolling (or dynamic) radius of the tyre on the wheel. For ease of analysis one value of the rolling radius can be used for all wheels if the same make and model of tyre is fitted to all wheels, i.e. $r_{r1} = r_{r2}$ etc. $= r_r$. In practice this

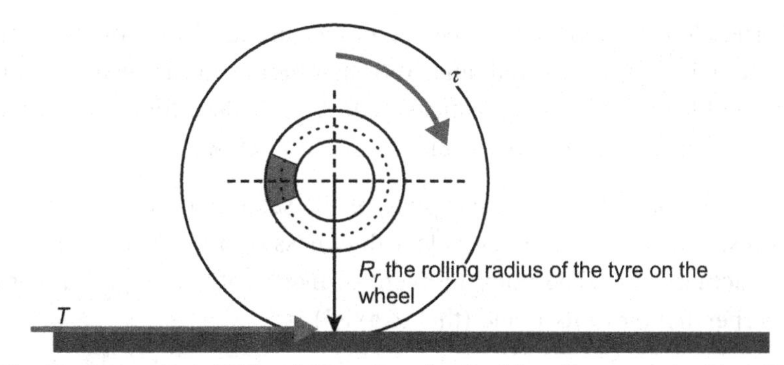

Figure 6.1: Disc Brake on Road Wheel Showing the Braking Force at the Tyre/Road Interface.

assumption may not be strictly correct, for example even with identical tyres the rolling radius is affected by inflation pressure and wheel load, which may not be the same on individual axles.

The design maximum torque required for each brake τ_{wimax} is therefore calculated using Equation (6.1a) for the maximum values of the braking forces T_{wi} for the required vehicle design parameters and deceleration using the braking ratio based on z_{crit} as shown in Equation (3.19b) for two-axle rigid vehicles such as cars and light vans:

$$\tau_{w1} = T_{w1}r_r = X_1Pzr_r/2 \tag{6.1b}$$
$$\tau_{w2} = T_{w2}r_r = X_2P_zr_r/2 \tag{6.1c}$$

and for example Equations (4.26)–(4.28) where for a vehicle with 'n' axles:

$$\tau_{wi} = T_{wi}r_r = X_iPzr_r/2 \tag{6.1d}$$

Alternatively the brake torque τ_{wi} required to decelerate a maximum axle load of N_{imax} at the required maximum deceleration z_{max} can be calculated:

$$\tau_{wi} = T_{wimax}r_r = N_{imax}z_{max}r_r/2 \tag{6.1e}$$

The maximum value of the torque τ_{wi} required from each brake on an axle to create the braking force T_{wi} depends upon the vehicle's mass and its deceleration, which are specified by the vehicle manufacturer. For passenger cars the worst case for the front brakes is usually at least gross vehicle weight (GVW) with the DoW braking ratio, at a target rate of braking of $z = 1$ (i.e. a deceleration of $J = 9.81$ m/s^2). The rear brakes worst case is also usually at least GVW but with the GVW braking ratio, at the same target rate of braking of $z = 1$. High-performance cars etc. may be designed to work to higher deceleration targets. Commercial vehicles cannot usually achieve such high decelerations, and may work to a target rate of braking of around $z = 0.8$. UN Regulation 13H requires an MFDD of 6.43 m/s^2, equivalent to $z = 0.655$ for M_1 vehicles while UN Regulation 13 requires an MFDD of 5.0 m/s^2, equivalent to $z = 0.509$ for commercial motor vehicles.

An additional torque load is imposed by the rotational inertia of the components of the axle, including the wheel assembly, hub and rotor assembly, and driveshafts, which can be estimated, calculated from CAD data, or measured. A procedure for estimating the moment of inertia of brake rotors and wheels is outlined below.

A brake disc can be modelled as a circular plate of the same outer diameter. To take account of the presence of vanes the equivalent thickness of a ventilated disc is assumed to be 80% of the actual value. The polar moment of inertia of a disc of radius r rotating about the axis perpendicular to its plane (the 'Z-axis') is:

$$I_{zz} = \tfrac{1}{2} mr^2 \tag{6.2}$$

For a ventilated disc of outer diameter 355 mm, thickness 20 mm, made from cast iron with a density of 7100 kg/m^3, the equivalent thickness is 16 mm and the estimated polar moment of inertia is 0.18 kg m^2.

A brake drum can be modelled in two parts: the flange as a circular plate and the wall as a cylinder, both of the same outer diameter. The polar moment of inertia of a cylinder rotating about its longitudinal axis (the 'Z-axis') is:

$$I_{zz} = mr^2 \tag{6.3}$$

For a large (commercial vehicle) cast iron brake drum of inner diameter 420 mm, rubbing path width 180 mm, flange and wall thickness 20 mm, the estimated polar moment of inertia is 1.92 kg m^2. For both these examples, modelling the brake disc and the flange of the brake drum as a circular plate includes an estimate of the moment of inertia of the hub.

Similarly, the moment of inertia of the wheel and tyre assembly can be estimated. The wheel and the rim can be modelled as a disc and a cylinder respectively. The tyre can be modelled as two annular rings to represent the sidewalls, and a cylinder to represent the tread. A steel wheel from a small passenger car is modelled as comprising a steel disc of outer diameter 350 mm and a cylindrical rim of 350 mm diameter and 130 mm wide, both 4 mm thick. The tyre for this wheel is modelled as two annular discs of 350 mm inner diameter, 550 mm outer diameter and thickness 10 mm to represent the sidewalls, and a cylinder of 550 mm diameter and 130 mm wide, 20 mm thick. The density of steel is 7800 kg/m^3, and for the tyre the rubber composite material is estimated as 1500 kg/m^3. The total polar moment of inertia of the wheel is estimated in this way as 0.75 kg m^2. The disc from this small car is solid and has an outer diameter of 220 mm, for which the estimated polar moment of inertia is 0.016 kg m^2, so on this car the wheel dominates the rotational inertia. On larger cars and commercial vehicles, the brake rotor represents an increased proportion of the rotational inertia, but the wheel still dominates, even if lighter weight alloy wheels are used. The effect of rotational inertia associated with the wheel and brake can be broadly estimated as 5%, so the brake torque can be specified as $1.05 \times \tau_{wimax}$.

However, this should always be checked accurately at the design stage for each vehicle brake system to ensure that the brakes are not under-specified.

Step 1.3 Design the Basic Braking System Parameters Based on the Vehicle Configuration – Calculate the Maximum Power Dissipation at Each Brake

In addition to being specified in terms of maximum braking torque, brakes may also have a specified power dissipation that relates to the ability of the brake to dissipate heat, which is covered in Chapter 7. The maximum power dissipation in the vehicle's brakes is usually calculated at this early stage of brake system design because it is needed to size the rotor (Step 2.2) and the stator components, in particular the brake pads or linings (Step 2.4).

As explained in Chapter 7, the instantaneous power dissipated in a brake 'i' is:

$$\dot{Q}_i = \tau_{wi}\omega \tag{6.4}$$

where ω is the instantaneous angular velocity (rotational speed) of the wheel and brake, and a simplifying assumption that is usually made at this stage is that there is no lateral variation in brake torque or wheel speed so $\dot{Q}_i$ is the same for two brakes on an axle. The maximum power dissipation occurs when ω is highest, i.e. at the start of a brake application to decelerate the vehicle. Over a single brake application from one speed (v_1) to another, lower, speed (v_2), the wheel rotational speeds are ω_1 (initial rotational speed) and ω_2 (final rotational speed) and the mean power dissipation is calculated from Equation (6.5):

$$\bar{\dot{Q}}_i = \tau_{wi}\frac{(\omega_1 + \omega_2)}{2} \tag{6.5}$$

The total energy Q_i dissipated in such a brake application can be estimated from:

$$Q_i = \bar{\dot{Q}}_i t \tag{6.6}$$

where t is the duration of the brake application. The energy dissipated by the brake (Q_i) can be compared with that estimated from the proportion of vehicle kinetic energy dissipated by the brake based on the braking ratio:

$$Q_i = ½ mV^2X_i/2 \tag{6.7}$$

This is used in Step 2.2 to calculate the single-stop temperature rise.

Step 2

Step 2.1 Specify the Foundation Brakes – Determine the Rotor Type and Size

Disc brakes with ventilated disc rotors are now almost universally used on the front axles of cars and light commercial vehicles because of their ability to dissipate more heat than solid rotors, especially at higher speeds. On the rear axles of these vehicles, drum brakes or disc brakes can used as discussed in Chapter 5. Disc brakes have been increasingly fitted to heavy commercial vehicles during the last 15 years, but drum brakes are still popular because of their robustness, low cost of ownership and ease of maintenance. Disc and drum brakes for commercial vehicles usually have standardised dimensions and the discs are usually ventilated because of the large amount of heat energy that needs to be dissipated (see Chapter 7).

Considerations that influence rotor size include:

1. Torque and power rating (see Step 1)
2. Packaging constraints
3. Single-stop temperature rise calculations
4. Fade and Alpine descent temperature prediction
5. Benchmarking.

The brake assembly must fit inside the wheel rim, which therefore limits its overall diameter. With a disc brake, the caliper has to fit over the disc, so the rotor outer radius is limited by the radial thickness of the caliper plus necessary clearance, as illustrated in Figure 6.2(a). For a drum brake, it is usual to allow more clearance between the drum outer circumference and the wheel rim inner surface to encourage air cooling and avoid any packing of road debris between the two, as this adversely affects brake cooling. Manufacturers of brakes for heavy commercial vehicles usually specify the wheel sizes that their brakes will fit, but for passenger car brakes it is usually the responsibility of the braking system designer to match the brake overall diameter with sufficient clearance inside the wheel.

For a disc brake, having determined the maximum disc outer diameter and thus the outer radius r_o, the inner radius r_i of the rubbing path can be determined from the 'good practice' rule of $r_o/r_i \leq 1.5$, as explained in Chapter 5. The disc effective radius (r_e) should

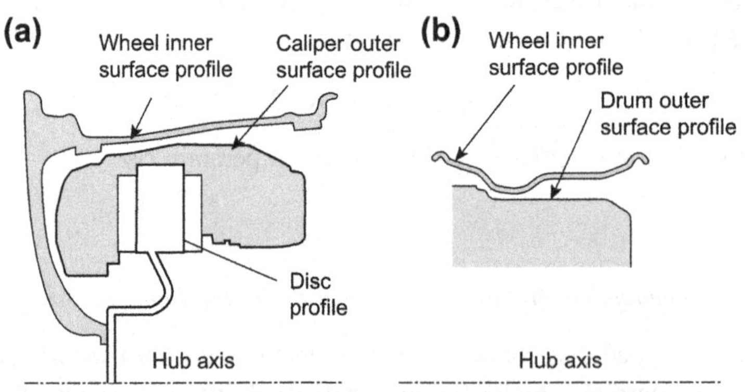

Figure 6.2: Brake Assembly and Wheel Packaging.
(a) Disc brake. (b) Drum brake.

be as large as possible, and the final design depends upon the pad surface area and aspect ratio (Step 2.4).

For a drum brake, r_e is the inner radius of the brake drum and depends upon the clearance between the drum outer circumference and the wheel as explained above, and the wall thickness of the drum, which is determined by thermal and mechanical strength and wear considerations. The friction material surface area (Step 2.4) is controlled by the lining arc lengths and width.

Step 2.2 Specify the Foundation Brakes – Determine the Thermal Mass of the Rotor

The thermal mass of the rotor is an important design parameter that should be included in the brake system design process at this early stage. It defines the 'single-stop temperature rise' (SSTR – explained in Chapter 7), which assumes that in a single brake application (under specified loading and speed conditions) the proportion of kinetic energy dissipated by any particular brake is entirely converted to heat and absorbed by the rotor assembly as there is no time for heat flow to the free surfaces for surface heat loss. Using Equation (7.7) the bulk temperature rise is calculated and compared with manufacturers' in-house standards for SSTR. These are typically as summarised in Table 6.1.

The rotor thermal mass also contributes to the temperatures reached under other test and usage conditions, which are considered in Chapter 7.

Step 2.3 Confirm the Brake Factor (or ηC^*)

In Chapter 5, the brake factor was defined for disc and drum brakes, based on the ratio of the total frictional force generated by the brake at the effective radius to the actuation force. The term 'brake factor' is usually used by passenger car and light commercial vehicle manufacturers for hydraulically actuated brake systems, while the parameter ηC^* (which includes the effect of actuator efficiency) is usually quoted by commercial vehicle manufacturers for pneumatically actuated systems. Here, the term 'brake factor' is used generically to indicate the relationship between the brake actuation force (P_a) and the braking force generated at the tyre/road interface (T_i). However, UN Regulation 13 (UN, Mar 2014) defines the brake factor as the ratio of output torque to the input torque as explained in Chapter 5.

Table 6.1: Typical Guidelines for Single-Stop Temperature Rise (SSTR)

Rotor Type	SSTR (°C)
Drum	350–400
Solid disc	550
Vented disc	600–650

For a disc brake, assuming that the actuation force P_a is the same as the resultant normal force at the pad/disc interface (N_c) (see Chapter 5), the torque τ_w generated by a disc brake is:

$$\tau_w = 2\mu P_a r_e \tag{5.4b}$$

Substituting $\mathrm{BF}_{disc} = 2\mu$, the brake force T_w at the tyre/road interface of one wheel is:

$$T_w = \mathrm{BF}_{disc} P_a r_e / r_r \tag{6.8a}$$

For axles with a disc brake on each end, the total braking force is:

$$T_i = 2\mathrm{BF}_{disc} P_a r_e / r_r \tag{6.9a}$$

Similarly, the torque generated by a drum brake is:

$$\tau_{drum} = (F_1 + F_2) r_e = (S_1 + S_2) P_a r_e \tag{5.21}$$

As discussed in Chapter 5, using the combined shoe factor ($S_1 + S_2$) as the brake factor (BF), Equation (6.8a) also applies to the drum brake:

$$T_w = \mathrm{BF}_{drum} P_a r_e / r_r \tag{6.8b}$$

For axles with a drum brake on each end, the total braking force is:

$$T_i = 2\mathrm{BF}_{drum} P_a r_e / r_r \tag{6.9b}$$

Note that equations (6.8a) and (6.8b) are the same, and so are equations (6.9a) and (6.9b).

As discussed in a commercial vehicle brake actuated by a pneumatic actuator via a slack adjuster (Figure 5.23), it can be very confusing because the brake factor (BF) is generally defined in the same way as the specific torque τ_s (see Equation (5.40)). The equivalent parameter to BF as used above is ηC^* (when the efficiency of the internal actuation system is included), and thus Equations (6.8) and (6.9) for commercial vehicle brakes and axles respectively become:

$$T_w = \eta C^* P_a r_e / r_r \tag{6.8c}$$

and the total axle braking force is:

$$T = 2\eta C^* P_a r_e / r_r \tag{6.9c}$$

Step 2.4 Confirm the Friction Material Size, Thermal and Mechanical Loading

Although friction force is, according to Amontons' law (Chapter 2), independent of the apparent friction area defined by the surface area of the brake pad or lining, the friction material may suffer mechanical or thermal failure if it is too small. The minimum surface area of friction material required to ensure satisfactory service is normally determined by consideration of two criteria, which may be specified by the friction material manufacturer or developed from experience by the vehicle manufacturer. Shear loading relates to the shear strength of the friction material or the pad/backplate bond, and power density (also called 'work rate') is a measure of the rate of heat transfer into the friction material that it can withstand before its structure breaks down. It is usually necessary to consider both criteria and design the friction material size to meet the higher value. Wear, which is also related to the size of the friction material surface area in frictional contact with the rotor, is evaluated later in the brake system design process (Step 5) but experience has shown that if the limits for shear loading and power density are not exceeded, the friction material wear is unlikely to be unsatisfactory.

Shear loading refers to the friction force per unit area of the brake pad or lining, as illustrated in Figure 6.3. Assuming that the wheels do not lock, this can be calculated from the braking force at the tyre/road interface (T_i; see Chapters 3 and 4). For an axle with two disc brakes each with two pads of friction surface area A_p:

$$\text{Shear loading(pad)} = T_i\left(r_r/r_e \cdot 4A_p\right) \tag{6.10}$$

For an axle with two Simplex drum brakes (floating actuator) each with leading shoe factor S_1, trailing shoe factor S_2, and brake factor = combined shoe factor (CSF = $S_1 + S_2$), the lining loading on each leading shoe (surface area A_l) is:

$$\text{Shear loading(leading shoe)} = T_i(S_1/\text{CSF})r_r/(r_e \cdot 2A_1) \tag{6.11}$$

(only the leading shoe is considered because $S_1 > S_2$).

Typical design target values for a front wheel drive passenger car are 1.4 MPa for front disc brakes and 0.3 MPa for rear drum brakes for a rate of braking $z = 0.5$ g at GVW. A high-performance car may reach 1.75 MPa at $z = 1$, at GVW. For an S-cam drum brake fitted to a heavy commercial vehicle axle, the design target value may lie in the region of 0.75 MPa for $z = 0.6$ g. The friction materials must retain sufficient shear strength to meet these targets at high temperature, e.g. repeated high-duty braking at GVW and high speed, or continuous (drag) braking.

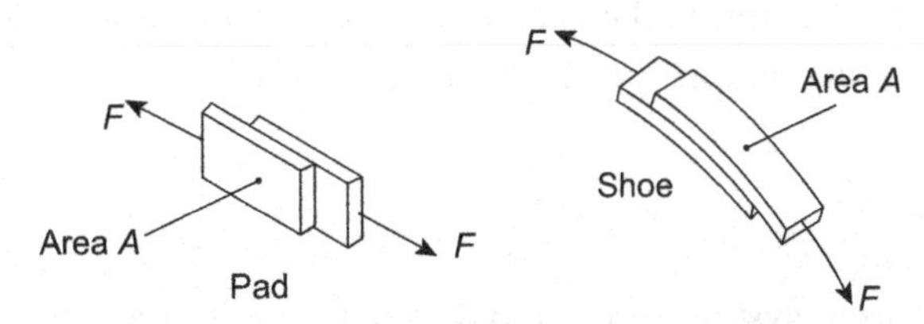

Figure 6.3: Friction Material Shear Loading.

The braking power generated during vehicle deceleration is greatest at the beginning of a brake application and decreases to a minimum (or zero if the vehicle comes to rest) at the end of the brake application. The mean power dissipation $\overline{\dot{Q}}_i$ associated with a vehicle axle (which has two brakes) may be calculated in terms of the mean angular velocity during the brake application, as previously explained from Equation (6.5). This may also be calculated in terms of the mean vehicle speed as indicated in Equation (6.12):

$$\overline{\dot{Q}}_i = T_i(v_1 - v_2) \tag{6.12a}$$

where v_1 and v_2 are in units of m/s. The mean power dissipation in decelerating a vehicle from road speed V_1 to V_2 km/h at a rate of braking of z is (the factor of 3.6 converts km/h to m/s):

$$\overline{\dot{Q}}_i = \frac{zPX_i(V_2 - V_1)}{3.6} \tag{6.12b}$$

Assuming again that the wheel does not lock, the mean power density for each individual pad of friction surface area A_p in a disc brake is $\overline{\dot{Q}}_i/4A_p$.

For an axle with two Simplex drum brakes (floating actuator) each with leading shoe factor S_1, trailing shoe factor S_2 and combined shoe factor $\text{CSF} = S_1 + S_2$, the mean power density on each leading shoe (surface area A_l) is $(\overline{\dot{Q}}_i/2A_l)\cdot(S_1/\text{CSF})$. Again, only the leading shoe is considered because $S_1 > S_2$.

Typical values of mean power density for a front wheel drive car during a 0.5 g deceleration to rest at GVW from 90% of vehicle maximum speed are:

8.5 MW/m^2 for front disc brakes;
1.2 MW/m^2 for rear drum brakes.

High-performance cars may reach higher power dissipations, e.g. at 1 g deceleration from V_{max} at GVW:

Mean pad work rate $\leq$ 20 W/mm^2;
Pad shear loading $\leq$ 1.75 MPa.

When steps 1 and 2 have been completed it is often useful to undertake 'benchmarking' to compare the proposed foundation brake specifications with those of competitor vehicles on the basis of size, weight, maximum speed, engine power, etc.

Steps 3 and 4

Step 3 (design the actuation system) and Step 4 (verify the design and check compliance with legislative requirements) both depend strongly on the type of vehicle, e.g. passenger cars and light commercial vehicles with hydraulically actuated brakes or commercial vehicles with pneumatically-actuated brakes. Although there are other types of actuation systems these are not considered here. From this point in the system design process onwards, the vehicle braking system layout design process is therefore split into these two categories of road vehicles, using examples of each to illustrate Steps 1–4. Aspects that are common to both types of actuation system may be covered for one and not repeated for the other.

Passenger Car and Light Commercial Vehicle Braking Systems with Hydraulic Actuation

The design of a hydraulic actuation system for a two-axle rigid vehicle such as a passenger car or light commercial vehicle is explained here, based on a passenger car with the specifications shown in Table 6.2.

From Step 1.1:

$$\text{Front/rear braking ratio}(X_1/X_2)\text{ for } z_{crit} = 0.6\text{: DoW 74/26, GVW 61/39}$$

The calculation of the maximum torque required at each wheel brake τ_{wimax} was explained in Step 1.2. For the front brakes on the example vehicle at $z_{crit} = 0.6$, the braking distribution X_1/X_2 is 74/26 (DoW loading condition), which is the worst case for the front brakes. The maximum worst case front brake torque at GVW and $z = 1$ is (including a correction factor of 1.05 for rotational inertia) $\tau_{w1max} = 3050$ Nm.

For the example vehicle at $z_{crit} = 0.6$, the braking distribution X_1/X_2 is 61/39 (GVW loading condition), which is the worst case for the rear brakes. The maximum worst case rear brake torque at GVW and $z = 1$ is (including a correction factor of 1.05 for rotational inertia) $\tau_{w2max} = 1600$ Nm.

The reason that the two different values of the front/rear braking ratio X_1/X_2 are used to calculate the maximum torque is that modern vehicles of this type can vary the ratio through, for example, electronic brake force distribution (EBD) to take account of loading

Table 6.2: Example Vehicle Data

	DoW	GVW
Wheelbase, $E = L_1 + L_2$ (mm)	2800	2800
Centre of gravity height, h (mm)	675	650
Position of centre of gravity behind the front axle, L_1 (mm)	1120	1486
Position of centre of gravity in front of the rear axle, L_2 (mm)	1680	1314
Vehicle mass, M (kg)	1750	2450
Tyre dynamic (rolling) radius, r_r (mm)	325	325

conditions. Therefore it is necessary to ensure that each brake has sufficient torque capacity to deliver the worst case brake force.

The maximum and mean power dissipations at each brake (Step 1.3) from an initial road speed of 145 km/h to rest for this vehicle are calculated from Equations (6.4) and (6.5):

Power Dissipation	Peak (kW)	Mean (kW)
Front brake	287	144
Rear brake	151	75

From Step 2.1, specifying the foundation brakes, rotor type and size, ventilated discs are selected for the front brakes, with the option of ventilated or solid discs on the rear axle. In the example vehicle the smallest wheel rim inner diameter is 200 mm (the car may be specified with different wheel and tyre sizes depending on the customer requirements, but the aim is to standardise the brakes as far as possible), and the maximum front brake disc outer radius to allow clearance for the caliper within the wheel rim is 150 mm, which is therefore specified as the outer radius, with a 100 mm inner radius (this is usually limited by the FWD hub size), giving an effective radius ($r_e = r_m$) of 125 mm. The ratio r_o/r_i is thus 1.50, which is just within the 'good practice' rule ≤ 1.5. The rear brake disc is also specified with 150 mm outer radius (usually for cosmetic reasons because alloy wheels tend to expose the brake), and 110 mm inner radius, giving $r_e = 130$ mm. The ratio r_o/r_i is 1.36.

As previously explained, the rotor size is also influenced by other factors, especially those relating to thermal performance and temperatures reached during braking. The theory and practice behind Step 2.2 is covered in more detail in Chapter 7, where the choice of solid or ventilated disc and the disc thickness are considered.

In Step 2.3, having designed the front and rear brake rotors in the example given, the brake factor of the disc brakes to be used is confirmed by specifying a suitable coefficient of friction μ. This may be based upon the vehicle manufacturer's in-house preferences, from previous experience, or as part of a recommendation from a friction material supplier. For this vehicle the front and the rear disc brake factors are initially specified as 0.76 ($\mu = 0.38$).

Step 2.4 confirms the friction material size based on the thermal and mechanical loading. Based on the typical limits of shear loading and mean work rate discussed earlier (1.4 MPa and 8.5 MW/m^2 respectively for disc brake pads in a 0.5 g stop at maximum GVW from 145 km/h, which is assumed to be 90% of this vehicle's maximum speed), the estimated pad areas are shown in Table 6.3. The limiting factor is the mean work rate.

Assuming the maximum road speed of this vehicle is 160 km/h, the design worst case for the front and rear brakes at 1 g deceleration from V_{max} at GVW is summarised in Table 6.4. Based upon the limits discussed for high-performance cars, pad mean work

Table 6.3: Estimated Values of Shear Loading and Mean Work Rate for Disc Brake Pads on the Example Vehicle in a 0.5 g Deceleration to Rest at Maximum GVW from 145 km/h

Axle	Pad Width (mm)	Pad Length (mm)	Pad Area (mm^2)	Shear Loading (MPa)	Mean Work Rate (MW/m^2)
Front	50	110	5500	1.06	8.6
Rear	40	75	3000	1.02	8.3

Table 6.4: Estimated Values of Shear Loading and Work Rate for the Same Area Disc Brake Pads on the Example Vehicle in a 1 g Deceleration to Rest at Maximum GVW from 160 km/h

Axle	Pad Width (mm)	Pad Length (mm)	Pad Area (mm^2)	Shear Loading (MPa)	Mean Work Rate (MW/m^2)
Front	50	110	5500	2.12	17.2
Rear	40	75	3000	2.0	16.6

rate $\leq$20 W/mm^2 and shear loading $\leq$1.75 MPa, the limiting factor is the shear loading. This may necessitate an increase in pad area, or a change of friction material if a higher strength material were available. Pad areas would have to be increased to 6500 and 3500 mm^2 respectively. Since the pad width is fixed by the inner and outer radii of the disc, this would mean an increase in pad length from 110 to 130 mm (front brake) and 75 to 87.5 mm (rear brake). As a result of this increase in the 'aspect ratio' (see Chapter 5), this may influence the design of the actuation system, which is discussed later.

At this stage it is also interesting to consider if a drum brake could be specified as an option for the rear axle of the vehicle and meet the mean work rate criterion. Taking a 250 mm diameter simplex drum brake as an example, the leading shoe factor (S_1 – see Chapter 5) might be 1.67 while the brake factor ($S_1 + S_2$) is 2.18. The arc length of the leading shoe might be 110° and the width 30 mm, giving a leading shoe lining area of 7200 mm^2. Under the worst case condition of the car loaded to GVW and decelerating to rest from 145 km/h, the brake lining mean work rate is predicted to exceed the previously discussed limit of 1.2 MW/m^2 for rear drum brakes at a rate of braking (z) less than 0.2. So whilst the drum brake might be thought to be satisfactory at lower speeds on the mean work rate criterion (where the car might see most of its usage), it would not be an option in a modern passenger car of this size and specification where consistent high braking capability under high-duty conditions is regarded as a prerequisite.

For actuation, the 'critical length' (L_c) of the disc brake pad assembly is calculated using Equation (5.13) to indicate whether one piston will satisfactorily distribute the actuation force at the pad/disc interface. Using the same dimensions and properties as in the example in Chapter 5, $L_c = 62$ mm, with the proposed pad lengths (110 mm front, 75 mm rear) the rear brake should be satisfactory with a single piston of diameter ≤ the pad width (40 mm), while the front brake may need two pistons. This is considered further in the design of the brake actuators (slave pistons).

Step 3 Design the Hydraulic Actuation System

Step 3.1 Specify the Actuation Mechanism (Hydraulic System)

Since the foundation brake design has now been defined for this vehicle, the brake system design can proceed to the design of the selected hydraulic actuation system. The brakes on most modern passenger cars and light commercial vehicles are actuated via a hydraulic actuation system that uses 'muscular' energy (i.e. the driver effort or force applied to the brake pedal) to create hydraulic pressure in the brake fluid to provide the required actuation forces at the foundation brakes. A servo (or booster) to amplify the driver effort is usually incorporated on modern cars as part of a 'servo-assisted' hydraulic brake actuation system (also termed 'power brake system' in the USA). (Although this type of actuation system is always described as a hydraulic system, its purpose is to transmit pressure without significant fluid flow so technically it is a hydrostatic system.) The basic components of the hydraulic actuation system are illustrated in Figure 6.4, and include the brake pedal, booster (if fitted), master cylinder, hydraulic lines (pipes, tubes, hoses, etc.) and slave cylinders at the brakes. To this may be added components that relate to safety or control systems, e.g. ABS, that are considered later in Chapter 11. Since the hydraulic pressures generated are high (a typical design maximum is 12 MPa or 120 bar), all the system components must be designed with high stiffness, otherwise the brake pedal travel and 'fluid consumption' would be increased by too much elasticity in the system. The brake pipes are thick-walled (usually steel) tube, and flexible hoses are required to connect parts that move relative to each other, e.g. from a subframe to a caliper; these are manufactured from composite

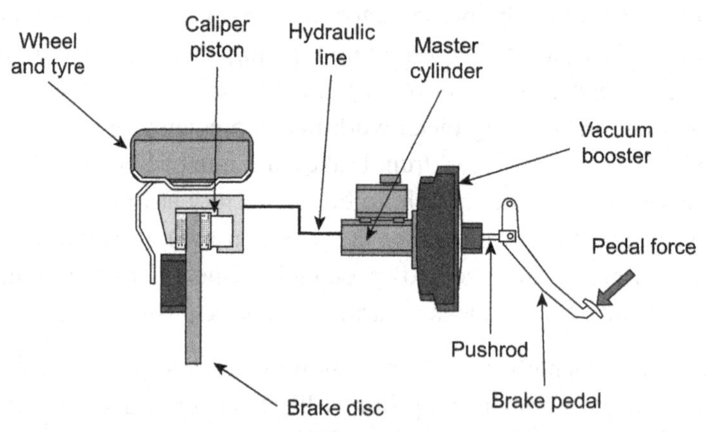

Figure 6.4: Basic Components of a Hydraulic Braking System for a Passenger Car or Light Van.

materials such as reinforced rubber, providing transverse flexibility while maintaining high radial stiffness. Pad compressibility is another parameter that contributes to elasticity in the system, influencing pedal travel, fluid consumption and ultimately 'pedal feel'.

It is a legislative requirement (UN Regulations 13 and 13H; see Chapter 8) that every road vehicle has at least two separate brake actuation circuits, as illustrated in Figure 6.5, so that in the event of the failure of one circuit, the capability remains for the driver to stop the vehicle safely, albeit at a reduced level of performance as specified by the relevant legislation. The simplest configuration for a two-axle rigid vehicle is a 'vertical split', where the front brakes are on one circuit and the rear brakes are on the other. The limitation of this design is that if the front circuit should fail, the adhesion available at the rear wheels may not be sufficient to provide the deceleration required by the legislation in the system part-fail condition (see Chapters 3 and 8). The vertical split can be satisfactory for rear wheel drive (RWD) and four-wheel drive (4WD) vehicles, but is usually not satisfactory for front wheel drive (FWD) vehicles because their weight distribution is concentrated on the front axle, especially when unladen.

For FWD cars and light commercial vehicles the 'diagonal split' is better. This ensures that braking is always available at one front and one rear wheel, each on opposite sides of the vehicle. Because front brakes generally contribute a higher proportion of the vehicle braking force than the rear brakes (i.e. $X_1 > X_2$), in the part-fail condition where only one circuit is providing the braking force, the vehicle is subject to a strong yaw moment, which can be minimised by appropriate steering geometry and suspension design, usually in the form of zero or negative offset steering. For vehicles where neither the vertical nor the diagonal split is satisfactory, other configurations can be used. For example, if hydraulic brakes have more than one actuating cylinder it is possible to actuate each piston

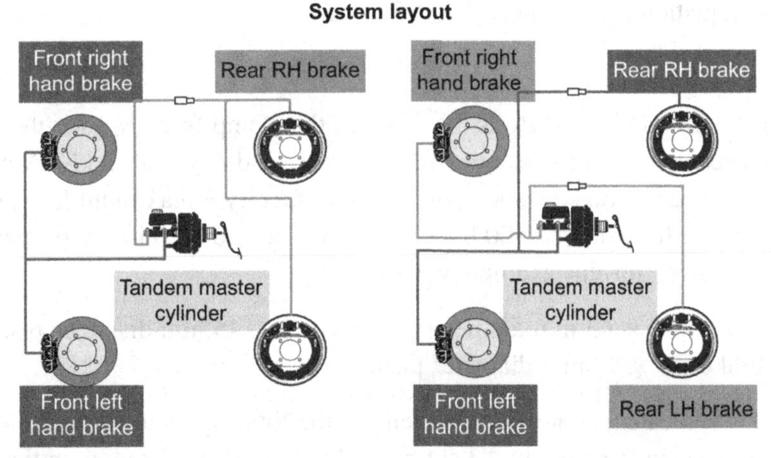

Figure 6.5: 'Split' Configurations for Passenger Cars and Light Commercial Vehicles with Hydraulic Braking Systems.

individually and connect each brake to both circuits. If all the brakes are connected to both circuits, the system is described as having a 'horizontal' split. Legislation does not require duplication of the brake pedal as it is assumed this cannot fail. The diagonal split is selected for the example vehicle here.

Step 3.2 Design the Brake Actuators (Hydraulic System)

Referring to Figure 6.4, the hydraulic pressure generated by the master cylinder is transmitted via the hydraulic brake lines to the slave cylinders, which provide the actuation force at each brake. The principle is the same for hydraulic actuation of a drum brake and a disc brake, with the difference that the disc brake requires a larger diameter slave cylinder than the drum brake because of the lower brake factor. In a drum brake hydraulic slave cylinder, the hydraulic seal is fitted to the piston and sliding contact occurs between the seal and the cylinder bore as the brake shoes are actuated. In a disc brake, the hydraulic seal fits in a groove machined in the cylinder bore. Under normal operation the displacement of the disc brake slave piston is small enough so that it is accommodated by deflection of the seal instead of sliding. This enables the hydraulic seal to provide positive pad retraction as the seal elastically recovers the deflection as 'rollback'; in comparison a drum brake requires 'pull-off' springs to retract the linings away from the brake drum. Slave cylinder pistons are also fitted with dust covers, which can provide a small additional retraction force.

For the example vehicle the required maximum brake torque calculated in Step 1.1 is $\tau_{w1max} = 3050$ Nm and $\tau_{w2max} = 1600$ Nm. From Equation (5.4b) the required maximum clamp force N_{cmax} is:

$$N_{cmax} = \tau_w/(2\mu r_e) \tag{6.13a}$$

On the basis of Equation (5.4e) where disc BF $= 2\mu$:

$$N_{cmax} = \tau_w/(r_e \mathrm{BF}) \tag{6.13b}$$

Assuming that theactuation force P_{ai} is the same as the clamp force N_{ci}, for the example vehicle the required actuation forces are $P_{a1} = 30.9$ kN and $P_{a2} = 16.3$ kN. A good starting point in sizing the brake slave cylinders is to specify a maximum line pressure in the hydraulic system, for example 100 bar (10 MPa). This would initially propose the following actuator sizes for this example vehicle:

Front disc brakes: 1 × 63 mm diameter piston (or 2 × 45 mm diameter pistons);
Rear disc brakes: 1 × 48 mm diameter piston.

Referring back to the calculation of critical length, the limiting factor in the selection of the number of pistons in the front brake caliper is likely to depend not upon the size and properties of the pad assembly, but upon the design and packaging of the caliper. This is because the single piston diameter proposed above is significantly larger in diameter than the radial width of the disc rubbing path and thus the pad assembly.

The braking performance of the whole vehicle with the initially proposed braking system must now be examined, using the analysis of vehicle braking distribution and adhesion utilisation presented in Chapter 3. The braking force exerted by the brakes on axle 'i' under normal braking conditions (T_i) is determined by the specification of the foundation brakes at each wheel (see Equations (6.9)). Assuming that the hydraulic actuation pressure at the front and rear brakes is the same (no valves, EBD, etc.), the ratio X_1/X_2 is calculated using Equations (3.19) (for the moment ignoring the effect of threshold pressure):

$$X_1 = T_1/Pz = \text{BF}_1 P_a r_{e1}/r_{r1} \tag{6.14a}$$

$$X_2 = T_2/Pz = \text{BF}_2 P_a r_{e2}/r_{r2} \tag{6.14b}$$

Using the proposed slave piston sizes (front disc brakes 2 × 45 mm or 1 × 63 mm diameter pistons; rear disc brakes 1 × 48 mm diameter piston):

$$X_1/X_2 = 65/35 = 1.83$$

which can be compared with the initial value (from Step 1.1) of $X_1/X_2 = 74/26$ (DoW) or 61/39 (GVW) based on $z_{crit} = 0.6$. As illustrated in Figure 6.6, this indicates that the rear wheels are 'over-braked' in the unladen condition so it is necessary to adjust the actuation design to meet the requirement of $z_{crit} \leq 0.6$ in this condition. Reducing the rear brake slave cylinder from 48 mm diameter to 38 mm achieves $z_{crit} = 0.6$, as illustrated in Figure 6.7.

This system would be satisfactory according to the criterion that $z_{crit} \geq 0.6$ (which of course depends upon the vehicle manufacturer's standards), i.e. the rear wheels will not lock before the front wheels at a deceleration of less than 0.6 g in the unladen condition, which is generally the worst case for an FWD passenger car or light van. This worst case might occur if all the parts of the braking system that might be incorporated to control the braking distribution between axles, e.g. valves, EBD, load sensing, etc., stopped working and the system defaulted to the DoW braking distribution. The worst case brake torques at GVW are approximately $\tau_{w1max} = 3050$ Nm and $\tau_{w2max} = 1040$ Nm for $z = 1$ (including 5% for rotational inertia), and values can be compared with the design maximum brake torques in Step 1.1, which were $\tau_{w1max} = 3050$ Nm and $\tau_{w2max} = 1600$ Nm for $z = 1$. The actuation pressure to achieve $z = 1$ (1 g deceleration) would be approximately 100 bar (10 MPa) compared with approximately 85 bar (8.5 MPa) in the first design. This might be too high for the vehicle manufacturer, in which case the slave cylinders could be increased in size while maintaining the ratio X_1/X_2. For example, the front brake slave cylinders could be increased from 2 × 45 mm to 2 × 49 mm diameter pistons, and the rear brake slave cylinders changed from 1 × 38 mm to 1 × 41 mm diameter pistons, which would give an estimated 82 bar (8.2 MPa) for $z = 1$. The other option of single actuation pistons on the

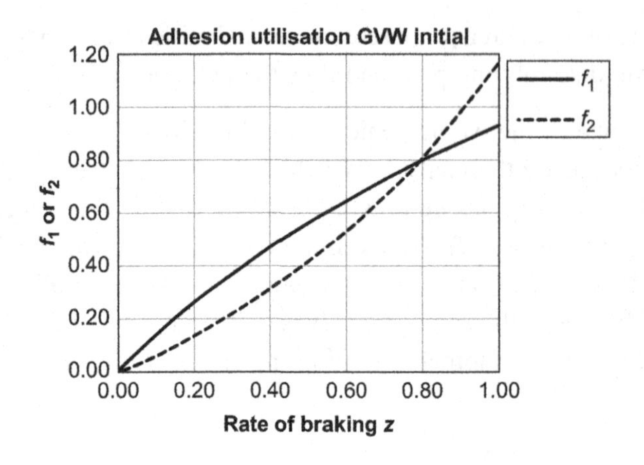

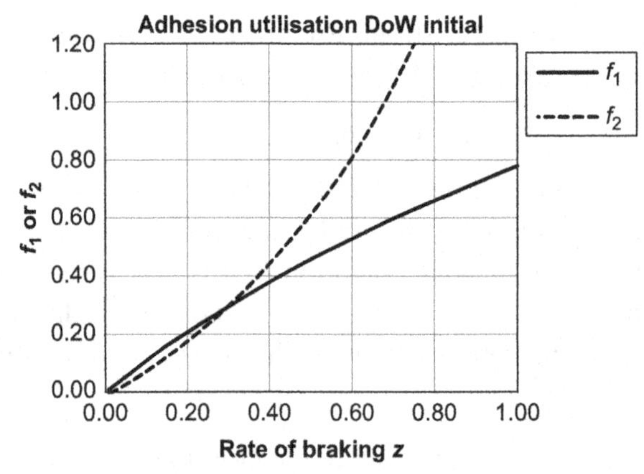

Figure 6.6: Adhesion Utilisation Curves for First Proposed Design Showing Premature RWL, $z_{crit} < 0.6$.

front brakes would require 1 × 69 mm diameter pistons with the rear brake having 1 × 41 mm diameter pistons.

There are two features of the adhesion utilisation characteristics shown in Figure 6.7 that require comment following on from the discussion of braking distribution in Chapter 3. First, the graphs illustrate the difficulty of designing a basic braking system that works efficiently under all usage conditions, especially at different loads, without any control of the braking distribution between axles, e.g. valves, EBD, load sensing. At DoW, the specification of $z_{crit} \geq 0.6$ is met, while at GVW the efficiency is low. For example, at a coefficient of adhesion between the tyre and road of $k = 0.6$, the maximum rate of braking before the front brakes lock is $z = 0.46$, which indicates unacceptably low efficiency in a modern passenger car or light commercial vehicle. Secondly, at DoW, the graph indicates that rear wheel lock occurs at $z = 0.65$, when the actuation pressure is approximately 42 bar (4.2 MPa). Since this is the worst case for premature rear wheel lock, it might be good

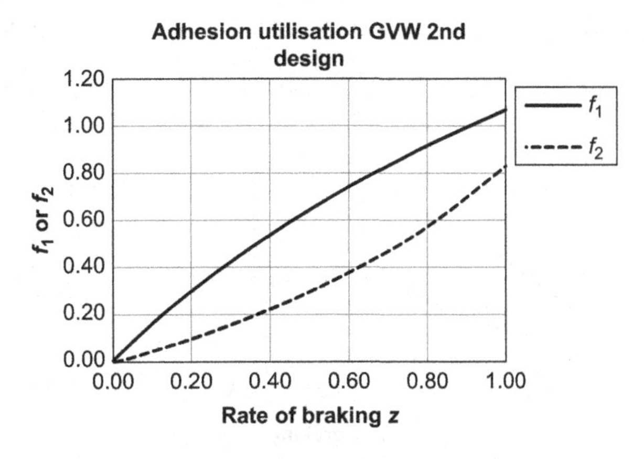

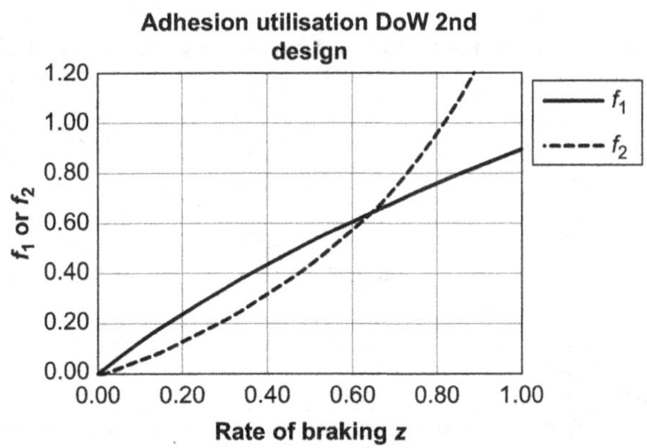

Figure 6.7: Adhesion Utilisation Curves for Second Proposed Design Showing $z_{crit} \geq 0.6$.

practice to limit the available pressure to the rear brakes to, for example, 40 bar, using a pressure-limiting valve, but the effect of this would be to make the braking efficiency at GVW unacceptable unless some form of proportional load sensing were incorporated. An alternative might be to install a pressure-reducing valve in which the pressure to the rear brakes is reduced above a specified pressure, or to use a deceleration-sensing valve that limits actuation pressure to the rear brakes. All these types of device have been utilised in the past, but since the fitment of ABS and more recently EBD, the active control of wheel lock (ABS) and actuation pressure distribution between axles (EBD) has enabled high-efficiency braking systems to be designed and installed.

As discussed in Chapter 3, to achieve the maximum braking efficiency ($\eta = 1$) over a wide range of vehicle loading and usage conditions, the 'ideal' braking distribution X_1/X_2 would need to be designed to be continuously variable and equal to the dynamic weight distribution for all values of z. Using Equations (3.38) and (3.39), the optimised actuation pressures for ideal braking distribution in the GVW and DoW loading conditions are shown in Figure 6.8. In the DoW condition the actuation pressure to the rear brake must

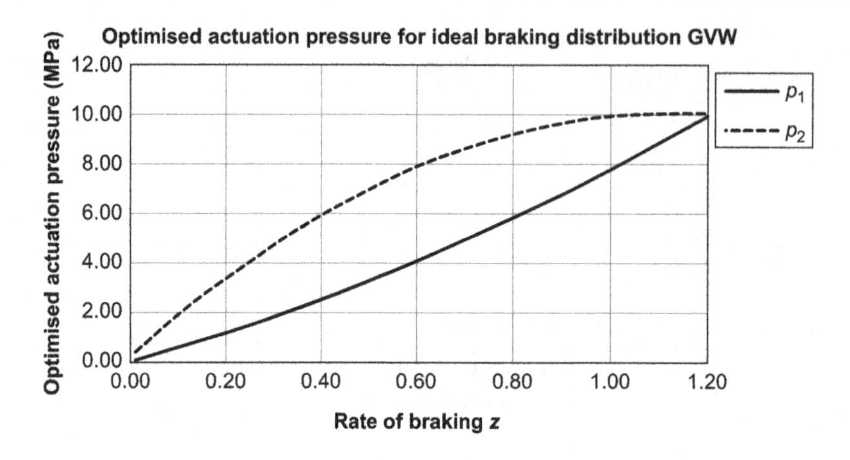

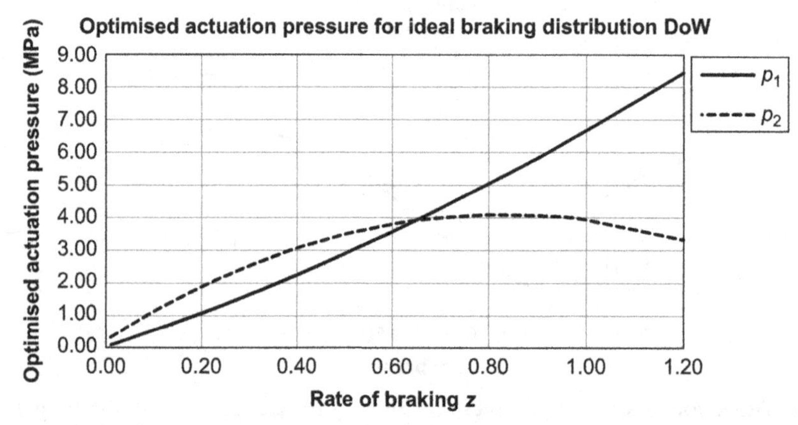

Figure 6.8: Optimised Actuation Pressures for Ideal Braking Distribution in the GVW and DoW Loading Conditions.

be lower than that to the front brakes, while in the GVW condition the actuation pressure to the rear brakes must be higher than that to the front brakes in order to take advantage of the greater adhesion on the rear axle in the GVW condition. However, to achieve the 'ideal' ratio X_1/X_2, the master cylinder pressure can only be reduced to give the required actuation pressure at the front or rear brakes. The result is that varying the actuation pressure at each brake to achieve the ideal braking distribution where the braking efficiency is always unity affects the relationship between the master cylinder pressure and the resulting deceleration of the vehicle.

Step 3.3 Design the 'Master Actuator' (Brake Pedal) System (Hydraulic System)

Step 3.3.1 is to design the brake pedal system, including the pedal configuration, master cylinder piston size and stroke, servo boost ratio and knee point, and fluid consumption. The design procedure starts with the calculation of the driver effort (the force applied by the driver to the brake pedal) that is required to decelerate the vehicle safely in the absence of any servo assistance, i.e. the 'booster fail' condition. The EU legislative requirement (UN, Feb 2014) is for a minimum vehicle deceleration in the booster fail condition of $z = 0.25$ with the driver effort (F_d) between 6.5 and 50 daN (see Chapter 8). The master cylinder can therefore only be designed in conjunction with the brake pedal design.

Brake Pedal

The brake pedal is one of the primary 'driver interfaces' between the customer and the vehicle. The response (deceleration) of the vehicle to the force applied to the brake pedal by the driver must be consistent and confidence-inspiring. First, the force system must be designed so that the force (F_d) applied by the driver's foot to the brake pedal (the 'driver effort') is multiplied by the brake pedal lever ratio (R_p) to give the pedal output force (F_p), which is also called the 'pushrod force':

$$F_p = R_p F_d \tag{6.15}$$

This ratio is initially specified as indicated in Figure 6.9, as the perpendicular distance from the pedal pivot to the centre of the pressure pad on which the driver's foot acts (L_1), to the perpendicular distance from the pedal pivot to the clevis pin where the pushrod is attached (L_2). Most drivers do not press centrally or squarely on the brake pedal pad, as indicated by wear of the rubber pad on the pedal, and a detailed and accurate analysis of the brake pedal mechanics is essential in the refinement of the braking system. The output force from the brake pedal is called the pushrod force, which in the absence of a brake servo or booster would directly actuate the piston in the master cylinder.

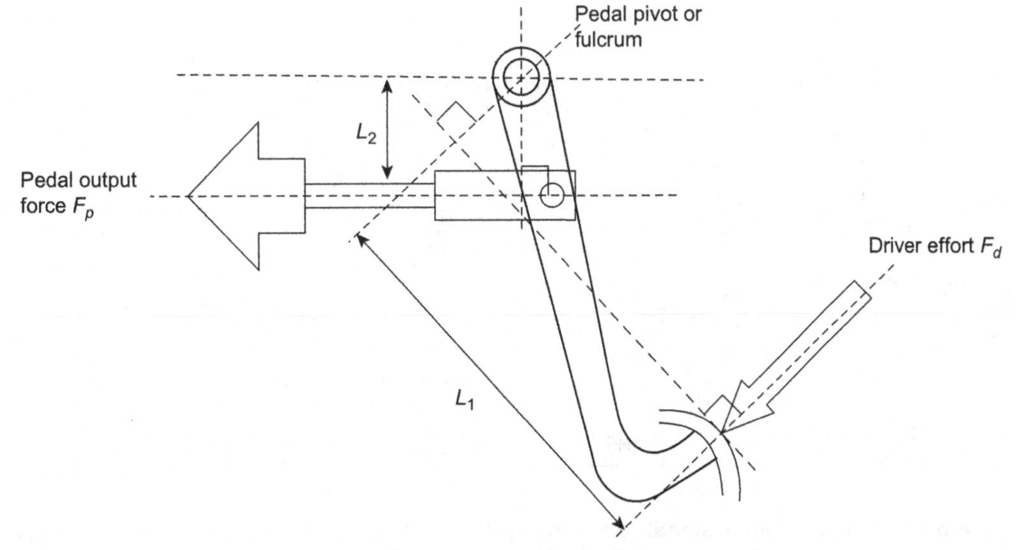

Figure 6.9: Brake Pedal.

The limiting factors on brake pedal design are:

- The available package space between the firewall, instrument cluster and the driver's legs for accessibility reasons.
- The effective pedal ratio in combination with the servo boost ratio, actuation design, and brake factor.
- Crashworthiness-related safety requirements, e.g. avoiding pedal intrusion into the driver's leg-room, which may require a pedal retraction feature.
- Ergonomic requirements to define the limits of travel and effort, the spacing between the pedals, and the trajectory of the brake pedal relative to the driver's seating position.

The last factor includes the 'step-over' height between the accelerator and brake pedals relative to the lateral separation of the pedals and any footrest that might be fitted, as illustrated in Figure 6.10. The driver may perceive different braking responses during brake pedal application depending on the driving situation, pedal apply speed and vehicle load condition. Because braking is a safety system, attributes such as consistency, 'feel' and responsiveness represent a high priority for drivers, although these seem to depend to some extent on the local market and the vehicle class. Parameters to be considered for a good pedal feel include:

- Initial 'dead' travel before deceleration is perceived
- Relationship between deceleration and pedal effort

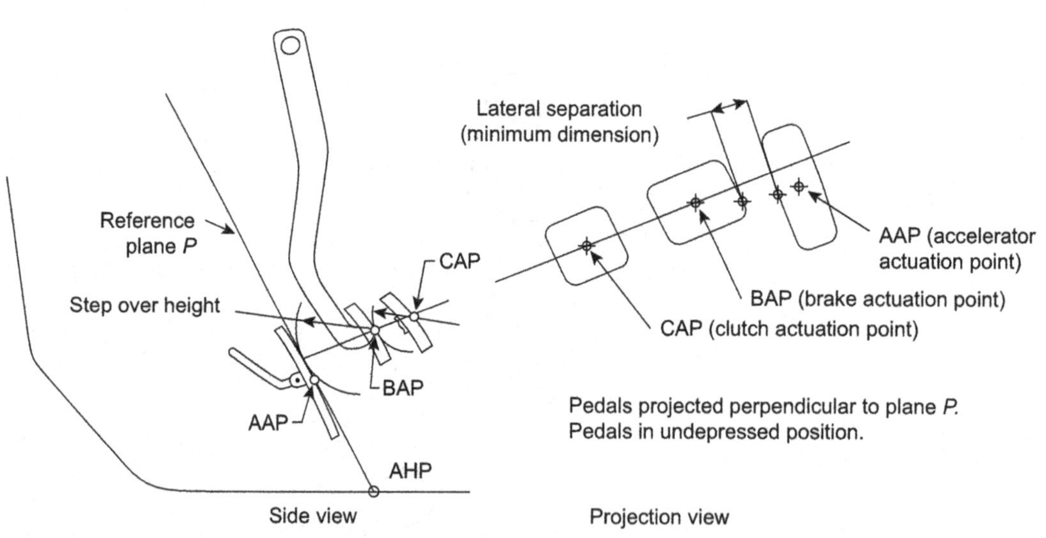

Figure 6.10: Nominal Dimensional Relationship Between Brake Pedal and Accelerator Pedal (Svensson, 2013).

- Relationship between deceleration and pedal travel
- Relationship between pedal travel and effort
- Deceleration consistency
- Static and dynamic hysteresis
- Time delay between the application of the driver effort and the deceleration response of the vehicle.

The pedal ratio (R_p) is usually designed to conform with the vehicle manufacturer's in-house standards, which are based upon the boundaries and restrictions explained above. Equation (6.15) indicates that a low pedal ratio (e.g. 2.5:1) would mean that the mechanical advantage is low and the driver would have to apply greater pedal force to achieve the same pushrod force (F_p) than if the pedal ratio were higher (e.g. 4:1). On the other hand, the pedal travel would be higher in the latter case. The brake pedal is usually part of a vehicle sub-assembly known as the 'pedal box', which includes the accelerator pedal, brake pedal and clutch pedal (for manual transmission), and is mounted on to the vehicle bulkhead. The brake pedal, pedal box and the bulkhead may deflect under actuation load, although such deflections should be minimised to avoid excessive pedal travel. The pushrod is designed for negligible compressive deflection. EU legislation assumes that the brake pedal is amply dimensioned and will not fail.

The maximum movement (travel) of the brake pedal (x_{pmax}) is determined by the distance between the 'off' position and the floor or bulkhead, including any deformation or deflection of the brake pedal, pedal box and the floor or bulkhead. The maximum pushrod travel or full stroke depends upon the pedal travel and the pedal ratio (R_p) as shown in Equation (6.16) and must not exceed the maximum length of piston travel of the master cylinder, which is usually specified by its manufacturer:

$$\text{Full stroke} = x_{pmax}/R_p \tag{6.16}$$

If the maximum pedal travel (x_{pmax}) is 100 mm and $R_p = 4$, the full stroke of the master cylinder would be 25 mm and the maximum length of piston travel of the master cylinder would in practice usually be specified as significantly more (e.g. 36 mm) to accommodate deformation and deflections of the system components under actuation forces.

Master Cylinder

The master cylinder converts the pushrod force (F_p) into brake circuit hydraulic pressure. Separate master cylinder functions are required for each circuit in case one fails. The necessary two functions are incorporated into a dual master cylinder, which is usually a 'tandem' design, in which two pistons share a single cylinder as illustrated in Figure 6.11(a). The 'primary' piston is actuated by the pushrod force and the hydraulic pressure generated actuates the 'secondary' piston, which floats in the cylinder bore.

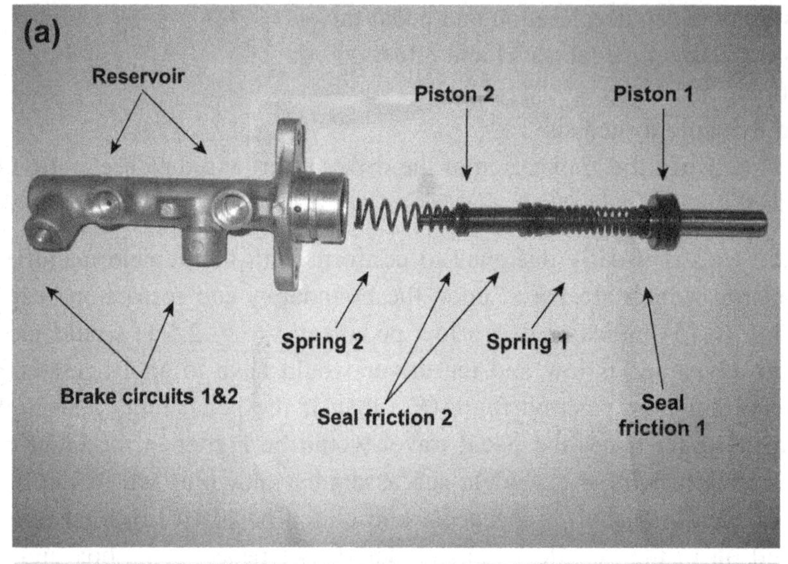

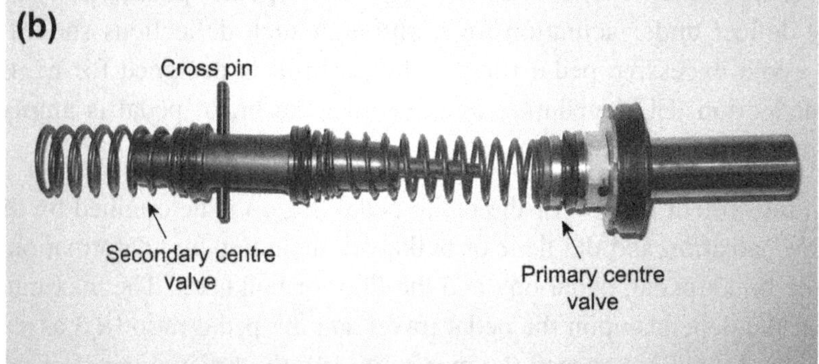

Figure 6.11: Master Cylinder Designs.
(a) Tandem master cylinder design. (b) Centre valve design (Ho, 2009).

Where the two pistons are the same diameter, the primary and secondary chamber pressures are in principle the same; in practice the secondary chamber pressure is slightly lower than the primary chamber pressure because of seal friction; usually the difference is less than 2 bar (0.2 MPa). Each chamber is supplied with fluid from separate parts of the fluid reservoir, and as the pistons move the supply ports are closed so that hydraulic pressure can be generated. Springs are provided to ensure that the pistons return to the 'off' position when the brakes are released, and together with the travel required to close the supply ports they contribute to the relationship between the pushrod travel and the force/pressure generated, which is so important in defining the pedal feel as discussed later. The level of hydraulic fluid in the reservoir should be maintained higher than the reservoir division or weir. If fluid is lost from one circuit the fluid level drops to the level of the weir, which is required (in the EU legislation) to trigger a visual warning to the driver, but enough fluid is retained behind the weir to provide for the needs of the circuit(s) that remain operational.

Many modern systems now use the 'centre valve' design of master cylinder as shown in Figure 6.11(b). The difference between this design and the standard tandem master cylinder is in the design of the hydraulic fluid flow from the reservoir to the primary and secondary pistons. In the centre valve design the fluid is supplied to the primary and secondary circuits through the open centre valve. When the primary piston is moved by the pushrod actuation towards the floating piston, it closes the valve inlet from the reservoir and at the same time opens the secondary piston valve to pressurise the secondary chamber from the primary chamber. The secondary piston valve remains shut if either circuit fails; if the primary circuit fails, pressure is generated in the secondary circuit by the mechanical transmission of the pushrod force to the secondary piston. If the secondary system fails, the circuit is closed so that pressure is generated in the primary circuit. The advantage of the centre valve design is that the piston seals do not act as valves, which reduces wear and damage of the seals, and also enables the relationship between the pushrod travel and the force/pressure generated to be more accurately designed.

If p_1 and p_2 are the hydraulic pressures generated in the primary and secondary chambers respectively by the input force P_{mc} (Gerdes et al., 1995):

$$p_1 = \left(P_{\mathrm{mc}} - P_{k1} - F_{f1} - P_{c1}\right)/A_{mc1} \tag{6.17a}$$

$$p_2 = p_1 - \left(P_{k2} + F_{f2} + P_{c2}\right)/A_{\mathrm{mc2}} \tag{6.17b}$$

where:

A_{mc1} and A_{mc2} are the cross-sectional areas of the primary and secondary pistons respectively;
F_{f1} and F_{f2} are the seal friction forces of the primary and secondary pistons respectively;
P_{k1} and P_{k2} are the spring pre-load forces on the return springs;
P_{c1} and P_{c2} are damping forces arising from the viscosity of the hydraulic fluid.

For braking system layout outline design, a useful approximation to Equations (6.17) is:

$$p = p_1 = p_2 = \eta_{mc} P_{mc}/A_{\mathrm{mc}} \tag{6.18}$$

η_{mc} is an efficiency factor associated with hydraulic actuators (cylinders); typically for a brake master cylinder it is approximately 0.95. It includes the friction force associated with hydraulic piston seals (and thus F_{f1} and F_{f2} in this case), which is non-linear, depending upon the speed of movement, fluid temperature, seal/bore contact area and hydraulic pressure. It can be divided into two parts: stiction (static friction) and dynamic (sliding) friction.

Brake Servo (Booster)

Drivers expect to be able to achieve high rates of braking at relatively low pedal efforts, which means that some form of servo or booster is usually incorporated to amplify the pushrod force (F_p) from the brake pedal. The most common type of servo device used on passenger cars and light vans is the 'vacuum' booster, although electrical boosters are becoming available. UN Regulation 13H (UN, Feb 2014) allows relatively high pedal effort to meet the braking performance prescribed for the booster fail condition ($z = 0.25$ equivalent to $J = 2.44$ m/s^2 for driver effort (F_p) between 6.5 and 50 daN) so back-up servo assistance is not usually designed into hydraulic braking systems.

When a servo or brake booster is incorporated into the actuation system, the pedal output force F_p (Equation (6.15)) is multiplied by the boost ratio (R_b) of the servo. This decreases the driver effort (F_d) required to decelerate the vehicle so that relatively low forces are required, thereby contributing to vehicle safety and driver comfort. The vacuum booster (Figure 6.12), requires air depression up to about 0.8 bar below atmospheric pressure. If the vehicle has a throttled gasoline engine this may be generated via an inlet manifold connection with a non-return valve. Where the engine inlet depression cannot be used, e.g. with a diesel engine, an electric powertrain, or for any other reason such as interference with the emissions control system, a separate 'vacuum' pump is required, which may be mechanically or electrically driven.

The vacuum booster has two chambers separated by a diaphragm; the 'apply' chamber is connected through a control valve to the atmosphere, and the 'vacuum' chamber is permanently connected to the depression source via a reservoir to allow for depletion from rapid repeated braking or failure of the vacuum pump. When the brake pedal is not actuated, both sides of the diaphragm are connected to the depression source, i.e. they are at the same pressure, which is below atmospheric pressure. When the brake pedal is actuated the control valve opens to allow a proportion of atmospheric pressure into the vacuum chamber, which creates a pressure difference between the two chambers.

The two basic design parameters associated with the vacuum servo are the boost ratio (R_b) and the knee point. R_b is the ratio of the vacuum booster total output force (P_{mc}) to the input force (F_p), and is determined by the design of the reaction disc and plunger in the air valve in the 'apply' chamber, which controls the relationship between the pedal force and the build-up of the pressure differential across the booster diaphragm. The vacuum booster total output force that actuates the master cylinder (P_{mc}) is therefore the sum of the input force from the brake pedal (F_p) and the servo assistance force generated by the pressure differential across the booster diaphragm (= pressure differential × booster diaphragm effective area), so that:

$$P_{mc} = F_p R_b \tag{6.19}$$

Brake off

Vacuum port

The diaphragm and input rod are fully retracted the vacuum port is open and there is a vacuum on each side of the diaphragm.

Brake applied

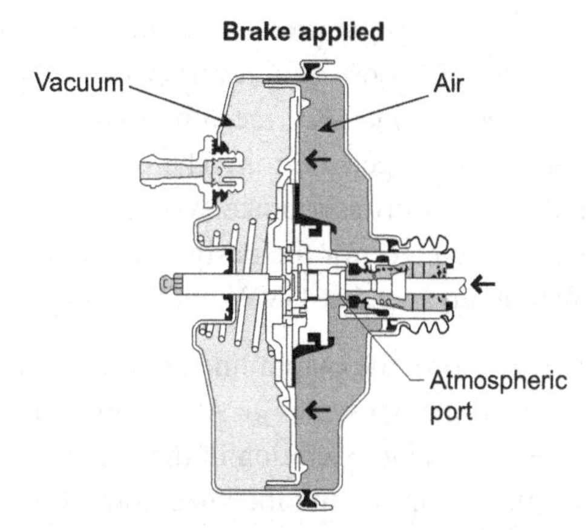

As the brake is applied the input rod assembly moves forward until the poppet valve closes the vacuum port. As the rod continues to move, the atmospheric port opens, and atmospheric pressure enters behind the diaphragm giving assistance to the input rod, thereby actuating the master cylinder.

Brake held on

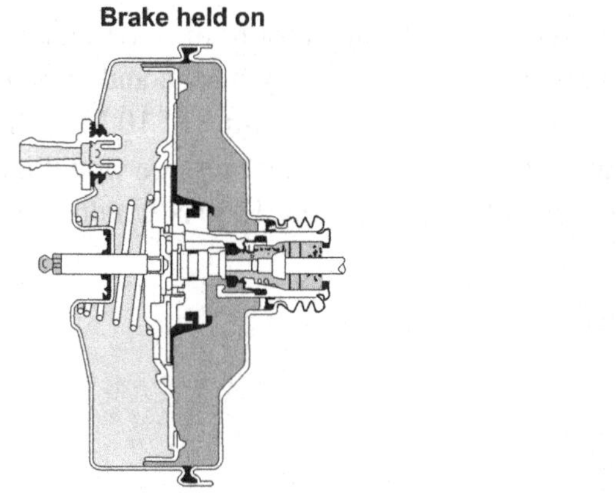

Brake released

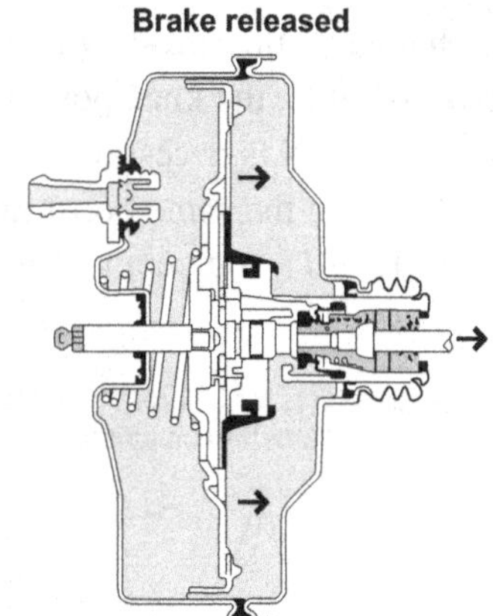

Figure 6.12: Vacuum Servo.

The total force amplification from the brake pedal to the master cylinder is the combination of the pedal ratio (Equation (6.15)) and the boost ratio (Equation (6.19)):

$$P_{mc} = R_p F_d R_b \tag{6.20}$$

In the event of failure of the vacuum booster for any reason, $R_b = 1$ because the pushrod force from the brake pedal (F_p) is transmitted to the master cylinder unchanged.

The knee point represents the maximum servo assistance force that can be generated by the vacuum booster. This depends on the size of the diaphragm over which the depression acts, since the force = pressure differential multiplied by the booster diaphragm effective area. The hydraulic line pressure at which the maximum available boost force is generated is called the knee point pressure, above which no further servo assistance can be generated, so the boost ratio reverts to unity and the system amplification reverts to that of the brake pedal alone (R_p), meaning that any further increase in deceleration would require substantially more pedal effort. The driver might interpret this significant change in the required pedal effort as a brake system failure and so the knee point pressure should be high enough to cater for the design maximum braking condition, usually at least $z = 1$ from maximum speed at GVW with hot brakes.

Figure 6.13 shows the output line pressure generated by the master cylinder (p) vs. the input force to the booster (F_p). At low input forces the pressure response is affected by the clearances and valve operation in the servo, over a region termed 'jump-in'. The gradient between the 'jump-in' and the knee point represents the boost ratio, so the lowest indicates the unboosted characteristic ($R_b = 1$), the next represents $R_b = 4$ and the steepest represents $R_b = 6$. A high boost ratio increases the apparent responsiveness of the brakes (MPa/N) but reduces the knee point pressure. To raise the knee point pressure it is necessary to increase the effective diaphragm area, by specifying a larger diameter booster or two diaphragms.

Four possible characteristics are shown in Figure 6.13. If the boost ratio is 4:1 and a 'small' diameter booster is fitted, the knee point pressure is approximately 8 MPa and an input force of approximately 1600 N is necessary to generate a brake line pressure of 10 MPa. If a larger booster were fitted with the same boost ratio of 4:1, the knee point pressure would increase to approximately 12 MPa so that an input force of approximately 1000 N would generate

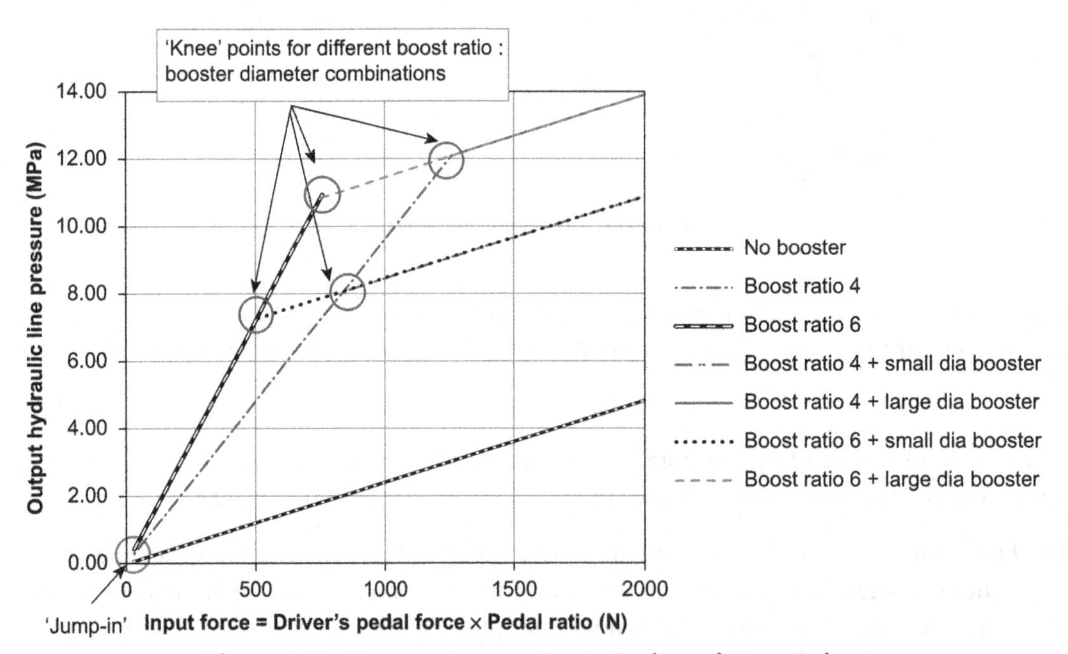

Figure 6.13: Vacuum Booster Boost Ratio and Knee Point.

10 MN/m^2. If the boost ratio were 6:1, with a small booster the knee point pressure would be approximately 7.3 MPa but the input force to generate 10 MPa would be approximately 1600 N, the same as with the 4:1 boost ratio and the small booster because this is above the knee point. With the larger booster the knee point pressure would be approximately 11 MPa and an input force of approximately 700 N would generate a line pressure of 10 MN/m^2. If the booster should fail completely, an input force of over 2000 N would be needed to generate 5 MPa compared with approximately 500 N with the 4:1 boost ratio or approximately 320 N with the 6:1 boost ratio, so the driver would experience a significant increase in the pedal force required to stop the vehicle. In the event of servo failure there is no back-up system; in consequence UN Regulation 13H requires that even in the booster fail condition the driver effort required to decelerate the vehicle at a mean fully developed deceleration (MFDD) of 2.44 m/s^2 is between 6.5 and 50 daN (65–500 N). This means that careful system design is required to avoid placing too much reliance on the servo assistance.

Combining Equations (6.18) and (6.20) defines the basic relationship between the driver effort (F_d) and the generated master cylinder pressure (p):

$$p = \eta_{mc} R_p F_d R_b / A_{mc} \tag{6.21a}$$

For the example vehicle an initial estimate of the required master cylinder size in the servo-fail condition, when the boost ratio $R_b = 1$, can be found from Equation (6.21b):

$$A_{mc} = \eta_{mc} R_p F_d R_b / p \tag{6.21b}$$

To meet the requirements of UN Regulation 13H in this condition ($F_{dmax} = 500$ N, $z = 0.25$, $R_b = 1$), an initial proposed design of the brake pedal (R_p) and the master cylinder diameter (D_{mc}) could be:

$$\text{Brake pedal lever ratio}\left(R_p\right) = 4 : 1$$

The braking system designed so far for the example vehicle has indicated that at GVW, the hydraulic line pressure required to achieve $z = 0.25$ with a safety margin to allow for operational factors including the effect of temperature on brake torque (fade) is approximately 3 MPa:

$$A_{mc} = 0.95 \times 4 \times 500 \times 1/3 \times 1\text{E}6 = 630 \text{ mm}^2$$

Master cylinder diameter $D_{mc} = 28$ mm

Many master cylinder manufacturers offer standard bore sizes, and 27 mm diameter is taken as the nearest standard size. The boost ratio and knee point can now be designed to achieve the desired relationship between the driver effort (F_d) and the vehicle deceleration. Equation (6.21a) can be rearranged to:

$$F_d = pA_{mc} / \left(\eta_{mc} R_p R_b\right) \tag{6.21c}$$

For a 27 mm diameter master cylinder, the actuation line pressure required for $z = 1$ at GVW is calculated at 8.3 MPa, in which case a boost ratio of 4 indicates a driver effort (F_d) of 310 N would be required for $z = 1$, GVW. Depending upon the vehicle manufacturer's requirements, this might be considered acceptable. In comparison a boost ratio of 6 would require a driver effort (F_d) of 210 N for $z = 1$, GVW. As previously explained (Figure 6.13) a larger diameter booster would be required to keep the knee point above 100 bar (10 MPa).

The system equation for the hydraulic brake actuation system can be formed by combining the individual component equations in Step 3.1, which are summarised below. It is assumed (see Chapter 5) that the pad actuation force (P_{ai}) is the same as the clamp force at the pad/disc interface (N_{ci}).

$$\text{Wheel brake torque: } \tau_{wi} = \text{BF}(p_i - p_{ti})A_{ai}\eta_i r_{ei} \tag{5.6b}$$

$$\text{Master cylinder pressure: } p_i = \eta_{mc} P_{mc}/A_{mc} \tag{6.22}$$

$$\text{Force amplification from the brake pedal to the master cylinder: } P_{mc} = R_p F_d R_b \tag{6.23}$$

Combining these gives the general brake hydraulic actuation system equation for the brake torque and the braking force generated at each wheel when the driver applies a force to the brake pedal (pedal effort F_d):

$$\tau_{wi} = \text{BF}_i\big((\eta_{mc} R_p F_d R_b/A_{mc}) - p_{ti}\big)(A_{ai}\eta_i r_{ei}) \tag{6.24}$$

The wheel braking force is:

$$T_{wi} = \text{BF}_i\big((\eta_{mc} R_p F_d R_b/A_{mc}) - p_{ti}\big)(A_{ai}\eta_i r_{ei})/R_r \tag{6.25}$$

and the axle braking force is:

$$T_i = 2\text{BF}_i\big((\eta_{mc} R_p F_d R_b/A_{mc}) - p_{ti}\big)(A_{ai}\eta_i r_{ei})/R_r \tag{6.26}$$

For the whole vehicle the total braking force ($T = T_1 + T_2$) is given by Equation (6.27). This equation applies for disc or drum brake combinations provided that the brake factor is correctly defined (see Chapter 5), and includes threshold pressures (p_{ti}) and cylinder efficiencies (η_i), which affect the actuation force generated at each brake. Estimated values of such efficiencies and losses are suggested; these are for guidance only as manufacturers derive and use their own detailed data. Equation (6.27) can also accommodate different hydraulic actuation pressures to the front and rear brakes (e.g. with pressure-reducing valves) by inserting a factor to modify the master cylinder pressure to the selected brakes.

$$\begin{aligned} T &= (T_1 + T_2) \\ &= 2\text{BF}_1\big((\eta_{mc} R_p F_d R_b/A_{mc}) - p_{ti}\big)(A_{sc1}\eta_{sc1} r_{e1})/r_{r1} + 2\text{BF}_2\big((\eta_{mc} R_p F_d R_b/A_{mc}) - p_{ti}\big) \\ &\quad \times (A_{a1}\eta_1 r_{e2})/r_{r2} \end{aligned} \tag{6.27}$$

Brake Pedal Feel and Fluid Consumption

Brake pedal feel is very important in modern passenger car brake system design, and is primarily determined by the relationship between the driver effort (force applied on the pedal by the driver's foot) and the pedal travel (movement), as indicated in Figure 6.14. It is such an important parameter in terms of the driver interface that the movement and deformation or deflection of the component parts of the system that affect the force/travel relationship must always be estimated or predicted (e.g. by FE analysis) at the design stage, or measured from previous similar installations.

The factors that influence pedal travel include:

- Brake pedal and mounting (bulkhead) clearance, deflection and deformation
- Servo (booster) valve clearances, spring settings and reaction disc deformation
- Master cylinder bore and seal deformation, and valve clearances (c_{mc})
- ABS/ESC system valve clearances and internal deformation
- Brake pipe and flexible hose deformation (c_{bp} and c_{bh})
- Slave cylinder bore and seal deformation (c_{sc})
- Brake fluid compression (c_{bf})
- Pad/disc or lining/drum clearances
- Brake pad or brake shoe assembly compression, deflection and wear
- Stator deformation – disc brake caliper and drum brake anchor plate
- Rotor deformation and deflection – disc and drum.

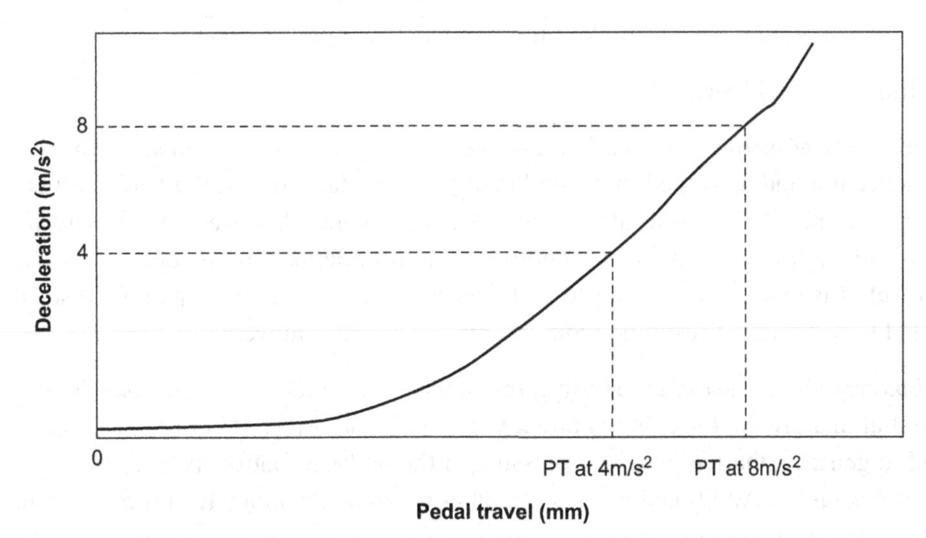

Figure 6.14: Brake Pedal Feel Example Characteristic.

When the driver applies the brakes, the initial brake pedal movement takes up clearances in the system, e.g. to actuate valves in the master cylinder and servo ('jump-in'), and move the pads or linings into contact with the disc or drum. Then, mechanical loading, thermal loading, internal hydraulic pressure and wear cause deformation and deflection of the actuation system components, which can only be accommodated by further movement of the brake pedal. The part of this extra movement that is directly caused by deformation of the hydraulic components under hydraulic pressure is termed 'fluid consumption', and this affects the relationship between brake pedal travel and the hydraulic pressure generated, which has to be allowed for in the system design. Vehicle manufacturers usually specify a design value of 'fluid reserve', which is defined as the ratio of the brake fluid displaced by the master cylinder piston to achieve a specified rate of braking to the brake fluid displaced over the master cylinder full stroke (x_{pmax}). The purpose is to avoid a situation where the extra movement to compensate for the fluid consumption causes the pedal to reach the limit of its travel before the required hydraulic pressure is generated. It should be noted that the term 'fluid consumption' does not imply that any hydraulic fluid is lost from the system, and the term 'fluid reserve' does not refer to the amount of brake fluid in the reservoir of the master cylinder, but to the volume (V_{mc}) of brake fluid of the master cylinder, as defined in Equation (6.22):

$$V_{mc} = A_{mc} \cdot \text{Full stroke} = A_{mc} x_{pmax} / R_p \qquad (6.28)$$

A typical design target could be 50% fluid reserve at $z = 1$, GVW (at a specified worst case operational condition, which may mean high brake operating temperatures after repeated braking), i.e.:

$$\frac{\text{Volume of brake fluid displaced to achieve } z = 1}{\text{Volume } (V_{mc}) \text{ of brake fluid in the master cylinder stroke}} \leq 0.5 \qquad (6.29)$$

This is illustrated in Figure 6.15.

Brake Pedal and Mounting (Bulkhead) Clearance, Deflection and Deformation As previously stated, deflection and deformation of the brake pedal, pedal box and the vehicle bulkhead should be minimised to achieve the stiffest system possible. This is achieved by CAE analysis. Although brake pedal and mounting clearance, deflection and deformation affects pedal travel, it is external to the master cylinder and the effect can be quantified separately and added to pedal travel resulting from the master cylinder movement.

Servo (Booster) Valve Clearances, Spring Settings and Reaction Disc Deformation The boost ratio (and ultimately the knee point) directly influence the driver effort (pedal force) required to generate the required line pressure in the brake actuation system and hence the vehicle deceleration. Additionally there are other components in a vacuum booster that affect the relationship between the pedal force and the line pressure, and thus the pedal feel; in particular, the valves must be sensitive and controllable enough to provide

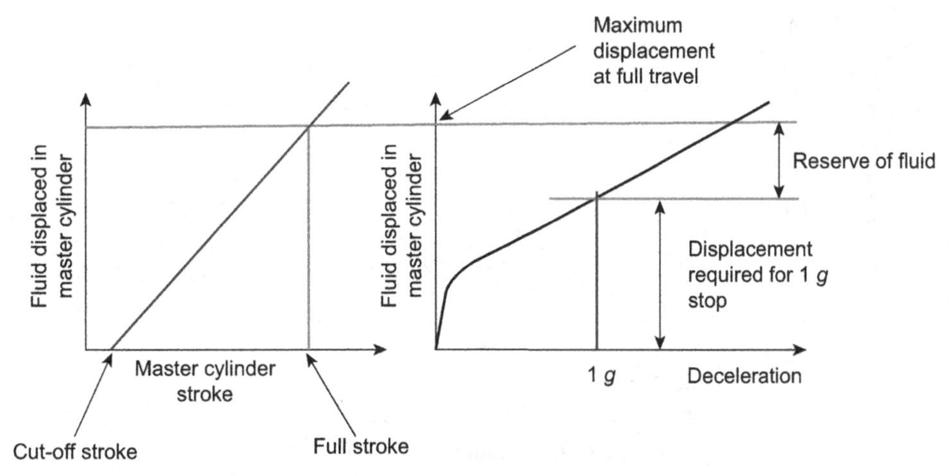

Figure 6.15: Brake Fluid Consumption Design Target.
(a) Fluid volume displaced vs. stroke. (b) Fluid volume displaced vs. deceleration.

proportional assistance even at very low pedal force under all driving conditions. Typically, around 20 N pedal force should start to generate boost assistance, and if this minimum force is higher or the brake pedal has to move too much, the result may be poor pedal feel. Similarly if a noticeably large amount of 'threshold' (or 'idle') pedal travel is required to actuate the control valves and initiate servo assistance, poor pedal feel would result because the pedal is moving without generating boosted hydraulic pressure and hence vehicle deceleration. Once the pedal force exceeds the threshold force and/or travel the vacuum servo starts to boost the hydraulic line pressure and this transition is called 'jump-in'. Day et al., 2009 and Ho (2009) found that the vacuum valve spring stiffness and the air valve spring stiffness had a great effect on the brake pedal feel. For example, the brake pedal feel could be improved by increasing the air valve spring stiffness because this reduces the idle travel.

An example characteristic of master cylinder pressure vs. pedal travel for a car brake actuation system (three datasets) is shown in Figure 6.16. The measurement was carried out in a passenger car (Ho, 2009) so the characteristic includes brake pedal and mounting clearance, deflection and deformation, the vacuum booster characteristic, and the full system fluid consumption.

Master Cylinder Bore and Seal Deformation, and Valve Clearances (c_{mc}) The radial expansion (u_r) of the master cylinder bore under hydraulic pressure p can be calculated from thick-walled pressure vessel theory (Lamé's equations):

$$u_r = \frac{r_a}{E\left(r_b^2 - r_a^2\right)}\left[(1-\nu)pr_a^2 + (1+\nu)pr_b^2\right] \tag{6.30}$$

r_a and r_b are the tube inner and outer radii respectively.

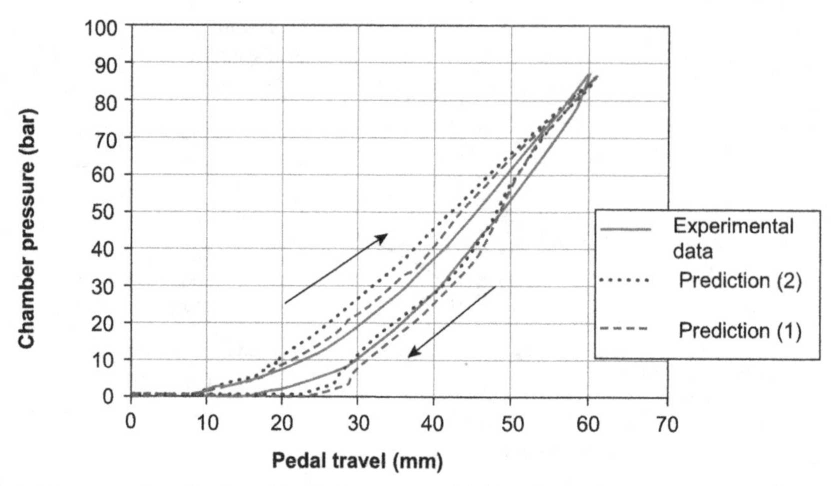

Figure 6.16: Vacuum Servo Characteristic Example: Master Cylinder Pressure vs. Pedal Travel (Three Datasets Shown).

Table 6.5: Comparison of the Estimated Expansion of the Bore of Different Master Cylinder Designs and Materials

Material	Cast Iron	Steel	Aluminium Alloy
Bore diameter (mm)	20.64	20.64	20.64
Wall thickness (mm)	5	3	7
Young's modulus (GPa)	175	200	75
Poisson's ratio	0.25	0.29	0.33
Hydraulic pressure (MPa)	10	10	10
Radial expansion (mm)	0.0013	0.0017	0.0028
c_{mc} (mm^3/MPa) (assuming a 25 mm stroke)	0.22	0.27	0.46
Fluid 'consumption' at 10 MPa (100 bar) (mm^3)	2.2	2.7	4.6

Using Equation (6.22) a comparison of the estimated expansion of the bore of different master cylinder designs and materials is shown in Table 6.5.

The measured pedal travel to actuate the valves and generate 1 MPa in a 20.64 mm diameter master cylinder was 4.5 mm, equivalent to 1.73 mm master cylinder piston travel ($R_p = 2.6$ on this particular installation) and a fluid consumption of 580 mm^3. The fluid consumption resulting from the deflection and deformation of the seals in the same master cylinder was predicted to increase linearly by approximately 10.5 mm^3 over the hydraulic pressure range 0–5 MPa (Ho, 2009), so the cylinder and seal deformation could be equivalent to approximately 2.2 mm^3/MPa. Together with the piston travel to actuate the valves, this suggests that the fluid consumption coefficient c_{mc} (mm^3 / MPa) for a 20.64 mm diameter centre valve master cylinder at hydraulic pressure p_{mc} (MPa) could be:

$$c_{mc} = 580 + (2.2p_{mc}) \tag{6.31}$$

This is an example that depends on the size, specification, design and manufacture of the master cylinder. Design improvements in master cylinder valves continue to be made to reduce the actuation movement needed.

ABS/ESC System Valve Clearances and Internal Deformation As explained in Chapter 11, ABS/ESC systems are designed to be neutral in terms of brake system performance including pedal feel when they are not operational. During operation they are designed to minimise transient dynamic effects (pulsing) on the brake pedal.

Brake Pipe and Flexible Hose Deformation (c_{bp} and c_{bh}) Hydraulic brake lines are thick-walled steel tubes, and thick-walled composite flexible hoses made from reinforced rubber where flexibility is required to connect parts that move relative to each other, e.g. from a subframe to a caliper. The radial expansion of thick-walled steel brake lines can also be predicted from Equation (6.21), and for a typical brake line at 100 bar the fluid consumption can be around 1.5 mm^3 per metre length. Hence $c_{bp} = 0.15$ mm^3/MPa · m. The composite hoses have a much higher flexibility, and a typical value of fluid consumption estimated from manufacturers' expansion data is $c_{bh} = 30$ mm^3/MPa · m.

Slave Cylinder Bore and Seal Deformation (c_{sc}) As explained in Chapter 5, slave cylinders and pistons provide the actuation force at the foundation brake (disc or drum) in a hydraulically actuated braking system. In drum brakes the slave piston acts directly on the shoe tip, and in disc brakes there may be one, two or more pistons acting on each pad to ensure that the actuation force is as uniformly distributed as possible over the pad/disc friction interface. The relationship between the applied hydraulic pressure and the actuation force generated by the slave cylinder piston is defined by Equation (5.5). As with the master cylinder the fluid consumption in a slave cylinder arises from deformation of the bore and deflection of the pistonseals. For a 57 mm diameter disc brake caliper piston, the fluid consumption resulting from the deflection and deformation of the seal was predicted using FEA to increase linearly by less than 2 mm^3 over the hydraulic pressure range 0–50 bar (Ho, 2009). The fluid consumption caused by expansion and deformation under 5 MPa hydraulic pressure in the slave cylinder was estimated to be 8 mm^3, giving an estimate for $c_{sc} = 2$ mm^3/MPa. Prediction of caliper deformation, especially the bore, by FEA is strongly recommended as an essential part of disc brake design as this depends strongly on the detailed design and construction of the caliper.

Brake Fluid Compression (c_{bf}) The compressibility of brake fluid is affected by temperature and the presence of trapped air and moisture. Vehicle manufacturers use vacuum fill techniques for charging the hydraulic brake system on the vehicle assembly line, and annual replacement of the fluid is recommended to minimise the effects of moisture (moisture content also affects vaporisation point). For braking system design purposes, the compressibility of brake fluid is specified in terms of a compressibility coefficient, c_{bf} MPa^{-1}, where the fluid consumption is the product of the working volume of fluid in the brake system and the compressibility coefficient c_{bf}. The physical and chemical properties of brake fluid are closely specified, and Figure 6.17 indicates the dependence of fluid compressibility (c_{bf}) on temperature for different types of brake fluid. The fluid compressibility (c_{bf}) can range from around 0.0007 MPa^{-1} at room temperature to around 0.003 MPa^{-1} at 200°C.

Pad/Disc or Lining/Drum Clearances In order to minimise residual drag forces (parasitic losses), running clearances between the rotor and stator friction elements must be maintained when the brake is not being used. As explained in Chapter 5, a threshold force is required to overcome the retraction forces generated by the 'pull-off' springs in drum brakes and the seal 'roll-back' in disc brakes. Then piston movement is required to close the running clearance. Since this occurs at low actuation pressures, the driver senses pedal travel without any significant deceleration response, so ideally the threshold pressures and running clearances will be kept to a minimum. A recent development in hydraulic brake actuation systems is to detect intended brake usage, e.g. by release of the accelerator, and use the ABS pump to lightly pressurise the actuation system to overcome threshold forces and take up running clearances in anticipation of a brake application.

Brake Pad or Brake Shoe Assembly Compression, Deflection and Wear Brake pad compression is influenced by the compressive stiffness of the friction material, and the flexure of the backplate as explained in Chapter 5 referring to friction interface pressure distribution. Modern brake pad friction material can be assumed to be linear in tension and compression over ±3% strain, and using a disc pad friction material compressive modulus of 3.5 GPa, assuming the pad is compressed uniformly by the actuation force, for a pad

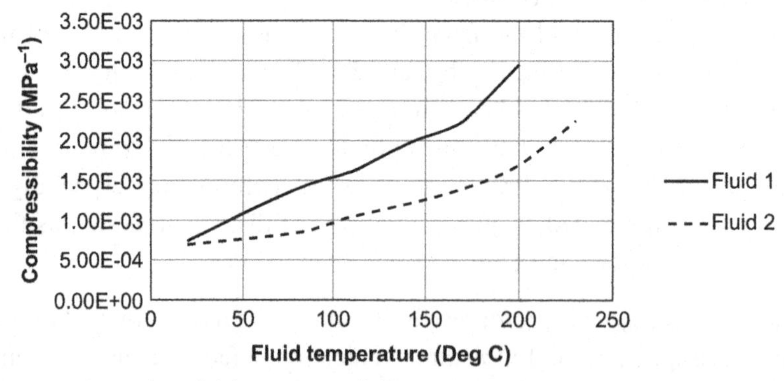

Figure 6.17: Brake Fluid Compressibility vs. Temperature (Internet Source).

area of 3500 mm^2, and thickness 12 mm, the compressive stiffness and compressibility are approximately 1.0×10^6 and 1.0×10^{-6} mm/N respectively. At a line pressure of 100 bar, the compression of the pad assembly, assumed uniform, is therefore 0.01 mm. The fluid consumption (for a 57 mm diameter piston) would be 17 mm^3. Because of flexural effects in the brake pad assembly (Chapter 5), disc brake pads do not experience uniform compression even where there is more than one actuation piston on each pad; therefore, this is a conservative estimate, which may be doubled to 34 mm^3 to provide a more realistic approximation. In sliding calipers, the outboard pad actuation force is usually applied via 'fingers' spaced apart to spread the load more uniformly. The process of wear causes the interface pressure distribution to tend towards uniform, but this does not necessarily reduce the flexure of the pad assembly when the actuation force is applied and removed because of the clearances generated between the friction material and the disc at the friction interface.

Drum brake shoes tend to run at larger clearances than disc brake pads, although sliding abutment shoes enable some of the clearance to be taken up by translation of the shoe rather than actuation displacement. At high actuation forces, flexure of the brake shoe can affect wear and pressure distributions (see Chapter 5), which can increase fluid consumption and thus actuator travel, but on modern cars and light vans, drum brakes are only fitted to rear axles and the braking duty is relatively low.

Stator Deformation — Disc Brake Caliper and Drum Brake Anchor Plate Disc brake caliper deflection was briefly discussed in Chapter 5; deformation, displacement and distortion of the caliper arise from caliper deflection under clamp load, caliper twisting under friction drag loading, and thermal deformation. Deformation and deflection of the anchor plate of the type of Simplex drum brake fitted to most passenger cars and light vans are not considered to affect fluid consumption significantly. In all designs of friction brake, stator deformation and deflection depend strongly on the detail design and construction, and the prediction of stator thermal and mechanical deformation by FEA at the detail design stage is essential.

Rotor Deformation and Deflection — Disc and Drum Good detail design of brake discs is essential to minimise fluid consumption arising from the deformation and deflection of brake rotors. A new brake can be manufactured and assembled with minimum deformation and deflection, but as soon as it is used, the disc will deform and deflect as a result of the thermomechanical loads imposed on it by braking friction. Some of the deformation will be permanent as a result of residual stresses created by thermo-elasto-plastic deformation during heating and cooling, taking the form of 'runout' arising from 'swash' and/or rotational distortion. Thermomechanical deformation (see Chapter 7) also causes 'coning' distortion and 'disc thickness variation' (DTV) may arise from uneven wear around the friction faces. As well as causing judder (see Chapter 10). This can cause increased clearance between the pad and disc, which increases fluid consumption and causes a deterioration of pedal feel over time, so it is important to minimise these effects.

Factors which influence the generation of DTV include:

- Disc design for minimal coning
- Stable disc material with temperature
- Disc and hub manufacture with tight tolerances on runout
- Assembly processes designed to minimise 'bolt-up' distortion
- Rigid stator (caliper) with minimum deflection and distortion under operational loads.

Heat generated during braking causes the friction ring to expand but it is constrained by the 'top hat' section at the hub, which causes coning deformation of the brake disc. The increased pad/disc clearance causes increased pedal travel, taper wear of the pads and non-uniform heat input.

Most vehicle manufacturers use fluid consumption factors that they have developed in-house based on measured data from previous designs and experience. Experimental measurement of brake line pressure vs. pedal travel from a static bench test rig is shown in Figure 6.18, which indicates a contribution of different parts of the hydraulic actuation system to fluid consumption. This system was a full vehicle system except for the vacuum booster and the ABS system on a static test rig. Pad/disc clearances were set nominally at 0.5 mm, and the pedal/mounting deformation at 5 MPa was measured at 5 mm. A comparison of estimated fluid consumption with the measured values for this set-up is shown in Table 6.6. The estimated values are not in full agreement, but are close enough to illustrate the method.

Step 4 Verify the Design and Check Compliance with Legislative Requirements (UN Regulation 13H)

Steps 1–3 above describe the brake system design procedure for the example passenger car defined in Table 6.2, and the typical design targets that must be verified for the braking system on this type of vehicle include:

- Compliance with all legal requirements.
- Brake pedal feel: the subjective perception of brake pedal force and travel to achieve the required vehicle deceleration should meet the manufacturer's targets.
- Stopping distance: the vehicle must be capable of achieving the minimum stopping distance and/or mean fully developed deceleration (MFDD).
- Stability and adhesion utilisation: the vehicle must decelerate without deviation or imbalance between axles.
- The vehicle must decelerate without generating noise, judder and vibration.

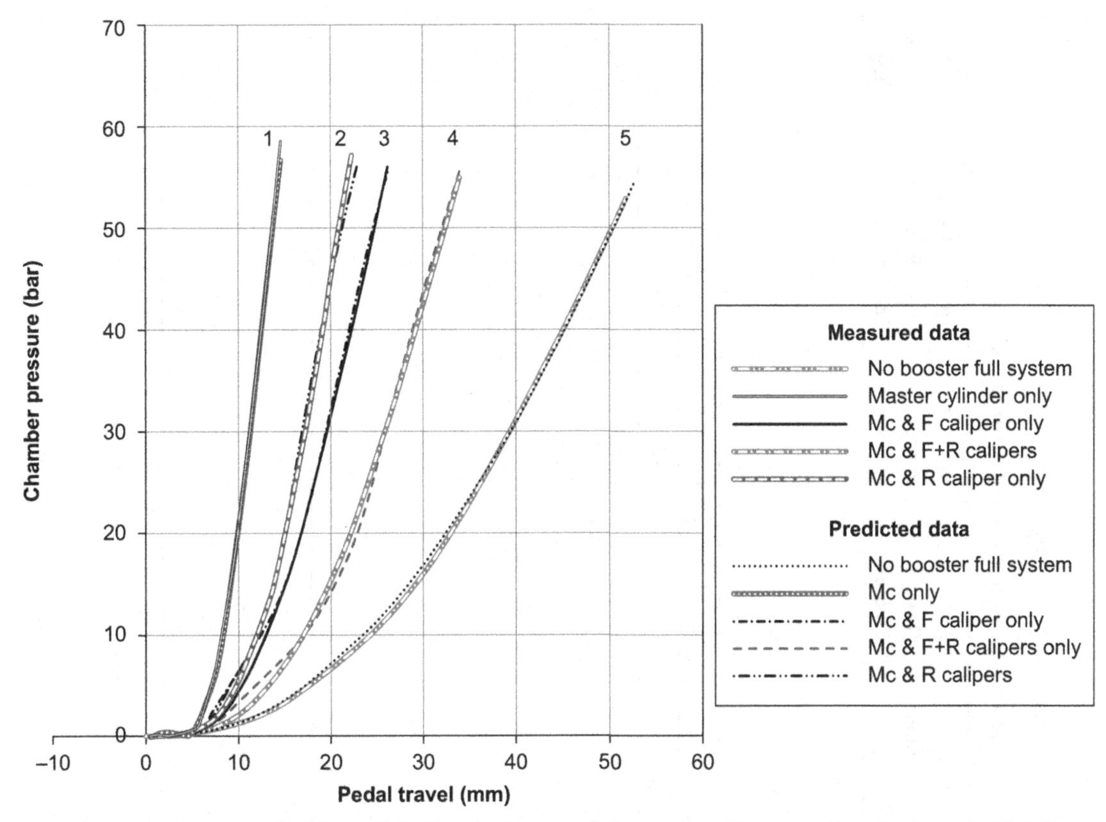

Figure 6.18: Contribution of Different Parts of the Hydraulic Actuation System to Fluid Consumption as indicated by the change in the relationship between master cylinder chamber pressure and brake pedal travel. (Ho, 2009).

- Wear and durability: the life of the brake rotors and stators must meet customer expectations.

Equation (6.27) is used to calculate the design braking performance of the vehicle specified in Table 6.2 with the additional designed system data summarised in Table 6.7.

Step 4.1 Intact System

Annex 3 of UN Regulation 13H (see Chapter 8) requires minimum levels of vehicle deceleration to be verified by tests. The prescribed performance is based on the stopping distance and the mean fully developed deceleration, MFDD. (Note that the symbol used in this book for MFDD is J_m, while that used in UN Regulations 13H and 13 is d_m) with the vehicle in the laden condition as specified by the manufacturer. For an intact system, i.e. the system is fully operating as intended, the vehicle is tested in the GVW condition at 100 km/h and J_m must be greater than 6.43 m/s^2 ($z = 0.655$) with the driver effort (F_d) not exceeding

Table 6.6: Comparison of Estimated Fluid Consumption with Measured Values

Fluid Consumption Estimate at 5 MPa (mm^3)			Master Cylinder Only	MC + 1 Rear Caliper	MC + 1 Front Caliper	MC + 1 Front and 1 Rear Caliper	MC + 4 Calipers
Brake pedal and mounting		2 mm	257	257	257	257	257
Servo (booster)			0	0	0	0	0
Master cylinder	c_{mc}	591 mm^3/MPa	591	591	591	591	591
ABS/ESC system			0	0	0	0	0
Brake pipe	c_{bp}	0.15 mm^3/MPa · m	0.00	0.83	1.65	2.48	4.95
Flexible hose	c_{bh}	30 mm^3/MPa · m	0.00	57.00	57.00	114.00	228.00
Slave cylinder	c_{sc}	2 mm^3/MPa	0	10	10	20	40
Brake fluid compression	c_{bf}	0.0007 MPa^{-1}	29	108	159	238	447
Pad/disc clearances	c_c	0.5 mm	0	570	905	1475	2950
Brake pad assembly compression and deflection			0	34	34	68	136
Stator deformation and deflection			0	0	0	0	0
Rotor deformation and deflection			0	0	0	0	0
Total estimated			878	1629	2015	2766	4654
Total measured			1094	2059	2509	3475	5791
Difference			216	430	494	709	1137

Table 6.7: Hydraulic Braking System Design Specification

Design Parameter	Design Value
Brake pedal ratio	4.0
Boost ratio	4.0
Booster diameter	Large
Brake master cylinder diameter	27 mm
Master cylinder efficiency	0.95
Brake master stroke	36 mm
Foundation brakes:	
Front brake factor (BF = 2μ)	0.76
Rear brake factor BF = 2μ for a disc brake (BF = combined shoe factor = S_1 + S_2 for a drum brake)	0.76
Front disc effective radius	125 mm
Rear disc effective radius	130 mm
Front slave cylinder diameter	1 × 63 mm
Front slave cylinder threshold pressure	0.05 MPa
Front slave cylinder efficiency	0.95
Rear slave cylinder diameter	1 × 38 mm
Rear slave cylinder threshold pressure	0.25 MPa
Rear slave cylinder efficiency	0.95

500 N. From Equation (6.21c) the predicted value of F_d at $z = 0.655$ is approximately 260 N, equivalent to a line pressure of 6.9 MPa. The larger diameter booster has been specified (see Figure 6.13) for which the knee point is >10 MPa, but the analysis indicates that 10.4 MPa line pressure is needed for $z = 1$, so this again suggests an adjustment to the system design as previously discussed relating to the front and rear brake piston size, which might be preferred to avoid reaching the knee point at $z \leq 1$.

Step 4.2 Partial System Failure

The braking system performance in the booster fail condition as specified in UN Regulation 13H requires vehicle deceleration (J_m) ≥ 2.44 m/s^2 ($z = 0.25$) from 100 km/h with the driver effort (F_d) not exceeding 500 N. A line pressure of approximately 2.65 MPa is required to achieve $z = 0.25$, and using Equation (6.21c) with R_b set to 1 for the servo-fail condition; this requirement is met with the required pedal force $F_d = 403$ N.

In the event of a single failure in the electronic control system, e.g. the loss of EBD or ABS, UN Regulation 13H requires vehicle deceleration ≥ 5.15 m/s^2 ($z \geq 0.525$) at $k \leq 0.85$. Since the system is based on $z_{crit} > 0.6$ this requirement is met (see Figure 6.7), where $z = 0.525$ at $k \approx 0.55$.

If one brake circuit fails, e.g. by rupture of the brake hydraulic pipe, UN Regulation 13H requires $J_m \geq 2.44$ m/s^2 ($z = 0.25$) from 100 km/h with F_d not exceeding 500 N. To simulate this in the calculation, the system failure can be simulated for each type of braking system split:

X-split: multiply BF_1 and BF_2 by 0.5: $F_d = 258$ N
Front–rear split: set the corresponding value of BF to 0
Front brake system failure ($BF_1 = 0$): $F_d = 384$ N
Rear brake system failure ($BF_2 = 0$): $F_d = 133$ N.

Step 4.3 Parking

UN Regulation 13H also requires that the parking braking system must be capable of holding a fully laden vehicle stationary on a 20% up or down gradient (angle of inclination = 11.3°) with a manually applied force to the control of ≤400 N. This requirement usually determines the design and specification of the parking brake, and very often the rear brake effective radius r_e is maximised in order to maximise the torque for any specified actuation system design.

Parking brakes can be mechanical or electrical, hand- or foot-operated systems; they must operate independently of the service (primary) brake system and require a separate adjustment device to compensate for brake wear. The parking brake can actuate the brakes of any axle(s) depending on the vehicle and brake configuration; for example a parking brake may be designed to operate on the front axle to meet the UN Regulation 13H requirements where rear axle adhesion utilisation does not permit the legal minimum performance to be reached. As mentioned in Chapter 5, on passenger cars, drum brakes are convenient as parking brakes and as such are widely used with separate mechanical actuation on the rear axle of mass-produced cars. On higher performance cars combined disc brake calipers may be fitted, which incorporate a separate mechanically actuated parking mechanism in sliding calipers. One problem with disc brakes for parking is that as they cool down from high operating temperatures, the pad and disc contract so the clamp force reduces which may lead to partial release of the parking brake. Electric parking brakes (EPB) may incorporate a 're-clamping' function to address this potential failure mode. Some 4WD vehicles may fit a parking brake on one of the transmission shafts to avoid parking actuation mechanisms and cable runs to the wheels, which can be vulnerable in off-road conditions.

A passenger car parking brake can be mechanically operated via a handbrake lever or by a foot lever, which must be positioned (UN Regulation 13H) so that the driver can operate the parking brake from the driving position. Both must incorporate a ratchet device to hold them in the 'on' position. There are several popular designs of handbrake levers for passenger cars, including the conventional straight lever, Z-shaped lever, U-shaped lever and the linear extending lever. The design of handbrake actuation levers for passenger cars is strongly influenced by styling and interior package concept and should always consider the ergonomic aspects of driver operation. An electric parking brake (EPB) system for a passenger car comprises an electronic controller and an electromechanical central actuator that actuates the parking brake function of the drum or disc brakes, usually via Bowden cables. EPB systems are becoming increasingly popular because of the advantages summarised below:

- Improved operational safety
- Simple operation, including hill hold capacity independent of driver skill
- Simpler packaging; no hand or foot lever to intrude into the occupant area
- Integral diagnostic system with an emergency braking function via the service brake system in connection with ESP if the service brake system fails
- Easier installation
- Enhanced safety by automatic application, reduced risk of vehicle 'roll-away'
- Potential integration into anti-theft system.

Any EPB is a safety-critical system that must be designed to prevent unintended dynamic application. Emergency release is mandatory in the case of electrical supply failure.

Comment on Verification Against Legislative Requirements — Hydraulic Braking Systems

Legislative requirements represent the minimum level of performance required of a road vehicle braking system, and vehicle manufacturers always aim to exceed this level of braking performance, often by a considerable margin. Additional 'in-house' braking performance standards are specified by individual vehicle and component manufacturers relating to many aspects of braking, e.g. the fluid consumption discussed earlier. For the example vehicle, the system pressure (assuming no valves or pressure differences between the front and the rear brakes) at $z = 1$ is predicted to be 10.5 MPa. Referring to Table 6.7, the master cylinder diameter is 27 mm and the stroke is 36 mm, so the working volume of fluid (v_{mc}) in the master cylinder is 20,600 mm^3. The fluid consumption can be estimated as explained earlier, and to meet the criterion defined in Equation (6.29) must not exceed approximately 10,000 mm^3 for this example. Based on the experimental data shown in Figure 6.18, the fluid consumption for a full car hydraulic brake actuation system (no booster) at 10 MPa is calculated to be approximately 8600 mm^3. As previously discussed, fluid consumption influences brake pedal feel, and Svensson (2013) stated that the pedal feel is strongly influenced by the relationship (Wheel cylinder area)/(Master cylinder area) $\cdot\, r_p$, which can be used to guide the selection of master and slave cylinder sizes.

Svensson also considered two alternative brake actuation system designs for a medium-sized passenger car, which are summarised in Table 6.8. He showed that with the same wheel cylinder diameter for each design (60 mm), the mechanical ratios for the two systems were the same:

$$\frac{\text{Wheel cylinder area}}{\text{Master cylinder area}} \cdot r_p = \left[\frac{60^2}{27^2}\right] \cdot 4 = \left[\frac{60^2}{22.2^2}\right] \cdot 2.65 = 19.7$$

Both designs satisfied the requirements of UN Regulation 13H, e.g.:

- Booster fail ($J_m > 2.44$ m/s^2): $J_m = 3.05$ and 2.70.
- Adhesion utilisation for $J \geq 5.15$ m/s^2 ($f_i \leq 0.85$): $f_i = 0.72$ (both designs).

The in-house requirement for the booster knee point to be above the actuation pressure required for $z = 1$ at GVW was also satisfied for both designs; the knee point came in at $z = 1.25$ and 1.83 respectively. He also identified additional benefits from Design 2 that included downsizing of the brake booster unit from an 8 & 9 (inch) Tandem to a 10 (inch) Single, representing a 1.5 kg weight reduction with a consequent cost saving. Brake pedal feel was practically unchanged:

- F_d at 0.3 g: 41/42 N
- F_d at 0.6 g: 84/86 N

Table 6.8: Comparison of Alternative Brake Actuation System Designs for a Medium-Sized Passenger Car (Svensson, 2013)

Design Specification	Design 1	Design 2
	Brake Pedal Ratio	
Actuation unit	4	2.65
Diaphragm size	8 & 9 inch (tandem)	8 & 9 inch (tandem)
Brake master cylinder diameter (mm)	27	22.2
Brake master stroke (mm)	36	46
	Foundation Brakes	
Front effective radius (mm)	125	125
Rear effective radius (mm)	130	130
Front disc effective mass (kg)	6.0	6.0
Rear disc effective mass (kg)	2.8	2.8

- Pedal travel for 0.3 g: 39/39 mm
- Pedal travel for 0.6 g: 52/53 mm

At $z = 1$ and GVW, the braking performance was also practically unchanged:

- Master cylinder pressure for 1 g: 97/99 bar
- Pedal travel for 1 g: 79/80 mm
- Fluid reserve for 1 g: 46/45%
- Pedal force for 1 g: 197/201 N.

Commercial Vehicle Braking Systems with Pneumatic Actuation

Commercial vehicle braking systems have several complicating features that are not present on passenger cars, and some of these are discussed here before dealing with the actuation system design. First, because of their large size and large mass, commercial vehicles present a very high demand on their foundation brakes to achieve safe and reliable braking. For this reason, many commercial vehicles are fitted with retarders that provide braking torque through the vehicle driveline and supplement the retardation provided by the foundation brakes, thereby reducing the braking duty on the foundation brakes. This is particularly important in long-duration braking, such as drag braking down long gradients, and retarders are thus generally categorised as 'endurance' brakes; for certain heavy commercial vehicles the fitment of a retarder is mandated in UN Regulations. Because retarders work through the vehicle driveline, the retarding torque they provide is reacted through the tyre/road adhesion at the driven wheels, and therefore the limit of stable braking is still controlled by the adhesion utilisation as explained in Chapter 4, and must be taken into account when designing the system (Markovic, 2002).

Engine braking alone provides negligible retardation of a laden commercial vehicle (unless the engine is designed or fitted with systems for enhanced engine braking, see below), so a retarder allows the driver to avoid using the foundation brakes for minor speed adjustments, e.g. when descending slight gradients, matching speed in a line of traffic or carefully approaching a hazard. This is different from passenger cars, where to make minor reductions in speed it is usually only necessary to lift off from the accelerator and rely on the engine braking. The benefit to the commercial vehicle operator is reduced wear of the foundation brakes. Retarder systems were first mentioned in European legislation in 1954, and Directive 71/320/EEC introduced brake fade tests that required all commercial vehicles to complete a repeated braking test (Type I test) and an additional downhill behaviour test, and these requirements still apply today. For load-carrying vehicles over 12 tonnes the vehicle must retain satisfactory hot brake performance after descending a 6% gradient at 30 km/h for 6 km, the Type II test. Passenger-carrying vehicles must be able to descend a 7% gradient at 30 km/h for 6 km, without the use of the foundation brakes, in such a gear that the maximum engine speed specified by the manufacturer is not exceeded, the Type IIa test. For both Type II and IIa tests the retarder may be used. Certain dangerous load-carrying vehicles are also required to fulfil the Type IIa test specified in Annex 5 of UN Regulation 13 (UN, Mar 2014). There are two basic types of retarder: engine brakes, which increase the overrun braking effect of the vehicle engine; and transmission retarders, which are additional devices incorporated in the transmission. These retarders convert the potential or kinetic energy of the vehicle into heat, which is then dissipated into the environment.

Engine brakes convert the braking energy absorbed into heat, which is dissipated by the engine cooling system. There are two categories: exhaust brakes and decompression brakes. Exhaust brakes consist of a valve fitted into the exhaust system, downstream of the turbocharger, which when operated severely restricts the exhaust flow and results in a high exhaust back pressure. This pressure, acting on the engine pistons during the exhaust cycle, when the exhaust valves are open, provides a retarding torque on the engine crankshaft, and being flow dependent, this type of brake is more effective at high engine speeds. Decompression brakes function on the compression and expansion cycles of the engine. By allowing the compressed air to escape into the exhaust port as the piston approaches top dead centre, a proportion of the air compression energy is lost and the net retarding torque is increased. This 'escape' is achieved either by partially opening the exhaust valve or by operating an additional small valve in the cylinder head. As these two types of engine brake act during different parts of the engine cycle, they can be used together and their effects are approximately additive.

Transmission retarders can be electromagnetic or hydrodynamic. Electromagnetic retarders are effectively an electrical generator where the inductive currents convert braking energy into heat at the rotors. Heat is dissipated into the surrounding air by ventilating fins in the rotating rotor. A primary hydrodynamic retarder is mounted ahead of the gearbox and has a power dissipation proportional to engine speed, and a secondary retarder is mounted to the gearbox output shaft or the transmission propeller shaft, so power dissipation is dependent upon road speed. Retardation is achieved by hydraulic fluid flowing between the rotating and stationary elements, the degree of retardation being controlled by the amount of fluid in the system. Heat transmitted to the hydraulic fluid during braking is normally transferred to the engine cooling system.

The design of an air actuation system for a multi-axle commercial vehicle combination (towing vehicle with trailer; see Chapter 4) is explained here without including retarder systems. The example on which the design is presented is a commercial articulated vehicle combination comprising a towing vehicle (tractor) with a semi-trailer (Figure 4.14) of the specification shown in Table 4.6. The calculation of the maximum torque required at each wheel brake τ_{wimax} was explained in Step 1.2; for a commercial vehicle the laden condition is the worst case for all axles. Equation (6.1d) requires the braking distribution

X_i to be calculated for each axle as explained in Chapter 4, which for the laden case at $z = 0.5$ is 0.254/0.263/0.483. Equation (6.1d) then gives the following maximum brake torque requirements at each axle at the design maximum rate of braking of 0.8:

$$\text{Tractor front axle: } \tau_{w1max} = T_{w1max} r_r = X_1 Pz r_{r1}/2 = 19.8 \text{ kNm}$$
$$\text{Tractor rear axle: } \tau_{w2max} = T_{w2max} r_r = X_2 Pz r_{r2}/2 = 20.6 \text{ kNm}$$
$$\text{Tractor rear axle: } \tau_{w3max} = T_{w3max} r_r = X_3 Pz r_{r3}/2 = 12.6 \text{ kNm}$$

(based on a three-axle bogie; see below)

As previously explained, for commercial vehicles the brakes on each axle should provide sufficient braking force to provide the required maximum deceleration for the maximum weight being carried on that axle. On this basis the maximum brake torque requirements at each axle at a design maximum rate of braking of 0.8 are comparable:

$$\text{Tractor front axle: } \tau_{w1max} = 19.4 \text{ kNm}$$
$$\text{Tractor rear axle: } \tau_{w2max} = 22.3 \text{ kNm}$$
$$\text{Semi-trailer axles: } \tau_{w3max} = 19.4 \text{ kNm}$$

In this example, the static loads on the semi-trailer in the fully laden condition are P_3 (axles) = 220 kN and P_4 (coupling) = 110 kN, where the GVW of the combination is 39.62 kN (40.4 tonnes), so to comply with the axle loading limits, three axles would be required on the trailer (9 tonnes per axle; the maximum permitted axle load on a tri-axle semi-trailer in the UK is 27 tonnes, i.e. 9 tonnes per axle). The three axles would be in the form of a bogie, and comprise six brakes, as previously explained in Chapter 4.

The maximum and mean power dissipations at each brake (Step 1.3) from an initial road speed of 96 km/h (legal maximum throughout the EU) to rest for this vehicle are calculated from Equations (6.4) and (6.5). (Generally, vehicle manufacturers test to the vehicle design speed irrespective of any speed limits as these may change and therefore influence an existing type approval.)

Power Dissipation	Peak (kW)	Mean (kW)
Tractor front brake	1072	536
Tractor rear brake	1114	557
Trailer brake	671	335

From Step 2.1 (specify the foundation brakes, determine the rotor type and size), ventilated discs are selected for the front and rear brakes of the tractor, with S-cam drum brakes on the trailer axles (disc brakes would also be an option for semi-trailers). For commercial vehicles there is a standard range of wheel rim sizes, e.g. 17.5, 19.5, 22.5 and 22.5 inches, and brake manufacturers specify brakes for each of these. This example vehicle uses the largest wheel size (22.5 inches) for which a commercially available disc brake of outer diameter 430 mm capable of delivering 26 kNm torque according to the manufacturer's specification would be suitable for the front and rear axles of the tractor. The brake disc has an effective radius $R_e = 173$ mm. A commercially available S-cam drum brake of 420 mm diameter ($R_e = 210$ mm) capable of delivering 18 kNm would be suitable for the trailer axles.

As previously explained, the thermal performance of the rotor during braking and the size and design of a commercial vehicle brake disc size is critically important. Step 2.2 is covered in Chapter 7.

For Step 2.3, the brake factor (or ηC^*) of the brakes is specified by the brake manufacturer (see Chapter 5). From Equation (5.47) the brake factor in the form of the specific torque (τ_s) of a cam actuated disc brake was estimated to be 26, and for an S-cam drum brake the brake factor was estimated to be 11 (Equation (5.45)).

Step 2.4 is intended to confirm the friction material size based on the thermal and mechanical loading. The typical limits of shear loading and mean work rate discussed earlier relate to cars and light vans, not commercial vehicles. Manufacturers of brakes for commercial vehicles provide a complete foundation brake unit including pads or lined shoes, for which the shear loading and work rate have been extensively proven, so commercial vehicle designers do not usually use thermal and mechanical loading – this is the responsibility of the brake manufacturer.

Step 3 Design the Pneumatic Actuation System

Step 3.1 Specify the Actuation Mechanism (Pneumatic System)

The brake system design for this commercial vehicle can now proceed to the design of the pneumatic actuation system, which for most modern heavy commercial vehicles uses compressed air up to a maximum of about 8.5 bar (0.85 MPa) to create an actuation force at the foundation brake, which is converted to an actuation torque via a lever. With drum brakes, especially the S-cam type specified here for the semi-trailer axles, this lever incorporates a 'slack adjuster' that provides automatic adjustment to cater for wear (see Figure 5.23). The internal mechanism of the foundation brake then converts the torque back into an actuation force, as explained in Chapter 5. The layout of the pneumatic actuation system is illustrated in Figure 6.19, and includes the brake actuator and lever components, as illustrated in Figure 6.20.

Step 3.2 Design the Brake Actuators

For a pneumatically actuated braking system the brake actuator design includes the pneumatic actuator size and stroke and the lever length. The input torque to an air-actuated brake was defined in Equation (5.32):

$$\tau_a = P_a l = (p - p_t) A_a \eta l \qquad (5.32)$$

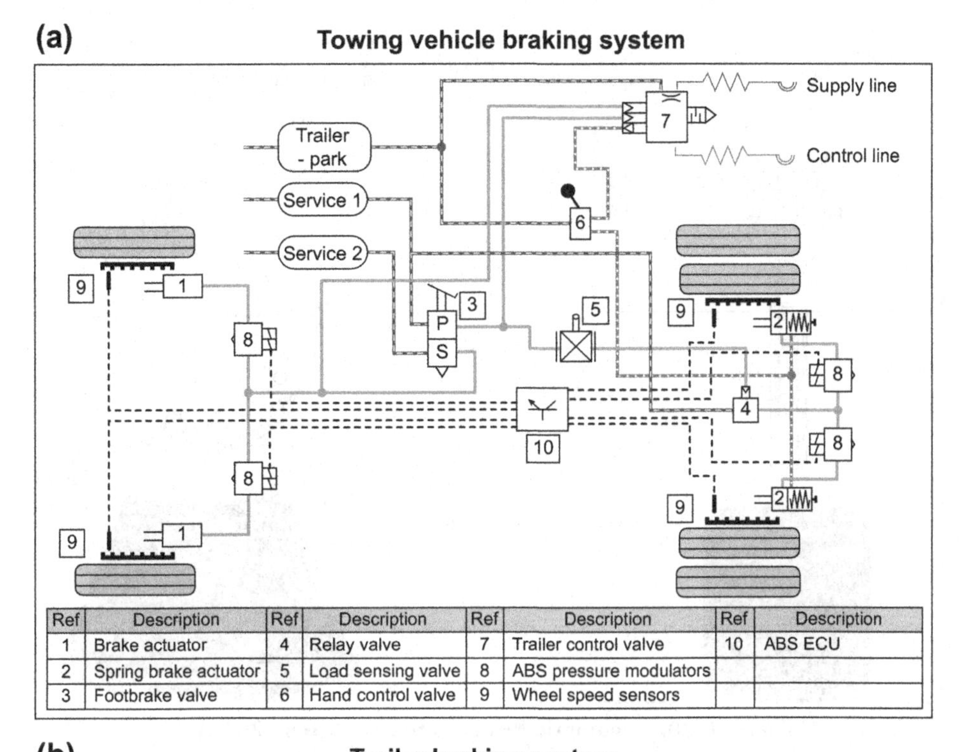

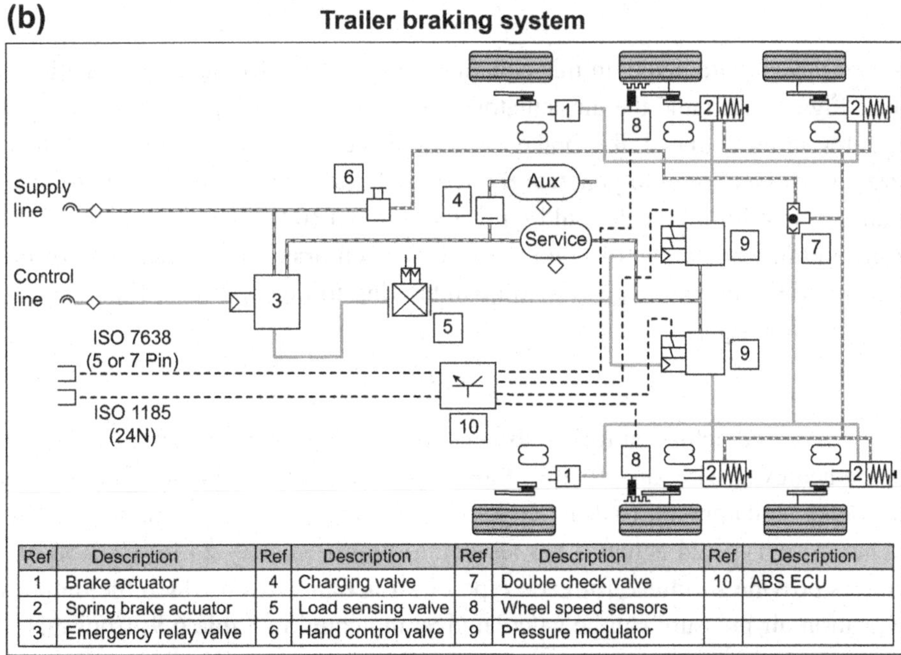

Figure 6.19: (a) Two Line/Dual Circuit Pneumatic Braking System with ABS for a Two-Axle Rigid Commercial Vehicle. (b) Two Line/Dual Circuit Pneumatic Braking System with ABS for a Semi-Trailer (Ross, 2013).

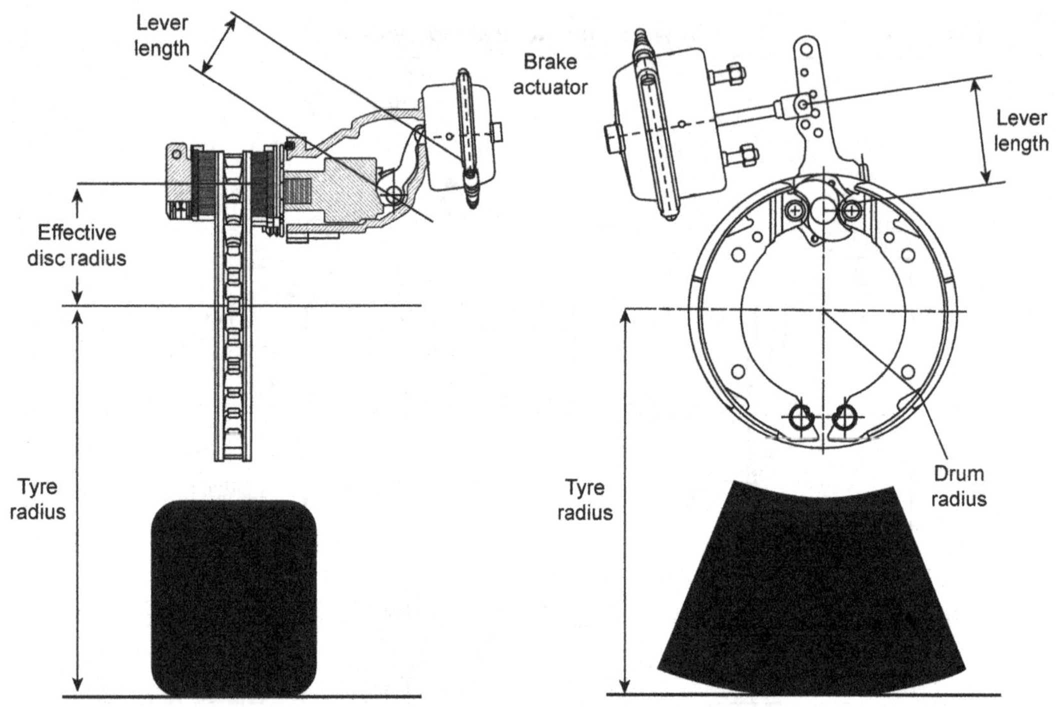

Figure 6.20: Pneumatic Brake Actuation (Ross 2013).

There are two design parameters in this equation that must be decided: the actuation lever length l (see Figure 6.20) and the air actuator size (area) A_a (see Figure 6.21). A typical lever length for an air-actuated disc brake of the selected size is $l = 76$ mm, and the air actuator size, must be chosen to generate the required actuation torque within the available effective air pressure limit, which could typically be 8 bar (0.8 MPa), to achieve a designed maximum vehicle rate of braking $z = 0.8$. Each brake on the rear axle of the tractor is required to generate $\tau_{max} = 20.6$ kNm braking torque and from Equation (5.40) (ignoring threshold torque):

$$\tau = \tau_a \mathrm{BF} \tag{5.40}$$

Therefore the required actuation torque to generate $\tau_{2max} = 20.6$ kNm is $\tau_{2a} = 792$ Nm, which can be achieved with an actuator of area approximately 0.014 m^2. Commercial vehicle air brake actuators are traditionally specified in terms of the actuation area in square inches, so a Type 24 actuator has an area of approximately 24 in^2 (0.0155 m^2). Manufacturers provide formulae for each type of air actuator from which the output force for any actuation air pressure can be calculated as indicated in Table 6.9; from this it can be seen that a Type 24 actuator is appropriate for this tractor rear disc brake. The tractor front disc brakes have a slightly lower design value $\tau_{1max} = 19.8$ kNm for which the required actuation torque is $\tau_{1a} = 762$ Nm, but the Type 24 actuator is appropriate for

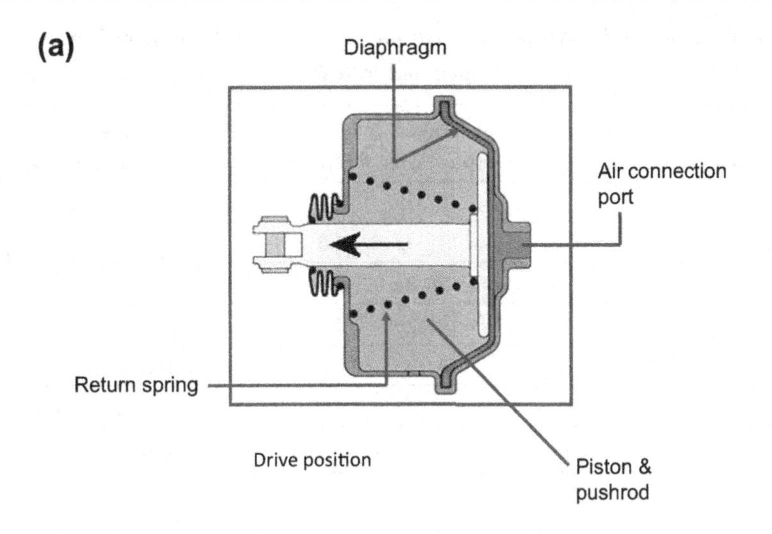

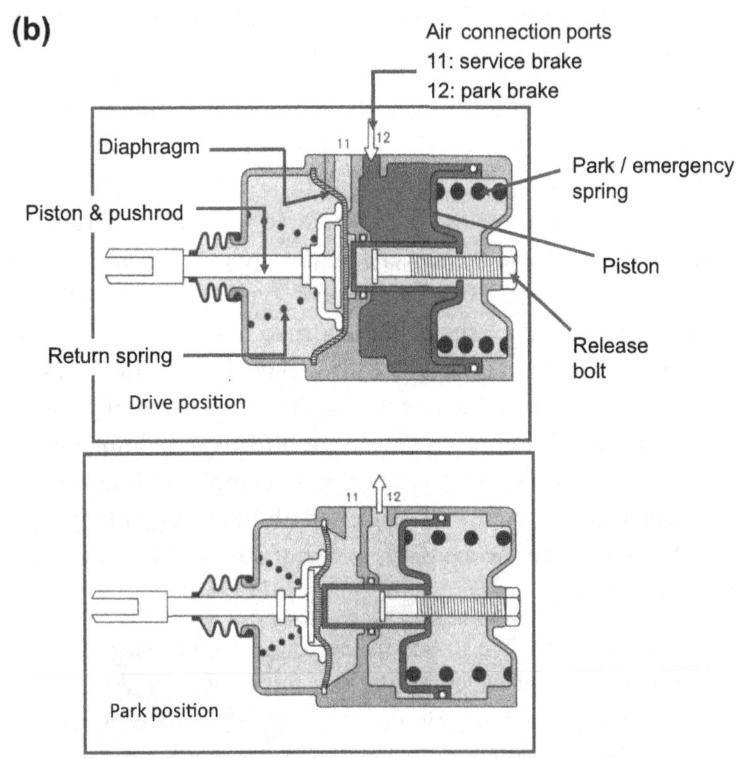

Figure 6.21: Pneumatic Actuators (Courtesy Arfesan).
(a) Standard actuator. (b) Spring brake actuator: (i) drive position; (ii) park position.

Table 6.9: Air Brake Actuator Formulae (Pneumatic Actuation Pressure p is Specified in MPa)

Actuator Type/Size	Disc Brakes Output Force P_a (N)	Drum Brakes Output Force P_a (N)
16	$P_a = 107.5p - 34.5$	$P_a = 104.5p - 19.0$
18	$P_a = 112.1p - 37.6$	
20	$P_a = 120.8p - 35.7$	$P_a = 124.0p - 24.0$
22	$P_a = 128.2p - 37.1$	
24	$P_a = 141.6p - 31.2$	$P_a = 143.0p - 28.5$
27		$P_a = 169.0p - 34.0$
30		$P_a = 196.0p - 39.0$

them as well. (The formulae shown in Table 6.9 are equivalent to the term $P_a = (p - p_t) A_a$ in Equation (5.32), shown above.) For the front and rear disc brakes on the tractor, the Type 24 actuator is paired with a 76 mm lever length.

For the S-cam drum brakes on the semi-trailer, the required actuation torque to generate $\tau_{3max} = 12.6$ kNm is $\tau_{3a} = 1143$ Nm. This is higher than for the disc brake because of the lower brake factor in the form of the specific torque (τ_s). A lever length of $l = 76$ mm is not appropriate because this would require it to be paired with a very large size air actuator to create the actuation torque, so a longer lever length is selected, e.g. 150 mm. With this an actuator area of approximately 0.010 m^2 is needed: Type 16 or Type 20 (which would also have a larger margin). In practice the actuator stroke may limit the lever movement, so it may be better to use a 135 mm lever length with a Type 20 actuator, or even a 127 mm lever with a Type 24 actuator. Type 24 actuators are the most commonly used actuator for drum braked semi-trailers and therefore the most cost effective; additionally a shorter lever provides more actuator stroke, which may be needed to comply with fade test requirements (Annex 11 of UN Regulation 13). Actuator stroke length becomes particularly important in high-duty drag or fade tests where the available actuator stroke is an essential element in fulfilling prescribed requirements especially for air consumption.

Because of the springs and other force losses in the foundation brakes and the pneumatic actuators, it is necessary to take into account the threshold torque required to overcome these in order to achieve the desired vehicle deceleration. The analyses above are estimates based on the effective actuation torque. The formulae for the air actuators shown in Table 6.9 include offsets for the threshold force in each actuator, which can typically range from 200 to 400 N, creating threshold torques of 24 Nm (Type 24, disc brake) and 32 Nm (Type 20, S-cam drum brake). To these must be added the brake threshold actuation torque (τ_t), which typically may be 10 Nm for a disc brake and 30 Nm for an

S-cam drum brake. The actuator manufacturer's formulae are used to calculate the output force net of the threshold force, and then the brake torque is calculated taking account of the brake threshold actuation torque according to Equation (5.40), which yields:

$$\tau_i = (\tau_{ai} - \tau_{ti})\text{BF} \tag{6.32}$$

The torque generated by each brake when the driver applies an air pressure of p (MPa) in the actuation system via the foot or hand valve is therefore:

$$\tau_{wi} = (((p_i - p_{ti})A_{ai}\eta_i l_i)\tau_{ti})\text{BF}_i \tag{6.33a}$$

where p_{ti} is the actuator threshold pressure (see Table 6.9) and τ_{ti} is the brake threshold torque. Alternatively the brake threshold force P_{ti} may be specified, in which case Equation (6.33b) is used:

$$\tau_{wi} = (((p_i - p_{ti})A_{ai}\eta_i) - P_{ti})l_i\text{BF}_i \tag{6.33b}$$

The wheel braking force is:

$$T_{wi} = (((p_i - p_{ti})A_{ai}\eta_i l_i)\tau_{ti})\text{BF}_i/r_{ri} \tag{6.34a}$$

or

$$T_{wi} = (((p_i - p_{ti})A_{ai}\eta_i - P_{ti})l_i\text{BF}_i/r_{ri} \tag{6.34b}$$

Commercial vehicles may have more than one axle to carry the required normal load, e.g. in the form of a bogie on a semi-trailer. Where 'n' is the number of axles carrying the normal load at any specific location on the vehicle or trailer, the 'axle braking force' is:

$$T_i = 2n_i T_{wi} \tag{6.35}$$

Step 3.3 Design the 'Master Actuator' (Brake Pedal) System (Pneumatic System)

In a pneumatically actuated braking system, the brake pedal takes the form of a proportional valve that increases the pneumatic actuation control pressure in response to the driver's demand. Although the valve may basically depend upon movement to determine the pneumatic pressure, modern footvalves for commercial vehicles are designed with force feedback to provide pedal feel to the driver in the same way that a hydraulically actuated car system does. The design of footvalves is not considered here as this is a specialist topic. 'Air consumption' is an important consideration in the design of pneumatically actuated braking systems, and a reserve of travel and operation in the event of failure in the energy source is required in all legislation as mentioned in the next section and in Chapter 8. Because this depends so strongly on the characteristics of individual system components for which the manufacturers' data and knowledge are required, the analysis air consumption and operating reserves are beyond the scope of this book.

Step 4 Verify the Design and Check Compliance with Legislative Requirements (UN Regulation 13)

Steps 1–3 above describe the brake system design procedure for the commercial vehicle defined in Table 4.6, and the typical design targets that must be verified for the braking system on this vehicle include:

- Compliance with all legal requirements including a reserve of travel in the brake actuation system.
- Pneumatic actuation system pressure and reserve, e.g. UN Regulation 13; the capacity of the energy reservoirs must be sufficient to enable the vehicle to be brought to a halt after the eighth brake application as prescribed for secondary braking.
- Stopping distance: the vehicle must be capable of achieving the minimum stopping distance and/or MFDD.
- Stability and adhesion utilisation: the solo vehicle and the vehicle combination must decelerate without deviation or imbalance between axles.
- The vehicle must decelerate without generating noise, judder and vibration.
- Wear and durability: the life of the brake rotors and stators must meet customer expectations.

For the whole vehicle, the total braking force ($T = T_1 + T_2 + T_3 + T_4$ as applicable) is given by Equation (6.36). This equation applies for disc or drum brake combinations provided that the BF is correctly defined (see Chapter 5), and includes thresholds and efficiencies that affect the actuation force at each brake. It can also accommodate different pneumatic actuation pressures to the different axles by inserting factors to modify the actuation pressure to the brakes (see Figure 6.23(a, b)).

$$\begin{aligned} T &= (T_1 + T_2 + T_3 + T_4 \text{ etc.}) \\ &= [2n_1(((p_1 - p_{t1})A_{a1}\eta_1 l_1)\tau_{t1})\mathrm{BF}_1/r_{r1}] + [2n_2(((p_2 - p_{t2})A_{a2}\eta_2 l_2)\tau_{t2})\mathrm{BF}_2/r_{r2}] \\ &\quad + [2n_3(((p_3 - p_{t3})A_{a3}\eta_3 l_3)\tau_{t3})\mathrm{BF}_3/r_{r3}] + [2n_4(((p_4 - p_{t4})A_{a4}\eta_4 l_4)\tau_{t4})\mathrm{BF}_4/r_{r4}] \text{ etc.} \end{aligned} \tag{6.36}$$

Equation (6.36) is used to calculate the design braking performance of the vehicle specified in Table 4.6 with the designed system data summarised in Table 6.10.

Step 4.1 Intact System

The design and functionality of pneumatically actuated braking systems must meet the legislative requirements for the braking performance of commercial vehicles, e.g. in UN

Table 6.10: Pneumatic Braking System Design Specification for an Articulated Tractor/Semi-Trailer Combination

Design Parameter	Design Value
Tractor: front axle(s) (#1)	
Number of axles (n_1)	1
Type of foundation brake	Air-actuated disc brake
Effective radius (r_{e1})	173 mm
Brake factor (BF_1)	26
Threshold force (P_{t1})	131N
Air actuator size	Type 24
Equation (Table 6.9)	$1416p - 312$
Lever length (l_1)	76 mm
Tractor: rear axle(s) (#2)	
Number of axles (n_2)	1
Type of foundation brake	Air-actuated disc brake
Effective radius (r_{e2})	173 mm
Brake factor (BF_2)	26
Threshold force (P_{t2})	131 N
Air actuator size	Type 24
Equation (Table 6.9)	$1416p - 312$
Lever length (l_2)	76 mm
Trailer: axle(s) (#3)	Semi-trailer
Number of axles (n_3)	3
Type of foundation brake	Air actuated S-cam drum brake
Effective radius (r_{e3})	210 mm
Brake factor (BF_3)	11
Threshold torque (τ_{t3})	23 Nm
Air actuator size	Type 20
Equation (Table 6.9)	$1240p - 240$
Lever length (l_3)	135 mm
Trailer: axle(s) (#4)	Not applicable to this example
Number of axles (n_4)	0

Regulation 13, which are summarised in Chapter 8, Tables 8.4 and 8.5. As for passenger cars, commercial vehicles must have at least two separate brake actuation circuits so that in the event of the failure of one circuit, the driver can stop the vehicle safely (residual braking performance). This can be achieved by two pneumatic circuits for a rigid commercial vehicle, usually in the form of a vertical split, i.e. the front and rear axles are on separate circuits (see Figure 6.19(a)) with a third circuit for a trailer. 'Spring brake' actuators (see Figure 6.21) are often fitted to rear axles of the towing vehicle and trailer axles to fulfil the parking brake requirements. In certain cases on the towing vehicle they also provide the means to fulfil the secondary braking performance; in all cases, air pressure is used to compress the spring and hold the brake off while driving.

The service brakes of a commercial vehicle are applied by a foot-operated control valve ('pedal'); the secondary and park controls may be via a hand-operated valve. The following performance requirement principles apply:

- 'Service braking performance': the defined performance is required when no faults are present and must be achieved when the driver has both hands on the steering control.
- 'Secondary braking performance': the defined performance is required when a 'single' defect is present and must be achieved when the driver has at least one hand on the steering control.
- 'Residual braking performance': the defined performance is required when there is a failure within a service braking circuit.

In practice the braking performance of a commercial vehicle combination is never evaluated, e.g. for Type Approval (UN Regulation 13) by testing the full vehicle combination, but by testing the towing vehicle and the trailer independently with appropriate compensation for the load transfer of the trailer on to the towing vehicle in the case of tractors for semi-trailers. To allow the safe interchangeability of tractors and trailers, UN Regulation 13 incorporates requirements to represent the dynamics of the vehicle combination when braking. A tractor coupled to an unladen semi-trailer is considered to be a solo tractor with an additional mass equivalent to 15% of the maximum allowable static load on the kingpin, located at the kingpin, to simulate the dynamic effects of the coupled semi-trailer on the tractor. As explained in Chapter 8, to simulate the effect of a laden semi-trailer coupled to a tractor, an additional mass M_s mounted at the kingpin is determined by Equation [8.4] and the height (h_{am}) of the centre of gravity of the additional mass is determined by Equation [8.5].

As explained in Chapter 8, compatibility between towing vehicles and trailers is ensured by examining the calculated values of T_m / P_m, where T_M is the total force developed by the towing vehicle's brakes and P_m is the simulated tractor weight (= (M + Ms) . g). When plotted against the coupling head air pressure, P_M, must lie within the compatibility bands as shown for a tractor / semi-trailer combination in Figure 6.22, with the proviso

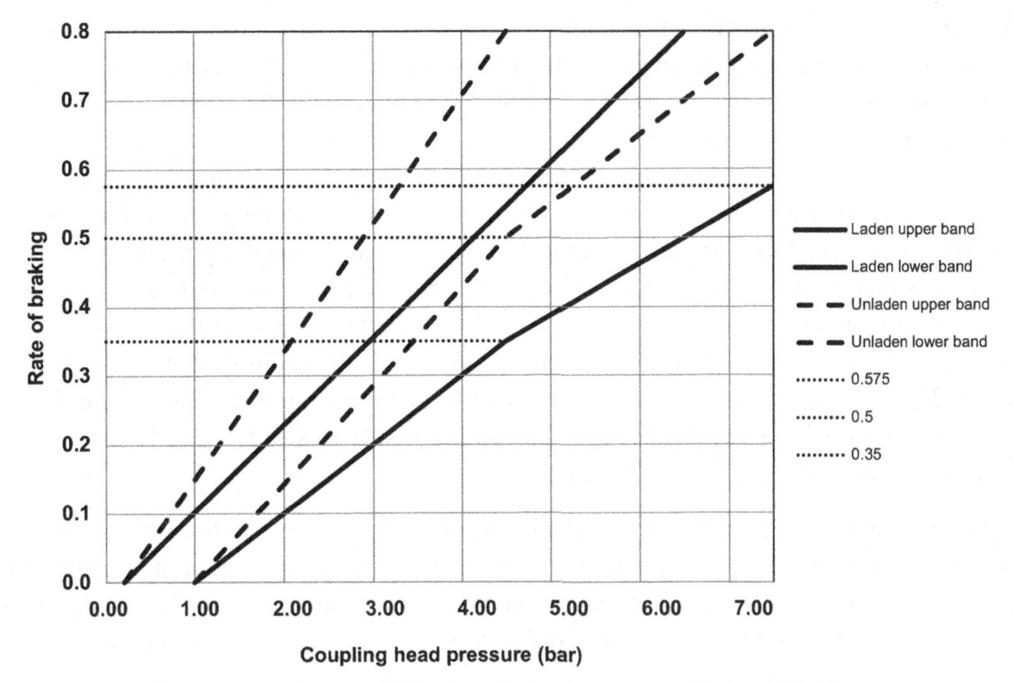

Figure 6.22: Compatibility Bands for Tractors with Semi-Trailers.

that the unladen requirement need not be met if the tractor is fitted with an approved anti-lock braking system.

For a semi-trailer, UN Regulation 13 (UN, Mar 2014) acknowledges that the trailer braking forces should not be sufficient to decelerate the total mass of the semi-trailer but only the load supported by the wheels of the semi-trailer. The load imposed at the coupling is effectively decelerated by the towing vehicle. The values of T_R/P_R for the semi-trailer, laden and unladen, when plotted against the coupling head pressure, must lie within certain bands, which are calculated for each particular type of semi-trailer to take account of dimensional and load differences. For example, comparing a short semi-trailer having a high centre of gravity with a long low semi-trailer having the same static axle load, the dynamic weight transfer from the axles on to the fifth wheel coupling will be greater for the short trailer, which should therefore require less braking force from the trailer brakes at any given combination of deceleration and coupling head air pressure. The calculated T_R/P_R compatibility band for the short, high trailer will therefore be lower than for the long trailer. As for all towing vehicles, only the laden case need be considered if approved ABS is fitted.

As discussed in Chapter 4, load sensing can be fitted to commercial vehicle combinations to adjust the actuation pressure supplied to the brakes on each axle or bogie to improve the braking distribution for different loading conditions (Shilton,

1969–70; Lindeman et al., 1997). It is also desirable to have the threshold torques or forces balanced between the towing vehicle and the trailer to avoid braking being unevenly distributed between the axles of the vehicle combination, especially at relatively low-duty operation. This is achieved by a trailer control valve mounted on the towing vehicle with an adjustable pressure bias to provide a positive predominance, i.e. increase the pressure in the control line to the trailer, thereby reducing the performance of the tractor relative to the trailer, the objective being to align the coupling head pressure at which braking commences on each element of the combination. A similar function is available in the emergency relay valve installed on the semi-trailer, as illustrated in Figure 6.19(b). The fully laden (GVW) condition is illustrated in Figure 6.23(a), and the effect of load sensing can be seen in Figure 6.23(b), where the air pressure to the tractor rear brakes and the semi-trailer brakes is reduced to control the braking distribution in the unladen condition.

The designed braking system performance for this example vehicle has been analysed including load sensing and predominance settings. The results indicate that the tractor and semi-trailer combination in the GVW condition achieves a deceleration of $J = 5$ m/s^2 with an actuation air pressure of about 5 bar under service braking, which is satisfactory.

Step 4.2 Partial System Failure

For partial system failure, secondary braking can be simulated, e.g. by setting the front brakes to be inoperative, in which case $J = 2.2$ m/s^2 is achieved with an actuation air pressure of about 3.3 bar, which is also satisfactory.

Step 4.3 Parking

As for passenger cars, the parking brake system on a commercial vehicle can actuate the brakes of any axles depending on the vehicle and brake configuration in order to meet the UN Regulation 13H requirements (see Chapter 8). Parking brake systems for commercial vehicles and their trailers with pneumatic brake actuation are usually based on the clamp force generated by spring brake actuators (Figure 6.21(a)), which, as indicated in Figure 6.19, are usually fitted to the rear axle(s) of the tractor, and in the example shown, two of the three axles in the semi-trailer bogie. The parking brake on a motor vehicle is actuated by a hand control valve in the driver's cab, but for a trailer the spring brakes can be manually operated by a separate valve mounted on the trailer side. In some cases the spring brakes are automatically applied, e.g. when the trailer is uncoupled and the supply line is exhausted.

Comment on Verification Against Legislative Requirements – Pneumatic Braking Systems

As for passenger cars, legislative requirements for the braking of commercial vehicles represent the minimum level of performance required of a road vehicle braking system,

(a)

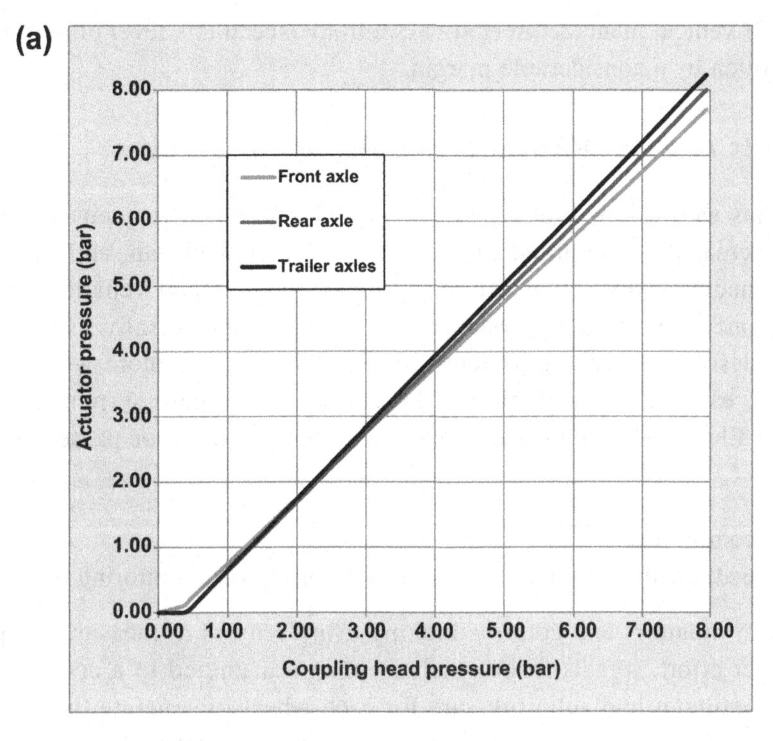

(b)

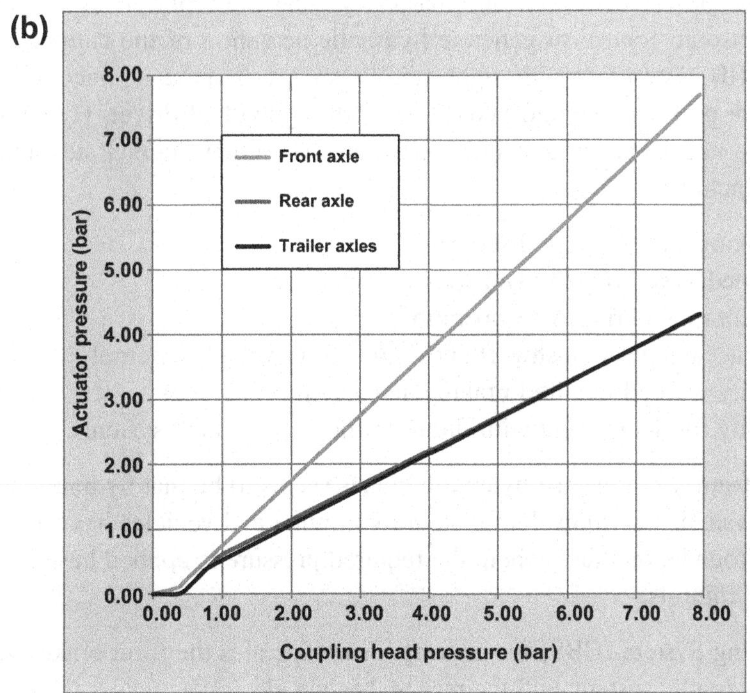

Figure 6.23: Articulated Commercial Vehicle Combination Braking System Pressure Characteristics (Ross, 2013).
(a) Laden. (b) Unladen.

and commercial vehicle manufacturers always aim to exceed this level of braking performance, often by a considerable margin.

Developments in Road Vehicle Brake Actuation Systems

This chapter has shown how braking systems can be designed for conventional hydraulically actuated brakes fitted to a passenger car or light van, and for conventional pneumatically actuated brakes fitted to commercial vehicles. Although hydraulic and pneumatic braking systems are still almost universally used on these vehicles, new designs of braking systems are likely to become more widely utilised in the near future, especially in electric and hybrid vehicles. These systems include (Moore, 2013) Electro-Hydraulic Brake (EHB) systems, which for passenger cars include:

- Electronic control unit
- Electronic pedal with pedal-feel simulator and sensors for monitoring driver settings.

An EHB system monitors and controls braking using sensors to measure the pedal travel and driver effort, signals from which are then transmitted to a control unit that calculates the optimum hydraulic pressure for each wheel to generate the amount of braking force needed. The booster and master cylinder are replaced by a hydraulic pump with electronic control to generate hydraulic actuation of the caliper piston. It is claimed that EHB systems provide light quick response, require no vacuum supply, and can adjust brake performance and pedal feel to the individual driver. Hydraulic braking is still available under part system fail conditions. The other claimed advantages of EHB systems include:

- Shorter braking and stopping distances
- Optimised pedal feel, braking and stability behaviour
- No pedal vibration during ABS operation
- Improved packaging for crashworthiness and easier vehicle assembly
- Compatibility with all required braking and stability functions
- Compatibility for networking with future traffic management systems.

The main disadvantage of electro-hydraulic brakes seems to be that hydraulic fluid is still needed in the system. The equivalent system for commercial vehicles has a pneumatic supply to each foundation brake where the required pressure is applied based on signals from the driver controls.

A Hybrid Braking System (HBS) for passenger cars operates the front brakes via a conventional hydraulic system, and the rear brakes via electric actuators with an integrated electric parking brake, so it is essentially a two-circuit system and an alternative to full

EHB. It avoids the need for long hydraulic lines and handbrake cables to the rear axle. The components of HBS include:

- Two hydraulic wheel brakes on the front axle
- Two electromechanical wheel brake modules at the rear axle
- Integrated Electric Parking Brake (EPB) at the rear axle
- Electronic Braking Systems (EBS) hydraulic unit including Electronic Control Unit (ECU)
- Pedal/booster unit with small booster.

The advantages of an HBS are claimed to include:

- Cheaper and cleaner electric actuation to the rear brakes, operating on a conventional 12 V system
- Easily combined with Electric Parking Brake (EPB) and Hill Launch Assist
- Easily combined with regenerative braking technology on rear wheel drive vehicles.

The disadvantages mainly relate to the increase of unsprung mass at the rear wheels.

An Electromechanical Brake (EMB) system for passenger cars is a full 'brake-by-wire' system, which eliminates brake fluid and hydraulic lines completely and generates the demanded braking force directly at each wheel. The components of the EMB include:

- Wheel brake actuation modules
- An ECU, which includes all brake and stability functions
- Electronic pedal module with a pedal feel simulator and sensors for interpreting driver demand.

The advantages of the EMB system are claimed to include:

- No hydraulic systems; and consequent packaging, environmental, assembly and maintenance benefits
- Easier implementation of braking and stability functions, including EPB, automatic hill hold and parking brake, and adjustable pedal feel characteristics
- Improved crash safety performance through elimination of the booster and master cylinder assembly and decoupling from pedal box
- Fully optimised braking giving shortest stopping distances and optimal stability
- Increased comfort with silent operation, adjustable brake pedal
- Can be easily networked with future traffic management systems, e.g. autonomous cruise control, and collision prevention systems.

The disadvantages include increased unsprung mass, and the size, mass and cost of the electric actuators needed to generate and hold the required brake force. A major limitation is that there is no secondary operation in the brake units in the event of electric power supply failure, which is needed under current legislation. A separate back-up actuation system would be required, adding cost, mass and complexity unless two batteries are used with independent electric control circuits. Electrical actuation of commercial vehicle braking systems is not considered practical because of the amount of actuation energy required (at each brake). Commercial vehicles may be fitted with endurance brakes (retarders) that are electronically controlled, but as previously explained, such retarders are generally designed for low-duty braking, and leave the foundation brakes to provide high-duty and emergency braking. Additionally it is not possible to bring a vehicle to rest by means of an endurance brake (retarder) as actual performance is speed dependent.

Chapter Summary

- A four-step procedure for the detailed design and layout of braking systems for road vehicles is presented, which includes the design of the basic braking system parameters, specifying the foundation brakes, designing the actuation system, and verification. Examples are worked through for a passenger car and a commercial vehicle, namely an articulated tractor and semi-trailer combination, which illustrate the important features of achieving a good basic system configuration. Only the friction braking is considered in the design, but the incorporation of other braking systems is briefly discussed.
- Two actuation systems are studied: the hydraulic system used on most passenger cars and light commercial vehicles (vans), and the pneumatic system used on most commercial vehicles. Steps 1 and 2 are common to both systems, but steps 3 and 4 are separated so that the detail of the actuation systems can be examined effectively.
- At all stages the design requirements of the vehicle braking system are considered in terms of the vehicle manufacturer's specification and the legislative requirements based on UN Regulation 13 or 13H. These include intact system and part-fail conditions for verification purposes. It is made clear that manufacturers' standards and specifications are designed to exceed legislative requirements, which are regarded as a minimum level of braking performance.
- The selection and sizing of the foundation brakes are considered based on worst case conditions, and considerations that influence disc and drum rotor size are explained. The importance of the 'z_{crit}' parameter, which defines the 100% braking efficiency point for the basic system of a two-axle rigid vehicle, especially a passenger car, is explained. Commercial vehicles and trailer combinations do not use this parameter in their basic system design, and a more useful alternative design approach is presented based on optimum adhesion utilisation. This includes the importance of load sensing.
- For passenger car hydraulic braking systems that rely on muscular energy to provide the required brake actuation force, the need for servo-assistance in the actuation system to

meet drivers' expectations of actuation force on the brake pedal (pedal effort) while at the same time meeting servo-fail requirements is explained. The importance of parameters such as pedal feel in terms of the driver interface and customer satisfaction with the vehicle are explained, and the parameters that define pedal feel are explained and discussed; these include compliance in the system arising from threshold pressures and forces, clearances, deflections, fluid compression, friction material compression, and hysteresis, among many others.

- For commercial vehicle combinations (towing vehicles and trailers) with pneumatic brake systems, the energy to actuate the brakes is provided by the pneumatic system. While aspects such as driver pedal effort and pedal feel are not so relevant in the system design, actuator forces and travel must be carefully designed so that the braking system works under all conditions of use, especially heavy duty and high speed braking. Compatibility of the trailer brake system with the towing vehicle brake system is discussed in terms of load sensing and predominance.
- Parking brakes and actuation systems are discussed for both passenger cars and commercial vehicles.
- Future developments in braking actuation systems are briefly discussed.

References

Day, A.J., Ho, H.P., Hussain, K., Johnstone, A., 2009. Brake system simulation to predict brake pedal feel in a passenger car, SAE 2009 Brake Colloquium, ISBN 978 0 7680 3335 9.

UN, Mar 2014. UNECE Regulation 13. Uniform provisions concerning the approval of vehicles of categories M, N and O with regard to braking, E/ECE/324/Rev.1/Add.12/Rev.8, March 2014.

UN, Feb 2014. UNECE Regulation 13H. Uniform provisions concerning the approval of passenger cars with regard to braking, E/ECE/324/Rev.2/Add.12H/Rev.3, February 2014.

Gerdes, J.C., Brown, A.S., Hedrick, J.K., 1995. Brake System Modeling for Vehicle Control. Advanced Automotive Technologies. ASME IMECE.

Ho, H.P., 2009. The influence of braking system component design parameters on pedal force and displacement characteristics. PhD thesis. University of Bradford, UK.

Lindeman, K., Petersen, E., Schult, M., Korn, A., 1997. EBS and trailer brake compatibility. SAE, 973283.

Markovic, T., 2002. Current position and development trends in air-brake systems for Mercedes-Benz commercial vehicles. IMechE Conference, Braking 2002.

Moore, I., 2013. Passenger car ESC systems, Braking of Road Vehicles. University of Bradford. ISBN 1 85143 272 1.

Newcomb, T.P., Spurr, R.T., 1989. A Technical History of the Motor Car. Adam Hilger. ISBN 0852 74074-3.

Ross, C.F., 2013. Commercial Vehicle case study, Braking of Road Vehicles 2013. University of Bradford. ISBN 1 85143 272 1.

Shilton, B.R., 1969–70. Braking system load sensing valves. Proc. IMechE. vol. 184 (12), 30–37.

Svensson, T., 2013. Brake system layout, Braking of Road Vehicles 2013. University of Bradford. ISBN 85143 272 1.

CHAPTER 7

Thermal Effects in Friction Brakes

Introduction

The rotor and stator components of a brake friction pair (disc and pads, or drum and linings) operate in sliding contact under the applied actuation forces to generate friction forces that oppose the direction of relative sliding to create the retarding torque required to decelerate the vehicle. Friction brakes are required to transform large amounts of kinetic energy into heat over very short time periods and in the process they create high temperatures, steep temperature gradients and substantial thermal stresses. So all parts of the foundation brake, especially the rotor and the friction material, must resist the temperatures and stresses arising from combined thermal and mechanical loadings, which can be extremely arduous. Although mechanical loading, e.g. contact forces and rotational stresses, on the brake rotor in themselves can be high, most critical brake design parameters are related to thermal loads. If the brakes become too hot, deterioration in brake performance and ultimately premature failure may result from a number of causes, such as:

- Brake fade, when the coefficient of friction between rotor and stator is reduced (see Chapter 2)
- Excessive thermal distortion, which can also affect the torque output from a brake (see Chapter 5)
- High thermal stresses at the metal rotor surface resulting in surface cracking
- Reduction in mechanical strength and other property changes in the rotor material
- Increased wear of the friction material
- Increased risk of fluid vaporisation and deterioration of rubber seals in the brake hydraulic actuation cylinders.

An understanding of the factors that influence the thermal behaviour during braking is therefore essential as a sound basis for foundation brake design, especially with ever-increasing vehicle speeds and axle loads placing higher demands on brake equipment. The relative importance of these factors depends on the type of braking duty being applied. The main factors that limit performance in single brake applications are high surface temperatures and consequent high thermal gradients that may give rise to thermal stresses sufficient to cause damage to the disc or drum. During repeated or drag braking, factors including the thermal capacity of the brake and surface heat transfer coefficients influence

Braking of Road Vehicles.

the peak temperatures attained, which need to be limited to avoid operational problems. Consequently it is of great importance to know what temperatures can be reached in brake discs, drums, pads and linings under all conditions of use. Methods of doing this are outlined for the cases of a single brake application, a series of repeated brake applications, and continuous drag braking. Numerical examples are given to illustrate the use of the analysis described, and the practical implications involved are also discussed.

As explained in Chapter 6, thermal analysis is an essential part of the vehicle braking system layout design process (Steps 2.2 and 2.4 of the four-step brake system design procedure) and require calculations or predictions of the following parameters:

- Friction pair surface temperatures and temperature distributions
- Thermal and thermomechanical stresses
- Thermal deformations and deflections
- Cooling characteristics
- Brake fluid temperatures
- Temperatures of seals, bearings and associated brake components.

Brake thermal analysis is also essential for design optimisation, material specification, brake friction performance prediction (e.g. fade at higher temperatures), component fatigue and wear life, cooling, and noise and judder (brake NVH). Because of the extremely complex physical nature of braking friction, it is difficult to predict many of these parameters to a very high degree of accuracy, and experimental validation is essential. However, brake thermal modelling and analysis have been developed to a level that currently is adequate for the successful design of foundation brakes for conventional friction pairs.

The heat generated by friction in a brake must be transferred away from the friction interface and dissipated as quickly as possible to avoid the generation of very high temperatures at the friction surfaces. As explained in Chapter 2, conventional friction materials have low heat transfer properties and tend to act as insulators or barriers to heat flow (see Table 2.4); therefore, most of the heat transfer takes place through the rotor. Thus, for good thermal performance a brake rotor material should have high thermal conductivity (k), a high specific heat capacity (C_p) and high density (ρ) in order to facilitate heat flow and absorption. Although high thermal mass (ρC_p) is important for heat absorption in a friction brake, there has to be a compromise between brake thermal mass and the unsprung mass that has to be accommodated in the design of vehicle suspensions. Once the frictional heat has been transferred into and through the rotor it is dissipated either to the surrounding air by surface convection or radiation, or to other brake components usually by conduction, often via an interface that has a thermal contact resistance. Automotive disc brakes have the friction surfaces of their rotors exposed, which is good for cooling by radiation heat transfer and the high surface heat transfer coefficients associated with forced convection from the airflow over the disc. The friction surface of a drum brake is inside the drum and thus the frictional heat has to be transferred through the drum material to the outside surface before it can be dissipated by convection or radiation. The result is that disc brakes exhibit better thermal performance than drum brakes, which is why they have become very widely used in high-duty vehicle braking systems. An important defining parameter of a friction brake rotor material is its maximum operating temperature (MOT), which should be higher than the maximum anticipated surface temperature under the most severe duty conditions. The rotor material should also have a low coefficient of thermal expansion to minimise thermal deformations and distortions, and should be resistant to wear.

Heat Energy and Power in Friction Brakes

Drivers use the brakes on their vehicle in many different and unpredictable ways. In previous chapters, the maximum torque from any brake on a vehicle was calculated and used to design the vehicle's braking system based on the vehicle manufacturer's standards and legislative requirements relating to the required vehicle deceleration. The parameters that affect the thermal loading of a vehicle's brakes include initial and final road speeds, initial brake temperature, deceleration, environmental conditions (ambient temperature, wind), resistance to vehicle motion (engine braking, retarder, aerodynamic drag, tyre/transmission rolling resistance), topography (downhill braking) and additional loads (trailer). The three types of braking conditions that can be defined for brake thermal performance analysis in road vehicles are:

1. Single application of the vehicle's brakes; often termed 'stops' (to rest) or 'snubs' (from an initial road speed to a final road speed).
2. Repeated application of the vehicle's brakes; often associated with fade tests.
3. Continuous application of the vehicle's brakes, often called drag braking; usually associated with the maintenance of a constant speed on a long downhill gradient.

These are incorporated into the legislative requirements, e.g. UN Regulations 13 and 13H, but as such they only represent the minimum performance requirements. Vehicle manufacturers set their own standards based on these three types, which generally represent much higher braking duties, and in some cases consumer expectations can lead to the acceptance of very demanding brake usage tests based particularly on repeated brake applications to evaluate braking performance (e.g. the 'AMS' test, which is explained later).

A single application of the vehicle's brakes from vehicle maximum road speed to zero at high deceleration represents a single very severe braking duty cycle, dissipating maximum kinetic energy. Although this type of brake usage can be relatively rare (it tends to be more common with high-performance cars) it is essential that the vehicle can perform such braking safely. Therefore, this loading case is often modelled in brake design analysis.

However, many other single brake applications from lower speeds and at lower decelerations are very common in vehicle operation and should also be examined. The main characteristic of such single brake applications from the thermal analysis point of view is that braking times are relatively short, meaning that the heat is absorbed by the brake components and the heat dissipation rate from them is quite low and can be neglected.

Repeated application of the vehicle's brakes often occurs at relatively close intervals, e.g. during urban driving or high-speed driving on motorways with busy traffic conditions. From the thermal analysis point of view, although each brake application may be relatively short, the successive brake applications build up a substantial amount of heat energy that has to be dissipated into the environment, so brake cooling must be included in these braking events. Brake cooling generally depends upon vehicle road speed, so urban driving at low speed may actually represent a worse case than motorway driving at high speed despite the higher kinetic energy involved. In legislative and development testing, one of the most important reasons for testing under repeated brake applications is the evaluation of the braking system's fade performance (see Chapter 9) to ensure that deceleration and stability are adequate even when the brakes are hot. Such tests include a series of time-spaced individual brake applications, often with the time spacing determined by a representative interval to allow some cooling 'off-brake', or in the extreme by the time taken for the vehicle to accelerate back up to the required initial speed.

Continuous application of the vehicle's brakes, termed drag braking, occurs when a vehicle has to maintain a safe constant speed down a long gradient. Although the equivalent deceleration and duty level are usually relatively low, the continuous generation of frictional heat over a long time with the friction pair constantly working at quite low road speeds where cooling is restricted represents a very high thermal demand on the brakes. This type of braking is also covered by legislation and is usually more of a problem on heavy commercial vehicles, which for this reason since 1991 have been required to be fitted with endurance brakes (retarders; see Chapter 6) to reduce the thermal load on the foundation brakes. Again, from the thermal analysis point of view, brake cooling must always be included in the thermal analysis of drag brake applications.

The energy associated with a moving road vehicle of mass M (kg) travelling at a speed of v (m/s) is the 'translational kinetic energy' Q_{ke} (J), where:

$$\text{Translational kinetic energy} = Q_{ke} = \frac{1}{2}Mv^2 \tag{7.1}$$

Note that in this book, 'vehicle road speed' is defined as V (km/h), and 'linear speed' is defined as v (m/s).

To this must be added the rotational kinetic energy of the rotating components of the vehicle engine and driveline, plus the wheels and brake rotors. In Chapter 6 the effect of the rotational inertia associated with the wheel and brake was broadly estimated as 5%, so the total kinetic energy could be approximated as 1.05 × Q_{ke}, although this depends on the vehicle design and should always be checked accurately at the design stage. This needs further consideration; the kinetic energy of the powertrain and drivetrain can be significant, but it is difficult to approximate the effect on the 'braking energy' because it depends strongly on the powertrain/drivetrain design and is to some extent offset by engine braking and other losses described below. If the vehicle decelerates from an initial speed v_1 to v_2 (m/s), the change in total kinetic energy is given by Equation (7.2). Note that although deceleration by a vehicle during braking is negative acceleration, for convenience in this book deceleration (J m/s^2) is always taken as positive, and $v_1 > v_2$:

$$\Delta Q_{ke} = \frac{1}{2}M\left(v_1^2 - v_2^2\right) \tag{7.2}$$

The 'braking energy', ΔQ_b, that is dissipated by the brakes of this road vehicle during a single brake application when it decelerates from v_1 to v_2 (m/s), is the part of the total kinetic energy that is not absorbed or dissipated by other retarding forces on the vehicle. This can include wind and tyre rolling resistance (Q_{wr} and Q_{rr}), transmission losses (Q_{tr}), engine braking (Q_{eng}), retarder (Q_{ret}) and regenerative braking (Q_{reg} – if fitted) estimated from the work done by the forces associated with each of these over the distance travelled by the vehicle during the brake application. Additionally, the effect of road gradient must be included; if driving uphill, kinetic energy is converted into potential energy (Q_g), which is effectively reducing the duty on the brakes, and vice versa if driving downhill when the duty on the brakes is increased. Thus, the braking energy can be estimated as:

$$\Delta Q_b = \Delta Q_{ke} - \Delta Q_{rr} - \Delta Q_{wr} - \Delta Q_{tr} - \Delta Q_{eng} - \Delta Q_{ret} - \Delta Q_{reg} \pm \Delta Q_g \tag{7.3}$$

Resistance to motion from wind and tyre rolling resistance and transmission losses (see Chapter 3) can be predicted for different road speeds from established formulae, but for the purposes of braking energy determination it is often evaluated experimentally by a 'coast-down' test in which the vehicle is driven up to a selected speed and allowed to decelerate without the use of the brakes in neutral gear (i.e. without engine braking). Such a test is carried out on a straight and level test track and repeated in both directions to neutralise the effect of wind. The initial and final road speeds, with the time interval between them, are used to calculate the mean deceleration. Typically the coast-down deceleration of a medium-sized passenger car, e.g. between 80 and 20 km/h, is around 3%g (~0.3 m/s^2), but depends on many features of the car's design and mechanical configuration, especially the tyres. From this value of coast-down deceleration (J_{rr}), the average retarding force from wind and tyre rolling resistance and transmission losses can be calculated. Similarly the effect of engine braking can be assessed by the same vehicle test with the vehicle in gear, but alternatively it can be calculated from the powertrain design data. Assuming constant deceleration, the energy dissipated by wind and tyre rolling resistance and transmission losses during the brake application at deceleration J (m/s^2) can be calculated from:

$$\Delta Q_{rr} + \Delta Q_{wr} + \Delta Q_{tr} = \frac{1}{2}m\left(v_1^2 - v_2^2\right)(J_{rr}/J) \tag{7.4}$$

The effect of a gradient can similarly be calculated from Equation (7.4), substituting the road gradient for the term (J_{rr}/J). Road gradients are defined as a slope; e.g. a 10% gradient is equivalent to an incline of $\sin^{-1}$ (0.1), i.e. $\alpha = 5.7°$ to the horizontal, and the longitudinal component of the vehicle weight is $m\sin\alpha$.

Common practice is to take the worst case for the thermal analysis of brakes, and this means ignoring any effect of engine braking that may offset the effect of the rotational inertia of the powertrain, drivetrain, and wheels, hubs and brake rotors as explained above. Many drivers naturally disengage the drivetrain by depressing the clutch pedal (of a manual transmission) during high-duty braking. The effect of engine braking through an automatic transmission depends upon the transmission design and control logic. Based on the example passenger car previously examined in Chapters 3 and 6, using Equation (7.3) the braking energy has been calculated for a rate of braking $z = 0.5$ from a road speed of $V_1 = 80$ km/h to $V_2 = 20$ km/h, ignoring engine braking and assuming that no retarder or regenerative braking is fitted to the vehicle. In the laden condition (GVW = 2450 kg):

Total KE $(Q_{ke}) = 567$ kJ
$(\Delta Q_{rr} + \Delta Q_{wr} + \Delta Q_{tr}) = 34$ kJ
$\Delta Q_g = 113$ kJ
(For a 10% uphill gradient $\Delta Q_b = 448$ kJ)
(For a 10% downhill gradient $\Delta Q_b = 675$ kJ)

In this example ΔQ_g is approximately 6% of Q_{ke} and since this represents the kinetic energy that is not dissipated through the brakes, it approximately offsets the rotational kinetic energy of the vehicle engine and driveline, plus the wheels and brake rotors, which was previously broadly estimated as 5%. The result is that for initial design purposes brake thermal calculations can assume that the braking energy dissipated during any brake application is the same as the change in kinetic energy, i.e. $\Delta Q_b = \Delta Q_{ke}$, and this is transformed into heat energy, which is calculated using Equation (7.2), and enters the vehicle brakes from the friction interfaces. Detailed prediction of temperatures, deformations, stresses, etc. in brakes for design prediction under critical conditions must, however, always review any such assumptions to ensure that the necessary detail is included to achieve the required accuracy. It is also assumed that the kinetic energy is 100% converted into heat energy via the friction process. This is widely accepted as a robust assumption, but it must be remembered that there are other mechanisms that absorb energy during braking friction, e.g. associated with chemical and phase changes in the friction pair, wear, and the generation of noise and vibration; these are considered to be very small (<1%) in comparison with the main mechanism of frictional transformation of kinetic energy into heat energy.

In a drag braking application when a constant vehicle speed is maintained on a downward slope for a distance L (m) at a constant gradient of angle α using the friction brakes and including an appropriate value of engine braking and rolling resistance, the potential vehicle energy change ΔQ_{pe} is:

$$\Delta Q_{pe} = Mg\Delta H = MgL \sin \alpha \tag{7.5}$$

where M is the vehicle mass, g is the acceleration due to gravity (= 9.81 m/s^2) and ΔH (m) (= $L \sin \alpha$) is the vertical height difference between the start and finish of the drag brake application.

Using the same example passenger car as above travelling at constant speed down a constant 10% gradient for 5 km, the potential energy change (ΔQ_{pe}) calculated from Equation (7.5) is 12 MJ. Since there is no change in translational or rotational speed in this case, rolling resistance forces must be included, and assuming that rolling resistance and engine braking combine to create a residual deceleration of 0.29 m/s^2 $^{(z = 0.03)}$ 3%g (most long drag brake usage occurs in mountainous areas where care is needed and drivers usually use low speeds so wind resistance is low), the retarding force from rolling resistance and engine braking is approximately 720 N and over the 5 km distance the work done is 3.6 MJ. The friction brakes are therefore required to dissipate 8.4 MJ. If the car is being driven safely and steadily at 50 km/h the duration of this drag braking is 6 minutes, and the power dissipation is a constant 33.4 kW. In comparison, the mean power dissipation for the single brake application shown above is 167 kW over a period of 3.4 s. This confirms that the contribution of any rolling resistance, engine braking and other residual forces that oppose residual motion is important in defining the braking force required for safe braking down long inclines. Where commercial vehicles are required to be fitted with retarders in addition to their foundation brakes, the effect of the retarder on reducing the energy dissipation required of the friction brakes is also significant.

The above analyses and examples consider the total vehicle braking energy, and discuss various assumptions and approximations to arrive at the energy dissipation requirements for the vehicle's brakes. To predict the thermal performance of individual brakes, the proportion of the total vehicle braking energy must be evaluated at each brake. This can be done by using the vehicle braking distribution (see Chapter 6), which defines the brake force generated by each brake (i) to achieve the desired vehicle deceleration, and hence the work done by each brake during vehicle braking. A convenient way to do this is to define the equivalent mass M_i associated with each wheel:

$$m_i = X_i M \tag{7.6}$$

From Equation (7.2) therefore:

$$\Delta Q_{kei} = \frac{1}{2} X_i M \left(v_1^2 - v_2^2 \right) \tag{7.7}$$

Braking power defines the rate of frictional energy transformation and is needed to calculate heat flux input for brake thermal analysis. The power $\dot{Q}_i$ (W) developed by a brake at any instant during a brake application was introduced in Chapter 6, Equation (6.4), where τ_{wi} is the instantaneous brake torque (Nm) and ω is the instantaneous angular velocity (rad/s). In a single brake application to bring a vehicle to rest it can be assumed that if τ_{wi} remains constant then the angular velocity ω of the wheel at any instant (time $= t$) decreases linearly with time to rest from an initial angular velocity ω_1. The rate of heat generation $\dot{Q}_i$ at time t is then as shown in Equation (7.8):

$$\dot{Q}_i = \tau_{wi} \omega \tag{6.4}$$

$$\dot{Q}_i = \tau_{wi} \omega = \tau_{wi} \omega_1 \left(1 - \frac{t}{(t_2 - t_1)} \right) \tag{7.8}$$

In practice the brake torque τ_i is limited by the adhesion between the tyre and the road. If the wheel locks, the angular velocity decreases to zero and the braking power dissipated by the brake also becomes zero. When this occurs, the braking power is then dissipated in the tyre/road contact, not in the brake. When there is braking slip at the tyre/road interface the angular velocity of the wheel and thus the brake is reduced and the actual power generated in the brake will be reduced. Correspondingly a proportion of the braking power is generated in the tyre/road interface, which is dissipated as heat, causing temperature rise in the tyre. However, in brake thermal analysis the worst case for the brake is always considered, namely the braking power is assumed to be 100% generated at the friction interface of a brake and transformed into heat to be dissipated by heat.

For a single brake application from an initial speed v_1 (> 0) to a final speed v_2 ($= 0$) with constant braking torque, the deceleration can be assumed to be constant. The braking torque takes time to develop when the brake is actuated, so the braking power change during braking has a typical form as shown in Figure 7.1. This indicates that maximum braking power $\dot{Q}_{imax}$ develops a short time after the start of the brake application (depending on the response time of the system). When established, the brake torque (and thus vehicle deceleration) can be assumed to remain constant, resulting in a linear

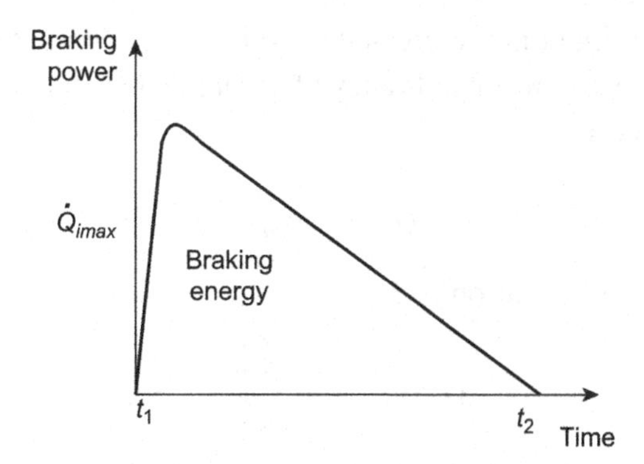

Figure 7.1: Braking Power in a Single Brake Application.

reduction of vehicle speed and braking power. At the moment the vehicle comes to rest (time $= t_2$), the braking power becomes zero. This is a simplified approach but is very useful for brake thermal analysis. Where the braking power is not constant with time, the area under the braking power curve represents the braking energy for brake i, which can be determined by integration:

$$\int_{t_1}^{t_2} \delta E_i = \int_{t_1}^{t_2} \dot{Q}_i \delta t \tag{7.9}$$

The brake power generation or heat input to the brake may be further simplified as shown in Figure 7.2(a, b). Figure 7.2(a) shows the maximum braking power as being instantly developed at the start of the brake application and then declining linearly to zero at the end. In Figure 7.2(b) the braking power is ramped up at a rate determined by the system response characteristics and is closer to the real brake application (Figure 7.1). A benefit of using the characteristic shown in Figure 7.2(b) is that it avoids a step heat input, which

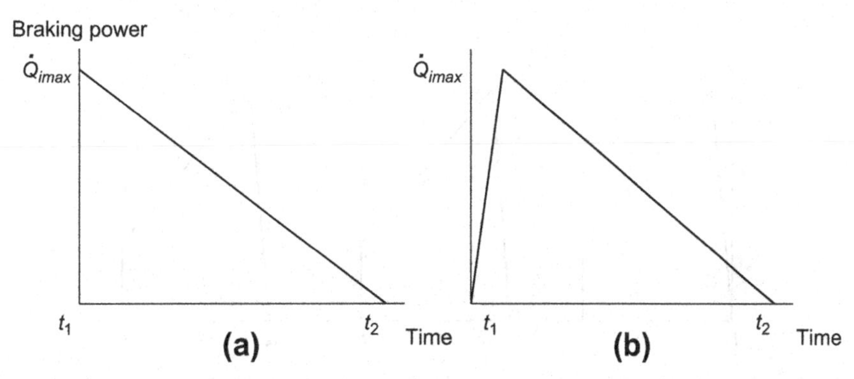

Figure 7.2: Idealised Braking vs. Time Characteristic.

allows more accurate heat input and more stable solutions in computer simulation or numerical modelling of heat transfer in brakes (Tirovic, 2013). The mean power dissipation $\overline{\dot{Q}}_i$ is calculated from:

$$\overline{\dot{Q}}_i = \frac{1}{2}\dot{Q}_{max} \tag{7.10}$$

or as indicated in Chapter 6, Equation (6.5):

$$\overline{\dot{Q}}_i = \tau_{wi}\frac{(\omega_1 + \omega_2)}{2} \tag{6.5}$$

The braking energy dissipated during the brake application is calculated from Equation (7.11), where t_1 is the time of the start of the brake application and t_2 is the time of the end of the brake application:

$$Q_i = \frac{1}{2}\overline{\dot{Q}}_{imax}(t_2 - t_1) \tag{7.11}$$

which is the same as Equation (6.6), where the time period $t = t_2 - t_1$.

Even if the wheel rotational speed at the end of the braking application is not zero, the braking power will decrease to zero when the brake is released since $\tau_{wi} = 0$. The braking power vs. time for a series of repeated brake applications with non-zero final speeds is shown in Figure 7.3. For drag braking at constant speed (Figure 7.4) the braking torque is constant and thus braking power $\dot{Q}$ remains constant after the initial rise. Ignoring the time taken to actuate the brake, the braking energy in a long drag brake application from time t_1 to t_2 (s) can be calculated from:

$$Q_i = \dot{Q}_i(t_2 - t_1) \tag{7.12}$$

Maximum braking power $\dot{Q}_{imax}$ and total braking energy (Q_i) are used to define brake thermal loading or duty. The heat flux $\dot{q}_i$ (W/m^2) at the brake friction interface is also an important parameter for brake thermal analysis; it is not usually uniformly distributed

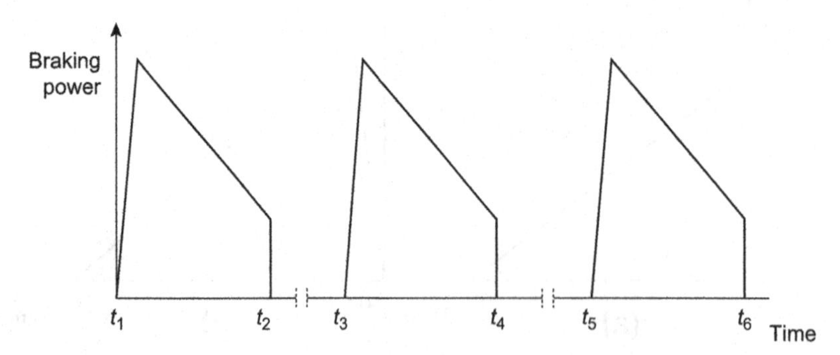

Figure 7.3: Repeated Braking vs. Time Characteristic with Non-Zero Final Speeds.

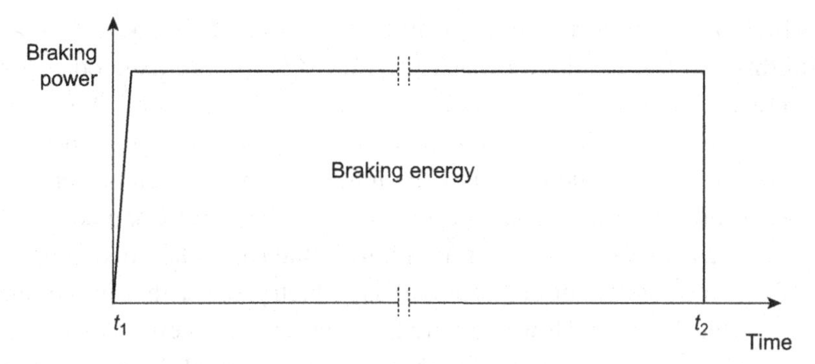

Figure 7.4: Drag braking vs. Time Characteristic, Constant Speed.

across the friction surfaces, but as indicated in Chapter 5 the average heat flux is used to quantify the thermal loading (power density) of the friction material.

Braking Energy Management and Materials

The kinetic energy that a brake will be required to convert to heat and dissipate on a vehicle determines first whether a disc or drum brake should be selected, and then its size. The advantages of disc brakes over drum brakes have already been discussed in Chapter 5, and brake sizing has been presented in Chapter 6 in terms of brake torque and braking force requirements on vehicles. However, the size of friction brakes, especially rotors, is very important in terms of their thermal capacity, i.e. their ability to absorb and dissipate very large amounts of heat energy over a very short time without overheating and premature failure. The heat must be allowed to flow through the brake components to be dissipated by heat transfer from exposed and contacting surfaces by convection, conduction and radiation, but most of the braking heat energy is dissipated by heat transfer from the rotor, disc or drum. A drum brake has a relatively large thermal mass (mC_p) compared to a disc and can therefore absorb more heat energy, but it dissipates the heat more slowly because it has to flow from the friction interface through the drum wall to be dissipated from the outside surface of the drum (which is usually shrouded by the wheel). A disc brake has a lower thermal capacity but can dissipate the heat more quickly because it has exposed friction surfaces over which airflow is good, and can be enhanced by the 'ventilated' (or 'vented') disc design in which air can flow between the friction surfaces.

Ventilated brake discs have two solid friction rings (often termed 'cheeks') separated by the vents that are formed between vanes that support and separate the two cheeks. Curved or angled straight vane designs can give improved cooling compared to tangential straight vane designs but are more expensive to manufacture and ideally have to be right- and left-handed, which increases costs (although some vehicle manufacturers do not differentiate in this way). It is important that the vane design permits good casting flow during manufacture. 'Pillar' designs of ventilated brake discs have been found to be effective for high-duty disc brakes, e.g. commercial vehicles and large high-performance cars. Bryant (2010) found that the design of the inner vent profile can not only aid the flow of cooling air, but can also reduce hot judder propensity (see Chapter 10). The design of the ventilated disc has also been found to influence brake noise propensity (see Chapter 10) by introducing circumferential asymmetry, e.g.with a prime number of vanes to decouple 'pair' modes. However, circumferential asymmetry can also affect frictional heat distribution, temperature distributions, and thermal deformations and stresses in brake rotors.

For conventional brakes (resin-bonded composite friction materials with cast-iron rotors) on high-speed, high-performance road cars, the rotor must be designed to have as much thermal mass as possible while at the same time it must dissipate the heat as quickly as possible, so a ventilated brake disc is the obvious choice. The mass of the brake disc is of course limited by constraints, e.g. unsprung mass, explained in Chapter 5. Commercial vehicles do not travel at such high speeds and can accommodate much larger brake rotors than passenger cars but still have to dissipate huge amounts of braking heat energy so disc brakes are fitted to a significant proportion of commercial vehicles. Despite the performance advantages of disc brakes, drum brakes remain an attractive option, combining large thermal mass with robustness and ease of maintenance. One approach to reducing the weight of a brake rotor is to use lighter materials while maintaining the thermal mass, which means that the lower density must be matched by a higher specific heat, or by increased thermal diffusivity, which enables the heat energy to flow through the rotor more quickly (thermal diffusivity $= k/\rho C_p$). Ideas that have been proposed include 'metal matrix composite' (MMC) discs, which have been found to have limited high-temperature capability. Another approach is to use lightweight designs, e.g. thin solid discs of high-strength steel perhaps with wear-resistant surface coatings to increase high-temperature durability when paired with high-duty, ceramic-type friction materials. Carbon–carbon brake friction pairs offer substantial weight savings, but are very expensive and can be too inconsistent in performance for the average road vehicle and driver, so their use tends to be limited to motorsport.

The stator components of a brake do not generally influence braking energy management, but if the overall mass of a foundation brake needs to be reduced, it may be possible to reduce, say, the caliper mass in order to retain the thermal mass of the disc. Aluminium calipers offer a moderate weight reduction but because of its lower modulus, the section needs to be increased to maintain sufficient stiffness, and the net effect is usually a decrease in the maximum disc size.

For conventional designs of brakes the thermomechanical properties of brake rotor materials are very important; the ability of the rotor material to absorb and transfer braking heat energy is as important as high mechanical strength at operating temperatures. Grey cast iron has been used for brake rotors on road vehicles for many years; these cast irons typically have low ductility and moderate strength, high thermal conductivity and specific heat characteristics, and excellent vibration damping properties. Additionally they are low cost and easy to machine in manufacture. Their characteristic microstructure is a result of slow solidification rates, which allow graphite flakes to be formed coupled with silicon alloying that promotes graphite formation. The wear and abrasion resistance of grey cast irons depend on the microstructure. Brake rotor design must take account of the limited mechanical properties:

- Low ductility – maximum tensile strain is less than 0.5%
- Low impact strength and high notch sensitivity because the graphite flakes act as stress raisers
- Low fatigue strengths because of the effects of the graphite flakes on crack initiation.

Grey cast irons maintain their mechanical properties up to approximately 500°C, above which the mechanical properties decrease significantly. The thermal properties for a typical automotive grey cast iron (e.g. ISO Grade 250, ultimate tensile strength 250 MPa; see e.g. Ihm, 2003) are indicated below (usually the properties reduce as bulk temperature increases):

- Density, 7200 kg/m^3
- Specific heat, 450–550 J/kg K
- Thermal conductivity, 45–54 W/m K
- Coefficient of thermal expansion, 12×10^{-6} K^{-1}
- Melt temperature, 1145°C.

Grey cast-iron brake rotors can be stress relieved after machining. This has been found to be beneficial to minimise runout of new rotors and the generation of runout, which can lead to judder (see Chapter 10). Stress relieving is typically performed in the temperature range of 500–650°C for up to 24 hours and has no significant effect on microstructure or mechanical properties. Castings are generally not annealed because annealing can adversely change the microstructure.

Medium carbon grey cast irons, e.g. Grade 250, tend to be used for small-diameter discs on small and medium-sized passenger cars. High-carbon grey cast irons, e.g. Grade 150, tend to be used for larger vehicles where strength retention at high temperature is not as critical and larger rotors can be accommodated. Alloying elements can be included in all grades of cast iron to improve mechanical strength but tend to reduce thermal properties (and manufacturability).

Brake Thermal Analysis

Brake thermal analysis is needed for the foundation brake design and for the brake system layout design stage (see Chapter 6), in Steps 1.3, 2.2 and 2.4. Additionally brake thermal analysis is an essential part of Step 5 (part of the brake system layout design process in Chapter 5, and covered in detail here): to evaluate operational effects including loading and usage, heat and temperature, wear and durability, consistency and stability. Bulk thermal effects relate to large-scale global temperature rise, which can be analysed as follows:

1. Prediction of the bulk temperature rise of brake components, including the brake disc or drum, brake pad or shoe and lining assembly, and other associated components such as the caliper, wheel bearings and suspension components.
2. Prediction of temperature distributions within the friction pair (disc/drum and pad/lining) and the prediction of thermal stresses.
3. Detailed thermomechanical analysis usually by computational methods such as the finite element (FE) method.

Bulk temperature rise $\Delta\theta$ can be estimated from the energy Q entering the brake component, the portion of the brake component mass (m) that is expected to be heated during braking, and its specific heat (C_p):

$$\Delta\theta = Q/mC_p \tag{7.13}$$

An example is the calculation of the 'single-stop temperature rise' (SSTR) of a brake disc or drum, as introduced in Chapter 6. The kinetic energy input to each brake on the vehicle during a single-stop brake application from its designed maximum speed to rest is calculated using Equation (7.7) with $v_2 = 0$. Using Equation (7.13) the bulk temperature rise is calculated for the part of the disc or drum that is in contact with the stators; in the case of a brake disc this is the mass of the friction ring, i.e. the mass of the disc less the hub flange and hat region, and in the case of a brake drum it is the mass of the friction cylinder, i.e. the mass of the drum less the flange (see Figure 7.5). The car front brake disc specified in Step 2.1 in Chapter 6 has 150 mm outer radius and 100 mm inner radius. A ventilated disc would normally be specified for the front brakes of this type of passenger car, with a 'cheek' thickness (i.e. the thickness of each of the two solid rings separated by the vents) of 5 mm, and separated by a 5 mm gap with vanes giving a 30% volume ratio. The mass of the 'friction ring' is therefore approximately 3.2 kg, and the SSTR from 145 km/h is approximately 400°C, which is less than 600–650°C as illustrated for a ventilated disc in Table 6.1. On the basis of this, the initial thermal performance of the disc based on the SSTR from the maximum speed of this vehicle of 160 km/h is satisfactory.

Experience has shown that if the SSTR value is below the limits illustrated in Table 6.1, service problems are unlikely to be encountered in disc brake designs. However, SSTR is

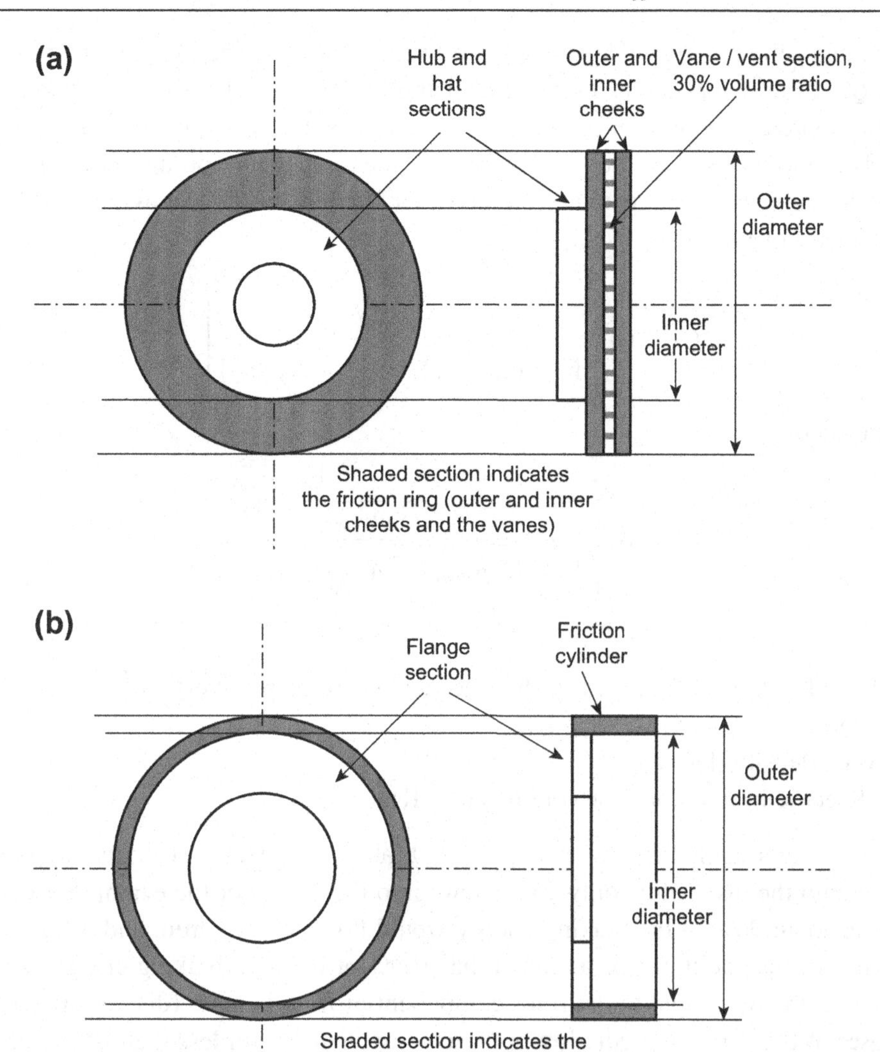

Figure 7.5: (a) Friction Ring of a Brake Disc. (b) Friction Cylinder of a Brake Drum.

a 'quick indicator' calculation and makes no allowance for brake cooling so it should not be relied upon in isolation for new brake designs. It is also useful as a basic check for all other thermal analyses of brakes.

The SSTR approach assumes that all the frictional heat generated is transferred into, and dissipated by, the rotor, but in practice some of the heat flows into (and through) the rotor and some into the stator. The 'partition' of the braking energy between the friction pair components (disc and pad, drum and lining) must therefore be determined in order to be able to calculate more accurately the temperature distributions in each component. The determination of the braking energy flowing into each part of a friction pair (Q_r to the rotor and Q_s to the stator) can be established using Equations (7.14) (Newcomb and Spurr, 1967). This approach to the heat partition at a sliding contact was originally presented by Jaeger (1942) and is based on the assumption that the temperature on the rotor friction surface is the same as the temperature on the stator friction surface, an assumption that is discussed later. For the rotor:

$$Q_{\mathrm{r}} = Q_i \left[\frac{A_{sr}\sqrt{k_r \rho_r c_{pr}}}{A_{\mathrm{sr}}\sqrt{(k_r \rho_r c_{pr})} + A_{\mathrm{ss}}\sqrt{(k_s \rho_s c_{ps})}} \right] \tag{7.14a}$$

and for the stator:

$$Q_{\mathrm{s}} = Q_i \left[\frac{A_{ss}\sqrt{k_s \rho_s c_{ps}}}{A_{ss}\sqrt{(k_s \rho_s c_{ps})} + A_{sr}\sqrt{(k_r \rho_r c_{pr})}} \right] \tag{7.14b}$$

where:

A_{sr}, A_{ss} = Friction surface area of the rotor and stator respectively (m^2);
k = Thermal conductivity (W/m K);
ρ = Mass density (kg/m^3);
C_p = Specific heat at constant volume (J/kg K).

This equation suggests that for the example disc brake about 98.4% of the braking energy would flow into the disc, while only 1.6% flows into the pads. For the example S-cam drum brake, about 96% of the braking energy would flow into the drum and 4% into the linings. This means that it can be assumed that 100% of the total braking energy $Q = Q_s$ can be used as the heat input to the rotor component of friction pairs (disc or drum) for the purposes of temperature prediction in these components, with less than 5% error. However, a small change in the area or thermal properties of the friction material can make a significant change to the heat energy input into the stator components (shoes and pads), and the prediction of temperature distributions in these components may consequently be subject to considerable error. This question of heat partition is considered in more detail later in this chapter.

Friction Surface Temperature Prediction Using Analytical Methods

Analytical methods for predicting brake friction surface temperatures have been used for many years, usually treating the friction pair components as semi-infinite bodies. The most widely used formulae are those proposed by Newcomb and Spurr (1967) and Limpert (2011), which are suitable for basic design calculations. Assuming that the frictional contact area (rubbing path) is a continuous band on the disc surface and the frictional heat is generated uniformly over this band, a 2D heat equation can be formulated to find the temperatures within a disc or drum. Considering a brake disc that has axial symmetry the heat equation in cylindrical coordinates is:

$$\frac{\partial \theta}{\partial t} = \alpha_1 \left(\frac{\partial^2 \theta}{\partial r^2} + \frac{1}{r}\frac{\partial \theta}{\partial r} + \frac{\partial^2 \theta}{\partial z^2} \right) \tag{7.15}$$

where α_1 is the thermal diffusivity of the rotor material $= k/\rho c_p$ (m^2/s), and the domain $r_i \leq r \leq r_o$ and $0 \leq z \leq d_1$ where d_1 is the semi-thickness of the (solid) rotor. The boundary conditions are that the temperature gradient is zero at the centre plane of the disc ($z = d_1$):

$$\frac{\partial \theta}{\partial z} = 0 \quad \text{at} \quad z = d_1 \tag{7.16}$$

and over the other surfaces the heat flux input ($\dot{q}_{i1}$) is:

$$k_1 \frac{\partial \theta}{\partial r} + h(\theta - \theta_o) = -\dot{q}_{i1}(t) \tag{7.17}$$

where h is a heat transfer coefficient (W/m^2 K) that defines the heat transfer from the exposed surfaces of the rotor, as explained later in this chapter. For the non-rubbing path surfaces of the disc the value of $\dot{q}_{i1}(t)$ is zero.

A similar analysis can be used to determine the temperature distribution in a brake drum, where Equation (7.15) applies within the domain $r_i \leq r \leq r_o$ and $0 \leq z \leq L_o$ (r_i is the drum inner radius, r_o is the drum outer radius and L_o is the depth of the brake drum). Solving these equations analytically is difficult, and computational methods are now used, as described later in this chapter. However, the metal rotor (which absorbs most of the generated heat) can be considered to be a 1D body with no edge effects and uniform heat flux over the full width of the rubbing path. In braking to rest (e.g. the SSTR calculation) heat transfer from the rotor can be neglected and the resultant heat conduction equation can be solved analytically. The form of the analytical solution depends on the value of the parameter $\lambda = d_1/\sqrt{\alpha_1 t_s}$, where d_1 is the brake drum thickness or the semi-thickness of the brake disc and α_1 is the thermal diffusivity of the rotor. If $\lambda \leq 1.21$ the body can be considered to be infinitely thick and the mean temperature rise θ at the friction surface is given by:

$$\theta = \frac{2\dot{q}_1 t^{1/2}}{\pi^{1/2}(k_1 p_1 c_1)^{1/2}} \left(1 - \frac{2}{3}\frac{t}{t_s}\right) \tag{7.18}$$

The temperature rises rapidly and reaches a maximum value at $t = t_s/2$, and then reduces because of the reduction in braking energy input as the vehicle slows down. The maximum temperature (θ_{max}) is given by Equation (7.19), which shows the importance of having rotor materials with high thermal property values (k_1, p_1, c_1) to reduce maximum temperatures:

$$\theta_{max} = 0.53\dot{q}_1 \left(\frac{t_s}{k_1 p_1 c_1} \right)^{1/2} \tag{7.19}$$

If $\lambda < 1.21$ the body cannot be considered to be infinitely thick and the temperature is given by:

$$\theta = \frac{\alpha_1 \dot{q}_1}{d_1 k_1} \left\{ \left(1 - \frac{t}{2t_s} \right) + \frac{d_1^2}{3\alpha_1} \left(1 - \frac{t}{t_s} \right) + \frac{d_1^4}{45\alpha_1^2 t_s} \right\} \tag{7.20}$$

The maximum temperature calculated from Equation (7.20) is greater than that calculated from Equation (7.19) and occurs nearer the end of the brake application. This is because the thermal capacity of the body is finite, and because there is no surface heat transfer from it, the resultant temperature is higher. In reality, peak temperatures on the rotor friction surface vary dramatically because of the continually changing contact between the friction pair. The magnitude of the disc surface temperature depends on the ratio between the real and apparent area of contact, which should be kept as high as possible through good bedding and burnishing, as discussed in Chapter 1. Because high temperatures may be attained at the rubbing path, temperature gradients normal to the friction surface cause thermal stresses that can be high enough to cause surface failure in the form of 'thermal cracks'. It is therefore necessary to predict as realistically as possible temperature distributions and associated thermomechanical deformation and stress behaviour in any brake drum or disc. This cannot be achieved to the required accuracy or detail by analytical methods, so computational modelling techniques have to be used, of which the most frequently used is finite element (FE) analysis.

Brake Temperature, Stress and Deformation Prediction Using Computational Methods

Brake thermal analysis using computational methods such as the FE method enables the prediction of bulk component temperature, stress, and deformation profiles and distributions to a much higher accuracy and realism than by using analytical methods. The advantages of FE analysis include the realistic modelling of component geometry, material properties and boundary conditions. Even so, brake friction interface contact and pressure distribution and frictional heat generation remain complex thermophysical phenomena and simplifications are still required in the modelling. Thermomechanical properties and boundary conditions must be accurately specified, and results must be validated and verified. FE analysis can also be used to model 'macroscopic' effects resulting from variation in contact and pressure across the friction interface, arising from local deformation and distortion of the rubbing surfaces (see Chapter 2). Such deformation and distortion is usually transient in nature, and is associated with the modelling of thermoelastic instability (TEI). Since the work done and therefore the frictional power generated at any point in the brake friction interface is proportional to the local friction coefficient (μ), contact pressure (p) and linear sliding speed (v) as indicated in Equation (7.21), as a result of local μ variation with temperature and other factors, interface pressure variations with position and time, and sliding speed variation in the radial direction across the rubbing path, different regions of the friction surfaces will generate heat at different rates. In the regions where the heat generation is higher, the temperatures will be higher and these regions will thus experience more thermal expansion, causing an increase in local pressure and a further increase in local heat generation. The pressure distribution will therefore change across the friction surface, further increasing pressure in high-pressure areas and reducing interface pressure in other areas, and the increased local heat generation may result in the formation of hot bands and/or hot spots. However, the increased temperature in localised areas will eventually cause a reduction of local friction coefficient and increased wear of the friction material (see Chapter 1). The combined effect will be a reduction of local heat generation and, consequently, interface pressure in those areas, transferring contact and pressure increases to previously less loaded areas. Ultimately the hot areas may substantially reduce local interface pressures and even lose contact in a cyclic process that is broadly known as 'thermoelastic instability' (Barber, 1969).

$$q = \mu p v \tag{7.21}$$

In hot spots and hot bands, friction surface temperatures can be much higher than those predicted using any assumption of uniform heat flux distribution across the friction interface. These high temperatures can cause significant changes in the local material properties at the friction surfaces and can also affect bulk material properties such as elastic and plastic modulus, Poisson's ratio, thermal conductivity, specific heat, thermal expansion coefficient, density, wear rate and friction coefficient, and cause degradation of the friction material in the friction surface (Day, 1983; Koetniyom et al., 2000; Tirovic and Sarwar, 2004), all of which must be adequately represented in FE modelling.

Thermal modelling of the friction pair enables the problem of heat partition to be avoided. More recent research has shown that the previously mentioned assumption that the friction surface temperatures are the same for the rotor and the stator is not realistic for a brake friction pair. Experimental evidence indicates that the friction surface temperature of a resin-bonded composite friction material is higher than the friction surface temperature of a cast-iron rotor, and there is a 'temperature jump' across the friction interface. This is discussed later in the context of predictive simulations involving thermal contact resistance at the interface. The 'five-phase' model of a brake friction pair (resin-bonded composite friction material and cast iron; see Figure 2.2) indicates how a temperature jump can occur. The surface of the friction material comprises a 'char' layer of friction material that has experienced high temperature arising from frictional energy transformation at the friction interface. The surface of the rotor is coated by a transfer film that is deposited during the process of frictional sliding, from the friction material. In between these two surfaces is wear debris. The physical mechanism of frictional energy transformation is that heat is generated by work done as the friction material slides over the rotor surface. As indicated in Chapter 2, the three main mechanisms of friction include adhesion, abrasion and deformation, and since the friction material is the weaker of the friction pair, it can be assumed that the majority of the work is done by adhesion, abrasion and deformation of the friction material at or near the friction surface, and the frictional heat energy is therefore generated in the surface layers of the friction material. The heat flows back into the friction material and across the friction interface into the rotor, and because the surface layers and the interfacial wear debris create a contact resistance to heat transfer, a temperature difference or 'jump' is created. The partition of heat between the friction material and the rotor is therefore determined by the thermal properties of the materials and the boundary conditions, but in practice for FE-based brake thermal analysis, realistic results have been obtained by assuming the heat is generated at the surface nodes of the friction material and transferred to the rotor through an 'interface contact conductance' (i.e. the reciprocal of contact resistance) of typically 2 kW/m^2 K.

Frictional contact at the microscopic scale is fundamental to the study of friction and wear and relates to the basic tribological characteristics of the selected friction pair. Much research is being undertaken to better understand and model these 'micro' effects, which are thus a research topic and not considered further here.

FE analysis starts with the generation of a 'mesh' of 'elements', which in modern FE analysis software systems is generated automatically. The type of element used must follow the FE system supplier's recommendations and verification examples should be completed before predictions are started. The element size and refinement of the mesh must be appropriate to the geometry of the part(s) being modelled, and their thermophysical properties, and the quality of the mesh must always be checked. For example, because the friction material has a low thermal conductivity, smaller elements (in the predominant direction of heat flow) must be specified to accommodate the steeper temperature gradient. This can be checked using the 'Fourier number' (*Fo*):

$$Fo = \alpha \Delta t / \Delta x^2 \tag{7.22}$$

where α is the thermal diffusivity of the friction material, Δt is the FE solution time increment and Δx is the element size thickness in the predominant direction of heat flow, usually normal to the friction interface. It is recommended that in FE mesh designs for heat transfer analysis and transient temperature prediction, $Fo \leq 0.5$ for a 1D heat transfer model and $Fo \leq 0.17$ for 3D models. As a general rule, for metal rotors the element size in the proximity of the friction surface should be approximately 1 mm thick, which with a time increment of 0.1 s usually gives good results. Setting the mesh design and the associated time increment values is much more difficult for the friction material, e.g. for FE modelling the complete friction pair. There is a large difference in the diffusivity (α) values between the rotor (disc/drum) material and the friction material (pad/lining). Because the solution time increment must be the same for the whole model, large differences in the preferred element size (Δx) can be required, and it is usually necessary to work to a satisfactory compromise.

Correct and meaningful material thermophysical properties are essential for successful and efficient FE modelling. The starting point is to assume constant linear, isotropic elastic properties for all FE thermal modelling, and then moving to include more advanced features such as temperature and time dependence, non-linearity and anisotropy as the detail of the exercise requires.

Boundary conditions in the form of surface heat transfer coefficients must be specified on all free surfaces. The heat energy input is specified (as calculated using the methods explained above). It is usual to specify an initial temperature distribution in the components that is the same as the surrounding or ambient temperature. Two types of temperature prediction analysis can be generated:

- Steady-state temperature prediction, which calculates the temperature distribution when equilibrium has been reached.
- Transient temperature prediction, which calculates the temperature distributions at different times as the heat flows through the components.

The most important is transient thermal analysis because single and repeated brake applications do not normally reach equilibrium conditions. Steady-state analysis is useful for drag braking where thermal equilibrium is reached. 2D axisymmetric or 3D approaches to FE modelling can be adopted. 2D axisymmetric FE modelling presents a mesh of a radial section through the disc, which is mathematically 'swept' through 360°; thus, any circumferential detail or variation such as the vanes in a ventilated disc or localised contact or heat flux cannot be included and so this approach is generally only useful for basic thermomechanical predictions. In 3D FE modelling, a sector of a ventilated brake disc bounded by planes of symmetry may be modelled to reduce the computational time requirement. In the past, computer power and availability were limited, and 2D axisymmetric modelling was preferred, but modern systems are sufficiently powerful that 3D FE modelling has become the standard.

The method for the prediction of brake disc temperatures, stresses and deformations using the FE method was demonstrated by Day et al. (1991). For a single brake application on a solid disc rotor of a passenger car front brake (vehicle and brake data are summarised in

Table 7.1: Vehicle and Brake Data

Vehicle Characteristics	
Vehicle mass	1300 kg
Brake force distribution X_1/X_2	66:34
Brake Application Characteristics	
Initial vehicle speed	100 km/h (28 m/s)
Final vehicle speed	0
Duration of brake application	4 s
Average deceleration	7 m/s^2 (0.7 g)
Initial brake temperature	20°C
Front Brake Discs (Solid)	
Disc outer diameter	227 mm
Disc ring inner diameter	127 mm
Disc thickness	11 mm
Total disc mass	3 kg
Disc ring mass	2.2 kg

Table 7.1), the calculated maximum heat flux over the whole rubbing path (both sides of the disc) is approximately 1.5 MW/m^2 and the SSTR is 171°C. This is a simple design that was quickly modelled using the 2D axisymmetric FE mesh design illustrated in Figure 7.6 (Day et al., 1991). Detailed geometric features such as the 'undercut' where the friction ring joins the hat section can usually be ignored for the purposes of temperature prediction, being only required for stress analysis where they are known to be important, e.g. in areas of high stress concentration. However, it is usually convenient to use the same FE mesh for thermal and stress analyses as the computational time costs are relatively insignificant, so in practice detailed FE models as illustrated in Figure 7.7 are now used for these purposes.

The frictional heat input for this brake was modelled assuming 98.4% of the total braking energy enters the disc. The braking power profile with time for a single brake application was modelled as in Figure 7.2(b), from which the local heat flux was calculated assuming uniform interface pressure (and thus frictional heat generation) on the friction surfaces of the disc. Surface heat transfer was neglected for a single brake application. Starting from an initial uniform temperature (the ambient temperature 20°C), the peak temperature predicted on the disc surface (rubbing path) was about 215°C after approximately 3 s of the 4 s brake application, which compares with 127°C calculated using Equation (7.19). The reason for the difference is that FE analysis predicted a significant radial temperature gradient from 215°C near the outer periphery to around 80°C at the inner radius of the

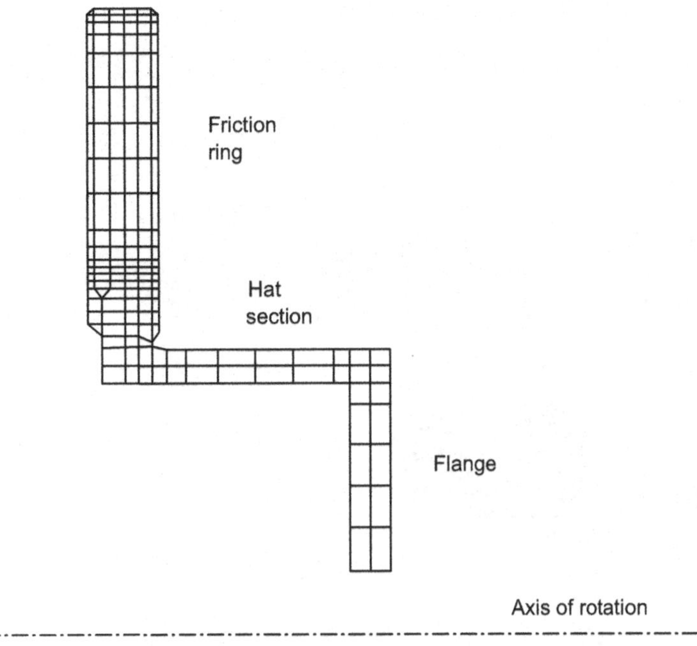

Figure 7.6: 2D Axisymmetric FE Brake Disc Model (Day et al., 1991).

rubbing path, while the analytical method assumes a uniform distribution of heat. High thermal gradients from the friction surfaces into the thickness of the disc were predicted by the FE analysis, and entirely disappeared towards the end of the brake application. Negligible temperature rise in the hat section of the disc indicates that the assumption that only the disc ring mass need be considered for the SSTR calculation was justified. At the end of the brake application the temperature predicted by the FEA was approximately 200°C compared with the calculated SSTR of 137°C.

Once the temperature distribution (steady-state or transient) at any point in time has been predicted, the FE analysis can be continued to predict the thermal stresses, deformations and displacements of a brake disc or drum in use. For this it can be assumed that the disc or drum flange (which is bolted to the hub) is a rigid boundary, i.e. there is no displacement of any node on this flange. Usually no mechanical loading is included in thermal deformation predictions for brake discs (clamp/actuation, friction or centrifugal forces) because the thermal stresses are the most significant, but for brake drums it is wise to check the effect of centrifugal forces. In all thermal stress and deformation analyses the thermal analyses are performed first and the predicted temperature data are usually stored for subsequent thermal stress and deformation analyses at selected braking times.

Example results from the FE analysis of a ventilated brake disc (from the front axle of a medium-sized passenger car) are shown in Figure 7.7, based on a 3D FE modelling procedure that utilises a 'sector' model, with symmetrical boundaries on radial planes either

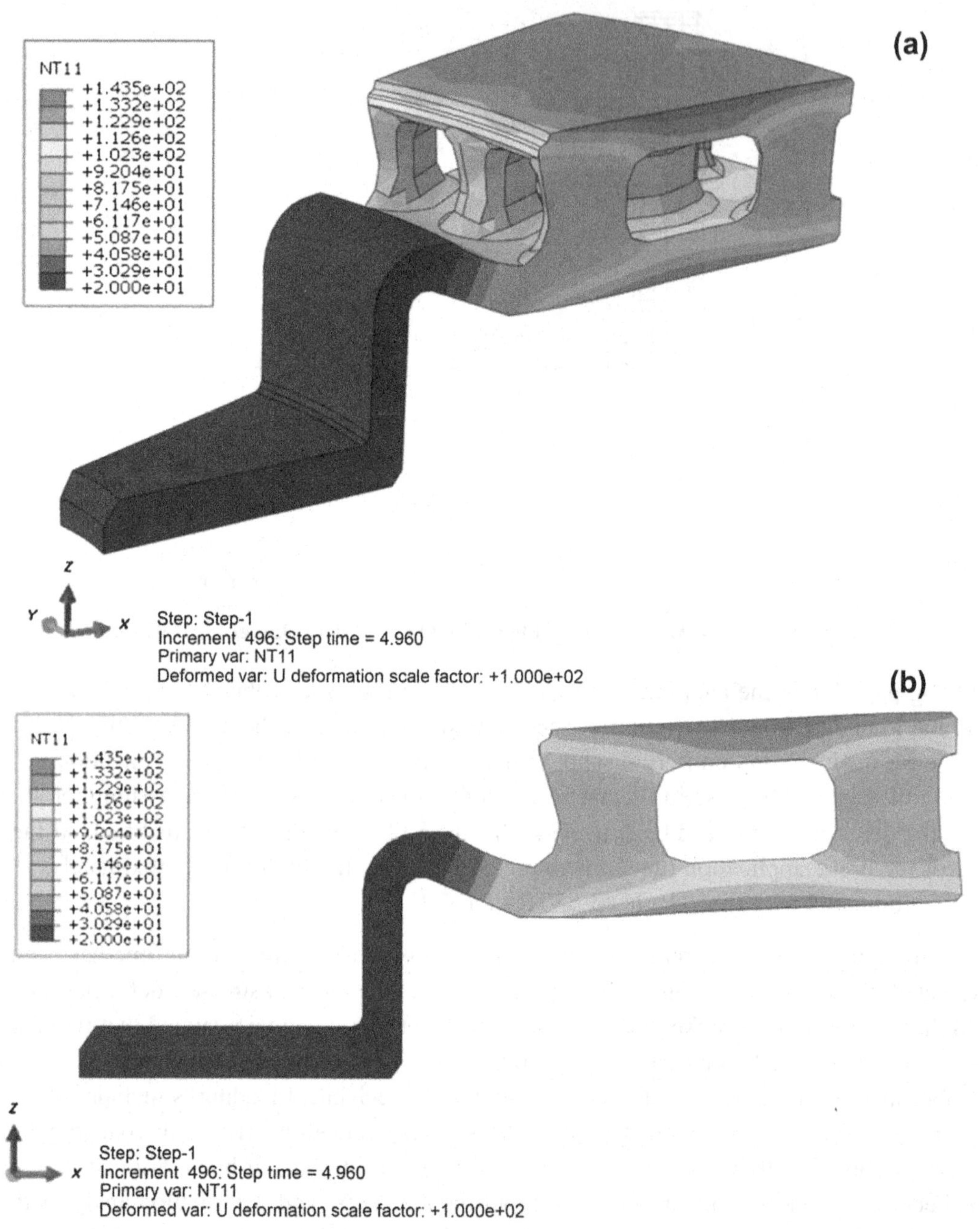

Figure 7.7: (a) 3D Sector Model of a Brake Disc Showing the Predicted Transient Temperature Distribution for a 0.4 *g* Brake Application. (b) The Same Transient Temperature Distribution Showing the Coning Distortion of the Friction Ring (Tang, private communication).

side of a complete vane region. For thermal analysis the assumption is that there is no heat flow across the boundary planes, and for stress analysis a plane strain boundary condition applies. Figure 7.7(a) shows the FE model with the predicted transient temperature distribution halfway through a single 0.4 g brake application, and Figure 7.7(b) illustrates the same temperature distribution superimposed on the deflected shape, i.e. the shape of the thermal deformation predicted for this temperature distribution. The detail of the junction between the friction ring joining the hat section is modelled in detail, and 'coning' distortion of the friction ring is clearly visible. These results illustrate the advantages of computational modelling methods like the FE method over classical analytical methods.

Although this analysis uses a 3D FE sector model, the thermal loading and boundary conditions assume circumferential symmetry, so full 3D modelling, which includes circumferential variation of contact and pressure at the friction interface, frictional heat generation, temperature distribution, thermal stresses and displacement, is needed to provide complete realism in FE-based thermal and thermomechanical predictions for the competitive design of high-duty brake rotors. There are many examples of this type of sophisticated modelling in published research; the frictional power can be modelled as being generated at the pad friction surface so that heat flows across the friction interface (via the interface contact resistance/conductance) to the disc, and as explained previously this avoids the need for heat partition. Because the heat flow into the disc varies circumferentially as it rotates, the disc surface temperature also varies circumferentially, as illustrated in Figure 7.8 (Tirovic, 2013).

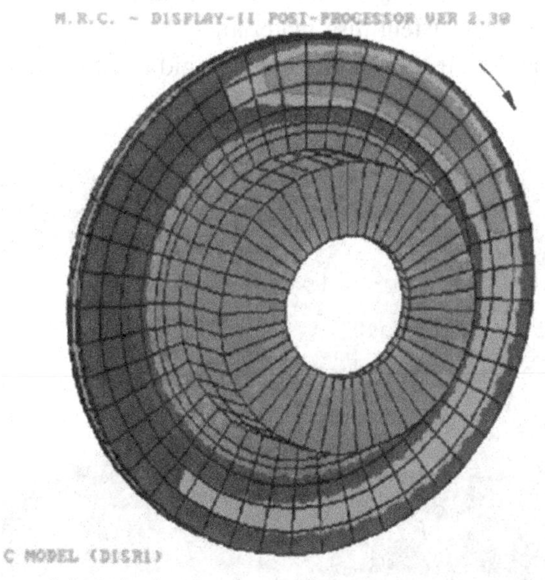

Figure 7.8: Circumferential Variation in the Surface Temperature of a Rotating Brake Disc (Tirovic, 2013).

Results from thermal stress prediction by FE analysis are often quoted in terms of von Mises stress values. Although the von Mises criterion is not representative for cast-iron materials, the resultant stress values can give a broad indication of where temperature gradients may cause high thermal stresses and thermal cracking on the surface of the brake disc. Radial cracks or crazing on the rubbing path of a brake disc is termed thermal cracking (see Figure 7.9) and is a brake disc failure mode resulting from stresses created by elastoplastic deformation of the cast iron during the heating cycle, which in turn generates residual tensile stresses during cooling. Even if thermal cracking is avoided, experience has indicated that hot and cold judder can result from high temperature and high thermal stress gradients normal to the disc friction surface (see Chapter 10).

Thermal deformation and distortion of brake discs is undesirable for many reasons. Coning can cause the pad/disc contact and pressure distribution to be biased towards the outer radius on the outboard side and the inner radius on the inboard side, which causes increased pedal travel, taper wear and non-uniform heat flux input into the disc. This leads to an increased risk of the formation of hot bands on the disc and hot spots or 'blue spots' associated with localised surface overheating and the localised change in phase of the cast-iron disc (or drum) material from pearlite to martensite. The overall effect is an increased incidence of thermal cracking and hot judder (Chapter 10). Good disc brake design should therefore aim to keep the braking energy input as even as possible, radially and circumferentially, so particular attention should be paid to the following factors:

- Design the disc for minimal coning.
- Design the disc and hub assembly for minimal runout and bolt-up distortion.
- Use a good quality and consistent disc material.
- Design the calipers and their mounting to be as rigid as possible.

Figure 7.9: Disc Thermal Cracking.

As previously mentioned, high temperature gradients are a major cause of thermal cracking in brake discs. In a ventilated brake disc, in order to allow airflow through the vanes, only one cheek is connected to the hat region, which can therefore be either the inner or the outer cheek. When the disc is designed so the outboard face is connected to the hat section (as in Figure 7.7) it has been observed that cracking occurs more on that (outboard) face. Conversely, if the inner disc cheek is attached to the hat section, then cracks occur on the inboard face. This may be the result of the additional heat transfer from the outer cheek to the hub causing higher temperature gradients through its thickness. To minimise coning distortion the hat section and the undercut should be designed to be as flexible as possible, consistent with structural integrity and the casting process, and since this also has the effect of restricting heat flow to the hub then disc cracking propensity may also be reduced. When the inner disc cheek is attached to the hat section this is known as 'reverse ventilated' disc design, which provides a longer hat section with consequent increased flexibility and reduced coning distortion of the disc. However, cooling of the disc can be noticeably worse because the airflow path is more restricted between the wheel and the disc.

Brake disc materials have recently seen a trend towards higher carbon contents in order to maximise rotor conductivity at the expense of tensile strength, and this demands even more sophisticated methods of design prediction because of the safety-critical nature of the brake disc. The FE analyses shown here assumed linear elastic properties, but more advanced non-linear elastic–plastic thermomechanical FE models can be used to predict the thermal performance more accurately, e.g. thermal cracking under extreme braking conditions. Grey cast iron is much weaker in tension than in compression, which needs to be included in FE analysis to take account of the cyclic loading behaviour of the material. Koetniyom et al. (2000) generated a model based on uniaxial cyclic tests, which was used in a non-linear 3D FE thermomechanical analysis of two variants of a ventilated brake disc, conventional and reverse ventilated. The results indicated which areas of the disc might experience high temperatures and plastic stresses that might lead to thermal cracking under repeated thermal cycling.

For a full FE simulation and prediction of temperatures and thermal stresses in brakes, both the rotor and the stator should be included, together with some consideration of thermoelastic instability (TEI). The maximum temperatures where the frictional contact is concentrated at 'hot spots' on the rotor surface can be expected to be much higher, perhaps by an order of magnitude, depending on many factors. Such advanced modelling is well established but is beyond the scope of this book. TEI modelling will inevitably predict increased stresses at the surface of a brake disc or drum as a result of hot spotting, even causing local material yielding. A brake disc has frictional heat flux input from both sides and away from the friction surface; the effects tend to 'even out' so that TEI has relatively little influence on the stresses in the friction disc ring–hat transition area where the detail design is critically important. Therefore, 2D circumferentially symmetric disc models can still be very useful and quick for the optimisation of this region, before more complex analyses are performed.

Heat Dissipation in Brakes

It has already been established that heat dissipation is essential in brake system design. Heat is dissipated from brakes by three modes of heat transfer: conduction, convection and radiation. For brake designs and braking duties on most road vehicles, convection is the prime mode of heat dissipation; however, because modern brakes are designed to operate at high temperatures the other two modes (conduction and radiation) should also be considered in brake thermal analyses to enable the accurate prediction of temperatures. As previously explained, for temperature predictions in single or short, low-duty brake applications, heat dissipation and cooling can be neglected but in longer brake applications, such as drag braking and repeated braking, it is important to model heat dissipation as accurately as possible.

Brake cooling can be predicted using analytical and computational methods. The predictions from analytical methods can be limited in accuracy, but the results can be very valuable, in particular for initial design estimates. Newcomb and Spurr (1967) explained how brake cooling could be analysed using Newton's law of cooling (Equation (7.23)), which states that heat loss is proportional to the temperature difference between the cooling body and its surroundings, assuming the principal mode of heat transfer from the brake rotor is convection to the environment. Conduction and radiation from brake rotors and other components are not always negligible and are considered later. The flow of air around the brakes is fundamentally important to effective heat dissipation and cooling. Cooling ducts to gather and direct airflow to the brakes are often placed in the front valance of a vehicle but if they are situated away from the centre line of the vehicle they can be ineffective, as the airflow can be parallel to the air intake, or stagnation can occur as it meets the turbulent regime around the wheel and brake region. Effective brake cooling ducts can be designed, but normally increase aerodynamic drag and therefore are not always acceptable to vehicle manufacturers.

$$(\theta_t - \theta_0) = (\theta_{t_1} - \theta_0)e^{(-bt)} \tag{7.23}$$

where:

θ_0 = ambient temperature (°C);
θ_t = bulk temperature of rotor at time t (°C);
θ_{t_1} = bulk temperature of rotor at the start of cooling (°C);
t = time (s);
b = cooling rate (s^{-1}).

The standard procedure for determining brake cooling rates is to drive the vehicle at constant speed and monitor the decay in rotor bulk temperature over time. Equation (7.23) can be written as the following:

$$\ln\frac{(\theta_t - \theta_0)}{(\theta_{t_1} - \theta_0)} = -bt \tag{7.24}$$

The slope of a graph of ln (natural log) of $(\theta_t - \theta_0)/(\theta_{t_1} - \theta_0)$ vs. time (t) gives the cooling coefficient (b), which depends upon the vehicle, the type of brake, the wheels and many other parameters, including the road speed. Newcomb and Spurr (1959, 1960) found that the effect of speed could be included as indicated in Equation (7.25), where b_0 is the rate at which heat is conducted through the hub and $Cv^{0.8}$ is the rate of convective heat dissipation from the rotor. The parameters b_0 and C also depend upon the vehicle, the type of brake, the wheels and many other parameters, and in the past have been best determined experimentally.

$$b = b_0 + Cv^{0.8} \tag{7.25}$$

The braking power (heat) dissipated from the exposed surface of the rotor ($\dot{Q}$) can be estimated from Equation (7.26), where A_s is the exposed surface area of the rotor. The parameter h is a heat transfer coefficient (W/m^2 K), which defines the heat transfer from the exposed surfaces of the rotor as previously described; in this case negative h indicates heat loss rather than heat input, i.e. cooling.

$$\dot{Q} = -hA_s(\theta_t - \theta_0) \tag{7.26}$$

During a brake application, if the braking energy flows into the rotor and creates a bulk temperature rise of $\Delta\theta$, and is then transferred through the rotor to be dissipated by surface heat transfer (e.g. convection) from its exposed surfaces:

$$Q = mC_p\Delta\theta \tag{7.27}$$

Combining Equations (7.26) and (7.27):

$$\dot{Q} = mC_p\frac{\mathrm{d}\theta}{\mathrm{d}t} = -hA_s(\theta_t - \theta_0) \tag{7.28}$$

and thus

$$\frac{\mathrm{d}\theta}{\mathrm{d}t} = -\frac{hA_s}{mC_p}(\theta_t - \theta_0) \tag{7.29}$$

which when integrated has the solution:

$$\theta(t) = \theta_0 + \mathrm{e}^{-\left(\frac{hA_s}{mC_p}\right)t}(\theta_{t_1} - \theta_0) \tag{7.30}$$

Equation (7.30) is the same form as Equation (7.23) and thus:

$$b = \left(\frac{hA_s}{mC_p}\right) \tag{7.31}$$

and therefore:

$$h = \frac{bmC_p}{A_s} \tag{7.32}$$

Some experimental data for disc brake cooling at constant speed are shown in Figure 7.10 in terms of $\ln(\theta_t - \theta_0)/(\theta_{t_1} - \theta_0)$ vs. time (t) for a solid and a ventilated disc. The cooling rate increases with the rotational speed of the disc, for the solid disc (Figure 7.10(a)) from $1.3 \times 10^{-3}\ \text{s}^{-1}$ at 450 rev/min to $2.6 \times 10^{-3}\ \text{s}^{-1}$ at 1200 rev/min (equivalent road speeds of 55 and 150 km/h). For the ventilated disc (Figure 7.10(b)) the cooling rate increases from 2.88×10^{-3} to $4.44 \times 10^{-3}\ \text{s}^{-1}$ between the same speeds, demonstrating the advantage of a ventilated disc design over a solid disc.

For repeated braking of a regular time interval Δt, Newcomb and Spurr (1967) showed that the average temperature of a brake rotor before the 'nth' brake application is:

$$\theta_{t_n} = \theta_0 + \Delta\theta\left\{\frac{e^{(-b\Delta t)} - e^{(-nb\Delta t)}}{1 - e^{(-b\Delta t)}}\right\} \tag{7.33}$$

and after the 'nth' brake application is:

$$\theta_{t_{n+1}} = \theta_0 + \Delta\theta\left\{\frac{1 - e^{(-nb\Delta t)}}{1 - e^{(-b\Delta t)}}\right\} \tag{7.34}$$

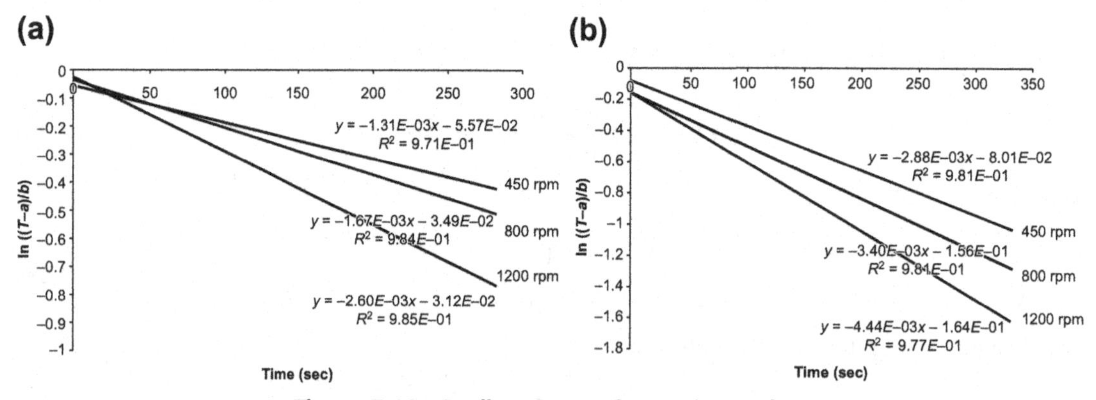

Figure 7.10: Cooling Curves for a Disc Brake.
(a) Solid. (b) Ventilated.

When a steady state is reached with repeated braking:

$$\theta_{t_{n+1}} = \theta_0 + \frac{\Delta\theta}{b\Delta t} \tag{7.35}$$

These analytical methods can only provide approximations but are useful for basic design purposes. The use of computational, especially FE, and also computational fluid dynamics (CFD), methods for brake thermal analysis has enabled predictions of temperatures in brakes on road vehicles to be made with great accuracy, provided that the thermophysical properties of the materials involved are known, and the boundary conditions are realistically applied. The main difference between FE and CFD methods in terms of brake heat transfer is that FE analysis models the brake components and the heat dissipation from them using surface heat transfer coefficients representing all three modes of heat dissipation as introduced above. The surface heat transfer coefficients for each mode must be defined and the resulting thermal or combined thermomechanical stresses and deflections can be calculated using the same models as explained previously. CFD analysis models the fluid (air) in which the brake operates and thus can predict how the convective heat dissipation varies around the brake (and the vehicle) depending on the airflow. It is possible to combine these two methods to create a very sophisticated simulation that includes conductive and radiative heat dissipation as well, but the conditions and laws of conductive and radiative heat dissipation must be defined by the user. In general, CFD analyses are much more complicated and require more time and more computing resources than FE analyses, but they do enable details of the brake installation in the vehicle and the influence of the real brake environment to be effectively modelled. To make brake thermal modelling more efficient, CFD analyses are often used to calculate convective heat transfer coefficients for typical brake design and operating conditions, which are then used in FE analyses. Experimental verification is important in all modelling and simulation to build confidence in the predictive techniques.

Convective Heat Transfer

Heat transfer from the exposed surfaces of a body such as a brake rotor to the environment is governed by Equation (7.26), where the local heat transfer coefficient from a surface of temperature θ_{st} is h_{conv} (W/m^2 K). Airflow is the main influencing factor on convective heat dissipation; a stationary brake disc or drum in still air dissipates heat only by natural convection, but if it is rotating, the heat dissipation increases as the convective cooling becomes 'forced' because of the airflow over the free surfaces, and this increase is more pronounced if the disc or drum is exposed to cross flow of air, e.g. from a side wind. Much work has been done in the study of convective heat transfer from brake rotors, and the factors influencing brake cooling and surface heat transfer coefficients have recently been studied in detail, e.g. by Tirovic and Voller (2002), who studied convective heat transfer from a commercial vehicle disc brake. By dynamometer testing in still air the influence of different disc, wheel carrier and wheel designs was investigated. Convective heat dissipation from the brake disc was predicted by FE and CFD analyses using the type of model shown in Figure 7.11. The CFD flow analysis predicted the airflow through the disc vanes and disc cooling, e.g. as indicated in Figure 7.12 for a disc at 100°C rotating at 450 rev/min in an environment at ambient temperature of 20°C. The maximum air exit temperature was 30°C at the exit from the vanes at the disc outer periphery, so the maximum air temperature rise was only 10°C. Flow stagnation and increase of air temperatures at the hat/friction ring transition region can also be seen. Predicted local convective heat transfer coefficients (h_{conv}) for the same disc rotating in the same environment ranged between 4 and 40 W/m^2 K, and were lowest in the hat area and highest at the air inlet into the channels (at the disc inner radius). High values were also predicted at the disc outer periphery. At the friction surfaces of the disc, the heat transfer coefficient increased in the radial direction outwards, from approximately 10 to 30 W/m^2 K. The average values of heat transfer coefficients were also calculated from the CFD analyses, and the relationship between average heat transfer coefficients and disc rotational speed is shown in Figure 7.13.

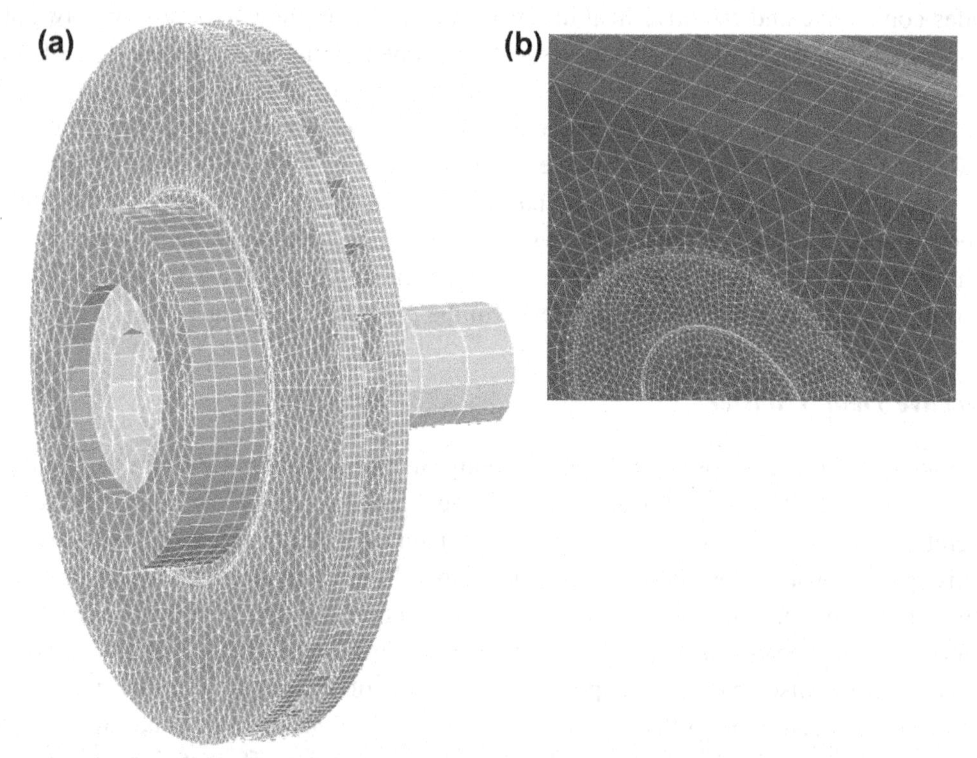

Figure 7.11: Examples of FE Mesh for Brake Disc Thermal Analysis (a) and CFD Mesh for Airflow and Convection Analysis (b) (Tirovic, 2013).

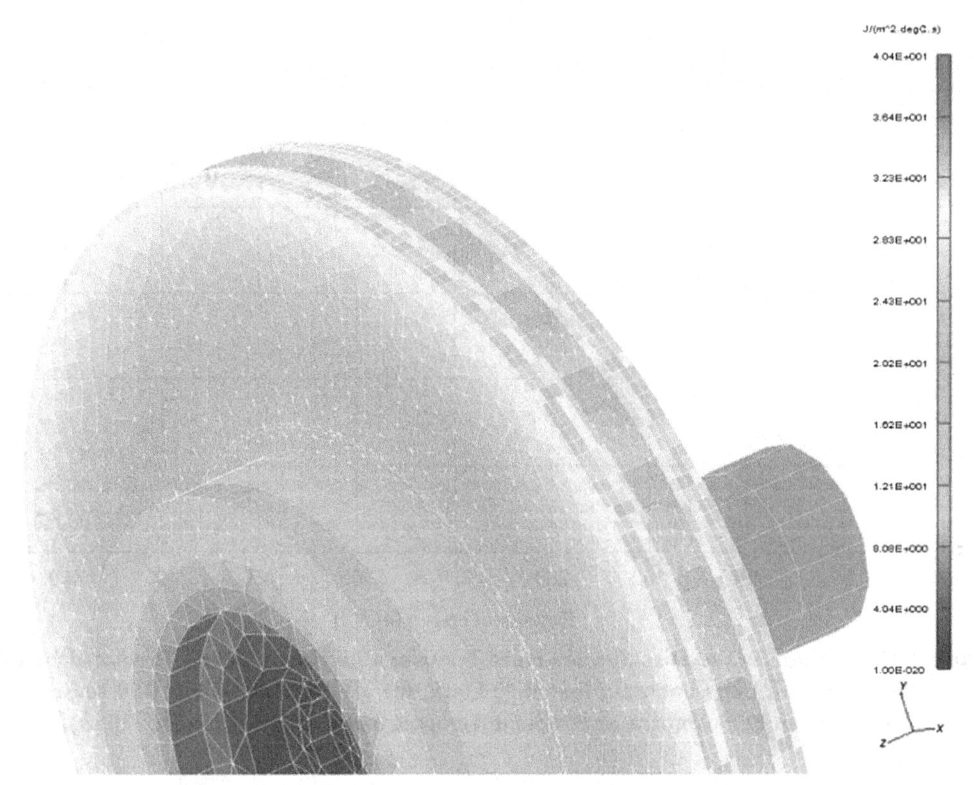

Figure 7.12: Brake Disc Surface Heat Transfer Coefficient Values Predicted by CFD for a Disc Rotating at 450 rev/min at 100°C in an Environment at Ambient Temperature of 20°C (Tirovic, 2013).

The influence of the wheel on convective heat transfer coefficients was also investigated by Voller et al. (2003) for the same environment (still air). A brake disc design in which the outboard friction cheek was attached to the hat showed very little change in the convective heat transfer coefficient with the wheel in place. This was because the air flowed into the vents from the inboard side, as indicated in Figure 7.14(a). A second design of brake disc, a reverse ventilated design with the inboard friction cheek attached to the hat (Figure 7.14(b)), showed similar cooling characteristics for the disc alone to the first design of brake in still air. However, when assembled with the wheel, the convective heat transfer coefficients reduced substantially because the wheel assembly restricted the supply of air into the vents from inside the wheel, not from the outside.

The second design of ventilated brake disc (Figure 7.14(b)) is desirable because it is known to reduce coning deformation, so, in order to improve cooling, the airflow into the vents was increased by piercing the wheel carrier, as illustrated in Figure 7.15. The result was convective heat transfer coefficients that were only slightly lower than for the disc alone, rotating without the wheel in still air.

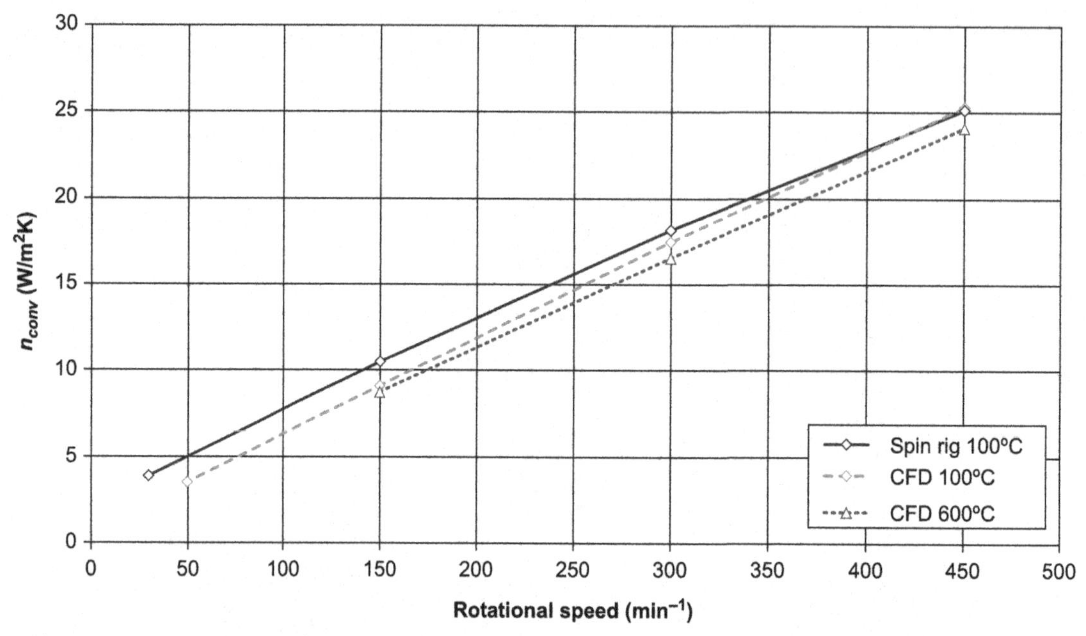

Figure 7.13: Average Brake Disc Surface Heat Transfer Coefficient Values Predicted by CFD and Compared with Experimental (Test Rig) Data, for a Disc Rotating at 450 rev/min at 100°C, in an Environment at Ambient Temperature of 20°C (Tirovic, 2013).

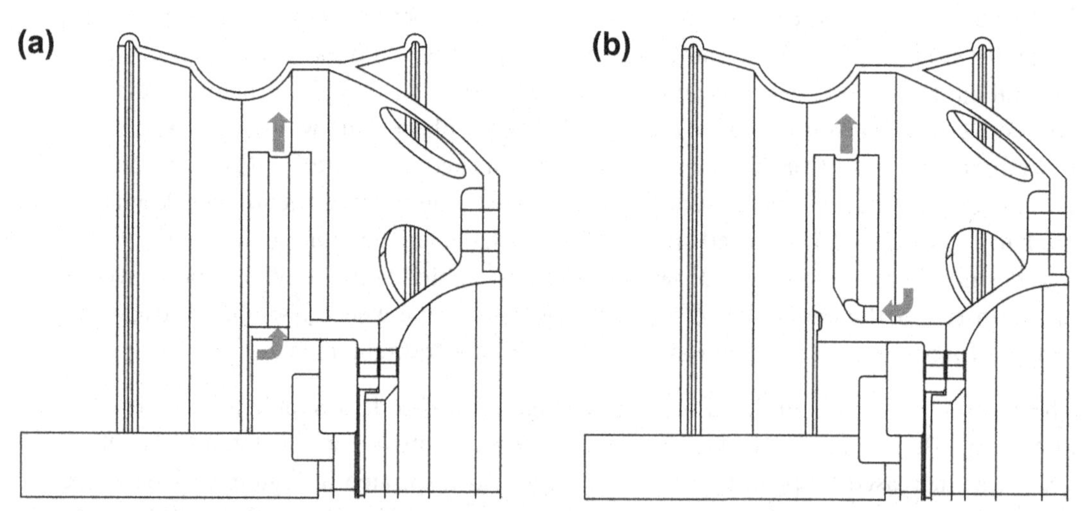

Figure 7.14: Cooling Airflow for Two Different Designs of Brake Disc (Tirovic, 2013).
(a) Outboard attachment. (b) Inboard attachment (reverse ventilated).

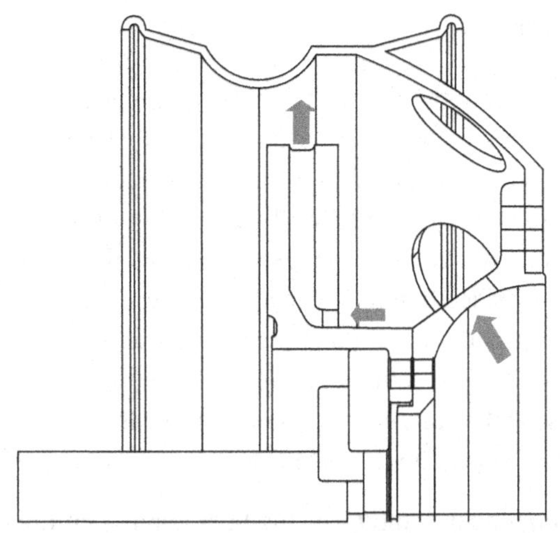

Figure 7.15: Anti-Coning Brake Disc and Pierced Wheel Carrier (Tirovic, 2013).

Figure 7.16 summarises the predicted heat transfer coefficients for the second design of brake disc and the two different wheel carriers. The brake disc with the pierced carrier wheel assembly had cooling reduced from the disc alone (no wheel), but is significantly higher than for the solid carrier. Wheel design was found to have practically no influence on disc convection cooling: steel or alloy wheels, with and without wheel ventilation holes. The reason seemed to be that on exit from the disc vents the air flows in the inboard direction.

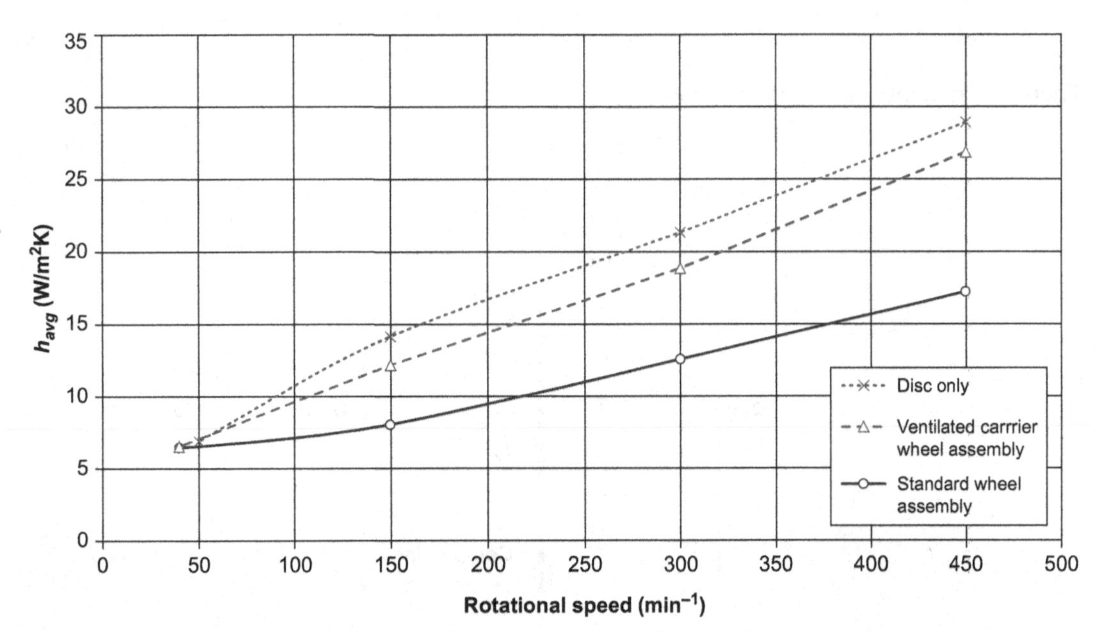

Figure 7.16: Average Heat Transfer Coefficient Values for the Second Design of Brake Disc Rotating at 100°C, in an Environment at Ambient Temperature of 20°C (Tirovic, 2013).

Conductive Heat Transfer

Heat transfer by conduction from the disc to the hub and the wheel is important in brake cooling, and involves heat transfer across interfaces between such components that are in contact, e.g. the disc is bolted to the hub flange, and the wheel is also bolted to this assembly. Conductive heat transfer also occurs at the interface of all other brake components, such as the pad, piston and caliper, and across all of these interfaces there is additional resistance to heat flow because of the 'interface contact resistance'. The interface thermal contact resistance is a result of the actual area of contact being a fraction of the nominal or apparent contact area A (see Chapter 2). The actual conductive heat transfer is achieved only over the contacting asperities. A typical steady-state temperature profile in two contacting solid bodies of cross-sectional area A (m^2) with heat flow through them is shown in Figure 7.17.

Q_1 is the heat entering component 1 and Q_2 is the heat leaving component 2. It is assumed that both components are insulated and steady-state conditions have been established, so that:

$$Q_1 = Q_2 \tag{7.36}$$

The rate of heat transfer through the solid ($\dot{Q}_{cond}$) is governed by the material thermal conductivity (k), the cross-sectional area (A) and the thermal gradient, and can be defined as:

$$\dot{Q}_{cond} = -kA\mathrm{d}\theta/\mathrm{d}x \tag{7.37}$$

Since in the steady state the conductive heat transfer is identical for both components:

$$Q_1 = Q_2 = Q_{cond} \tag{7.38}$$

Thermal gradients can be determined as:

$$d\theta/dx = -Q_{cond}/kA \tag{7.39}$$

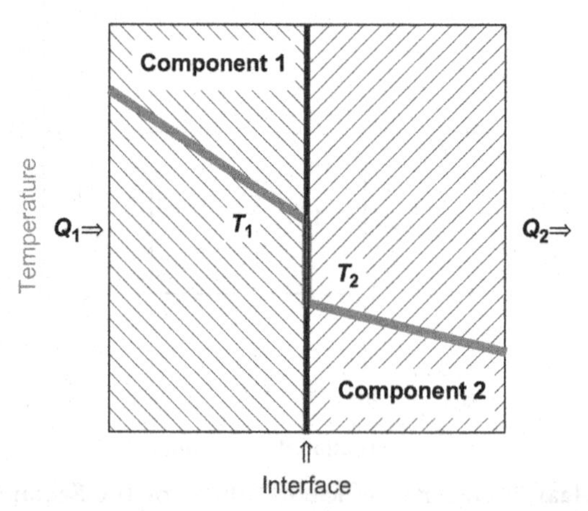

Figure 7.17: Thermal Contact Resistance (Tirovic, 2013).

The contact area (A) is identical for both components, and the difference in the thermal gradients through components 1 and 2 in Figure 7.17 is the result of the difference in the materials' thermal conductivities. In Figure 7.17 component 1 material has lower thermal conductivity, causing a higher thermal gradient. The step change at the interface caused by the interface contact resistance is similar to the thermal contact resistance at the friction interface discussed earlier and illustrated in the 'five-phase' model of a brake friction pair (see Figure 2.2), with the obvious difference that the interface illustrated in Figure 7.17 is static while the friction interface includes relative sliding motion.

Figure 7.18 and Equations (7.37)−(7.39) assume conduction-only heat transfer at the interface. Where there is no intimate contact, i.e. between the asperities, heat transfer may be achieved by convection and radiation, which may not be as effective as conduction, and hence the overall interface contact resistance to heat flow may be increased. In FE modelling it is more convenient to use thermal conductance (the reciprocal of thermal contact resistance) expressed in the form of (conductive) heat transfer coefficient h_{cond} (W/m^2 K), with the additional benefit that by using the same nominal heat transfer coefficients and units (m^{2}K/W) for all three heat transfer modes,

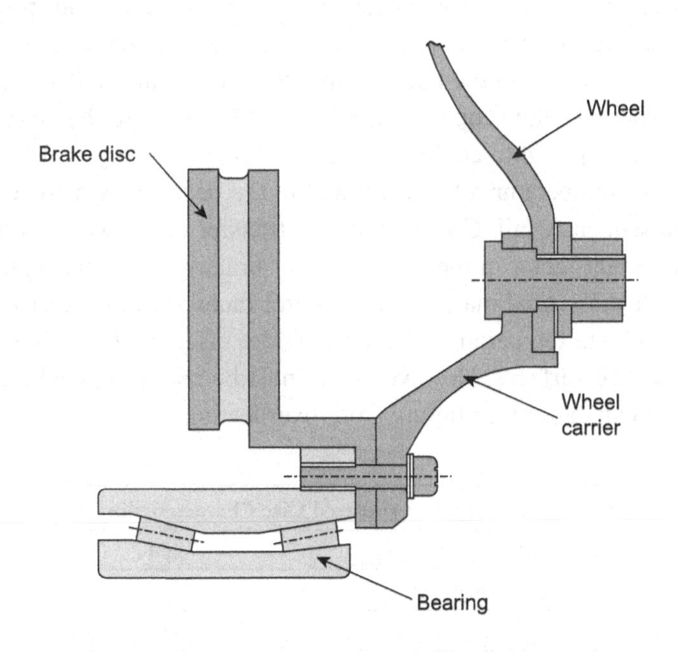

Figure 7.18: Cross-Sectional View Though a Typical Commercial Vehicle Front Wheel, Hub and Brake Assembly (Tirovic, 2013).

the rate of heat transfer for each mode can be directly compared. The conductive heat transfer can be calculated using:

$$Q_{cond} = h_{cond} A \Delta\theta_{int} \tag{7.40}$$

where:

$$\Delta\theta_{int} = \theta_1 - \theta_2 \tag{7.41}$$

The value of thermal conductance (h_{cond}) (and/or thermal contact resistance) at an interface between two bodies through which heat is being transferred can only be evaluated experimentally. It depends on the materials in contact, surface finish and condition, ambient medium, interface pressure, temperature, and many more parameters. In actual vehicle operation brake rotors and hubs are exposed to the environment (rain, snow, salt, etc.) and can be dismantled for maintenance, all of which will affect the interface thermal conductance. Some typical values are given below from the work of Tirovic and Voller (2002).

Figure 7.18 shows a cross-section view though a typical commercial vehicle front wheel, hub and brake assembly in which the disc flange is sandwiched between the hub and the wheel carrier (on commercial vehicles with disc brakes the wheel is not bolted directly to the hub but to the intermediate 'wheel carrier', which in this case is not pierced). The heat generated at the disc friction surfaces is conducted through the hat section of the disc to the flange, where it can flow into the hub and into the wheel carrier. The heat conduction path into the hub provides little scope for heat dissipation because the bearing ring mass is relatively small and unexposed to cooling air. This path is also very undesirable because of the possibility of high temperatures being created in the bearings, which could cause them to overheat and subsequently fail. Conductive heat transfer to the wheel carrier and then the wheel is an important factor in the dissipation of braking heat energy; the carrier and wheel usually have a substantial mass and their outer faces are in contact with fresh and fast flowing air at ambient temperature, which provides effective heat dissipation by forced convection from the free surfaces. However, care must be taken to avoid high wheel rim temperatures, which can cause tyre failure from overheating.

Commercial Vehicle Ventilated Disc Characteristics	
OD (mm)	434
Thickness (mm)	45
Mass (kg)	33
Number of vanes	30
Internal surface area (m^2)	0.24
Friction ring thermal capacity (kJ/K)	9.5

Average values of thermal conductance (h_{cond}) for the disc/carrier interface were evaluated by Tirovic and Voller (2002). Around 12 kW/m^2 K was measured for new components and for used components that exhibited corrosion at the interface, the thermal conductance was reduced by about 40%, to about 7 kW/m^2 K. With high thermal conductivity paste or thin aluminium foil inserted in the interface, the h_{cond} values significantly increased to 60 kW/m^2 K or more. Temperature was found to have a relatively small influence; thermal conductance increased slightly with increased temperature. The effect of interface pressure in the form of bolt-up torque was more significant; in the used condition the relationship shown in Equation (7.40) can be used to calculate average thermal conductance as a function of interface pressure:

$$h_{cond} = 150P + 1560 \tag{7.42}$$

where the units of thermal conductance h_{cond} are W/m^2 K and of interface pressure P are MPa.

Radiative Heat Transfer

The heat transferred by radiation from the surfaces of a brake disc and other brake components (Q_{rad}) is determined by Equation (7.43), which predicts a substantial increase of radiative heat dissipation at elevated temperatures:

$$Q_{rad} = \varepsilon \sigma A_s \left(\theta_D^4 - \theta_0^4 \right) \tag{7.43}$$

where:

- ε is the emissivity of the surface;
- σ is the Stefan–Boltzmann constant (W/m^2/K^4);
- A_s is the area of radiative heat emission (m^2);
- θ_D is the surface temperature of the disc (K);
- θ_0 is the ambient (air) temperature (K).

Equation (7.43) for radiative heat transfer can be rewritten in a form similar to Equation (7.26) for convective (or conductive) cooling:

$$\dot{Q}_{rad} = -h_{rad} A_s (\theta_{st} - \theta_0) \tag{7.44}$$

The radiative heat transfer coefficient h_{rad} is defined as:

$$h_{rad} = \frac{\varepsilon \sigma \left(\theta_{st}^4 - \theta_0^4 \right)}{(\theta_{st} - \theta_0)} \tag{7.45}$$

which can be further reduced to:

$$h = \varepsilon \sigma \left(\theta_{st}^3 + \theta_0^3 + \theta_{st} \theta_0^2 + \theta_0 \theta_{st}^2 \right) \tag{7.46}$$

The most difficult aspect of radiation heat transfer modelling and analysis relates to the value of emissivity; the material, the condition of the surface and its temperature all influence the value of emissivity. These parameters are affected by wear, transfer and/or surface films or deposits, time and temperature, and the emissivity of a brake disc friction surface can be evaluated by measuring the surface temperature using an infrared (IR) temperature sensor and a contact device such as a thermocouple, with the emissivity value being adjusted until the same temperature reading with both is obtained. Using an average emissivity of 0.55 in Equation (7.45), radiative heat transfer coefficient (h_{rad}) values were calculated and are shown in Figure 7.19 (Tirovic, 2013). This indicates that radiative heat transfer occurs even at temperatures as low as 100°C, when $h_{rad} = 5$ W/m^2 K, which is comparable to convective heat dissipation by natural convection. At 400°C, radiative heat dissipation becomes much more significant with h_{rad} equal to 16 W/m^2 K, while at 600°C h_{rad} rises to 31 W/m^2 K. This illustrates the importance of using correct emissivity values for reliable brake cooling prediction.

Modern FE analysis systems can model radiative heat transfer from surfaces including emissivity variation with temperature, and can also model cavity radiation. Although useful scientifically, this can substantially increase computational requirements with limited effect on the accuracy of the results. When modelling the cooling of a ventilated brake disc it can safely be assumed that there is no heat dissipation by radiation heat transfer from the internal surfaces of the vanes, since the heat is reflected back. Therefore,

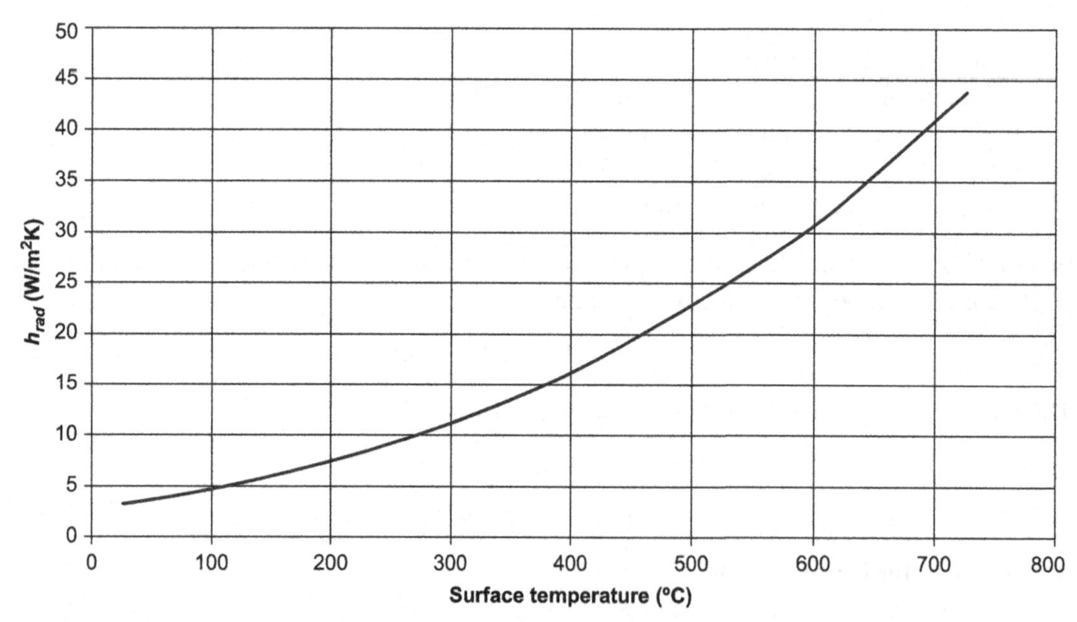

Figure 7.19: Radiative Heat Transfer Coefficient h_{rad} Increase with Surface Temperature for Emissivity $\varepsilon = 0.55$ (Tirovic, 2013).

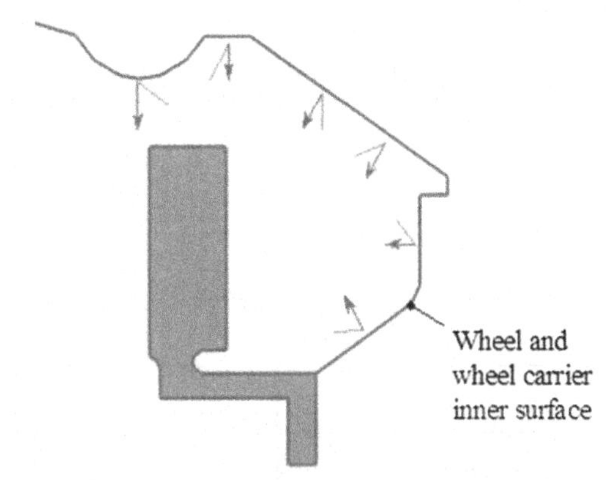

Figure 7.20: Radiative Heat Reflection in a Brake Disc, Wheel Carrier and Wheel Environment (Tirovic, 2013).

radiation heat transfer from a ventilated disc for any given surface temperature would be expected to be closely similar to a solid disc of the same size and surface area. As shown in Figure 7.20, heat radiated from the brake disc is partly reflected back from the wheel and wheel carrier (on a commercial vehicle disc brake assembly) to the disc, and under heavy-duty, repeated or drag brake applications, radiative heat reflection can result in an increase of over 20°C in brake disc temperatures.

Tirovic et al. (2002) determined the influence of individual heat dissipation modes from a commercial vehicle brake undertaking a drag brake application similar to the UN Regulation 13 Type 2 test (see chapters 8 and 9) for the effectiveness of cooling hot brakes after drag braking down long inclines. The braking energy in this test corresponded to 6 km of drag braking of a 16,500 kg rigid two-axle commercial vehicle travelling at a constant speed of 30 km/h down a 6% incline. Only the front brakes were considered ($X_1/X_2 = 50/50$), and the effects of rolling and air resistance, retarder and engine braking were not included, to represent the worst case. The contribution of individual modes of heat transfer to disc cooling was investigated at a disc surface temperature of 600°C, which was achieved at approximately 430 s into the drag braking application. Figure 7.21 shows the contribution of individual modes of heat transfer to the braking power heat dissipation from the disc at this temperature; a wheel rotational speed of 150 rev/min corresponds to a road speed of 30 km/h.

At a disc temperature of 600°C (Figure 7.21(a)), the total braking power heat dissipation was about 11.5 kW and the contributions of the individual modes of heat dissipation were approximately:

Conduction 2 kW (18%);
Convection 4.5 kW (39%);
Radiation 5 kW (43%).

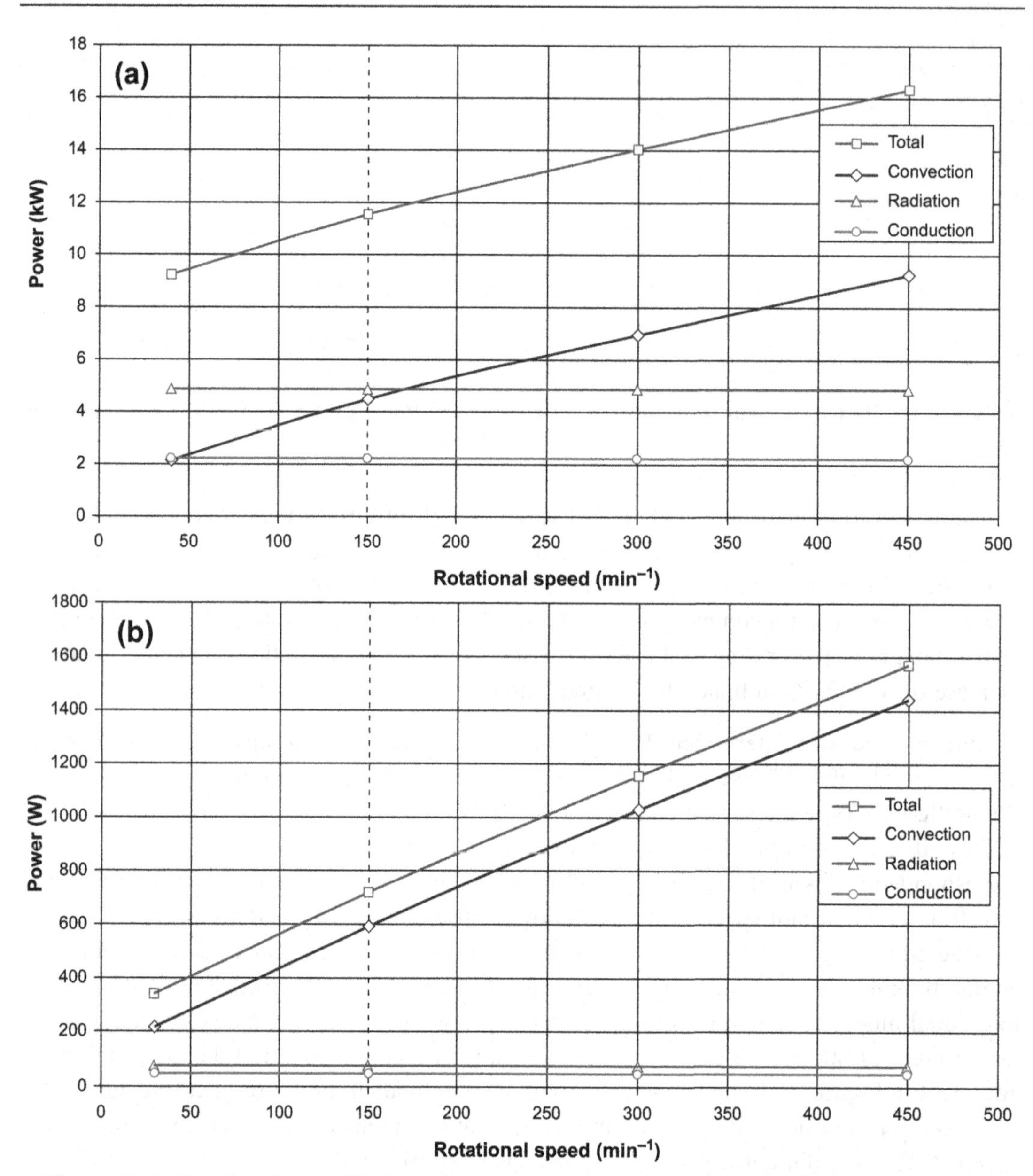

Figure 7.21: Braking Power Dissipated by Each Mode of Heat Transfer for the Brake Disc at 600°C (a) and 100°C (b) (Tirovic, 2013).

Conductive and radiative heat dissipation were not affected by speed but convective cooling varied from 2 kW at 40 rev/min to over 8 kW at maximum speed (450 rev/min). At a disc temperature of 100°C (Figure 7.21(b)) the total heat dissipation was about 700 W, and over 80% is by convective heat transfer.

The analysis, modelling and simulation of heat transfer in friction brakes is complicated and requires an increasingly sophisticated approach that FE methods are well developed for. The design characteristics of the wheel, wheel carrier (if fitted) and the foundation brake substantially influence brake cooling. It is important to include all modes of heat transfer, although convection remains the dominant cooling mode for most brake designs and duties for which FE and CFD are essential predictive methods. Conduction is probably the least studied heat dissipation mode from friction brakes, and radiative heat dissipation should be included at all temperatures, because even at the relatively low temperature of 100°C, its contribution to brake cooling is similar to that of convective cooling by natural convection. The main difficulty in predicting radiative cooling remains the determination of the surface emissivity, which changes with surface condition and temperature.

Chapter Summary

- The conversion of kinetic energy into heat energy in friction brakes and the dissipation of that heat energy are explained and discussed. The factors that influence the thermal behaviour of brakes during braking and the importance of thermal effects in brakes in terms of braking performance are also discussed and explained.
- Three different types of braking application are presented and discussed: single application braking, repeated application braking and continuous application or drag braking, all of which must be considered in the thermal design of braking systems for road vehicles. Basic methods of analysis for these are presented and explained.
- The management of the heat energy in brakes is discussed in terms of brake design; this adds to the design procedure based on brake torque and braking forces already presented in Chapter 6. Rotor materials are discussed, with a focus on conventional grey cast iron.
- The calculation of temperatures and temperature distributions in brakes starts with a review of analytical methods for rotor bulk temperature prediction, e.g. the 'single-stop temperature rise' (SSTR), which is a very useful starting point for brake thermal sizing.
- The generation of heat at the friction interface and the heat flow into the rotor, the 'heat partition', is discussed and explained.
- Computational methods for brake thermal analysis, namely finite element analysis (FEA) and computational fluid dynamics (CFD), are introduced and explained. Examples of temperature and stress predictions, together with thermal deformation predictions generated by FEA, are presented for transient and steady-state conditions.
- The importance of heat dissipation and cooling is discussed in the context of heat transfer in brakes, and 'brake cooling coefficients' that define how the temperatures of the brakes on a vehicle change with time during driving are explained.

- Surface heat transfer coefficients for three principal modes of heat transfer (convection, conduction and radiation) from brakes to dissipate the heat to the environment are explained. The use of CFD modelling to predict local heat transfer coefficients is demonstrated with examples.

References

Barber, J.R., 1969. Thermoelastic instabilities in the sliding of conforming solids. Proc. Royal Soc. A312381-394.

Bryant, D., 2010. Thermo-elastic deformation of a vented brake disc. University of Huddersfield. PhD thesis.

Day, A.J., Tirovic, M., Newcomb, T.P., 1991. Thermal effects and pressure distributions in brakes. Proc. IMechE. vol. 205, 199–205.

Day, A.J., 1983. Energy Transformation at the Friction Interface of a Brake. PhD thesis, Loughborough University, UK.

Ihm, M., 2003. Introduction to gray cast iron brake rotor metallurgy. SAE/TRW Automotive [internet source].

Jaeger, J.C., 1942. Moving sources of heat and the temperature at sliding contacts. Proc. R. Soc. N.S.W vol. 76, 222–235.

Koetniyom, S., Brooks, P.C., Barton, D.C., 2000. Finite element prediction of inelastic strain accumulation in cast-iron brake rotors. Braking 2000 Conference, Leeds 2000, UK.

Limpert, R., 2011. Brake Design and Safety, 3rd edn, SAE, ISBN 978 0 7680 3438 7.

Newcomb, T.P., Spurr, R.T., 1959. Transient temperatures reached in brake drums and linings. Proc. Auto. Div. IMechE. (No. 7), 227–244.

Newcomb, T.P., Spurr, R.T., 1960. Temperatures reached in disc brakes. J. Mech. Eng. Sci. vol. 2 (No. 3), 167–177.

Newcomb, T.P., Spurr, R.T., 1967. Braking of Road Vehicles. Chapman & Hall. ISBN 0412092700.

Tirovic, M., 2013. Thermal effects in brakes, Braking of Road Vehicles 2013. University of Bradford. ISBN 978 1 85143 269 1.

Tirovic, M., Sarwar, G.A., 2004. Design synthesis of non-symmetrically loaded high-performance disc brakes. Part 2: finite element modelling. Proc. IMechE. Part F, vol. 218, 89–104, 2004.

Tirovic, M., Voller, G., 2002. Optimisation of heat dissipation from commercial vehicle brakes. FISITA 2002 World Automotive Congress, Helsinki, Finland.

Tirovic, M., Voller, G.P., Morris, R., Gibbens, P., 2002. Heat dissipation from commercial vehicle brakes. BremsTech 2002 Conference, Munich, Germany.

Voller, G.P., Tirovic, M., Morris, R., Gibbens, P., 2003. Analysis of automotive disc brake cooling characteristics. Proc. IMechE. vol. 217. Part D.

CHAPTER 8

Braking Legislation

Introduction

Effective and efficient braking systems are fundamental to the safety of all road vehicles, and for this reason the braking of road vehicles is very closely controlled by law in most countries across the world to ensure that road vehicles are designed and constructed to brake safely and efficiently under all conditions of operation.The purpose of braking legislation is to ensure the provision and maintenance of safety standards for road vehicles and their braking equipment in terms of minimum safety standards and performance requirements in order to ensure that the public is protected against unreasonable risk of accidents. The most basic requirements are that each and every road wheel of the vehicle must be fitted with a controllable brake device to contribute to the overall deceleration of the vehicle, and that wheel lock and wheel lock sequence must not cause dangerous vehicle instability. Braking legislation is always advancing to keep up with technological developments in the road vehicle industry, and ensure that strategic targets, e.g. relating to road safety, are addressed.

Braking legislation defines the minimum performance standards required of a modern road vehicle in terms of its deceleration, stability and controllability, taking account of potential failures in parts of the system. Drivers expect very high standards of performance from the braking systems of modern vehicles, and manufacturers design their vehicle braking systems to exceed the legislative requirements, regarding them as a minimum performance level. The main way in which braking legislation is applied in practice is through Regulations and Directives, which are part of a comprehensive legal framework for all aspects of road vehicles in terms of their construction and use. The major world regions that enforce road vehicle legal standards and the associated legislative bodies are:

- Europe: EC (European Community), EU (European Union) and UN (United Nations).
- United States and Canada: Federal Motor Vehicle Safety Standards (FMVSS) and Canadian Motor Vehicle Safety Standards (CMVSS).
- Australia: Australian Design Rules (ADR).
- Japan and eastern Asia: type approval test procedures (TRIAS) (refer to the Japanese National Traffic Safety and Environment Laboratory (JNTSEL)).

Across these regions there are two basic approaches to ensuring that all road vehicles comply with the legislative requirements, which are 'type approval' and 'self-certification'.

Type approval is the procedure that is used in Europe and requires manufacturers to obtain formal approval for a vehicle 'type' and any associated 'systems', 'components' and 'separate technical units', which verifies compliance with UN Regulations and thus that all the legislative requirements are met. An Approval Authority, or a 'Technical Service', carries out the required testing using defined test procedures and methodologies to ensure a consistent approach. Once a product has gained type approval, it can be sold and used in the market without any further testing or verification, but if the manufacturer wishes to make a significant change, it may be re-submitted for type approval, which is, of course, a major endeavour.

Self-certification is the system that is used in countries such as the USA, Canada and Australia, and requires manufacturers carry out their own tests to certify compliance with the regulations to prove that each product on the market conforms to the legal requirements as specified in the relevant regulation(s). Compliance is then verified by the administrative authorities of the legislative body by inspection based on random sampling of the products. In the USA, manufacturers must confirm with the National Highway Traffic Safety Agency (NHTSA) that their products conform to the relevant Federal Motor Vehicle Safety Standards (FMVSS), and products that are certified as compliant in this way have a certification label and can be sold for use in the USA. Again, the required tests have rigorously defined test procedures and methodologies to ensure a consistent approach. If verification inspection reveals non-compliance with the regulations, the manufacturer is allowed to view it and provide an explanation, but in extreme cases, the NHTSA can require the manufacturer to remove the product from sale and recall all products already in the field for rectification. The self-certification system means that manufacturers can in principle more easily make changes to their products, verify compliance and introduce them to the market.

The information and legislative details in this chapter are, to the author's best knowledge, correct at the point of publication. However, the legislation relating to road vehicle braking is extensive and detailed, and the reader is advised to refer to the Regulations, Standards and associated regional documentation for full details and verification, and to ensure that they have full and correct understanding.

European Legislation for Road Vehicle Braking

Within the European Community, legislation relating to the braking of road vehicles is defined in terms of Regulations that are mandatory, and Directives that are not mandatory but may be adopted into national law by member states to which they are addressed. In the EU, a Regulation is binding in its entirety and is directly applicable in all member states, and this chapter concentrates on UNECE Regulations. The legislation covers three broad classes of vehicles with a design road speed exceeding 25 km/h:

- Road vehicles and their trailers
- Two- and three-wheel road vehicles and quadricycles
- Agricultural and forestry vehicles.

Road vehicles and their trailers are categorised as follows:

- Category M, passenger vehicles that are defined as motor vehicles with at least four wheels designed and constructed for the carriage of passengers:
  - Category M_1: passenger vehicles comprising no more than eight seats in addition to the driver's seat.
  - Category M_2: passenger vehicles comprising more than eight seats in addition to the driver's seat, and having a maximum mass not exceeding 5 tonnes.
  - Category M_3: passenger vehicles comprising more than eight seats in addition to the driver's seat, and having a maximum mass exceeding 5 tonnes.
- Category N, goods vehicles that are defined as motor vehicles with at least four wheels designed and constructed for the carriage of goods:
  - Category N_1: goods vehicles having a maximum mass not exceeding 3.5 tonnes.
  - Category N_2: goods vehicles having a maximum mass exceeding 3.5 tonnes but not exceeding 12 tonnes.
  - Category N_3: goods vehicles having a maximum mass exceeding 12 tonnes.
- Category O, trailers, including semi-trailers:
  - Category O_1: trailers with a maximum static combined wheel loading not exceeding 0.75 tonnes.
  - Category O_2: trailers with a maximum static combined wheel loading exceeding 0.75 tonnes but not exceeding 3.5 tonnes.
  - Category O_3: trailers with a maximum static combined wheel loading exceeding 3.5 tonnes but not exceeding 10 tonnes.
  - Category O_4: trailers with a maximum static combined wheel loading exceeding 10 tonnes.
- Note that for full (chassis) trailers the combined axle load is equal to the trailer mass but not for centre axle trailers and semi-trailers, as the load transmitted to the towing vehicle is not included.
- European legislation for two- and three-wheel road vehicles and quadricycles, agricultural and forestry vehicles is not included here, but the general principles of road vehicle braking legislation can be regarded as good practice in the design of their braking systems.

- Category L, motor bicycles, tricycles and quadricycles:
  - Category L_1–L_5: two- and three-wheel vehicles.
  - Category L_6, L_7: quadricycles.
- Categories T and C: agricultural tractors.
- Categories R and S: agricultural trailers.

For many years European braking legislation was basically defined by EEC Directive 71/320 and its subsequent amendments, but from November 2014 this (along with 49 others) will be repealed by the EC General Safety Regulation (GSR) EC 661/2009 (EC, 2009, and Gaupp, 2013). In the GSR the overriding EU legislation is defined by the United Nations Economic Commission for Europe UN Regulation 13 for commercial vehicles (categories M_2, M_3, N and O), and UN Regulation 13H for passenger cars and light vans (categories M_1 and N_1) (UN, Mar 2014 and UN, Feb 2014). As explained later, category N_1 vehicles are covered by both UN Regulations 13 and 13H, even though some category N_1 vehicles may have the same braking systems as some category M_1 vehicles, e.g. light vans derived from cars. These braking regulations are now widely accepted by many other world regions and countries, and the harmonisation of braking standards for passenger cars and light vans across the world has been incorporated into UN Regulation 13H. Harmonisation of braking standards for heavy commercial vehicles is still being pursued. Within the UN the 'Working Party on Braking and Running Gear' (GRRF) technical group is a group of experts that meets biannually to consider research and analysis to develop the active safety requirements of road vehicles. It is one of the six 'Permanent Subsidiary Working Parties' of the 'World Forum for Harmonisation of Vehicle Regulations' that prepares regulatory proposals on active safety, specifically on road vehicle braking and running gear.

The General Safety Regulation (GSR) the approach of mandatory European legislation based on UN Regulations 13 and 13H, and removes complexity from having UN Regulations as alternatives to the separate EU Directives with the same scope. The dates that the braking-related provisions apply from (or will apply from) are summarised in Table 8.1. The GSR also introduces new definitions relating to braking as listed below, and defines fundamental provisions for the type approval of motor vehicles relating to the following new braking-related requirements.

- Electronic Stability Control Systems (ESC)
- Advanced Emergency Braking Systems (AEBS)
- Lane Departure Warning Systems (LDW)
- Tyre Pressure Monitoring Systems (TPMS)
- Tyres, with regard to wet grip, rolling resistance and rolling noise.

Table 8.1: Application Dates for ESC and AEBS Provisions in the EU General Safety Regulation

Vehicle Category	Description	New Types of Vehicle	New Vehicle Registrations
	ESC		
M_2	Passenger vehicles	11 July 2013	11 July 2015
M_3	Class III and <16 tonnes passenger vehicles with pneumatic systems, and all others except the two shown below	1 November 2011	1 November 2014
M_3	Class II, III and B vehicles with hydraulic systems	11 July 2013	11 July 2015
M_3	Class II and III passenger vehicles with 'air-over-hydraulic' systems	11 July 2014	11 July 2016
N_1	Light vans	1 November 2011	1 November 2014
N_2	Vehicles other than the two shown below	11 July 2012	1 November 2014
N_2	Vehicles with hydraulic systems	11 July 2013	11 July 2015
N_2	Vehicles with 'air-over-hydraulic' systems	11 July 2014	11 July 2016
N_3	Vehicles other than the three shown below	1 November 2011	1 November 2014
N_3	Two-axle tractors for semi-trailers, including pneumatic systems, EBS and ABS/non-ABS	1 November 2011	1 November 2014
N_3	Three-axle vehicles with EBS and ABS	1 November 2011	1 November 2014
N_3	Two- and three-axle vehicles with pneumatic systems and ABS	11 July 2012	1 November 2014
O_3	Trailers with a combined axle load between 3.5 and 7.5 tonnes	11 July 2012	1 November 2014
O_3	All trailers other than those above	1 November 2011	1 November 2014
O_4	All types with up to three axles and air suspension	1 November 2011	1 November 2014
	AEBS		
M_1, N_1	Not required	–	–
M_2, M_3	Level 1* Level 2*	1 November 2013 1 November 2016	1 November 2015 1 November 2018
N_2, N_3	Level 1* Level 2*	1 November 2013 1 November 2016	1 November 2015 1 November 2018

*Note that AEBS Levels 1 and 2 refer to Appendix 1 of Annex II to EC Regulation 347/2012; see Table 8.11.

EU vehicle type approval may be sought under the following procedures.

EU type approval of road vehicles is based around a 'whole vehicle' framework. EC Whole Vehicle Type Approval (ECWVTA) is used for most road vehicles in Europe, and allows a vehicle design to be type approved for sale, registration and entry into service across all member states in the EU with no need for further tests or checks. It is aimed primarily at manufacturers of vehicles and bodywork producing large numbers of the same vehicle type or product each year and can be applied to complete or partially completed vehicles. Before the adoption of ECWVTA in 2007, EC type approval allowed a vehicle approved in one member state to the relevant Directive or Regulation to be sold across the EU, and a member state may still have and enforce additional requirements on vehicles approved nationally. ECWVTA requires each member state to introduce the respective provisions into national law, which means that the ECWVTA provisions replace any pre-existing national requirements.

EC Small Series Type Approval (ECSSTA) is for low-volume car producers. As for full ECWVTA it allows Europe-wide sales but with technical and administrative requirements that are more adapted to smaller businesses. The number of units of one type of vehicle to be registered, sold or put into service per year in the EC is limited to 1000 category M_1 vehicles per year. To avoid time-consuming and disproportionately expensive testing only the separate Directives or Regulations listed in the Appendix to Part I of Annex IV (e.g. no destructive testing) are required.

National Small Series Type Approval (NSSTA) allows the number of units of one type to be registered, sold or put into service per year in one member state to be determined by that member state provided that it does not exceed the maximum limits of categories M_1 (75), M_2 and M_3 (250), N_1 (500), N_2 and N_3 (250), O_1 and O_2 (500), O_3 and O_4 (250). The type approval is restricted to the territory of the member state that granted the approval, but another member state should not refuse the type approval unless it has reasonable grounds to believe that the technical provisions according to which the vehicle was approved are not equivalent to its own.

Individual Vehicle Approval (IVA) allows member states to grant individual approvals in accordance with their national rules without destructive tests. Again, approval is restricted to the territory of the member state that granted the approval and other member states may refuse individual vehicle approvals. IVA is usually used for the type approval of:

- Vehicles for use on construction sites or in quarries, ports or airports.
- Vehicles for use by the armed services, civil defence, fire services and forces responsible for maintaining public order.

- Mobile machinery.
- Vehicles intended exclusively for racing on roads.
- Prototypes of vehicles used on the road under the responsibility of a manufacturer to perform a specific test programme provided they have been specifically designed and constructed for this purpose.

Multi-Stage Type Approval (MSTA) is for the type approval of a partially completed type vehicle where different manufacturers are responsible for the approval and conformity of production of the systems, components or separate technical units added at that specific stage of vehicle completion. Many vehicles are built as part of a Multi-Stage Build (MSB) process, where a base vehicle (e.g. a chassis or chassis/cab) is produced and approved as an incomplete vehicle, and then another manufacturer (normally a body builder or converter) finishes the vehicle so it becomes a completed vehicle.

Braking Regulations

The European UN Regulations 13 and 13H (UNECE March 2014, UNECE February 2014) form the basis of the road vehicle braking legislation covered in this chapter. The US FMVSS Rules and Regulations Standard No. 135 (Light vehicle brake systems) (US FMVSS 135) and Standard No. 121 (Air brake systems) (US FMVSS 121) are also considered. The UN Regulation 90 for replacement or 'aftermarket' brake system components (UNECE, 2012) is also discussed. UN Regulations 13 and 13H concern the 'adoption of uniform technical prescriptions for wheeled vehicles, equipment and parts which can be fitted and/or be used on wheeled vehicles and the conditions for reciprocal recognition of approvals granted on the basis of these prescriptions.' They specify procedures for type approval including specifications and tests, and many Annexes, which for example relate to:

- Braking tests and performance of braking systems, including energy consumption (pneumatic systems), the procedure for monitoring the state of battery charge, and provisions relating to energy sources and energy storage devices.
- Braking distribution between vehicle axles, including a wheel-lock sequence test procedure and a torque wheel test procedure.
- Test requirements for vehicles fitted with ABS systems, including symbols and definitions and the utilisation of adhesion (see Chapter 11), performance on differing adhesion surfaces, and method of selection of the low-adhesion surface.
- Inertia dynamometer test method for brake linings (see Chapter 9).
- Electronic stability control systems and brake assist systems, and special requirements that apply to the safety aspects of complex electronic vehicle control systems (see Chapter 11).

UN Regulation 13H covers vehicles of categories M_1 and N_1 and is 'harmonised' with other countries' regulations, e.g. FMVSS 135. UN Regulation 13 covers vehicles of categories M, N and O; requirements for category M_1 vehicles are not included in UN Regulation 13. Category N_1 vehicles are covered by both UN Regulations 13 and 13H, even though some category N_1 vehicles may have the same braking systems as some category M_1 vehicles (e.g. light vans derived from cars). UN Regulations 13 and 13H are intended to be as design-neutral as possible; they do specify requirements for the braking system and various components but do not necessarily specify how these requirements are to be achieved. Some design-specific items are included for features such as commercial vehicle tractor/trailer compatibility and indicator lights for warning signals. In comparison with UN Regulations 13 and 13H, FMVSS 135 specifies requirements for light vehicle braking systems and associated park brake systems, while FMVSS 121 establishes performance and equipment requirements for braking systems on vehicles equipped with air (pneumatic) brake systems. As discussed later, in general the requirements of UN Regulations 13 and 13H are equal to or more stringent than the comparable requirements of FMVSS 121 and 135.

The requirements of EU type approval to UN Regulations 13 and 13H are verified by independent bodies or 'Technical Services', organisations that are designated by the Approval Authority of an EU member state as testing laboratories to carry out tests on its behalf, and/or as conformity assessment bodies to carry out the initial assessment and other tests or inspections. Examples are TUV in Germany and UTAC in France. In some countries the approval authority does the testing and conformity assessment itself, e.g. VCA in the UK and RDW in the Netherlands. The manufacturer may also act as a technical service for 'self-testing', where the tests involved are relatively simple. In order to maintain a high level of confidence between the approval authority and the technical services all documentation must be consistent and transparent; the format of the test reports and the information in them is fully specified in the legislation. There is provision in the EU type approval process for 'virtual testing', i.e. calculations and computer simulations that can demonstrate whether a vehicle, a system, a component or a separate technical unit fulfils the technical requirements. A virtual method does not require the use of a physical vehicle, system, component or separate technical unit and thus reduces costs for manufacturers because there is no need to build prototypes for the purposes of type approval. Virtual testing allows a manufacturer to conduct its own tests for approval (which implies a designation as a Technical Service), which must provide the same level of confidence in the results as a physical test.

The performance required by EU legislation for road vehicle braking systems is based on the stopping distance and/or the mean fully developed deceleration (MFDD), which together represent one of the most important aspects of active vehicle safety, namely the ability of the vehicle to stop quickly. It is required that the performance of a braking system is determined by measuring the stopping distance in relation to the initial speed of the vehicle and/or by measuring the MFDD during the test. The braking performance tests required for type approval are performed on vehicles, components and separate technical units that are representative of the type to be approved. Type approval is usually completed for the 'worst case' vehicle, i.e. a variant or version from the vehicle specification (which may be hypothetical) that represents the type to be approved under the worst conditions for a specific test. The decisions and the justification for them must be recorded in the approval documentation. The manufacturer may also select, in agreement with the Approval Authority or Technical Service, a vehicle, a system, a component or a separate technical unit which, while not representative of the type to be approved, combines unfavourable features with regard to the required level of performance. Virtual testing methods may also be used to aid decision-making during the selection process. By considering the worst case from a vehicle range for all of the test parameters the number of tests required and the number of vehicles needed can be minimised.

Table 8.2 suggests the main factors that might be considered when selecting the vehicle to be tested for type approval on the braking system. It is not possible to specify a worst case for all possible kinds of brake tests for vehicle approval, brake line integrity, ABS, rig testing and different vehicle types, so in some cases all variants of equipment or parameters have to be tested while in other cases only the worst case or the extremes (minimum or maximum) have to be considered.

UN Regulation 13H (Category M_1 and N_1 Vehicles)

Road vehicle brakes must comply with the requirements of UN Regulation 13H (UNECE, February 2014) in normal use, including being resistant to deterioration, e.g. by wear and corrosion, and the effects of magnetic or electrical fields. The braking system must provide the basic functions summarised below, and the reader is referred to the Regulation documents for full details (UNECE, February 2014).

- Service braking. To control the movement of the vehicle and to bring it safely and quickly to rest under all conditions of operation in a 'graduable' action by the driver without the removal of his/her hands from the steering control. Automatic wear adjustment is required for the service braking system.
- Secondary braking. If a failure occurs in the service braking system, it must be possible by application of the service brake control to bring the vehicle safely to rest within a reasonable distance under all conditions of operation in a 'graduable' action. The driver must be able to obtain this braking action from the driving seat without the removal of his/her hands from the steering control. It is assumed that not more than one failure of the service braking system can occur at one time.
- Park braking. To hold the vehicle stationary on an up or down gradient even in the absence of the driver, with the brakes held in the locked position by a purely mechanical device. Again, the driver must be able to achieve this braking action from the driving seat.

Table 8.2: Worst Case Parameter Selection for EU Vehicle Type Approval (Gaupp, 2013)

Parameter	Worst Case Selection
Wheelbase	The shortest wheelbase maximises braking weight transfer effects under dynamic testing. Additional actuation transmission lines are included to simulate the longest reaction time and air usage or fluid consumption.
Vehicle mass and centre of gravity	The heaviest laden vehicle is the worst case for stopping distances and fade tests. The unladen case usually gives the highest front to rear weight ratio. The highest centre of gravity gives greatest braking weight transfer.
Speed (maximum)	Any vehicle from the range that is capable of achieving 80% of the maximum design speed of the fastest vehicle in both laden and unladen conditions. The selected vehicle must be capable of achieving the Type-0 (engine connected) test speeds and the necessary acceleration for the Type-I test (explained later) of the fastest variant in the range.
Number of axles	All variants.
Foundation brakes	The vehicle with the greatest braking thermal load, e.g. smallest brake, solid disc, small brake diameter, worst cooling, maximum wheel dynamic (rolling) radius with respect to deceleration/brake force and minimum with respect to unladen wheel lock and fade tests (Type-I, Type-II and Type-III). Drum brakes are more critical than disc brakes with respect to temperature, friction material durability and brake balance (side-to-side of the axle). The extremes are the lowest vehicle weight with the largest brake and the heaviest vehicle with the smallest brake. For commercial vehicles with pneumatic actuation, automatic adjusters are considered a worse case than manual adjustment. For many tests the requirement is that brakes shall be closely adjusted, which requires overriding the automatic adjuster to optimise the actuator stroke.
Front/rear braking ratio X_1/X_2	With respect to secondary and residual braking performance.
Brake force distribution	Unfavourable brake force distribution, e.g. associated with the wheelbase, centre of gravity position and adhesion utilisation at each axle.
Brake linings and pads	All variants must be tested, but some fade testing can be omitted if the vehicle manufacturer has an approval report for the friction material performance measured, e.g. on an inertia dynamometer.
Brake transmission components	All variants, but where alternative makes of valve are used the worst case selection can be made from comparisons of valve manufacturers' data or actual tests.
Compressor (pneumatic systems)	Lowest output performance.

Table 8.2: Worst Case Parameter Selection for EU Vehicle Type Approval (Gaupp, 2013)—cont'd

Parameter	Worst Case Selection
Lever length, chamber size, brake actuation transmission lines (pneumatic systems)	The combination that gives the greatest air usage per stroke or maximum pressure reduction time at the ABS control valve.
Reservoir capacity (pneumatic systems and vacuum servo in hydraulic systems)	Largest for charging time, smallest for depletion and dynamic tests. For pneumatic systems, the smallest ratio of reservoir volume to air usage per application.
Engine	Worst case with reference to weight distribution. Lowest vacuum performance (static pedal effort to line pressure can be used to establish this). Lowest engine speed for compressed air build-up times (if different compressors are to be covered). Minimum engine braking effects, but if this is not compatible with the speed and acceleration requirements of the test, an additional vehicle may be required. Lowest engine power output with respect to Type-II test.
Transmission	Manual or automatic gearbox, which contributes to the lowest engine braking. Type approval fade tests (Type-I and Type-II) are completed with the vehicle in gear so transmission contributes to the overall retardation of the vehicle. However, commercial vehicles commonly have many gearbox and final drive options so in practice one way to cover all options is to carry out the fade test heating procedure in neutral as a worst case.
Overall transmission ratio	The combination that contributes the least engine braking effect at the critical speeds for the Type-I and -II tests. If this is not compatible with the speed and acceleration requirements, an additional vehicle may be required.
Suspension	With commercial vehicles, normally rubber suspension is a worst case, being stiffer with less deflection and less stability. Leaf springs give the next stiffest suspension and air suspension is usually the most compliant and stable, although the associated requirement for more auxiliary air should be considered in terms of air build-up time for static tests. All significant variants that may affect adhesion, weight distribution or the control of load-sensing devices. Multi axle suspension, i.e. the lowest ratio of the brake pressures required to lock the wheels of the unladen vehicle on a good surface when all brakes are in operation.

The service, secondary and park braking systems may have common components provided that there are at least two controls, independent of each other and readily accessible to the driver from the normal driving position provided that the requirements for secondary braking are met in the event of a failure. The service braking system control must be independent of the park braking system control and all brake controls should return to the fully off position when released except for the park brake control when it is mechanically locked in an applied position. The park braking system must be designed so that it can be actuated when the vehicle is in motion and may include the partial actuation of the service braking system through an auxiliary control, i.e. the electric park brake (EPB) switch. (For comparison, UN Regulation 13 states 'If the service braking system and the secondary braking system have the same control, the park braking system shall be so designed that it can be actuated when the vehicle is in motion. This requirement shall not apply if the vehicle's service braking system can be actuated, even partially, by means of an auxiliary control.')

If the service braking system incorporates servo assistance, e.g. a 'booster' (see Chapter 6), which requires an energy reserve, if this fails then secondary braking must be within the driver's muscular energy capability as defined by the prescribed maximum force applied to the service brake control. If the service braking system cannot achieve the minimum required secondary braking performance without the use of assistance from a stored energy system, e.g. in a 'power hydraulic' system, a warning signal is required to activate before the stored energy is insufficient to achieve secondary braking performance after four full-stroke brake applications. However, in the event that the service braking performance cannot be reached with such a stored energy system, a red warning signal must activate. The secondary braking performance must always be guaranteed in any failure condition; the overall intention is that the vehicle can be brought safely to rest in the event of failure of any part of the vehicle, including the engine.

The primary, secondary and park braking performance requirements are detailed in the UN Regulation 13H document, and are outlined here. Tests to verify the braking performance must be carried out on the vehicle in the laden and unladen conditions at the speeds prescribed for each type of test, and if the maximum design speed of a vehicle is lower than the prescribed speed the vehicle's maximum speed is used. Where the vehicle is fitted with an electric regenerative braking system, the test requirements depend upon the category of the system (A or B). Good practice in brake experimental testing (see Chapter 9) is written into the Regulation, and the experimental conditions are closely defined to ensure that the results are as consistent as possible, e.g. there should be no wind liable to affect the results, the track or road should be level, and at the start of the tests, the tyres must be cold and inflated to the pressure prescribed for the load actually borne by the wheels when the vehicle is stationary. The tests may include up to six stops, and the general behaviour of the vehicle during braking must be checked, especially at high speed. The tests include:

- Type-0 test. Ordinary performance test with engine disconnected with cold brakes (Table 8.3(a)).
- Type-0 test. Ordinary performance test with engine connected with cold brakes (Table 8.3(a); this test is not run if the maximum speed of the vehicle is ≤125 km/h).
- Type-I test: fade and recovery test (Table 8.3(b)). The pedal force must be sufficient to attain a mean deceleration of 3 m/s^2 during every brake application; during brake applications, the highest gear ratio (excluding overdrive, etc.) must be continuously engaged, and for regaining speed after braking the maximum acceleration allowed by the engine and gearbox must be used to attain the speed v_1 in the shortest possible time. Wear adjustment must be automatic for the service brakes and must ensure effective braking and the capability for normal running after the Type-I test.

Table 8.3(a): Summary of Test Conditions and Minimum Service and Secondary Braking System Performance Requirements for Type-0 Tests for Category M_1 and N_1 Vehicles in UN Regulation 13H

		Service Braking	Secondary Braking
Type-0 test with engine disconnected	V	100 km/h	100 km/h
	$s \leq$	$0.1V + 0.0060V^2$ (m)	$0.1V + 0.0158V^2$ (m)
	$J_m \geq$	6.43 m/s^2	2.44 m/s^2
Type-0 test with engine connected	V	80% $V_{max} \leq$ 160 km/h	
	$s \leq$	$0.1V + 0.0067V^2$ (m)	
	$J_m \geq$	5.76 m/s^2	
	F_d	6.5–50 daN	6.5–50 daN

Table 8.3(b): Summary of Test Conditions and Minimum Service and Secondary Braking System Performance Requirements for Type-I Tests for Category M_1 and N_1 Vehicles in UN Regulation 13H

Type-I Test (Fade and Recovery Test)	V_1 (km/h)	V_2 (km/h)	Number of Brake Applications (n)
	80% $V_{max} \leq$ 120 km/h	$0.5V_1$	15

V = vehicle road (test) speed (km/h).
s = stopping distance (m).
J_m = mean fully developed deceleration (m/s^2).
F_d = Force applied by the driver to the brake pedal, also termed force applied to foot control (daN).
V_{max} = maximum vehicle road speed (km/h).

- The Type-I test must be followed by a hot performance test (Type-0 engine disconnected) and recovery, which can include further reconditioning of the brake linings before a second cold performance (Type-0) test is made to compare with that achieved in the hot test. The recovery performance must not be less than 60% of the original Type-0 test with the engine disconnected.
- Immediately after the hot performance test, a recovery test must be completed: four stops from 50 km/h with the engine connected, at a mean deceleration of 3 m/s^2, allowing an interval of 1.5 km between the start of successive stops having immediately accelerated at maximum rate to 50 km/h after each stop.

The service braking system performance is verified using these tests, and then the performance of the secondary braking system is tested using the Type-0 test with the engine disconnected. The test conditions and minimum service and secondary braking system performance requirements for category M_1 and N_1 vehicles in UN Regulation 13H are summarised in Table 8.3(a). The response time of a hydraulic braking actuation system must not exceed 0.6 s.

Stopping distance (s) is defined as the distance travelled from the moment the brake pedal is actuated until the vehicle comes to rest. MFDD (J_m) is calculated over part of the brake application as the deceleration averaged with respect to distance over the interval V_b to V_e, according to:

$$J_m = \frac{\left(V_b^2 - V_e^2\right)}{25.92\left(s_g - s_b\right)} \text{ m/s}^2 \tag{8.1}$$

where:

V_1 = initial vehicle road speed (km/h);
V_b = vehicle speed at $0.8V_1$ (km/h);
V_e = vehicle speed at $0.1V_1$ (km/h);
s_b = distance travelled between V_1 and V_b (m);
s_e = distance travelled between V_1 and V_e (m).

The use of MFDD to determine braking performance is valuable because it is easily measured and thus provides an objective braking force value, e.g. to prescribe minimum performance requirements for periodical technical inspection (PTI). Stopping distance is valuable because measuring from the instant the brake control is applied includes the distance travelled during the system response phase. Additionally, variations in response time may affect stopping distance when deceleration remains constant. Prior to the addition of stopping distance into the European braking legislation, only the time-related deceleration was used, but this is not now regarded as an objective measure of the MFDD because by averaging the deceleration over time, the effect of a low deceleration at high speed at the beginning of the braking event is proportionally under-weighted and a high deceleration at low speed at

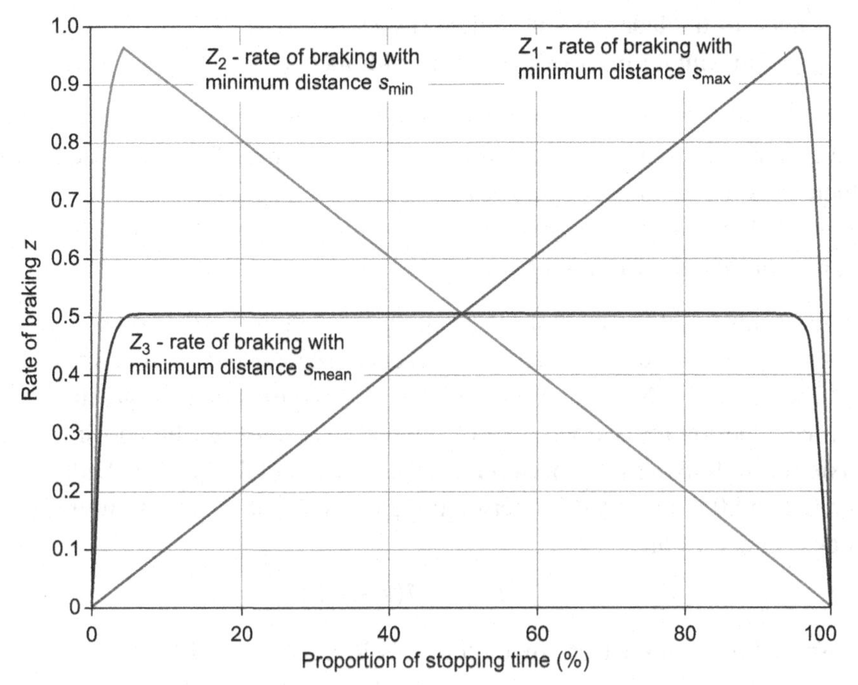

Figure 8.1: Different Examples of Rate of Braking vs. Time (Gaupp, 2013).

the end of the braking is proportionally over-weighted, as illustrated in Figure 8.1. This demonstrates in theory that the three rate-of-braking curves z_1, z_2 and z_3 result in very different stopping distances, although their time-averaged decelerations are all 50% g:

z_1: the braking force is linearly increasing until the vehicle comes to rest, which gives the longest stopping distance.
z_2: the braking force is linearly decreasing until the vehicle comes to rest, e.g. because of fade, which gives the shortest stopping distance.
z_3: the braking force is constant until the vehicle comes to rest, which gives a stopping distance between z_1 and z_2.

The minimum Type-0 service braking performance of a motor vehicle towing an unbraked trailer must not be less than 5.4 m/s^2 in both laden and unladen conditions, and must be verified by calculation using:

$$J_{M+R} = J_m \frac{M_{laden}}{M_{trailer\ max}} \tag{8.2}$$

where:

J_{M+R} = the calculated MFDD of the motor vehicle coupled to an unbraked trailer (m/s^2);
J_m = maximum MFDD of the solo motor vehicle achieved during the Type-0 test with engine disconnected (m/s^2);

M_{laden} = mass of the laden motor vehicle (kg);
$M_{trailer\ max}$ = maximum mass of the unbraked trailer.

The park braking system must be capable of holding the laden vehicle stationary on a 20% up or down gradient. If coupled to an unbraked trailer, the park braking system of the motor vehicle must be capable of holding the combination stationary on a 12% up or down gradient. Where the park brake control is manual, the operating force must not exceed 40 daN, or 50 daN if it is foot operated.

One of the most important features of UN braking legislation is adhesion utilisation and basic braking stability based on wheel-lock sequence (see Chapter 3). For vehicles that are not equipped with ABS, UN Regulation 13H Annex 5 requires that for all states of load of the vehicle, the adhesion utilisation curve of the rear axle must not lie above the curve for the front axle for all braking rates between 0.15 and 0.8, and for tyre/road adhesion values (k) between 0.2 and 0.8. For k values between 0.2 and 0.8, the required minimum rate of braking is determined from:

$$z \geq 0.1 + 0.7(k - 0.2) \tag{8.3}$$

(This is illustrated in diagram 1 of Annex 5 of UN Regulation 13H.)

Braking stability is verified by a wheel-lock sequence test in the laden and unladen conditions with the engine disconnected at a test road speed of 65 km/h for $z \leq 0.50$ or 100 km/h for $z > 0.50$. Neither rear wheel must lock before both front wheels lock for braking rates between 0.15 and 0.8, as indicated by:

1. No wheels lock
2. Both wheels on the front axle and one or no wheels on the rear axle lock
3. Both axles simultaneously lock.

If wheel lock-up commences at $0.15 > z > 0.8$ the test is invalid and should be repeated on a different road surface. If the test indicates that the rear wheels do lock before the front wheels, additional tests are required. If a valve or similar device is incorporated into the vehicle's braking actuation system to achieve braking stability, in the event of its failure it must be possible to stop the vehicle under the conditions of the Type-0 test with the engine disconnected to give a stopping distance $\leq 0.1V + 0.0100V^2$ (m) and MFDD $\geq$ 3.86 m/s^2.

For vehicles fitted with ABS (UN Regulation 13H specifies three categories based on the control strategy; see Chapter 11) the adhesion utilisation requirement as described above and in Equation (8.3) is not applicable since the ABS is designed to avoid wheel lock. However, in the event of a single electrical functional failure that only affects the ABS function, the subsequent service braking performance must not be less than 80% of the prescribed performance according to the Type-0 test with the engine disconnected. This corresponds to a stopping distance of $0.1V + 0.0075V^2$ (m) and an MFDD of 5.15 m/s^2. A test procedure for evaluating ABS braking performance based on UN Regulations 13H and 13 requirements is described in Chapter 11. Additional requirements on ABS braking systems include that wheels must not lock when full braking actuation force is suddenly applied, or when the vehicle passes from a high-adhesion surface to a low-adhesion surface while braking. Similarly, when the vehicle passes from a low-adhesion surface to a high-adhesion surface while braking, the vehicle deceleration must rise to the appropriate high value within a 'reasonable' time.

UN Regulation 13H also allows for a modification of vehicle type resulting from the fitting of alternative friction material (brake linings) to vehicles that have already been type approved by dynamometer testing and comparison of the alternative friction material without the need to test the whole vehicle. Five sample sets of the brake lining are required to be compared with five sets of linings conforming to the original components in the first type approval. The inertia dynamometer equipment and set-up are defined, and the brake lining equivalence is based on a comparison of the results from the Type-0 cold performance test and the Type-I fade test procedures summarised below. Brake linings must also be visually inspected on completion of the tests to check that they are in satisfactory condition for continued use in normal service.

- Type-0 cold performance test. Three brake applications with the initial temperature below 100°C from an initial rotational speed and at a mean torque equivalent to the deceleration in the Type-0 test with the engine disconnected (see Table 8.3(a)). Additional tests are carried out at rotational speeds ranging from 30% to 80% V_{max}.
- Type-I fade test. According to the vehicle Type-I procedure for hot performance, when the mean braking torque must be within ±15% of that recorded with the original brake linings conforming to the component.

The final part of UN Regulation 13H relates to special requirements to be applied to the safety aspects of complex electronic vehicle control systems (see Chapter 11).

UN Regulation 13 (Category M, N and O Vehicles)

As with UN Regulation 13H, road vehicle brakes must comply with the requirements of UN Regulation 13 (UNECE, March 2014) in normal use, and must provide the basic functions summarised previously for UN Regulation 13H in terms of service braking, secondary braking and park braking with the addition of residual braking (in UN Regulation 13H requirements for residual braking performance are not explicitly stated, being part of the secondary braking performance requirements). The braking must be controlled by the driver from the driving seat, and must be graduable, including the braking of the trailer in a vehicle combination. In the event of failure of the towing vehicle's service braking system, where that system consists of at least two independent parts, the part or parts not affected by the failure must be capable of partially or fully actuating the brakes of the trailer. Again, the reader is referred to the UN Regulation 13 document for full details (UNECE, March 2014). Where the braking power is derived from or assisted by a source of energy independent of the driver, UN Regulation 13 (and UN Regulation 13H) defines the braking actuation system in terms of two independent functions, namely the control transmission and the energy transmission. The control transmission refers to the combination of components that control the operation of the brakes, including the control function and the necessary reserve(s) of energy, and the energy transmission refers to the combination of components that supply the brakes with the necessary energy for their function, including the necessary reserve(s) of energy. The majority of commercial vehicles and their combinations with trailers utilise pneumatic or compressed air actuation systems, as explained in Chapter 6. The control transmission connections between towing vehicles and trailers can be a combination of a pneumatic supply line, a pneumatic control line and an electric control line (refer to the latest UN Regulation 13 documentation for up-to-date information (UNECE, March 2014)). These connections and their connector hardware are closely defined. Coupling force control to reduce the difference between the dynamic braking rates of the towing and towed vehicles (see Chapter 6) is only permitted in the towing vehicle and is checked at the time of type approval. The braking performance requirements for commercial vehicles are detailed in the UN Regulation 13 document, and were introduced in Chapter 6. Service, secondary and residual braking performance are defined below, and the reader is referred to the UN Regulation 13 document for full details (UNECE, March 2014).

- 'Service braking performance'. The defined performance is required when no faults are present and must be achieved when the driver has both hands on the steering control.
- 'Secondary braking performance'. The defined performance is required when a 'single' defect is present and must be achieved when the driver has at least one hand on the steering control.
- 'Residual braking performance'. In UN Regulation 13 this is defined as the braking performance required when there is a failure within one service braking circuit (service braking performance is when no faults are present and secondary braking performance is when a single defect is present).

Tests to verify braking performance for EU type approval must be carried out in the laden and unladen conditions at the speeds prescribed for each type of test. In the laden condition the distribution of mass among the axles must be that stated by the manufacturer. Where several arrangements of the load on the axles is possible, the distribution of the maximum mass among the axles must ensure that the load on each axle is proportional to the maximum permissible load for each axle. The prescribed speeds in the tests depend on the vehicle category, as indicated in Tables 8.4, 8.5 and 8.6. Where the vehicle is fitted with an electric regenerative braking system, the test requirements depend

Table 8.4(a): Service and Secondary Braking Performance Requirements for Solo Vehicles (Categories M and N) Based on UN Regulation 13 (Type-0 Test with Engine Disconnected; Refer to the UN Regulation 13 Document for Full Details)

	Service		Secondary			
	N_1	M_2, M_3, N_2 and N_3	M_2 and M_3	N_1	N_2	N_3
Initial vehicle road speed, V_1 (km/h)	80	60	60	70	50	40
Stopping distance, s (m)	$0.15V + (V^2/130)$		$0.15V + (2V^2/130)$	$0.15V + (2V^2/115)$		
	≤ 61.2	≤ 36.7	≤ 64.4	≤ 95.7	≤ 51.0	≤ 33.8
MFDD (m/s²)	5		2.5		2.2	
Control force, F_d (daN)	≤ 70		≤ 70			

upon the category of the system (A or B). A brief overview of some of the tests for motor vehicle categories covered in UN Regulation 13 is provided below (refer to the Regulation 13 documentation for full details (UNECE, March 2014)):

- Type-0 test. Ordinary performance test with engine disconnected (see Table 8.4) with cold brakes.
- Type-0 test. Ordinary performance test with engine connected (see Table 8.5) with cold brakes.

Table 8.4(b): Residual Braking Performance Requirements for Solo Vehicles (Categories M and N) Based on UN Regulation 13 (Type-0 Test with Engine Disconnected; Refer to the UN Regulation 13 Document for Full Details (UNECE, March 2014))

	Residual				
	M_2	M_3	N_1	N_2	N_3
Initial vehicle road speed, V_1 (km/h)	60	60	70	50	40
Laden Stopping distance, s (m)	$0.15V + ((100/30) \times (V^2/130))$		$0.15V + ((100/30) \times (V^2/115))$		
	≤ 101.3	≤ 101.3	≤ 152.5	≤ 80.0	≤ 52.4
Laden MFDD (m/s²)	1.5	1.5	1.3	1.3	1.3
Unladen Stopping distance, s (m)	$0.15V + ((100/25) \times (V^2/130))$	$0.15V + ((100/30) \times (V^2/130))$	$0.15V + ((100/25) \times (V^2/115))$		$0.15V + ((100/30) \times (V^2/115))$
	≤ 119.8	≤ 101.3	≤ 180.9	≤ 94.5	≤ 52.4
Unladen MFDD (m/s²)	1.3	1.5	1.1	1.1	1.3
Control force, F_d (daN)	≤ 70				

Table 8.5: Service Braking Performance Requirements for UN Regulation 13 Type-0 Test with Engine Connected, for Category M and N Vehicles (Refer to Regulation 13 Document for Full Details (UNECE, March 2014))

		Category				
	Type of Test	M_2 0, I	M_3 0, I, II, or IIA	N_1 0, I	N_2 0, I	N_3 0, I, II
Type 0 test with engine connected	$V = 0.80V_{max}$ but not exceeding:	100 km/h	90 km/h	120 km/h	100 km/h	90 km/h
	Stopping distance (s)	$\leq 0.15V + \frac{V^2}{103.5}$				
	J_m	$\geq 4.0\ m/s^2$				
	Control force (F_d)	≤ 700 N				

The Type-0 braking performance of a trailer (category O_2 and O_3) equipped with a pneumatic braking system can be calculated either from the braking rate of the towing vehicle plus the trailer and the measured thrust on the coupling or, in certain cases, from the braking rate of the towing vehicle plus the trailer with only the trailer being braked. The engine of the towing vehicle is disconnected during the braking test. The service brakes of category O_2 and O_3 trailers are tested under continuous (drag) braking with the vehicle laden and the energy input to the brakes equivalent to that recorded in the same period of time with a laden vehicle driven at a steady speed of 40 km/h on a 7% down gradient for a distance of 1.7 km (refer to the UN Regulation 13 document for full details (UNECE, March 2014)).

- Type-I test. Fade test with repeated braking (see Table 8.6) followed by a hot performance test (measured at the same conditions as the Type-0 test), which must demonstrate braking performance not less than 80% of that prescribed for the motor vehicle category, or less than 60% of the Type-0 test performance with the engine disconnected. For trailers, the hot brake force at the periphery of the wheels when tested at 40 km/h

Table 8.6: Test Conditions for UN Regulation 13 Type-I Test (Fade Test with Repeated Braking), for Category M and N Vehicles (Refer to Regulation 13 Document for Full Details (UNECE, March 2014))

	Conditions			
Type-I Test (Fade and Recovery Test) Vehicle Category	V_1 (km/h)	V_2 (km/h)	Time Interval Δt (s)	Number of Brake Applications (n)
M_2	80% $V_{max} \leq 100$ km/h	$0.5V_1$	55	15
N_1	80% $V_{max} \leq 120$ km/h	$0.5V_1$	55	15
M_3, N_2, N_3	80% $V_{max} \leq 60$ km/h	$0.5V_1$	60	20

Table 8.7: Braking Performance Requirements for Trailers Based on EC Regulation 13 (Refer to Regulation 13 Document for Full Details (UNECE, March 2014))

	Semi-Trailer	Full (Chassis) Trailer	Centre Axle Trailer
	Minimum Braking Force is Based on a Percentage of the Static Axle/Bogie Load		
Initial vehicle road speed, V_1 (km/h)	60	60	60
Braking force (%)	≥45	≥50	≥50
Maximum coupling head pressure (bar)	6.5	6.5	6.5
Supply line pressure (bar)	7.0	7.0	7.0

must not be less than 36% of the maximum stationary wheel load, or less than 60% of the figure recorded in the Type-0 test at the same speed (see Table 8.7). The service brakes of trailers of categories O_2 and O_3 can be tested laden under drag braking conditions with energy input to the brakes equivalent to a laden vehicle driven at 40 km/h down a 7% gradient for 1.7 km.

- Free-running test. For motor vehicles equipped with automatic brake adjustment devices, after completing the Type-I tests and allowing the brakes to cool it must be verified that the vehicle is capable of free running by checking that the wheels can be rotated by hand, or that rotor temperatures do not increase by more than 80°C when the vehicle is driven at a constant speed of 60 km/h with the brakes released.
- Type-II test (downhill behaviour test). Laden motor vehicles are tested with an energy input equivalent to that recorded in the same period of time with a laden vehicle driven at an average speed of 30 km/h on a 6% down gradient for a distance of 6 km, with the appropriate gear engaged and the endurance braking system (if fitted) being used. At the end of the test, the hot performance of the service braking system as measured in a Type-0 test with the engine disconnected (control force < 70 daN) should be:
  - Category M_3: $s \leq 0.15V + (1.33V^2/130)$ (the second term corresponds to an MFDD $J_m = 3.75$ m/s^2).
  - Category N_3: $s \leq 0.15V + (1.33V^2/115)$ (the second term corresponds to an MFDD $J_m = 3.3$ m/s^2).
- Type-IIA test (endurance braking performance). The endurance braking system of category M_3 long-distance motor coaches and category N_3 vehicles authorised to tow a category O_4 trailer is tested with an energy input equivalent to that recorded in the same period of time with a laden vehicle driven at an average speed of 30 km/h on a 7% downhill gradient for a distance of 6 km.
- Type III test (fade test for vehicles of category O_4). Twenty brake applications from 60 km/h each of 60 s duration, with the control force required to attain an MFDD of 3 m/s^2 at the

first brake application (refer to the UN Regulation 13 document for full details (UNECE, March 2014)). At the end of the test, the service braking hot performance is measured under the same conditions as the Type-0 test (with different temperature conditions and starting from an initial speed of 60 km/h) and the hot brake force at the periphery of the wheels must not be less than 40% of the maximum stationary wheel load, and not less than 60% of that in the Type-0 test at the same speed. This is followed by a free-running test.

Secondary braking performance may be achieved by means of the footbrake control or a hand-operated control. If the secondary braking performance is achievable by the footbrake control this also satisfies the requirements for residual braking performance. Referring to Table 8.7, compensation is made for the unbraked mass of the towing vehicle when determining the trailer braking performance, and no general requirements are defined for residual or secondary performance. However, 30% of the prescribed performance is required when certain specific failures are present, namely ABS failure or electric control transmission failure. Prescribed performance, whether service, secondary or residual, must be achieved without wheel lock. For trailers that are equipped with a single circuit braking system (note that in the case of trailers with EBS the electrical transmission cannot be a purely 'single circuit braking system'), two additional requirements apply:

- 'Automatic braking performance'. Braking performance of 13.5% of the maximum laden static axle load is required when a failure in the supply line is present.
- 'Auxiliary equipment'. The braking system may supply air for the function of auxiliary systems, e.g. air suspension, but the braking system must be protected so that in the event of a failure within the auxiliary system the braking system must be able to achieve a performance of 80% of that prescribed for the specific vehicle.

As explained in Chapter 6, the braking performance of a vehicle combination is not evaluated for EU type approval by testing the full vehicle combination, but by testing the towing vehicle and the trailer independently with appropriate allowance for the load transfer of the trailer on to the towing vehicle in the case of tractors for semi-trailers. UN Regulation 13 incorporates requirements to represent the dynamics of a vehicle combination when braking; e.g. an 'unladen' combination of tractor (with driver) and unladen semi-trailer is represented by a solo tractor with an additional mass located at the kingpin equivalent to 15% of the maximum allowable static load on the kingpin. A laden combination is a tractor (with driver) coupled to a laden semi-trailer, where the dynamic load of the semi-trailer on the tractor is represented by an additional mass M_s mounted at the fifth wheel coupling, which is determined from Equation (8.4), and the height (h_{am}) of the centre of gravity of the additional mass is determined from Equation (8.5). The effect of Equations (8.4) and (8.5) is therefore to simulate the effect on the tractor of the vertical and horizontal forces imposed on the fifth wheel coupling by a laden semi-trailer; both M_s and h_{am} increase as z increases. It should be noted that any retarding force produced by an endurance brake (retarder) is not taken into consideration when defining the compatibility performance.

$$M_s = M_{so}(1 + 0.45z) \tag{8.4}$$

$$h_{am} = (hM + h_3M_s)/(M + M_s) \tag{8.5}$$

where:

> M_{so} represents the difference between the maximum laden mass of the tractor and its unladen mass (M), i.e. the maximum allowable static load on the kingpin;
> h is the height of centre of gravity of the solo tractor;
> h_3 is the height of the coupling (kingpin or fifth wheel);
> M is the unladen mass of the solo tractor.

Requirements for the distribution of braking between vehicles' axles, and requirements for compatibility between towing vehicles and trailers, are detailed in UN Regulation 13. First, the 'threshold' actuation must avoid erratic brake operation at low actuation pressures, and it is required that the development of braking force on the axles of each independent axle group is within the pressure ranges summarised below:

(a) Laden vehicles. At least one axle must start to develop a braking force when the pressure at the coupling head is 20–100 kPa, and at least one axle in every other axle group must start to develop a braking force when the pressure at the coupling head is $\leq$120 kPa.
(b) Unladen vehicles. At least one axle must start to develop a braking force when the pressure at the coupling head is 20–100 kPa.

UN Regulation 13 requires that for all states of loading of two-axle motor vehicles that are not equipped with ABS, the rate of braking shown in Equation (8.6) must be achieved for k values between 0.2 and 0.8. (A tractor without ABS that is designed to operate with a semi-trailer would also need to meet the adhesion utilisation requirements outlined above when operating solo.) For all states of load of the vehicle, the adhesion utilisation curve of the rear axle must not lie above the curve for the front axle, but with qualifications depending on the vehicle category. This means that the requirements for adhesion utilisation and wheel-lock sequence depend upon the type of vehicle. These relate to the state of load of the vehicle, and include for example that in the case of category N_1 vehicles with a laden/unladen rear axle loading ratio not exceeding 1.5 or having a maximum mass of less than 2 tonnes, in the range of z values between 0.3 and 0.45, an inversion of the adhesion utilisation curves is permitted provided that the adhesion utilisation curve of the rear axle does not exceed the line defined by $k = z$ by more than 0.05 (UN Regulation 13 Annex 10 para 3.1.2.1 for two-axled vehicles). This is the line of ideal adhesion utilisation shown in Diagram 1A of Annex 10 (see Figure 8.2). Refer to UN Regulation 13 (UNECE, March 2014) for full information.

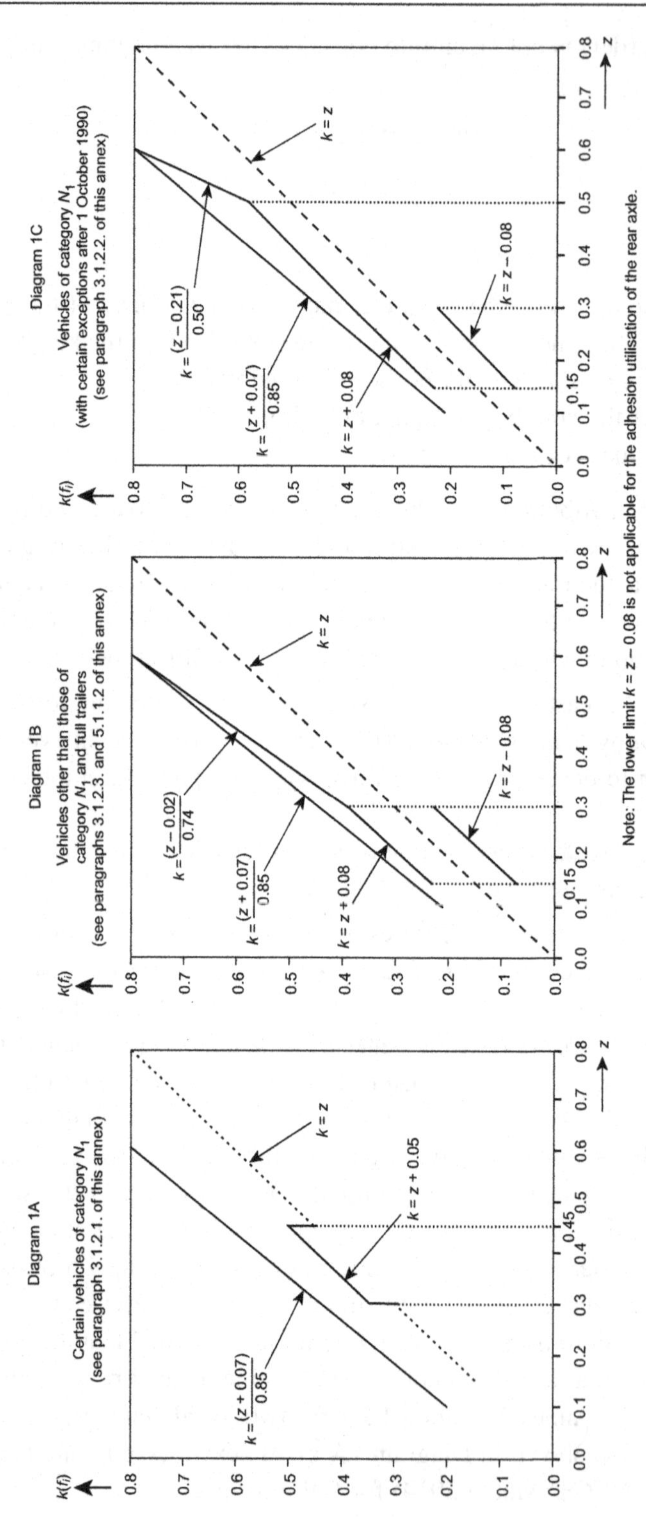

Figure 8.2: Diagrams 1A to 1C of Annex 10 of UN Regulation 13 (UN, March 2014).

$$z > 0.10 + 0.85(k - 0.20) \quad (8.6)$$

For category O_2, O_3 and O_4 trailers, the total braking forces at the tyre/road interface of the wheels must be at least the proportion of the maximum static normal wheel load as indicated below (category O_1 trailers are generally unbraked but if they are equipped with a braking system they must fulfil the requirements of a category O_2 trailer):

Full trailer, laden and unladen:	50%
Semi-trailer, laden and unladen:	45%
Centre-axle trailer, laden and unladen:	50%

As discussed in Chapter 6, the braking forces developed by the towing vehicle and the trailer of a commercial vehicle combination should be balanced so that the towing vehicle is not braking the trailer and vice versa. Compatibility between a towing vehicle and its trailer is therefore important. For safety reasons the aim should be that the adhesion utilisation at each axle of the vehicle combination is the same, and there is no significant delay in the response time. The brake compatibility requirements in UN Regulation 13 ensure that the total adhesion utilisation of the vehicle is within upper and lower bounds over a range of the pressure at the coupling head of the control line as explained below. For commercial reasons the brake compatibility requirements are intended to avoid badly matched vehicle combinations that can have poor braking performance and create in-service problems of under-used ('glazed') or over-used ('denatured') brake linings. UN Regulation 13 includes coupling force control to reduce the difference between the dynamic braking rates of towing and towed vehicles; the action of the coupling force control is only permitted to reduce the absolute coupling force values. Any coupling forces resulting from endurance braking systems must not be compensated by the service braking system of either the motor vehicle or trailer. Thus the coupling force control only takes account of the deceleration from the service braking system, and the retarding force from the endurance braking system therefore increases the coupling forces for a given combination deceleration. This avoids a potentially dangerous situation that might occur, e.g. with a commercial vehicle and trailer combination during a long downhill descent if the towing vehicle used the endurance brake while the trailer used the foundation brakes. In the event of an emergency brake application the combination of cold towing vehicle brakes and hot trailer brakes could result in a dangerous 'jack-knife' instability due to the over-utilisation of the towing vehicle's rear brakes. Vehicles equipped with an integrated endurance braking system must also be equipped with ABS acting at least on the service brakes of the endurance braking system's controlled axle and on the endurance braking system itself. This means that either the ABS controls the braking force generated by the endurance brake or switches it off during a brake-induced ABS intervention.

To ensure compatibility and interchangeability between different towing vehicles and trailers, UN Regulation 13 requires that the ratio of the total braking force developed at the wheels of the towing vehicle to its weight (defined in UN Regulation 13 as T_m/P_m), and the ratio of the total braking force developed at the wheels of the trailer to its weight (defined in UN Regulation 13 as T_r/P_r), when plotted against the coupling head pressure, must lie within certain defined bands that are calculated for each type of trailer to take account of dimensional and load differences. However, for tractor/semi-trailer combinations, UN Regulation 13 acknowledges that the semi-trailer braking forces should not be sufficient to decelerate the total mass of the semi-trailer but only the load supported by the wheels of the semi-trailer. The load imposed at the coupling is effectively decelerated by the towing vehicle. For example, comparing a short semi-trailer having a high centre of gravity with a long low semi-trailer having the same static axle load, the dynamic weight transfer from the axles on to the fifth wheel coupling will be greater for the short semi-trailer, which should therefore require less braking force from its brakes at any given deceleration and coupling head air pressure. The calculated compatibility band for the short, high semi-trailer will therefore be lower than for the long, low semi-trailer. For all towing vehicles, only the laden case need be considered if approved ABS is fitted. The compatibility bands for a tractor/semi-trailer combination are illustrated in Chapter 6 (Figure 6.22), and the reader is referred to Figures 2, 3 and 4 of Annex 10 of UN Regulation 13 for the full range of compatibility bands for commercial vehicles and their trailers.

UN Regulation 13 (and 13H) requires that the service braking system 'shall distribute its action appropriately among the axles'. ABS is now effectively standard on most road vehicles, being intended to be used to avoid locking of the wheels in extreme braking conditions, but as previously explained (see Chapter 4) any road vehicle should have a fundamentally sound braking distribution for all loading conditions. Most new road vehicles are exempt from the adhesion utilisation requirements of UN Regulation 13 (Annex 10) relating to the distribution of braking among the axles of vehicles and requirements for compatibility between towing vehicles and trailers because they are equipped with ABS. So vehicle manufacturers do not have to submit adhesion utilisation curves for type approval when the vehicle is equipped with ABS that fulfils the requirements of Annex 13 of UN Regulation 13. However, for vehicles towing a category O_2 trailer, paragraph 1.1 of Annex 13 does require them to meet the requirements for unladen compatibility in Annex 10. For commercial vehicles, especially in combination with a trailer or semi-trailer, the ability to adjust the braking distribution (see Chapter 4) is the only way to accommodate the very large variations in axle loading between the two extremes of loading, unladen to laden, and achieve a maximum efficiency braking system under all operating conditions. This means that irrespective of whether ABS is fitted or not, load sensing is necessary for most commercial vehicles. Load sensing for commercial vehicles (category N) was introduced in Chapter 4 in terms of the optimum utilisation of tyre/road adhesion when all wheels simultaneously approach their point of locking.

The park braking system of a motor vehicle must be capable of holding the laden vehicle stationary on an 18% up or down gradient. The park braking system of a motor vehicle coupled to a trailer must be capable of holding the combination of vehicles stationary on a 12% up or down gradient. If coupled to an unbraked trailer, the park braking system of the motor vehicle must be capable of holding the combination stationary on a 12% up or down gradient. Where the park brake control is manual, the operating force must not exceed 60 daN, or 70 daN if it is foot operated, and a park braking system that has to be actuated several times (e.g. a ratchet device) to achieve the prescribed performance is admissible. For trailers that are required to be equipped with a service braking system, park braking must be provided even when the trailer is separated from the towing vehicle and must be capable of holding the laden trailer stationary on an 18% up or down gradient. The force required to be applied to the control device must not exceed 60 daN.

UN Regulations 13H and 13, depending on vehicle category, require a minimum deceleration capability of $z = 0.66$ g ($J_m = 6.43$ m/s^2) and $z = 0.51$ g ($J_m = 5.0$ m/s^2) respectively. These values are below the tyre/road adhesion limits of most passenger car and commercial vehicle tyres respectively on normal dry road surfaces. The minimum prescribed secondary braking performance in UN Regulation 13H for a category M_1 vehicle is $J_m \geq 2.44$ m/s^2 (see Table 8.3). For commercial vehicles, the minimum prescribed residual performance of the service braking system in UN Regulation 13 in the event of failure in a part of its actuation system transmission depends on the vehicle category and load condition (laden or unladen), and varies between 1.1 and 1.5 m/s^2 (see Table 8.4b). The residual braking performance is the required minimum remaining braking performance 'under foot' when a transmission failure occurs in the brake actuation system, and this means that a two-circuit brake system is indirectly required by UN Regulation 13. In comparison, UN Regulation 13H requires higher secondary braking performance ($J_m \geq$ 2.44 m/s^2) in the event of failure in a part of its transmission. UN Regulation 13H requires a recovery test after the hot performance test, and the recovery performance must not be less than 70% or more than 150% of the performance recorded in the Type-0 test with the engine disconnected. The comparative bench test procedure 'Inertia dynamometer test method for brake linings' according to Annex 15 of UN Regulation 13 or Annex 7 of UN Regulation 13H can only be applied by the vehicle manufacturer when road performance tests have already been carried out with a vehicle. As explained in Chapter 2, the philosophy that underpins both UN Regulations 13 and 13H is that the wheels on the rear axle(s) of a road vehicle should never lock up before those on the front axle during a braking manoeuvre, thus maintaining stability. Whereas UN Regulation 13 allows this to be verified by a mathematical calculation, UN Regulation 13H requires it to be verified by a wheel-lock sequence test and if this test fails an additional torque wheel test is required.

However, this verification is not required if the vehicle is equipped with ABS, which applies to the majority of all type approved vehicles in Europe.

US Road Vehicle Braking Legislation

Braking legislation in the USA requires that manufacturers must confirm with the National Highway Traffic Safety Agency (NHTSA) that their products conform to the relevant Federal Motor Vehicle Safety Standards (FMVSS) through the process of self-certification. The NHTSA can inspect any product at any time to evaluate whether a vehicle or equipment item conforms to the performance requirements. The relevant standards include FMVSS 135 for light vehicle brake systems (US FMVSS 135), FMVSS 121 for air brake systems (performance and equipment requirements for braking systems on vehicles equipped with air brake systems) (US FMVSS 121), and FMVSS 105 for hydraulic and electric brake systems (requirements for hydraulic and electric service brake systems, and associated park brake systems) (US FMVSS 105). Specific components are also covered, e.g. FMVSS 106 for hydraulic brake hoses. The self-certification process is different from the EU type approval system but has much the same purpose to 'insure safe braking performance under normal and emergency driving conditions' (US FMVSS 135, US FMVSS 121, US FMVSS 105). FMVSS 135 applies to passenger cars (manufactured on or after September 2000) and to multi-purpose passenger vehicles, trucks and buses with a 'gross vehicle weight rating' (GVWR) $\leq$ 3.5 tonnes (manufactured on or after September 2002). FMVSS 121 applies to trucks, buses and trailers equipped with air brake systems, but with some exceptions based on size, speed and weight, and FMVSS 105 applies to multi-purpose passenger vehicles, trucks and buses with a GVWR $\geq$ 3.5 tonnes that are equipped with hydraulic or electric brake systems. The classes of road vehicles specified by the US Department of Transportation, Federal Highway Administration (FHWA 13) are outlined in Table 8.8. These classes can be compared with the EU vehicle categories summarised earlier, noting the following differences:

- EU categories for passenger-carrying vehicles are dependent on vehicle mass and number of seats.
- All EU categories of commercial vehicles and combinations are dependent on vehicle mass.
- Vehicle combinations of more than one motor vehicle and one trailer of any type (FHWA classes 11, 12, 13) are not permitted in EU legislation (except in the case of special purpose vehicles).

The standards specify the test procedures that must be used for the purposes of self-certification, which include ASTM and SAE as well as procedures defined in the FMVSS document. An outline of the FMVSS 135 self-certification test procedure for passenger cars with hydraulic actuation is illustrated in Table 9.2 in Chapter 9, and Table 9.3 outlines

Table 8.8: Classes of Road Vehicles Specified by the US Department of Transportation, Federal Highway Administration (FHWA 13)

FHWA 13 Vehicle Class	Description
1. Motorcycles	Two- or three-wheeled motorised vehicles.
2. Passenger cars	Rigid vehicles from mini (>680 kg unladen mass) to heavy (>1690 kg) passenger cars manufactured primarily for the purpose of carrying passengers, sport utility vehicles (SUV), also includes pickup trucks and 'vans' (multi-purpose vehicles for the transport of more passengers than a normal passenger car). These vehicles can tow one- or two-axle trailers.
3. Pick-ups, panels, vans	Rigid vehicles with four wheels for the transport of goods and passengers ('vans'), can tow one- or two-axle trailers.
4. Buses	Full-length passenger-carrying buses with two or three axles.
5. Single-unit two-axle trucks	Rigid vehicles with two axles and six wheels (dual rear tyres) primarily for the transport of goods, but including larger pickup trucks and recreational vehicles.
6. Single-unit three-axle trucks	Rigid vehicles with three axles primarily for the transport of goods.
7. Single-unit four- or more axle trucks	Rigid vehicles with four or more axles primarily for the transport of goods.
8. Single-trailer three- or four-axle trucks	Vehicles primarily for the transport of goods with three or four axles consisting of a rigid motor vehicle (tractor) towing a semi-trailer.
9. Single-trailer five-axle trucks	Vehicles primarily for the transport of goods with five axles consisting of a rigid motor vehicle (tractor) towing a semi-trailer.
10. Single-trailer six- or more axle trucks	Vehicles primarily for the transport of goods with six axles consisting of a rigid motor vehicle (tractor) towing a semi-trailer.
11. Multi-trailer five- or less axle trucks	Vehicles primarily for the transport of goods with five or fewer axles consisting of a rigid motor vehicle (tractor) towing multiple trailers including a semi-trailer and a rigid (chassis) trailer.
12. Multi-trailer six-axle trucks	Vehicles primarily for the transport of goods with six axles consisting of a rigid motor vehicle (tractor) towing multiple trailers including a semi-trailer and a rigid (chassis) trailer.
13. Multi-trailer seven- or more axle trucks	Vehicles primarily for the transport of goods with seven or more axles consisting of a rigid motor vehicle (tractor) towing multiple trailers including a semi-trailer and a rigid (chassis) trailer.

the SAE J2522 inertia dynamometer global brake effectiveness test procedure. Table 9.4 outlines the FMVSS 121 inertia dynamometer test procedure for pneumatically actuated braking systems for commercial vehicles and trailers. Table 9.5 outlines the SAE 2430 dynamometer effectiveness characterisation test for passenger cars and light truck brake replacement friction materials (Agudelo and Ferro, 2005).

The requirements for brake system design and operation in FMVSS 135 are broadly similar to UN Regulation 13H (because they are 'harmonised'). There are some differences, e.g. vehicle braking performance tests are required to be carried out at 'lightly loaded vehicle weight' (LLVW), which is defined as the unloaded vehicle weight plus 180 kg (including driver and instrumentation), and 'gross vehicle weight rating' (GVWR), which is equivalent to GVW. The required braking performance in FMVSS 135 aligns with UN Regulation 13H (see Table 8.3), i.e. the Type-0 test with engine disconnected is comparable with the FMVSS 135 'cold effectiveness' test, which requires a stopping distance ≤ 70 m from a road speed of 100 km/h with the transmission position in neutral, equating to a calculated average deceleration of 5.51 m/s^2. If the vehicle cannot reach $V = 100$ km/h the required stopping distance performance is calculated from $s \leq 0.1V + 0.0060V^2$ (m), which is the same as in UN Regulation 13H. The FMVSS 135 test for high-speed effectiveness with the transmission position in gear (which is not run if $V_{max} \leq 125$ km/h) compares with the Type-0 test with engine connected and has the same performance requirement. Hot performance (fade) test requirements are discussed in Chapter 9; for the first hot stop, the stopping distance must be less than or equal to a calculated distance based on 60% of the deceleration actually achieved on the shortest GVWR cold effectiveness stop. In addition, for at least one of the two hot stops $s \leq 89$ m from 100 km/h, which approximates to 80% of the cold effectiveness deceleration requirement. Pedal force requirements are the same as in UN Regulation 13H. With a hydraulic circuit failure or booster failure, which may be equivalent to a partial system failure, the FMVSS 135 performance requirement is the same as for the UN Regulation 13H secondary braking requirement, namely a stopping distance of 168 m from 100 km/h. Park brake performance requirements are the same, namely 20% gradient.

For vehicles without ABS, the wheel-lock sequence requirements are that the adhesion utilisation curve for the rear axle must lie below the line defined by $z = 0.9k$ for rates of braking $0.15 \leq z \leq 0.80$. If the vehicle meets the performance requirement that three valid tests result in the front axle locking before or simultaneously with the rear axle, or the front axle locks up with only one or no wheels locking on the rear axle, the vehicle is considered to meet the adhesion utilisation requirements. If any one of the three valid tests on any surface results in the rear axle locking before the front axle or the rear axle locks up with only one or no wheels locking on the front axle, a specified torque wheel test must be completed.

The requirements for pneumatic brake system design and operation that are specified in FMVSS 121 indicate that all trucks, buses and trailers must be fitted with ABS, which on rigid vehicles directly controls the wheels of at least one front axle and the wheels of at least one rear axle of the vehicle. Wheels on other axles may be indirectly controlled by the ABS. This equates to a category 2 ABS in UN Regulation 13; towing vehicles are required to have the equivalent of a category 1 ABS. The ABS on full trailers must directly control the wheels of at least one front axle and at least one rear axle, and on semi-trailers the ABS must directly control the wheels of at least one axle. Wheels on other axles may be indirectly controlled by the ABS. This equates to a category B ABS in UN Regulation 13. Any wheel may lock at speeds below 32 km/h or at any speed depending on the axle, and above this speed 1 s duration wheel lock is permitted on certain axles. Provisions are included for warning of ABS failure, and any malfunction that affects the ABS must not increase the actuation and release times of the service brakes. Vehicles with ABS must satisfy an ABS functional failure test which, with the ABS power source disconnected, requires a stopping distance from 100 km/h of 85 m distance ≤ 70 m from a road speed of 100 km/h with the transmission position in neutral, which equates to a calculated average deceleration of 4.54 m/s^2.

The service and emergency braking performance requirements defined in FMVSS 121 depend upon the type of vehicle and road speed, and are specified in terms of stopping distance (FMVSS 121, Table 2). The control forces to achieve these are not defined. The calculated approximate average decelerations from these values (which show slight variations depending on the vehicle road speed) are summarised in Table 8.9, and the

Table 8.9: Service and Emergency Braking Performance Requirements in Terms of Calculated Average Deceleration from Table 2 of FMVSS 121 (US FMVSS 121)

Type of Vehicle	Calculated (Approximate) Average Deceleration (m/s^2)	
	Service Braking	Emergency Braking
Loaded and unloaded buses	4.2	1.9
Loaded single-unit trucks	3.8	1.9
Loaded tractors with two axles, or with three axles and GVWR $\leq$ 31.8 tonnes, or with four or more axles and GVWR $\leq$ 38.6 tonnes	4.7	1.9
Loaded tractors with three axles and GVWR > 31.8 tonnes or with four or more axles and GVWR > 38.6 tonnes	3.8	1.9
Unloaded single-unit trucks	3.5	1.9
Unloaded tractors (solo)	5.0	1.6

reader is referred to the FMVSS 121 document for full details of stopping distance requirements. These can be compared with the requirements of UN Regulation 13 in Table 8.4. No braking tests for trailers are defined, as trailer brake performance is determined by dynamometer testing. Park brake performance requirements are defined in terms of a drag test or 20% gradient hold.

There are several functions for commercial vehicle braking systems that are defined in FMVSS 121 but are not defined in UN Regulation 13, e.g. the preparation of brakes prior to testing, and trailer braking only (not permitted in Regulation 13). There are more functions that are defined in UN Regulation 13 but not in FMVSS 121, e.g.:

- Specific fade tests for vehicles (FMVSS defines less stringent dynamometer procedures)
- Towing vehicle and trailer compatibility
- Wheel lock (prescribed braking performance must be achieved without wheel lock although directly controlled wheels may lock below 15 km/h and for brief periods above this speed)
- Split-μ stability
- Vehicle stability control
- AEBS.

UN Regulation 90

UN Regulation 90 (UNECE, 2012) relates to the approval of replacement brake lining assemblies, drum brake linings and rotors (discs and drums). It was introduced in 1995 to control the manufacture, sale, distribution and installation of replacement brake pad assemblies and lined shoe assemblies, and sets minimum performance standards for replacement brake rotors and stators for vehicles of categories M, N and O. With the exception of replacement drum brake linings (i.e. linings without the metal shoe) for vehicles of categories M_3, N_2, N_3, O_3 and O_4, brake lining materials alone (i.e. not part of a stator assembly) are not approved. From November 2014 UN Regulation 90 will be mandatory in Europe as part of the GSR (EC, 2009) for all categories of vehicles including motorcycles (L), passenger vehicles (M), goods vehicles (N) and trailers (O).

UN Regulation 90 (UNECE, 2012) covers three key areas: performance, conformance of production, and marking and packaging. In terms of performance, replacement stators (pad or lining assemblies) are required to (a) satisfy the relevant braking prescriptions of UN Regulation 13, (b) display performance characteristics similar to those of the original part it is intended to replace, and (c) not contain asbestos. The frictional performance of a replacement stator is compared by performance testing to the original equipment (OE) assembly in a cold performance equivalence test. It must be within a tolerance band compared to the OE part as specified in the Regulation for that class of vehicle (e.g. within ±15% of the OE part). These performance tests may be carried out on either a dynamometer or a complete vehicle on a test track (depending on the category of vehicle and test). The replacement brake pads and lined shoes have to meet production tests to ensure that they are robust (e.g. shear strength according to ISO 6312, compressibility according to ISO 6310, material hardness according to ISO 2039-2), and must carry clear identification marks and be properly packaged in sealed boxes with fitting instructions and application information. The production plant or factory must conform to acceptable control standards (such as ISO 9002) and has to undergo a conformity of production (CoP) audit, which is a process carried out by the approval authority to check the quality systems and procedures. Regular tests are required after the replacement brake lining assembly has been approved, which include measurement of the friction coefficient, compressibility, shear and hardness of the brake lining material and assembly.

Complex Electronic Vehicle Control Systems

The incorporation of complex electronic vehicle control systems into road vehicle braking systems started with ABS and has been extended to electronic brake control (EBS) and vehicle stability control (VSC or ESC) – see Chapter 11. Automatic brake applications are defined by two functions: automatically commanded braking (the main purpose being to decelerate rather than to stabilise the vehicle) and selective braking (the main purpose being to stabilise rather than to decelerate the vehicle). These have brought an associated need for advanced legislation to ensure that such systems are effective and safe. Annex 8 of UN Regulation 13H and Annex 18 of UN Regulation 13 address the special safety requirements of complex electronic vehicle control systems. These do not specify the performance criteria for such systems, but cover the methodology applied to the design process and the information that must be disclosed to the technical service for type approval purposes. Complex electronic vehicle control systems are defined in the Regulations as 'those electronic control systems which are subject to a hierarchy of control in which a controlled function may be overridden by a higher level electronic control system/function.' It also states that 'a function which is overridden becomes part of the complex system', and this includes conventional ABS as its action overrides the driver's braking demand. All systems in which a controlled function, whether mechanical, pneumatic, hydraulic or electrical, may be overridden by a higher level electronic control system or function are covered, as are the special requirements for documentation, fault strategy and verification with respect to the safety aspects of complex electronic vehicle control systems. The associated information is required to show clearly that the complex electronic vehicle control system respects, under normal and fault conditions, all the appropriate performance requirements specified elsewhere in the Regulations. The information that the manufacturer must provide includes how the current operational status of the system can be checked, descriptions of the system functions including input and output variables and how the output variables are controlled, system diagrams, and statements of how the system is intended to operate safely with failures within it. Full use of tools such as failure mode and effects analysis (FMEA) and fault tree analysis (FTA) is essential.

Electronic Stability Control (ESC) systems are explained in Chapter 11, and apply selective braking with the main purpose of stabilising the vehicle. Under UN Regulation 13H, category M_1 and N_1 vehicles must be fitted with an ESC system that meets the functional and performance requirements as measured by prescribed test procedures. These include a 'sine with dwell test', which defines a steering pattern of a sine wave at 0.7 Hz frequency with a 500 ms delay beginning at the second peak amplitude, as shown in Figure 8.3. Category M_1 and N_1 vehicles of mass greater than 1735 kg may be equipped with a vehicle stability function that includes rollover control and directional control, and meets the technical requirements of UN Regulation 13 Annex 21.

For vehicles of categories other than M_1 and N_1, the GSR through UN Regulation 13 makes ESC mandatory for type approval and thus for registration, sale and entry into service of new vehicles (see Table 8.1). It defines a vehicle stability control function, which may include directional control and/or rollover control to improve the dynamic stability of the vehicle. The mandatory requirement includes directional control and rollover control for motor vehicles, and rollover control on trailers. Directional control means a function within a vehicle stability function that assists the driver, in the event of understeer and oversteer conditions, within the physical limits of the vehicle in maintaining the direction intended by the driver in the case of a power-driven vehicle, and assists in maintaining the direction of the trailer with that of the towing vehicle in the case

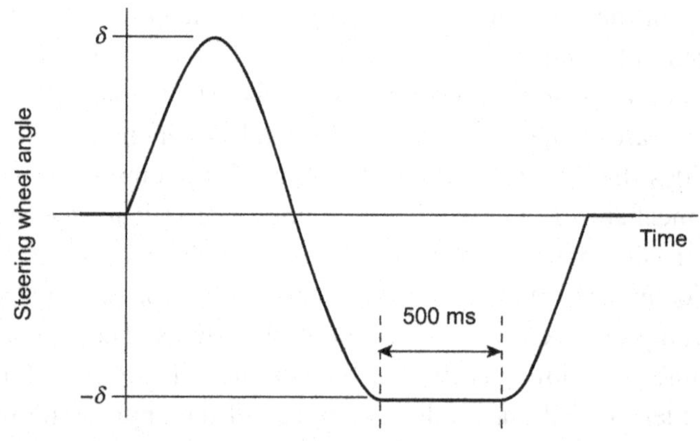

Figure 8.3: Sine with Dwell Test for ESC (UN Regulation 13H Annex 9 (UN, Feb 2014)).

of a trailer. Rollover control means a function within a vehicle stability function that reacts to an impending rollover in order to stabilise the power-driven vehicle or towing vehicle and trailer combination or the trailer during dynamic manoeuvres within the physical limits of the vehicle. UN Regulation 13 Annex 19 (Performance testing of braking system components) provides, in combination with Annex 20 (Alternative procedure for the type approval of trailers), an alternative method for the type approval of a trailer braking system. A vehicle stability function can be tested and evaluated in terms of a 'vehicle framework', covering for example:

- Trailer type (semi-trailer, centre-axle trailer, full trailer), axle configuration, steering axles, lift axles, suspension type, wheelbase, track and wheel type.
- Braking system, anti-lock braking system configuration and foundation brake type.
- Vehicle stability function components including sensors, controllers and modulators.

UN Regulation 13 Annex 21 defines the special requirements for vehicles equipped with a vehicle stability function, but does not prescribe a defined test procedure. Any of the dynamic manoeuvres shown in Table 8.10 may be used to demonstrate the vehicle stability function.

Electronic Braking Systems (EBS) are complex electronic vehicle control systems in which the control is generated and processed as an electrical signal in the control transmission, and the electrical output is in the form of a signal to control devices that produce brake actuation forces from stored pneumatic energy (in contrast to an electromechanical brake, which is actuated by electrical energy). As explained in Chapter 11, EBS offers advantages of faster braking control (which gives decreased stopping distance) and intelligent control of the braking force distribution between the front and rear axles of road vehicles and/or between the axles of the towing vehicle/trailer combination. Commercial vehicles with EBS

Table 8.10: Dynamic Manoeuvres to Demonstrate the Vehicle Stability Function for Commercial Vehicles (UN Regulation 13 Annex 21 (UNECE, March 2014))

Directional Control	Rollover Control
Reducing radius test	Steady state circular test
Step steer input test	J-turn
Sine with dwell	
J-turn	
μ_t-split single lane change	
Double lane change	
Reversed steering test or 'fish hook' test	
Asymmetrical one period sine steer or pulse steer input test	

are required by the European legislation to be fully compatible with conventional trailers and semi-trailers (with or without ABS). One of the most fundamental changes in the new EBS requirements in the EU legislation is to distinguish between 'control transmission' and 'energy transmission' in terms of the brake actuation system. In older designs of road vehicle braking actuation systems, both 'energy transmission' and 'control transmission' were provided by the same mechanism, e.g. the muscular energy used to control and actuate a conventional hydraulic braking system on a passenger car (see Chapter 6), where the service braking performance is generated only from muscular energy, namely by the action of the driver's foot acting on the brake pedal.

The EU Regulations require that the service braking system is able to generate a static total braking force at least equivalent to that required by the prescribed Type-0 test (see Tables 8.3 and 8.4), even when the vehicle's 'ignition' has been switched off and the key has been removed. In order to meet this requirement a motor vehicle with an electric brake control transmission might be fitted with a'wake-up'function activated by applying the brake pedal. Other requirements include:

- A control transmission failure must be indicated to the driver by a warning signal; a red warning signal is used to indicate when the prescribed service braking performance (attained without wheel lock) has not been reached.
- A red warning signal is required if excessive voltage drop of the battery is detected.
- A red warning signal is also required in the event of failure in a trailer braking system, failures due to insufficient energy and/or service brake performance.
- A test must be included to verify that the battery has enough energy capacity by requiring a given number of full brake applications.

For commercial vehicles the full actuation of the trailer brakes must remain ensured in the event of a failure in the electric control transmission of the service braking system of a towing vehicle equipped with an electric control line. This requirement is more stringent than for a towing vehicle with conventional air actuation, where in the case of a circuit failure only partial actuation of the trailer brakes is required. The vehicle manufacturer has two options to show a 'red signal failure' of the trailer on the dashboard:

- Two lamps option: red (towing vehicle signal) and yellow (trailer signal).
- One lamp option: separate red warning signal (to specifically indicate trailer faults that require a red signal).

These warning signals must illuminate when the electrical equipment of the vehicle and the braking system are first energised, and while the vehicle is stationary, the braking system must verify that none of the specified failures or defects are present before extinguishing the signals. If there are electrical failures or defects that are not detected under static conditions, these must be stored when they are detected and then must be displayed at all times when the ignition switch is on for as long as the failure or defect persists. This should form part of the vehicle's start-up warning signal sequence and memory function test for all electrical system failures with respect to the functional and performance requirements of the electrical braking equipment.

All road vehicles (including trailers) are required to be fitted with lights at the rear of the vehicle that indicate when the brakes are being actuated. For EBS vehicles equipped with endurance braking and/or regenerative braking system of category A, the stop light signal must be generated at a vehicle deceleration >1.3 m/s^2, but may be generated below this. Since the EU Regulations now include automatically commanded braking, which is intended to decelerate (rather than to stabilise) the vehicle, the stop lamp signal must be generated at a vehicle deceleration ≥ 0.7 m/s^2. Since the main purpose of selective braking is to stabilise (rather than to decelerate) the vehicle, the stop lights must not be illuminated. When a vehicle is equipped with emergency braking indicators (e.g. the existing brake lights can be controlled to operate at higher intensity or flash) activation is only permitted when the vehicle deceleration is above 6 m/s^2 for category N_1 vehicles (cars), and 4 m/s^2 for category M_2, M_3, N_2 and N_3 vehicles.

'Advanced Emergency Braking System' (AEBS; see Chapter 11) is a generic name for any system that can apply emergency braking independent of driver control, i.e. automatically commanded braking. It is designed to detect automatically an emergency situation and activate the vehicle braking system to decelerate the vehicle with the purpose of avoiding or mitigating a collision. A 'Collision Mitigation Braking System' (CMBS) is a type of AEBS that can autonomously apply emergency braking in order to mitigate the severity of a collision that has become unavoidable, and a 'Collision Avoidance Braking System' (CABS) can autonomously apply emergency braking in order to avoid a collision. AEBS uses sensors to monitor the proximity of an object in front of the vehicle to which it is fitted, and to detect conditions where the relative speed and distance between the two vehicles suggests a collision might occur. The system provides the driver with at least an acoustic and haptic warning and if the driver does not respond to the warning it will enter the emergency braking phase in which the AEBS activates the braking system to brake the vehicle to avoid or reduce the severity of the collision. No value of deceleration is defined; the requirements specify a minimum speed reduction that depends on the vehicle category. AEBS requirements are now defined in a new UN Regulation 131. During both phases, the driver can at any time take control and override the system. The AEBS should be robust in the sense that actual conditions and events should not result in false warnings or braking to the extent that they encourage the driver to switch the system off. A yellow optical warning signal is required to indicate when there is a failure in the AEBS or it is switched off. The GSR has set mandatory implementation dates for AEBS on some passenger and goods

Table 8.11: Two Level Approvals for Category M_2, M_3, N_2 and N_3 Vehicles with AEBS (Appendix 1 of Annex II to EC Regulation 347/2012) (see Table 8.1 for Implementation Dates Under the GSR)

	Level 1 Approval	Level 2 Approval
Scope	Vehicles of categories M_2 and M_3, and types of category N_2 vehicle with a maximum mass exceeding 8 tonnes, provided that these types of vehicle are equipped with pneumatic or air-over-hydraulic braking systems and with pneumatic rear axle suspension systems	Level 1 vehicles plus vehicles with hydraulic braking systems and non-pneumatic rear axle suspension systems. Will also include category M_2 and N_2 types of vehicle with a maximum mass not exceeding 8 tonnes
Speed reduction of subject vehicle	Not less than 10 km/h	Not less than 20 km/h
Target speed	32 ± 2 km/h	12 ± 2 km/h

vehicles and, because of the complexity of the technology, two levels of approval have been defined as summarised in Table 8.11.

Brake Assist Systems

Brake Assist Systems (BAS) were developed to improve the braking performance of drivers of M_1 vehicles in emergency situations. Many drivers do not make effective use of the full capability of the braking system in such circumstances, they do not apply sufficient pedal force to reach maximum braking during an emergency stop and they do not act quickly enough. The BAS detects when a driver might be intending to make an emergency stop and will reduce the brake actuation time and boost the brake pedal effort, e.g. by actuating the ABS pump. The potential for pedestrian safety and protection benefits from active safety systems, especially BAS, combined with changes to passive safety requirements, is part of EC Regulation No. 78/2009 ('On the type-approval of motor vehicles with regard to the protection of pedestrians and other vulnerable road users'), which defines a 'brake assist system' as a function of the braking system that deduces an emergency braking event from characteristics of the driver's brake demand and under such conditions assists the driver to deliver the maximum achievable braking rate, or is sufficient to cause full cycling of the ABS. The function of the BAS has to be verified (by testing) to confirm that the operation of the brake assist system is correctly triggered so as to apply the maximum achievable deceleration of the vehicle. UN Regulation 13H does not mandate the installation of a brake assist system, but the framework Directive 2007/46/EC requires that vehicles of categories M_1 and N_1 that are produced in unlimited series have to comply with the pedestrian protection EC Regulation No 631/2009. Accordingly UN Regulation 13H Supplement 9 was mandated for M_1 category vehicles from November 2011 for new types and will be from November 2014 for existing types. UN Regulation 13H defines two different BAS categories:

- Category A Brake Assist System. A system that detects an emergency braking condition based primarily* on the brake pedal force applied by the driver, e.g. it may be triggered at pedal force actuation rate of approximately 550 N/s.
- Category B Brake Assist System. A system that detects an emergency braking condition based primarily* on the brake pedal speed applied by the driver, e.g. it may detect an emergency braking condition triggered at a pedal speed ≥300 mm/s.

(*As declared by the vehicle manufacturer.)

A previous category C BAS (which detects an emergency braking condition based on multiple criteria including the rate at which the brake pedal is applied) has been deleted and is now covered by the above definitions of BAS categories A or B. A conventional dual-rate brake booster can also be capable of meeting the definition of a BAS and especially for heavier vehicles a dual-rate brake booster can also qualify (under the Regulations) as an acceptable category A BAS if its pedal force characteristic is similar to that shown in Figure 8.4. This indicates that the maximum braking force must be obtained with a pedal effort that is not less than 40% of what would be the case with a standard brake booster. Refer to UN Regulation 13H (UNECE, February 2014) for full information.

Regenerative Braking Systems

As described in Chapter 11, vehicles that incorporate regenerative braking are equipped with devices to convert the vehicle's kinetic energy into a form of energy that can be

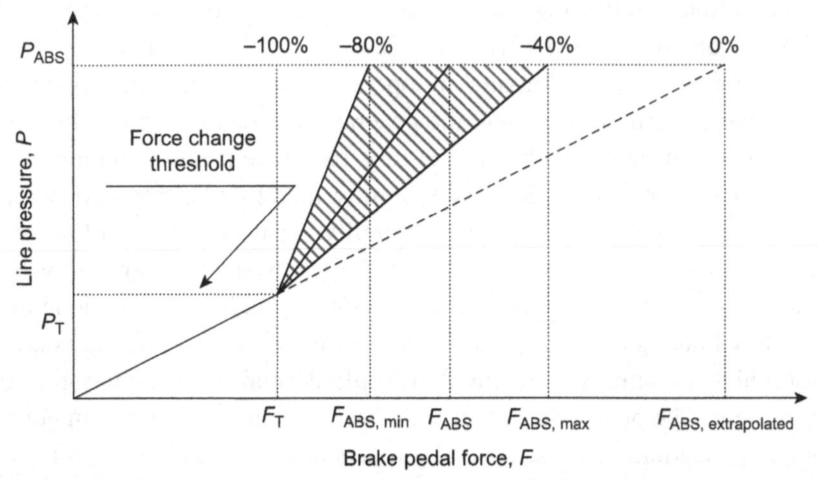

Figure 8.4: Pedal force Characteristic for Category 'A' BAS (UN Regulation 13H Annex 9 (UN, Feb 2014)).

stored onboard and reused. Typically such vehicles use a generator or pump to convert kinetic energy into another form of energy and store it in batteries, ultracapacitors, hydraulic reservoirs, flywheels, etc. This technology has introduced great demands on to vehicle designers and legislators for safe braking, and as far as the legislation is concerned, is potentially further complicated by the wide range of possibilities for the technologies involved. UN Regulations 13 and 13H base the legislative requirements on electric regenerative braking systems, the requirements for which were introduced in 2002 to allow vehicles to include the means to recover kinetic energy by regenerative braking. These braking regulations include electric vehicles and hybrid vehicles (with internal combustion engines and electric motor/generators). In the UN Regulations 13 and 13H, 'electric regenerative braking' means a braking system which, during deceleration, provides for the conversion of vehicle kinetic energy into electrical energy. 'Electric regenerative braking control' means a device that modulates the action of the electric regenerative braking system. Electric regenerative braking systems are separated into two categories:

- Category A: an electric regenerative braking system that is not part of the service braking system.
- Category B: an electric regenerative braking system that is part of the service braking system.

Since an electric regenerative braking system cannot produce sufficient braking force under all operating conditions, especially at low vehicle speed and under emergency braking (see Chapter 11), a combination of conventional friction braking and regenerative braking is always necessary to provide full braking capability. Regenerative braking requires two (or possibly more) modes of braking, e.g. friction braking and electrical braking, to be controlled so that together they create the required retarding force on the vehicle. In UN Regulations 13 and 13H this is described as 'phased braking', a term that describes how different modes of braking are operated from a common control. One of the modes may be given priority by delaying the generation of retarding force by another mode, e.g. increased movement of the control (e.g. the brake pedal) is required to bring it into operation. The regenerative braking system must therefore be 'blended' with the friction braking system to ensure that vehicle stability is ensured even on low-adhesion road surfaces. In order to permit the maximum recovery of kinetic energy in vehicles, the most common strategy is to use the electric regenerative braking system alone to provide as much light-duty braking as possible and in this case the friction braking control would be initiated at a higher control movement than would normally be adopted in a friction-only braking system. The precise design of the phasing should permit the majority of normal braking applications to be achieved by regenerative braking alone; this is typically achieved by controlling regenerative braking alone up to a preset level of vehicle retardation since this braking is usually applied through a single drive axle. Figure 8.5 illustrates the concepts.

The contribution from other sources of braking may also be phased to allow the regenerative braking system alone to be utilised, provided that both of the following conditions are met:

- Intrinsic variations in the torque output of the electrical regenerative braking system, e.g. from the state of charge (SoC) of the traction batteries, are automatically compensated by a suitably graduated reduction of the phasing setting, i.e. the friction braking is introduced to make up for any reduction in the regenerative braking torque.
- The braking distribution requirements on low-adhesion roads are met or the ABS operates to prevent wheel lock at all times.

It is a requirement of UN Regulations 13 and 13H that all road wheels are braked. A category B electrical regenerative braking system is part of the service braking system, but selective disconnection of the regenerative braking component is permitted provided that the friction braking source remains permanently connected and the braking stability requirements are met, as previously explained. Additionally the friction braking system must continue to operate with the prescribed degree of effectiveness. If the regenerative braking is disconnected the friction braking must be able to compensate for the loss of that component of braking, e.g. during gear changes (if the motor/generator is the engine side of the transmission) or if the regenerative braking is disabled during ABS operation. Otherwise regenerative braking can only be disengaged by moving the accelerator control and/or when the gear selector is in the neutral position. For vehicles of categories M_2 and N_2 (<5 tonnes) with category A electric regenerative braking the control can be a separate switch or lever. For vehicles with category B electric regenerative braking the service braking system can only have one control device. Vehicles fitted with both electric regenerative braking systems of categories A and B must not cause the action of the service braking control to reduce the electric regenerative braking effect generated by the release of accelerator control. Under UN Regulation 13H brake light illumination is required for vehicle decelerations $>1.3\ m/s^2$ whether under regenerative braking alone or not, but brake light illumination must not be generated at a vehicle deceleration $\leq 0.7\ m/s^2$. Between vehicle decelerations of 0.7 and $1.3\ m/s^2$ brake light illumination is optional. Under UN Regulation 13 brake light illumination is required for vehicle decelerations $>1.3\ m/s^2$ whether under regenerative braking alone or not, and brake light illumination at vehicle decelerations $\leq 1.3\ m/s^2$ is optional.

Vehicle control and behaviour tests must be carried out to prove the stability of the vehicle under braking is maintained. The ABS must control the regenerative braking system and during ABS operation the regenerative braking system, both category A and B, is typically disabled. For vehicles powered completely or partially by an electric motor (or motors),

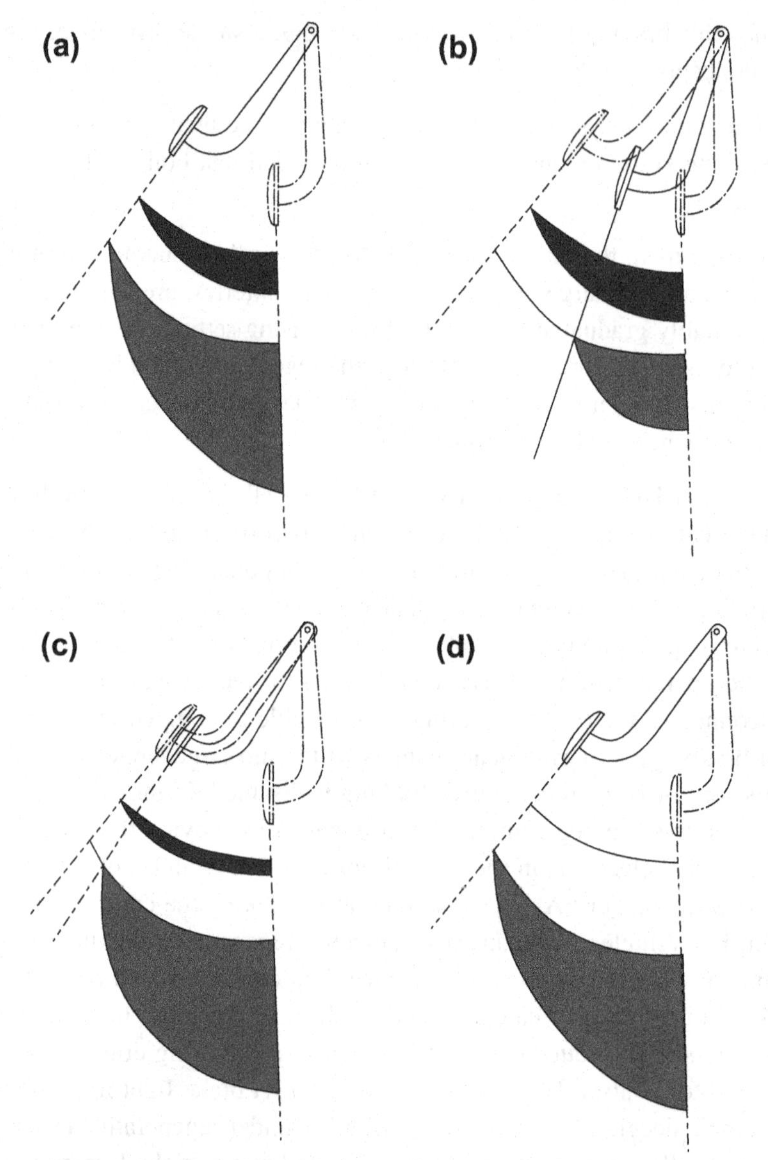

Figure 8.5: Examples of Phased Movement for Electrical Regenerative Brake Actuation Controls (Gaupp, 2013).
(a) Non-phased operation. (b) Maximum phased operation. (c) Intermediate phased operation. (d) Zero phased with no regenerative braking.

which are permanently connected to the wheels and fitted with an electric regenerative braking system of category A, behaviour tests must be carried out on a vehicle test track with a low adhesion coefficient ($k \leq 0.3$) at a speed of 80% V_{max} (but not exceeding 120 km/h) to check that stability is retained.

Chapter Summary

- The purpose of braking legislation for road vehicles worldwide is explained in terms of its importance in ensuring road safety standards and safe and stable braking so that the public is protected against unreasonable risk of accidents.
- The major world regions that enforce road vehicle legal standards and the associated legislative bodies are identified. The two basic approaches to ensuring that all road vehicles comply with the legislative requirements are described; these are type approval (in the EU) and self-certification (e.g. in the USA, Canada and Australia). Categories of motor vehicles and routes to type approval in the EU are outlined and global harmonisation of braking regulations is briefly discussed.
- The UN Braking Regulations 13 and 13H for passenger cars and commercial vehicles are separately explained and discussed. The importance of the EC General Safety Regulation (GSR) is explained, and the changes that will be introduced when it comes into force in November 2014 are discussed.
- The US braking regulations based on the Federal Motor Vehicle Safety Standards are outlined and compared with UN Regulations 13 and 13H.
- The braking performance technical requirements for service braking, secondary braking, park braking and residual braking in UN Regulations 13 and 13H are explained, including the importance of 'mean fully developed deceleration' (MFDD) and stopping distance and stability in terms of wheel-lock sequence. The test procedures associated with type approval are explained for road vehicles and trailers and the selection of the 'worst case' for type approval testing is explained.
- The need for compatibility between towing vehicles and trailers and the way this is covered in the EU Regulations are explained.
- US road vehicle braking legislation including FMVSS 135 for light vehicle brake systems and FMVSS 121 for air brake systems is outlined, including the US FHWA 13 vehicle classifications. A short comparison of the US standards with the UN Regulations is presented.
- UN Regulation 90 for the approval of replacement brake parts and components is briefly considered.
- New and recent EU legislation for complex electronic vehicle control systems including antilock braking (ABS), vehicle (electronic) stability control (VSC or ESC), electronic braking systems (EBS), automatically commanded braking (to decelerate the vehicle) and selective braking (to stabilise the vehicle) are explained and discussed. Methods for evaluating vehicle stability functions, including appropriate dynamic manoeuvres, are presented. Indicators and other warning functions for these complex systems to the driver and other road users are discussed. Electronic Braking Systems (EBS) and the separation of the brake control transmission and the brake energy transmission are described, together with the need to be able to actuate the vehicle's brakes when its control systems are deactivated.

- The legislative background to Advanced Emergency Braking Systems (AEBS) that can apply emergency braking independent of driver control is discussed. The way these are being included in the UN Regulations in terms of the Collision Mitigation Braking System (CMBS) and Collision Avoidance Braking System (CABS) is considered. Similarly the legislative background to Brake Assist Systems (BAS) is examined.
- Braking legislation for road vehicles with regenerative braking systems, e.g. electric and hybrid vehicles, is presented and discussed for two categories: A – an electric regenerative braking system that is not part of the service braking system; and B – an electric regenerative braking system that is part of the service braking system. The way the UN Regulations cover the control of blended braking (combined friction and regenerative braking) is explained.

References

Agudelo, C.E., Ferro, E., 2005. Technical overview of brake performance testing for original equipment and aftermarket industries in the US and European markets, Link Technical Report FEV205-01, Link Testing Laboratories, Inc.

Gaupp, W., 2013. Braking legislation, Braking of Road Vehicles 2013. University of Bradford. ISBN 1 85143 272 1.

UN, Mar 2014. UNECE Regulation 13. Uniform provisions concerning the approval of vehicles of categories M, N and O with regard to braking, E/ECE/324/Rev.1/Add.12/Rev.8, March 2014.

UN, Feb 2014. UNECE Regulation 13H. Uniform provisions concerning the approval of passenger cars with regard to braking, E/ECE/324/Rev.2/Add.12H/Rev.3, February 2014.

UN, 2012. UNECE Regulation 90. Uniform provisions concerning the approval of replacement brake lining assemblies, drum brake linings and discs and drums for power-driven vehicles and their trailers, E/ECE/324/Rev.1/Add.89/Rev.3, February 2012.

EC Regulation, 661/2009. Concerning type-approval requirements for the general safety of motor vehicles, their trailers and systems, components and separate technical units intended therefor. Official Journal of the European Union, L 200/1, July 2009.

US Department of Transportation, Federal Motor Carrier Safety Administration, Rules and Regulations Standard No. 135; Light vehicle brake systems (FMVSS 135).

US Department of Transportation, Federal Motor Carrier Safety Administration, Rules and Regulations Standard No. 121; Air brake systems (FMVSS 121).

US Department of Transportation, Federal Motor Carrier Safety Administration, Rules and Regulations Standard No. 105; Hydraulic and electric brake systems (FMVSS 105).

CHAPTER 9

Brake Testing

Introduction

Brake testing for road vehicles traditionally takes many forms, ranging from large fleet user trials of complete braking systems through the laboratory dynamometer testing of foundation brakes, to small sample component testing. However, the increase in the capability and sophistication of computer-based modelling and simulation over the last 20 years has improved the accuracy of predictive design analysis methods and has enabled many aspects of brake testing for design optimisation and verification to be carried out in the form of 'virtual' testing. Experimental brake testing is expensive and time-consuming, so the use of virtual testing methods is attractive and useful in the cost competitive and ambitious time scales of new vehicle design and development. Ultimately though, experimental testing can be the only way that a design can be finally proven, or information essential to accurate design prediction can be obtained, so it still has an important role in brake system design and development. Virtual testing using computer-aided engineering (CAE) methods and software usually takes the form of 'experiments' to predict the effect of a range of design changes to optimise the design performance. Many CAE systems feature inbuilt 'design of experiments' (DoE) routines to automate and increase the efficiency of the optimisation process. The accuracy of the computer predictions relies upon accurate data, e.g. relating to material properties, which can only be obtained from experimental testing. In particular, the accurate characterisation of friction materials in terms of mechanical, thermal and tribological properties is fundamentally important, and some of the difficulties associated with friction material thermophysical and tribological properties have been described in previous chapters. When designs of road vehicle brakes and braking systems have been completed, they have to be verified by vehicle manufacturers and their suppliers, which can only be achieved by experimental testing involving hardware, with the result that experimental testing has moved from the role of 'development' 20 years ago to that of 'design verification' now, saving time and cost as in all fields of engineering design. The ultimate design verification for road vehicles is legislative testing for type approval or self-certification, which requires experimental testing of road vehicle brakes and braking systems to demonstrate that the vehicle meets the legal requirements, but even this is increasingly underpinned by more sophisticated calculation and predictive methods, especially in areas such as complex electronic control systems.

Braking of Road Vehicles.

The primary function of any road vehicle braking system is to generate repeatable and controlled levels of vehicle deceleration under all conditions of operation and use, and experimental confirmation by the vehicle manufacturer that such function will always be achieved remains an important and essential part of the vehicle design process. Brake system, subsystem and component suppliers also have to confirm that their products consistently and reliably provide the required function, and may use experimental testing for this purpose and also to develop their products to perfection since there are still many aspects of brake performance and operation that cannot be predicted accurately. The reason is usually that data, properties and information about parts of the brake system are not available to the level of detail required for accurate prediction under actual operating conditions, which themselves might not be known precisely. For example, the mechanical properties of cast iron used for brake rotors are readily available for a limited range of operating loads and temperatures, but not for the extremes of temperature generated in a brake operating at peak duty. If it is necessary to measure the material properties under conditions that represent the extremes of brake operation this can be prohibitively expensive and time-consuming, so manufacturers often use experimental testing, e.g. of complete brakes, to generate data needed to complete the design. Another very important example is the friction material; the friction behaviour of disc brake pads and drum brake linings is fundamentally important information for brake and brake system design calculations. As explained in Chapter 2, it is almost impossible to predict the tribological (friction and wear) properties and behaviour of most friction materials to the required accuracy, so experimental testing is an essential part of the formulation and manufacture of suitable friction materials.

Once a vehicle has been manufactured and sold, it passes into the customer domain, where the manufacturer has no control over how it is used but retains product liability in the event of a failure that can be attributed to any deficiency in the design or manufacturing process. If a suspected deficiency is found, experimental testing will be required to confirm or deny its existence and, if confirmed, containment actions and countermeasures would usually be verified experimentally as part of a structured procedure for failure mode investigation. Any form of testing is a costly activity, and experimental testing, especially that which involves whole vehicles, is particularly expensive so there should always be compelling and valid purposes for brake testing. These could typically include:

- Design optimisation, e.g. computer-based virtual 'experiments' to iterate to an efficient and optimal design.
- Verification of system and component performance, including durability and reliability.
- Verification of vehicle braking performance for the vehicle manufacturer's and the legislative bodies' requirements, principally deceleration and stability, efficiency, actuation forces, system part-fail, etc., but also including durability and reliability.
- Measurement of material properties for brake design.

- Characterisation of friction material tribological properties for (a) material formulation and development purposes, and (b) brake design purposes.
- Investigation of vehicle braking characteristics for failure mode investigation and countermeasure development.

In this chapter the options for the experimental testing of brakes and braking systems ranging from whole vehicles to small samples are described and discussed. Test preparation, procedures and protocols, instrumentation, data acquisition, results analysis, and the interpretation and reporting of results are all considered. Predictive methods have been described and discussed in other chapters, but virtual testing methods as such are not considered in detail because these are a specialist topic beyond the scope of this book.

The safety of all personnel associated with experimental brake testing is paramount. In laboratory testing video cameras should be positioned to provide a visual image of the test being undertaken, so that personnel are not exposed to hazards such as noise and dust, and the effects of any potential failure. When undertaking vehicle testing, drivers should be comprehensively trained to drive safely and repeatably on test tracks and on public roads. Vehicle braking performance tests can be carried out on test tracks but are not permitted on public roads in the EU; however, a vehicle may be fitted with instrumentation and datalogging equipment and driven on public roads provided that the equipment does not interfere with the driver's control of the vehicle and it does not present a hazard to any other road users. Suitable permissions and insurance cover must also be in place, e.g. the installation of pressure transducers and brake circuit isolation valves can require special agreement from insurers and police forces before test vehicles can be driven on public roads. A minimum requirement is typically an interlocking system that prevents the actuation of brake circuit isolation valves on public roads, e.g. via a security key that is only available at a test track. For occupant safety, all instrumentation mounted within a test vehicle should be designed to minimise possible injury to the driver and any other occupants in the event of an accident, especially taking into account the consequences of protective safety devices such as an airbag.

Instrumentation and Data Acquisition in Experimental Brake Testing

Parameters that can be measured in experimental brake testing include:

- Temperature
- Deceleration
- Torque
- Force
- Pressure
- Displacement and distance
- Vibration frequency, amplitude and phase.

As explained below, modern transducer technology can be implemented to provide the sensor technology for these measurements, and then signal conditioning is needed to convert the transducer output signal into a suitable standardised form to be recorded as a time series on a data acquisition system, usually referred to as a datalogger. The ability to gather, record and process time series experimental data represents a significant advance in experimental capability over the last 25 years or so. Previously, data collection and processing was a time-consuming and labour-intensive task. Modern data acquisition systems are efficient, compact and flexible, and can record vast amounts of data, making it possible to record entire tests of several hours' duration, and not just specific test datapoints. They can be employed on all types of brake test rigs and equipment, whether on the vehicle for track or road testing, or in the laboratory on dynamometer or sample testing.

Temperature

Because friction brakes work by converting kinetic energy into heat energy, the single most important parameter to measure and control in experimental brake testing is temperature. As explained in Chapter 7, the effect of the heat generated is to create temperature distributions in the foundation brake, and because of the effect of the heat on the friction and wear performance of the friction material and the contact relationship between the rotor and stator components, changes in the temperature distribution and magnitude can dominate the measured brake performance. It is therefore essential that temperature is either controlled or monitored during all experimental brake testing. Within the braking industry, standard practice is to measure the temperature of the rotor, either at the surface or within the material of the rotor below the surface, to provide a direct indication of the amount of work the brake is doing, which can then be used in the control of the test. For example, it is most important to ensure that when a series of repeated brake applications is made to evaluate the brake performance, the 'start' temperature of the brake application is controlled to be the same, which is usually achieved by allowing the rotor to cool down between brake applications. As explained in Chapter 7, temperatures depend on the location on and in the brake components and the time at which the measurement is taken. The physical method of temperature measurement employed can also significantly influence the result, and accuracy in the temperature measurement requires the correct calibration of the sensor and instrumentation used. The most common methods of temperature measurement on a brake are rubbing thermocouples or non-contacting radiation sensors (usually infrared) to measure rotor surface temperature, and/or embedded thermocouples in the rotor to measure temperature below the rotor surface. Each method is capable of providing consistent and accurate temperature measurement, but the results may not be directly comparable. A most useful diagnostic tool for the initial experimental assessment of a brake system's performance is a hand-held digital thermometer, which enables a quick measurement of the temperature of each brake on a vehicle that has just completed a braking event. Similarly, thermal paints can be quickly and easily applied to brake components to indicate the maximum surface temperature that the component subsequently reaches. The disadvantage is that the point at which the maximum temperature was reached, or for how long, cannot be determined.

The thermocouple is the most commonly used sensor for brake temperature measurement. Two wires of dissimilar metals are joined to make a junction at each end; one end is held at a reference temperature and the other is exposed to the temperature being measured. The 'Seebeck' effect creates a potential difference (voltage) between the two wires that provides a proportional indication of the temperature difference between the two junctions. In modern systems the 'cold' junction (reference temperature) is automatically set by 'cold junction compensation' in the associated electronic signal conditioning unit (usually referred to as the 'thermocouple amplifier'). To improve the durability of the thermocouple, the measuring junction is usually attached to a sacrificial 'skid', which slides on the rotor surface, making it robust enough to be used on vehicle brake tests. A temperature difference across the contact will always exist, but careful experimental technique can minimise this. Usually the force holding the skid against the rotor surface is adjusted until the temperature measured is the same whether the rotor is moving or stationary; too low contact force will indicate a low temperature and too much contact force will cause the skid to generate frictional heat, which will indicate a high temperature during operation. Examples of skids are illustrated in Figure 9.1. The response of the thermocouple to a temperature change on the rotor surface is damped by the presence of contact resistance across the skid and the thermal mass of the skid and thermocouple. Highly transient temperatures or 'flash' temperatures, which can occur on the friction surface of the rotor (see Chapter 7), therefore cannot be measured by this type of thermocouple, but very fine thermocouple wires can be used to form a very low thermal mass sensor where the measuring junction touches directly on the friction surface. This arrangement can be useful for research purposes but is not sufficiently durable for most types of brake testing. 'Bare wire' thermocouples are seldom used, having been replaced by 'sheathed' thermocouples where the wires are encased in a protective metal sheath (and insulated from it) to avoid problems of electrical interference, especially when connected to datalogging systems on vehicles. A comparison of temperatures measured by two different types of thermocouple and skid is shown in Figure 9.1(c); the older 'bare wire' thermocouple unit with the copper button skid (Figure 9.1(a)) is less responsive than the modern sheathed thermocouple unit (Figure 9.1(b)) because of its higher thermal mass.

Where a rubbing thermocouple is set up to slide on the friction surface of a brake rotor, it can create a track that can affect the frictional performance of the brake, and may also be

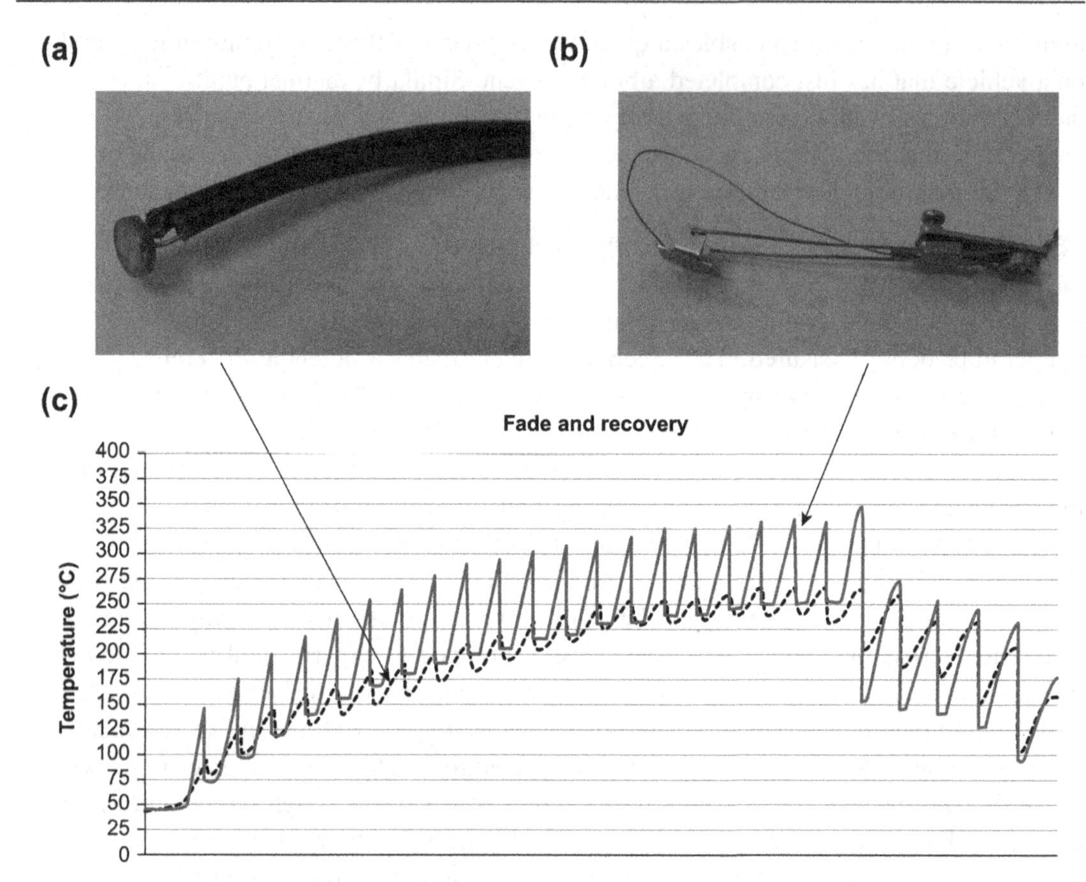

Figure 9.1: Comparative Performance of Two Types of Rubbing Thermocouple (McLellan, 2012). (a) Bare wire rubbing thermocouple with copper button. (b) Sheathed rubbing thermocouple with steel skid. (c) Comparative performance of the two types of rubbing thermocouple.

affected by temperature 'banding' across the rubbing path so the measured temperature is not fully representative of the typical rotor surface temperature under those operating conditions. Temperature measurement for European type approval specifies the rubbing thermocouple to be contacting the outside of a brake drum, and the periphery of a brake disc, which is intended to minimise variations arising from transient and contact effects at the friction interface. The result is that lower temperatures are measured because the sensor is further away from the friction interface, but provided that the test protocol is always the same, results are directly comparable.

'Embedded' thermocouples are widely used (especially in brake pads by the North American industry), but their setting up and interpretation of the measurements compared with those from rubbing thermocouples are more difficult. This is because (a) the embedded thermocouple introduces a tribological non-homogeneity in the surface of one part of the friction pair, (b) the local heat transfer is affected by the thermal mass of the embedded thermocouple, and (c) it is difficult to set the precise depth of the thermocouple sensor relative to the friction surface in which it is embedded, which means that it may be measuring the temperature at a different part of the temperature profile. The thermocouple can be embedded in the friction material (stator) or in the rotor and the temperature/depth sensitivity is especially critical in friction material because the temperature gradient normal to the friction interface is steep as a result of its insulating thermal properties. A thermocouple embedded in the friction material can only indicate friction surface temperature when the brake is applied and the friction pair is in full contact. If the thermocouple is embedded in the rotor it will only indicate temperature at that one location and therefore provides a cyclical measurement that depends on the depth of the thermocouple, the conductivity of the rotor material and any hot banding at the friction interface. Embedded thermocouples (and any other instrumentation that is mounted to the brake rotor) also require transmission of the transducer signal to the conditioning and recording devices. Historically this requirement has been met by the use of bulky and expensive slip rings, but modern telemetry devices are now available, which are much more convenient.

The availability of compact and lightweight sensors has made non-contacting infrared (IR) temperature-sensing technology very attractive for measuring surface temperatures in experimental brake testing. However, the reliable and accurate measurement of brake temperatures by IR is difficult because of the uncertainty associated with the emissivity of the surfaces being measured; the accurate interpretation of the IR reading depends upon the emissivity of the surface radiating the heat. This applies to all surfaces but especially the friction surfaces because continual changes in the physical state of the friction surface due to the transient nature of third body layers, corrosion and contamination, surface topography and temperature directly influence the emissivity of the rotor surface. Whilst temperature distributions over the surfaces may be consistent at any point in time and can be compared, the absolute temperature values may be inaccurate. It is essential to carry out a detailed calibration against a conventional method of temperature measurement under the precise conditions of operation, and to check this frequently during the testing.

Deceleration

Since the required function of a vehicle braking system is deceleration, its measurement is very important in the experimental testing of vehicle brakes and braking systems. By recording the instantaneous deceleration as a time series, variation during a brake application can be examined so that in-stop characteristics can be quantified. Traditional methods of measuring vehicle deceleration included portable decelerometers, which were mounted inside the vehicle and read by an observer. The two most popular were 'pendulum' devices: the 'Tapley' meter, which was adopted by the UK Ministry of Transport before the introduction of the rolling road test; and the 'Bowmonk' decelerometer, which was much more compact and could be mounted to the windscreen or dashboard to be read by the driver or observer. This latter device is still occasionally used to provide a quick check on a vehicle's braking performance. Electronic accelerometers are now readily available and are almost universally used in conjunction with portable datalogging systems on road vehicles to provide a full record of braking performance during vehicle experimental testing. Three-axis accelerometers are usually specified (sensing longitudinal, lateral and vertical acceleration/deceleration) since the extra cost is negligible. They should be mounted as near as possible to the vehicle's centre of mass to minimise the errors resulting from transient changes in vehicle pitch caused by weight transfer during braking (or acceleration). These can be avoided if the accelerometer is mounted on a gyroscopic self-levelling base, but this is not usually necessary for measuring longitudinal braking and acceleration in modern road vehicles since pitch changes are small. Calibration is easily achieved by mounting the accelerometer on a hinged plate, which can be raised to the vertical (equivalent to 100%g) via specified angular increments. An alternative to the decelerometer is measuring the rotational speed of a separate low-inertia wheel attached to the vehicle, usually trailing behind it. Vehicle deceleration can be calculated from the change in wheel angular velocity and can eliminate measurement errors relating to weight transfer and pitch change.

Visual display of deceleration to the test driver has been found to be important. When conducting vehicle braking tests, the driver is often required to generate a constant rate of braking, which requires modulating the pedal force in response to the deceleration display. For a driver to interact with and respond to a real-time display, an analogue-type dial display calibrated to provide a near 360° full-scale deflection has been found to be easiest to respond to. Ribbon or digital numeric displays are less effective for test driver interaction, although for reading steady-state or slowly changing data digital numeric displays help to eliminate reading errors. This applies equally to all measurements where driver interaction or manual reading is needed.

Torque

There are two methods used to measure the torque generated by a brake, namely a torque sensor and a 'torque arm'. The former is fitted 'in-line' with the shaft or component that is transmitting the torque, and works by sensing a torque-sensitive parameter such as the 'twist' of the shaft or the relative angular displacement of the sensing elements of the device. Such devices are able to sense dynamic torque up to frequencies in excess of 1 kHz, and may be used on laboratory equipment such as dynamometers or on actual vehicles. Where a brake is to be tested as part of a 'wheel corner', e.g. mounted on a brake dynamometer with the associated vehicle suspension components, an in-line torque transducer between the hub and the dynamometer drive is very effective. The torque arm method is usually associated with static test rigs, whether full-scale inertia dynamometers or small-sample friction test machines; the brake caliper or torque plate is mounted to a freely rotating shaft, which is restrained from rotating by a radial lever or 'torque arm' attached to it, the distal end of which is connected via a load cell to 'ground'. The product of the force measured by the load cell and the torque arm length indicates the torque generated. The relatively high inertia of this system precludes its use for dynamic torque measurement above a few hertz. By altering the effective length of the torque arm it is possible to use a standard range of load cells within a test laboratory to cover a wide range of measured torques, while retaining suitable sensitivity of measurement.

Where experimental determination of individual brake performance or braking balance between individual wheels on full vehicles is required, torque sensors in the form of 'torque wheels' are used. These are road wheels fitted with strain gauges or similar sensing devices, which provide a direct indication of wheel torque on a vehicle on a test track; torque wheels are generally not considered to be safe for use on public roads. As with embedded thermocouples, the signals from the rotating wheel require the use of telemetry to transmit the data to the signal conditioning and datalogger units. Torque wheels are sensitive to transient forces and torques caused by irregularities in the road surface, which can introduce substantial noise into the torque signal, so sophisticated filtering, signal conditioning and data processing are required. They are expensive and durability can be an issue, but the technology is evolving to address these. Brake torque at each wheel of a vehicle can also be measured in safety with the vehicle stationary using a chassis dynamometer, and with the addition of brake test instrumentation it is possible to investigate braking distribution between axles and basic compatibility on commercial vehicle combinations.

Force

Many different types of load cells and other force transducers are available for use in experimental brake testing. On vehicles the measurement of the actuation force at the brake is avoided if possible because of the high temperatures, restricted space and hostile environment. It is used only where necessary, e.g. to confirm actuation system function and efficiency, and instead brake actuation force is usually calculated from the actuation pressure, which is easier to measure accurately. Driver effort, i.e. the applied pedal force, is often measured using a force transducer between the brake pedal pad and the driver's foot, which in the past has taken the form of a portable instrument based on a hydraulic or pneumatic pad clamped to the pedal, which is piped to a calibrated pressure gauge. Such transducers are bulky and difficult to use, and tend to affect the ergonomic relationship between the driver's foot and the brake pedal so that the measured force can be unrepresentative. They should only be used by experienced drivers and should not be used on public roads. Another problem with brake pedal effort measurement is that drivers seldom press the brake pedal perpendicular to the pedal pad, while force transducers measure the perpendicular component. More up to date and convenient methods include an electronic load washer or strain gauging the pedal lever, both of which minimise any interference with driver ergonomics and can be used by any driver.

A recent development in the controlled measurement of driver effort is the 'robot driver', which is an actuation device that can be fitted to the driver's seat and connected to the brake pedal via an actuation rod, while still allowing a human driver to operate the vehicle. This device can achieve very accurate programmable force applications to the brake pedal at exactly the correct angle of force to represent the driver's foot and leg actuation. A load cell mounted on the actuation rod measures the applied force, which is fed back to the controller to generate a highly consistent pedal effort at the set level. The accuracy and repeatability of the device has been demonstrated to be much higher than can be achieved by even the most skilled human driver. Because the device sits on the driver's seat and a human driver is still required to drive the vehicle, packaging is intrusive and special training is required to use the equipment.

Pressure

Brake actuation pressure downstream of all valves, boosters, ABS pump, etc., whether pneumatic or hydraulic, is a fundamentally important parameter that must be measured accurately in experimental brake testing. A traditional Bourdon tube pressure gauge can be used to provide an immediate indication of line pressure, but the gauge and the associated piping does affect the actuation characteristics, especially in hydraulic system fluid consumption and pedal travel, and Bourdon tube pressure gauges should not be used when pedal feel assessment is required. Bourdon tube pressure gauges need to be mounted in the driver's line of sight and thus are intrusive in the vehicle's interior and present a potential safety hazard. In-line pressure transducers have negligible effect on system characteristics, and the signal output can easily be recorded on a suitable datalogger. Care must be taken when using such transducers on vehicles with hydraulically actuated braking systems to ensure that any seals within the transducer are compatible with the brake fluid used. In the case of commercial vehicles with pneumatic brakes, installation is simplified by the UN Regulation 13 requirement for all vehicles to provide suitable test points for pressure measurement to confirm various valve operations. A suitable test point connector can be found at each air chamber on the vehicle.

Displacement and Distance

Displacement measurement is required to investigate and confirm system actuation. Linear variable displacement transducers (LVDT) are used to measure large displacements such as brake pedal travel or pneumatic actuator displacement, and an alternative is a 'string potentiometer' where a linear displacement is converted to a rotary displacement for measurement and can be easier to fit into confined spaces. Very small displacements need to be measured to assess circumferential variations in rotational precision, e.g. the concentricity and 'ovalising' of brake drums and 'coning', 'runout'' and thickness variation of brake discs. These are important parameters in brake-induced vibrations, particularly judder, as explained in Chapter 10, and non-contacting displacement transducers, e.g. those that utilise capacitative principles, are useful for this type of small displacement measurement. Such transducers need to be robust to heat, temperature and mechanical vibration because they have to be located close to the disc surface, and are more suited to laboratory dynamometer use rather than on-vehicle measurement.

Distance travelled is an important parameter in determining braking system performance, e.g. for legislative testing, and if distance can be recorded on a time base, speed can also be determined. For long braking distances an accurate measurement of distance travelled can easily be achieved by marking the beginning and the end of the braking event by marks on the road, but this is seldom practical nowadays. A more accurate measure of vehicle stopping distance and also speed can be obtained from a 'fifth wheel'. This is a separate lightweight wheel of known diameter but should not be confused with the term 'fifth wheel' applied to the coupling between the tractor and semi-trailer of an articulated commercial vehicle, see Chapter 4. It can be mounted on to the test vehicle on a suspension system to maintain ground contact at all times throughout the braking event and the distance is calculated from the rotation of the wheel. Optical sensor systems are extensively used, for example based on an optical sensor reflecting off the road surface. Commercial systems are available that are generally unaffected by the physical nature of the road surface, although wet road conditions can be challenging. GPS sensor systems can now provide accurate measurement of vehicle position and speed, and because of the small size and convenient packaging GPS technology is very suitable for vehicle testing and is becoming widely used.

Vibration Frequency, Amplitude and Phase

Experimental techniques and instrumentation for the measurement of noise and vibration represent a very specialised area of activity, which is covered in more detail in Chapter 10. A portable 'noise meter' (which may actually measure sound pressure level) can be used to measure the frequency of specific noises and identify frequency components of complex waveforms. Noise can be recorded on a hand-held voice recorder for subsequent analysis, and both these are very useful during the experimental brake testing of a vehicle whether on a test track or on the public road, since they are completely self-contained and do not require installation on the vehicle. Vibration measurement usually requires some form of accelerometer transducer, which is attached to the component under investigation. For vehicle brake testing, the associated signal conditioning and datalogging equipment usually requires fitment into the vehicle. For laboratory use, very sophisticated techniques and instrumentation equipment can be employed to measure and analyse frequency, amplitude and phase; these include the use of lasers, which are subject to strict safety controls. More information about experimental techniques for measuring brake noise and vibration is presented in Chapter 10.

Experimental Design, Test Procedures and Protocols for Brake Testing

The fundamental performance of a friction braking system for road vehicles is defined by the friction pair, which forms the core of the foundation brake. The lowest level of brake component that would normally be considered for experimental testing is included in the foundation brake package illustrated in the boundary diagram in Figure 9.2. The ability of a friction pair to achieve optimal performance depends entirely upon the nature of the contact and pressure distributions achieved at the friction pair interface, but testing of the friction pair components in isolation is only considered for material design and research purposes, and this is covered later in this chapter. Inside the boundary the friction material and rotor (disc or drum) form the friction pair. The friction material is mounted to the stator to form the disc brake pad or drum brake shoe assembly, which in turn is located by the disc brake caliper or drum brake torque plate. The actuation force is applied to the stator as explained in Chapter 5. Outside the boundary are examples of other vehicle components and systems that interact with the foundation brake. Depending upon the purpose of the test, these components and systems may or may not need to be included (see Table 9.1).

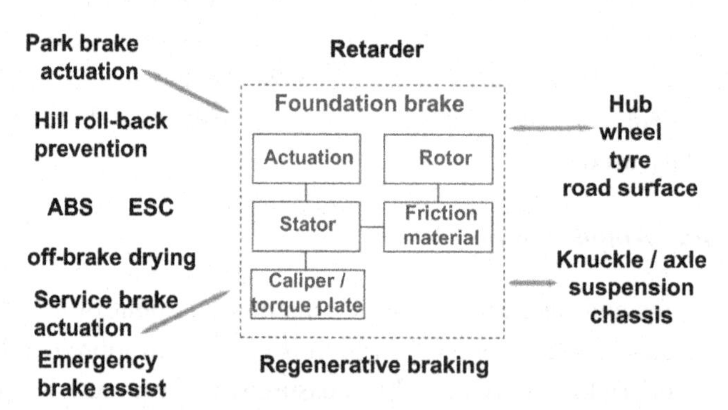

Figure 9.2: Foundation Brake Boundary Diagram (After McLellan, 2012).

Table 9.1: Generic Experimental Test Options for Brakes and Braking Systems (McLellan, 2012)

Brake and Braking System Test Options	Focus	Purpose of Testing
Vehicle fleet	Total system and components	Durability Wear 'Manners' Operating environment
Individual vehicle	Total system and components	Performance evaluation and design verification Product liability Legislative compliance Dynamic behaviour NVH qualification Roadworthiness
Vehicle axle (rolling road dynamometer)	Wheel end components	Design validation Legislative compliance Concern qualification Concern resolution
Vehicle corner (full-scale inertia dynamometer)	Wheel end/suspension components	NVH resolution
Wheel end (full-scale inertia dynamometer)	Wheel end components	Performance mapping Friction pair development Quality
Component rigs (cyclic torque and fatigue)	Individual components	Design validation Durability Environmental testing Quality
'Bench' test rigs using full-size brake components	Individual components	Friction and wear assessment, Friction material formulation and development Quality
Scale rigs (e.g. Chase machine, FAST machine)	Friction material	Friction and wear assessment, Friction material formulation and development Research Quality
Computer modelling and virtual testing	Whole vehicle system to individual components	Design Pre-test prediction

From the boundary diagram illustrated in Figure 9.2, a 'P-diagram' can be generated showing the driver's viewpoint as illustrated in Figure 9.3, which helps to identify some of the issues that might provide reasons for brake testing. It should be noted that pedal effort has no meaning for power brake systems including pneumatic systems because the force

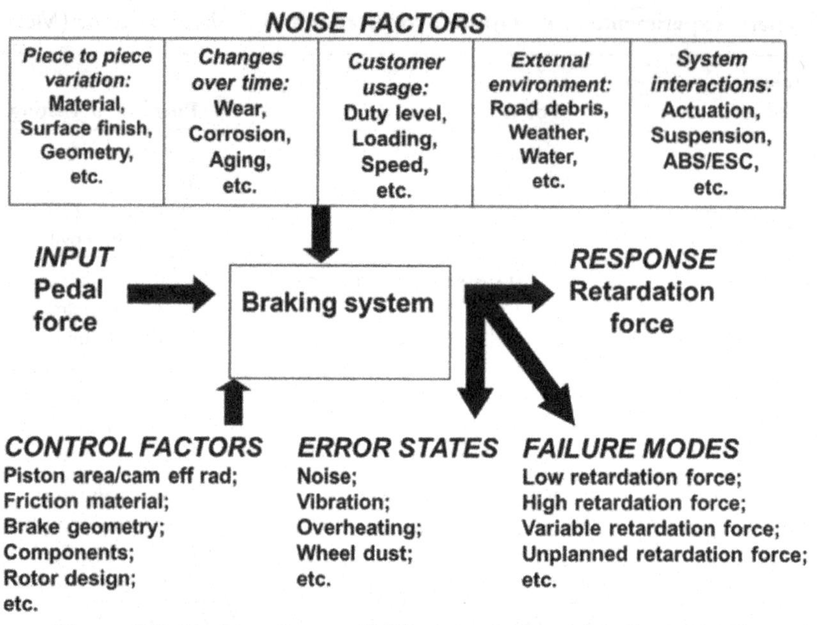

Figure 9.3: Braking System P-Diagram (After McLellan, 2012).

that needs to be applied by the driver is decided by the design of the operating valve and not related to the muscular energy needed to actuate the brakes.

A test procedure comprises a list of the steps that must be taken in order to complete the test and obtain the desired test data, results and outcomes. Most brake test procedures follow some form of standardised protocol so that results from different tests can be compared with those from similar tests carried out by other organisations. Considerable efforts have been made to develop standardised test protocols across the global brake industry, and the relevance and success of these seem to depend on the intended area of investigation. The simplest level of brake performance test is best suited to the development of a standard protocol, e.g. in the area of type approval or certification testing where the relationship between actuation force and output force is to be quantified. More complex types of performance test such as those required to assess brake noise are less suited to the development of a standard test protocol, but such protocols have been usefully designed, e.g. the SAE J2521 test procedure for high-frequency squeal noise. Test procedures for research purposes are least likely to be able to follow established protocols, but even so there is benefit in using them wherever possible, or at least referring to them. As a general rule, the identification and quantification of error states and robustness to noise factors in brakes require the development of sophisticated and dedicated test procedures. The generic test options available for the experimental testing of road vehicle brakes and braking systems are outlined in Table 9.1; the cost of the test options broadly tends to decrease down the table, from 'Vehicle fleet' to 'Computer modelling and virtual testing'. The basic test parameters that must be controlled and/or measured are common across the range of experimental brake testing. Many types of transducers are commercially available to measure each parameter as explained earlier, but occasionally new techniques need to be developed for which purpose-designed instrumentation is required. Such instrumentation may ultimately find its way into mainstream usage and availability.

Test Vehicles, Dynamometers and Rigs

As indicated in Table 9.1, there are many options for testing brakes and braking systems, and a wide range of machines, rigs and devices exists for the experimental testing of brakes, from full vehicles to 'scale' test machines. A complete vehicle can be instrumented with sensors as described above, fitted with data acquisition equipment, and driven on a test track to undertake specific test procedures, or on public roads as 'field testing' for general 'real-world' data gathering. The very large capability of modern data acquisition systems for on-board data collection, processing and storage means that data from entire journeys of several hours' duration can be recorded for subsequent analysis. Arrangements may be made with specific types of fleet users to gather more general data about the performance of the braking system, e.g. durability, wear and 'manners' as stated in Table 9.1. 'Manners' is a term that is used to describe the behaviour of a vehicle's braking system to an ordinary driver. Because the emphasis in vehicle fleet testing tends to be on high mileage over a short time coupled with significant duty levels, arrangements are often made to gather data from vehicle in police or taxi fleets. The vehicles are often not fitted with any specific instrumentation, the emphasis being on wear and condition assessment after an extended period of use in the 'field', but sometimes the monitoring of particular parameters necessitates a relatively low level of instrumentation and datalogging, which must be robust and invisible to vehicle users. Similar arrangements can be made for commercial vehicles, e.g. semi-trailer fleets are particularly useful for durability and wear evaluation in the field. Careful statistical analysis is required to interpret vehicle data obtained from fleet trials because of the limited sample and the non-random selection.

Brake testing on actual vehicles driven on test tracks or public roads is always associated with a high degree of variation arising from non-controllable factors such as driver style, weather, traffic conditions, etc. In the past, some vehicle and component manufacturers utilised closely controlled test routes on public roads, where the vehicle was driven over a specified route a number of times at the same time each day. Timed stages in the route controlled the speed, and in the event of excessive deviation from the schedule, e.g. because of traffic delays, the test would be aborted. Vehicles would be instrumented and drivers were required to apply controlled brake applications at intervals that, together with those needed for routine driving, were not supposed to exceed a defined number for each stage. Brake temperatures were also monitored and compared with defined values at each stage on the route. If any control parameters were exceeded the test would also be aborted. As vehicles, roads and types of driving have changed, this type of closely controlled road test schedule has been replaced first by track testing (and more recently by 'rolling road' testing). Test tracks have been built at 'proving grounds', which may be operated by vehicle manufacturers, although there are many independent proving grounds that offer a range of test tracks to simulate actual roads and conditions. For example, a proving ground may include a high-speed circuit to simulate motorway driving, a low-speed circuit to simulate urban roads, a long straight track for brake performance testing, tracks with rough surfaces, and low adhesion and 'split-μ' tracks for stability testing. Additionally there may be water troughs for wet recovery tests, salt-spray and humidity test facilities, etc. Proving grounds are designed and operated to provide for all aspects of vehicle testing, but are essential for brake testing away from the dangers of the public road with the added benefit that variability is reduced because the route is predetermined and the traffic is controlled.

Experimental testing of brakes and braking systems on complete vehicles on the road or test track is time-consuming and expensive, and the trend over the last 20 years has been for manufacturers to increase their utilisation of laboratory and virtual testing methods. The most obvious alternative to vehicle field testing is'rolling road' or 'chassis' dynamometers. These are widely used for vehicle performance testing, particularly fuel consumption, where a number of standard 'drive cycles' have been established for the purpose. Such drive cycles are not designed for brake testing on a chassis dynamometer, so special drive cycles for brake testing are required. Chassis dynamometers use electric motor/generator units connected to each wheel to simulate vehicle mass and inertia for the purposes of acceleration and deceleration. For brake testing, motor/generator torque is applied to represent the required energy dissipation over the defined braking event, and rolling resistance and gradients can be included in the control logic. The rolling road type of chassis dynamometer connects the motor/generator to the vehicle's brakes through large-diameter rollers on which the vehicle's wheels sit, which thus includes the effects of tyre/road contact and slip in the testing. In principle this is beneficial, but in practice the curvature of the tyre/roller contact patch can adversely affect tyre durability (the amount depends upon the diameter of the drum), and the surface (and curvature) of the drum is not necessarily representative of a road surface. Four-wheel rolling roads can be used for braking distribution evaluation, but unless the control system includes the simulation of the effect of braking weight transfer, two-wheel systems can be just as useful. Motor/ generators may operate one per axle or one per brake if side-to-side variation is required to be investigated. Because of the size of the rollers and the motor/generators, rolling road chassis dynamometers are specialist test facilities. A smaller and lower cost alternative is the direct drive chassis dynamometer, where the motor/generator is coupled directly to the wheel hub, and the tyre/road contact is bypassed.

Performance tests, wear tests and many other types of brake tests can be carried out in the laboratory using inertia dynamometers and specialist test rigs. The test procedures for these may either be designed to simulate brake duty levels and distributions similar to those experienced on actual vehicles, or may be purpose designed for specific investigative purposes. The most widely used laboratory-based machine in experimental brake testing is the inertia dynamometer. Shaft-type inertia brake dynamometers (see Figure 9.4) are designed so that the brake rotor is attached to a driven shaft (usually by an electric motor) on which flywheels are mounted to simulate the appropriate part of the vehicle inertia and kinetic energy. The brake stator (caliper or torque plate/anchor bracket) is mounted to the

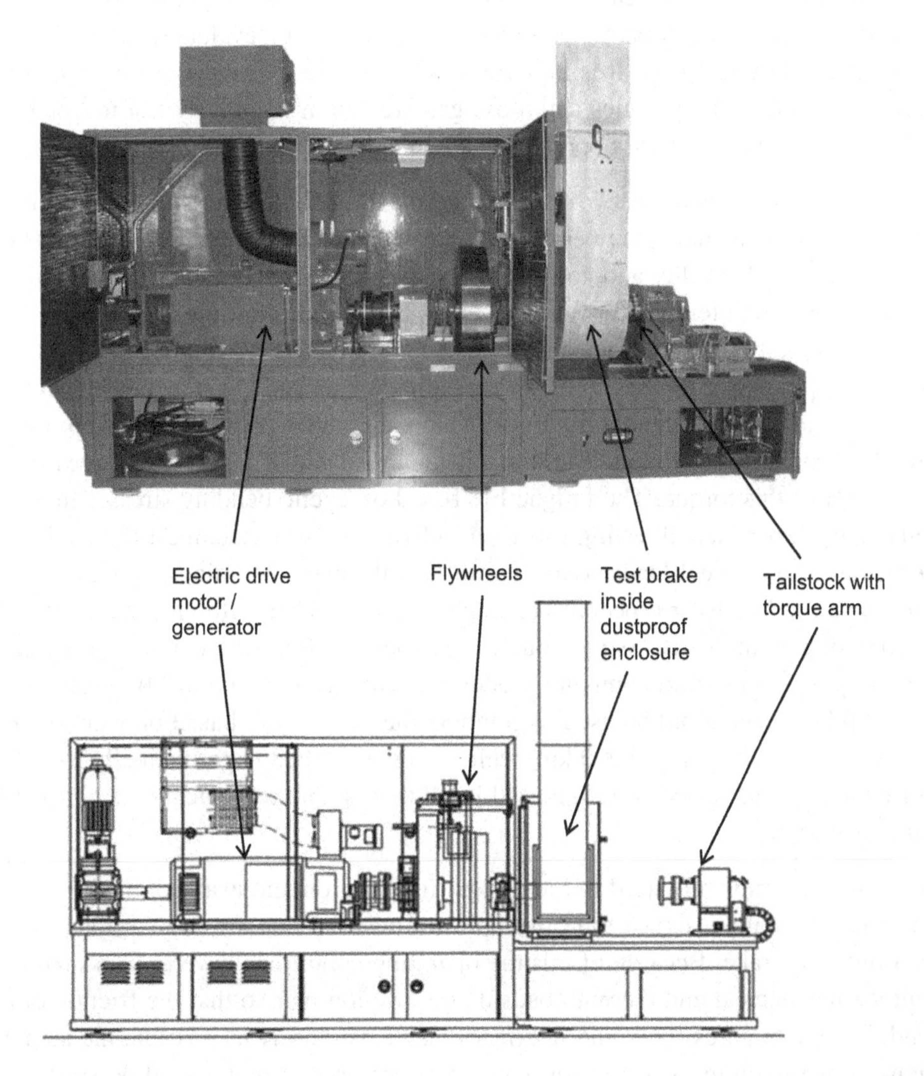

Figure 9.4: Inertia Dynamometer for Brake Testing (Courtesy of Link Engineering).

'tailstock' shaft of the machine; this shaft is free to rotate on bearings and is constrained by a torque arm with a load cell to measure torque. It is also possible to mount the stator on its suspension, e.g. for noise and vibration testing. The machine operates by engaging the motor drive to rotate the flywheels and rotor, and then when the required rotational speed is reached, the drive is disconnected and the brake is used to decelerate the flywheels. Inertia dynamometers are suitable for performance, durability, wear and noise and vibration brake testing, and may be installed with environmental control chambers and many other features to simulate the actual conditions experienced on the vehicle. Sophisticated control systems enable the dynamometer to simulate inertia in place of flywheels, making changes in vehicle weight conditions such as weight transfer easy to accommodate. Regenerative braking and brake blending can be evaluated on specially designed inertia dynamometers. It is important to correlate experimental test data between those gathered from actual vehicles and those gathered from laboratory test machines such as inertia dynamometers so that laboratory test data are known to be representative.

Brake component test rigs are used for design verification and for quality monitoring purposes, and environmental chambers may be provided to investigate the effect of factors such as temperature, humidity and corrosion. For example, in operation brake components are subject to very high temperatures and stresses, usually in the form of cyclic loading with substantial load reversals. Specialised servo-hydraulic test machines are used to generate and apply cyclic torques to the brake caliper or torque plate/anchor bracket using where possible an industry standardised peak value, e.g. the torque required from the brake for 1 *g* vehicle deceleration. The brake is then actuated at high enough pressure to clamp the rotor at this torque. The fatigue life based on cyclic bending stresses may also be evaluated by experimental testing using servo-hydraulic test machines; the loads may be based on dynamic wheel loads coupled with lateral forces including impact forces, e.g. from striking a road kerb, or especially on commercial vehicles with close-coupled axle bogies, cornering scrub loads. As previously explained, such testing is now mainly carried out to verify predictions from computer modelling; Tumbrink et al. (2000) explained how the 'strain life' concept could be used to compute the service life based on worst-case customer usage for a life of 300,000 km, which is then verified experimentally using a randomised mixture of scaled or normalised loads that is applied typically at 17 Hz and ambient temperature.

Scale rigs are predominantly used in the design, formulation and evaluation of friction and rotor materials, and are designed to test samples of material that are smaller than would be used in a full-size brake. Because of this the operating conditions have to be scaled to represent the mechanical and thermal loads on the friction pair so that the friction drag force and the temperatures (bulk and interface) match as closely as possible the actual conditions of operation in the brake for which the friction pair is intended. Resin-bonded composite friction material is operating temperature sensitive (see Chapter 2), and therefore the operating temperature is fundamentally important. Because the friction pair is much smaller than in a real brake, the same braking power density generates much less heat, and the temperature rise in the friction pair components may be lower as a result. Additionally, the smaller dimensions mean there is a reduced heat transfer path so the heat is conducted away at a different rate, and surface heat transfer is affected. If tests are being controlled by temperature measured by an embedded or sliding thermocouple on the rubbing path, the operating conditions will not be representative, so the results will not provide a true value for the friction coefficient or wear rate. Despite this there is no doubt that small-scale friction rigs can be useful and they are extensively used for research and development purposes.

One type of scale rig that is extensively used is the 'pin-on-disc' machine for the characterisation of the friction and wear properties of many different sliding friction pairs. These machines comprise a rotating disc driven at constant speed by an electric motor, against which a small 'pin' of material is pressed. The pin is usually mounted in a holder that is connected to a force transducer so that the friction force can be directly measured, but an in-line torque transducer can be used as an alternative. The rotating axis can be vertical or horizontal; the latter is preferred for research purposes as 'dead-weight' loading (gravity force) can be directly applied. For resin-bonded composite friction material evaluation, the pin size should not be less than 10 mm diameter so that a representative friction material surface is presented to the rotor. For more homogeneous material such as metals and polymers, smaller pin sizes can be used. Care must be taken to ensure that the actuation and restraining force vectors are aligned correctly and any servo effects from the pin carrier and actuation system are accounted for, otherwise the measured friction force will be inaccurate. Usually pin-on-disc machines are used under constant normal load conditions, the load being applied over a period of a few seconds and then removed to allow the friction interface to return to the required temperature. The information summarised in Table 2.1 was generated on a vertical axis pin-on-disc machine.

For commercial (rather than research) purposes, the minimum size of a scale friction material specimen was for a long time considered to be approximately 12.5 mm × 12.5 mm (1/2 inch square). The FMT (also known as 'Chase') machine has been a standard machine for friction and wear assessment since the 1950s, being used to provide a general assessment of the friction and wear of resin-bonded friction materials for rating purposes, quality control and friction material development. It utilises a 25 mm (1 inch) square specimen dead weight loaded against the inner surface of a 280 mm (11 inch) brake drum rotating about a horizontal axis, and its operation is defined in SAE J661. Electrical heaters are provided to maintain required operating temperatures. The standard test is $2\frac{1}{2}$ hours and covers different cycles to assess baseline friction performance, fade and recovery, and wear, providing μ−speed information. The Friction Assessment Screening Test (FAST) machine is intended for disc pad materials; it is also a horizontal axis machine but with a disc rather than a drum. The 12.5 mm (1/2 inch) square specimen is actuated by axial force generated by a calibrated (hydraulic) actuation system. The standard test is 90 minutes at constant torque and constant speed, providing μ–speed information. More information about these machines can be found in Rhee and Schwartz (1979). An alternative to the pin-on-disc configuration is the annular configuration small-sample tester, which is less representative of actual operational conditions in automotive disc and drum brakes, but is useful for annular disc brakes as used in aircraft and some special purpose vehicles. The limitations of small-sample test machines to represent the friction and wear characteristics of modern automotive friction materials under real-life operating conditions led to the development of bench test machines for the characterisation of the friction and wear properties of automotive materials using actual brake components. These machines comprise a rotating brake disc or brake drum driven by an electric motor usually at constant speed, while the brake is actuated; an example is the 'Krauss' test machine developed in Europe in the 1970s (Rhee and Schwartz, 1979).

An example of a purpose-designed small-sample scale friction research test rig is shown in Figure 9.5. This machine has two working heads, each comprising a 125 mm diameter disc manufactured from the desired rotor material and a sector-shaped specimen of approximately 625 mm^2 (1 square inch). The actuation force is applied via a hydraulic piston and measured by an electronic load cell. Torque is measured by a torque arm arrangement with another electronic load cell. The machine is fully computer controlled and can be fitted with an environmental chamber for ambient temperature control. It can be used for both friction material performance mapping and wear mapping at power densities up to 5 W/mm^2.

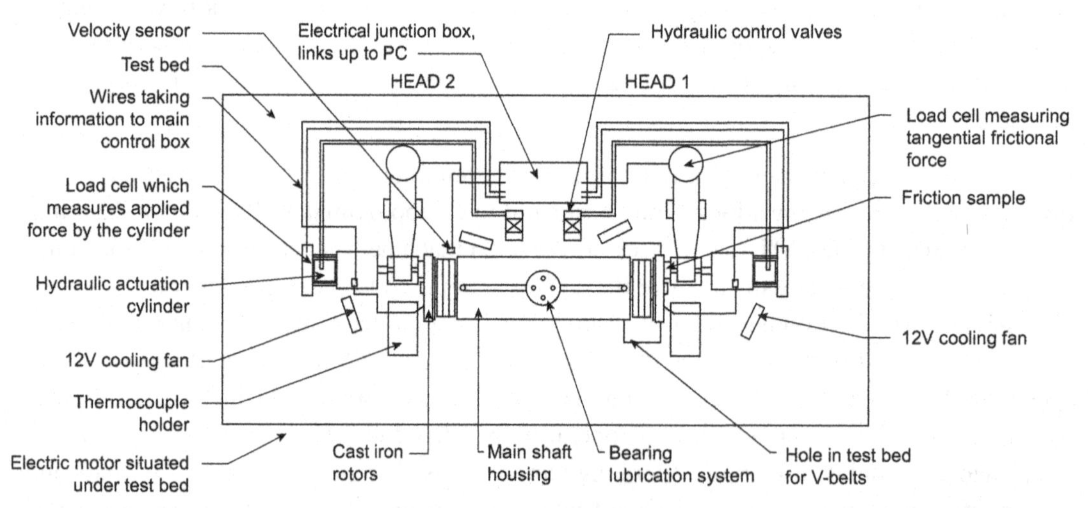

Figure 9.5: Purpose-Designed Small-Sample Scale Friction Research Test Rig.

Brake Experimental Test Procedures

Preparation, Bedding and Burnishing

With all brake experimental testing it is essential to take account of the preparation and conditioning history of the brake system and friction pair before critical measurements are taken. Unless the purpose of the test is to evaluate a brake installation in its operational condition from new, consistency and repeatability of results that are representative of the stabilised performance of brake friction pairs will only be obtained if the friction pair is carefully prepared through a process of bedding and burnishing (see Chapter 2). The goal of a bedding and burnishing procedure is to achieve full contact and tribological compatibility between the two mating surfaces of a brake friction pair while at the same time causing minimal alteration to the physical and chemical nature of the two surfaces. With resin-bonded composite friction materials this means keeping the maximum operating temperature of the friction material during the process well below the bulk temperature it has experienced during the manufacturing process. 'Green' performance is the term used to describe the performance of a brake friction pair before bedding; it is of critical safety importance on a newly produced vehicle and after servicing where friction components have been replaced. When new brake pads or lined shoes are fitted, care should be taken to ensure that their friction surfaces conform to the rotor surfaces, e.g. brake discs should not be significantly worn. In the case of drum brakes, the radius of the friction surface of the brake linings should match the inner radius of the drum, and if there is a significant difference they should be radius ground (see Chapter 5). This is particularly important for commercial vehicles where the bedding of a large drum brake with long-life linings can be a very protracted exercise. The friction surfaces of brake discs or drums can be prepared for bedding/burnishing before performance testing as follows:

1. Clean the rotor friction surface with a degreasing agent.
2. Use medium (grade 60) emery paper to lightly abrade the surface in a random fashion.
3. Wipe with a clean lint-free cloth to remove dust.
4. Clean again with a degreasing agent.
5. Wipe again with a clean lint-free cloth to remove all traces of preparation.

A bedding-in procedure for a typical passenger car could comprise the following sequence of brake applications, with the emphasis on conditioning the friction surfaces (burnishing) rather than wearing them. Operating temperatures should be kept low ($<100°C$ start of application temperature, less than 200°C maximum), and equivalent rates of braking should be limited to less than 0.25, although the low wear rate of some modern (NAO) friction materials extends the time and cost of bedding and so high deceleration or torque levels are more often employed during bedding nowadays (up to $z = 0.3$), and temperatures may be allowed to reach 250°C. It is always recommended that random and varied duty levels are incorporated into bedding cycles, and a few higher deceleration applications towards the end of the bedding-in sequence are useful to complete the friction material surface conditioning.

1. Start temperature 80°C (measured by rubbing thermocouple on the disc friction surface), vehicle road speed 50 km/h, apply 20 brake applications to decelerate to 30 km/h at 20%*g* deceleration, vehicle laden to midway between laden and unladen condition. Drive at the same constant road speed after each brake application to cool the brakes to the required start temperature.
2. Increase the road speed to 70 km/h and repeat the process of applying 20 brake applications to decelerate to 50 km/h at 20%*g* deceleration, vehicle laden to midway between laden and unladen condition. Drive at the same constant road speed after each brake application to cool the brakes to the required start temperature. At no time should the indicated temperature exceed 200°C.
3. Repeat 1 and 2 until a total of 200 brake applications have been completed. If possible disassemble the brakes to inspect the friction surfaces for completeness of bedding and quality of burnishing (see Chapter 2). Occasional high-duty applications may be inserted into the sequence as mentioned above.
4. Reassemble the brakes and repeat 2 from a start temperature of 100°C, after 10 of the 20 applications apply five brake applications from 70 to 50 km/h at 50%*g* deceleration.

The vehicle's brakes should then be ready for performance testing, although care should still be taken to compare the first set of test data with subsequent test data for consistency. Many test procedures, such as those required for legislative testing, specify the bedding and burnishing procedures.

Preparation must also include validating the test set-up, confirming brake actuation and control, health and safety provisions including risk assessments, checking and calibration of all sensors, transducers and instrumentation, and confirmation of the correct functioning of signal conditioning and datalogging equipment. The high level of sophistication and complexity in modern instrumentation and data acquisition systems can easily obscure failure or inaccuracy. Most brake test procedures include 'instrument checks' in the initial stages for these purposes.

Braking Performance

'Braking performance' (sometimes known as 'effectiveness') experimental testing may take many forms, including noise and vibration, water recovery, salt recovery, ambient temperature, humidity, static/parking, partial system fail and vehicle compatibility/overload. It is essential to determine the purpose of a braking performance test before a suitable test procedure can be selected or designed. To the driver, the most important braking performance characteristics are stopping distance and driver effort (or brake pedal force) for a muscle actuated braking system (hydraulic; see Chapter 6). Essentially this represents driver effort as the input force to the system, and stopping distance as the response, but as indicated in Figure 9.3, the response is usually associated with Newton's second law, i.e. the retardation force. Because pedal effort has no meaning for non-muscle actuated, e.g. pneumatically actuated, braking systems (see Chapter 6), the brake actuation force is usually considered as the system input force and this is conveniently represented by the actuation pressure at the foundation brake. The limits of a brake performance envelope are controlled by the physical characteristics of the rotor in terms of ultimate thermal capacity, maximum operating temperature, ambient temperature and operating environment, and the energy conversion capabilities of the friction pair (see Chapter 7). Power density also has a finite maximum value within the vehicle's capabilities in terms of mass, speed and tyre/road adhesion, and care should be taken not to exceed this if testing on a brake test dynamometer where tyre/road adhesion is not a limiting factor.

The basic brake performance test carried out on a road vehicle consists of a sequence of braking events controlled at set decelerations (or actuation pressures) and at specific load conditions (usually laden or unladen) from a range of initial road speeds to rest (or nearly to rest) with the vehicle out of gear to minimise the effect of transmission drag. Prior to any testing the vehicle should always be checked to ensure it is operating correctly and safely, and the instrumentation and datalogging is functioning correctly. The test procedure should specify whether the driver (human or robot) is required to apply the brake at constant actuation pressure or to achieve constant deceleration. The test should be conducted on a straight, level, test track with a high adhesion zero camber surface, and successive brake events should be driven in the opposite direction to balance out any effect of wind and incline. Each braking event should follow the procedure stated below, and be repeated at least twice, making a minimum of three for each setting. The vehicle should be equipped with instrumentation to measure longitudinal deceleration, actuation pressure at each brake (although some procedures only call for actuation pressure to be measured at one brake on each axle), deceleration, road speed and, if required, parameters such as pedal travel, preferably recorded continuously on a time base on a datalogger. ABS/ESC is usually disabled for basic brake performance testing. A basic brake performance test procedure is outlined below:

1. Commence each braking event only when the initial temperature of the brake (as indicated by a rubbing thermocouple or an embedded thermocouple as explained earlier) has reached the specified 'start of stop' temperature, e.g. 100°C. If the indicated temperature is too low, drive while gently applying the brakes to increase the temperature to the set value.
2. Drive the vehicle at constant road speed 5–10 km/h above the set speed, select neutral gear and coast until the set speed is reached, and then apply the brake to achieve the required deceleration or pressure.

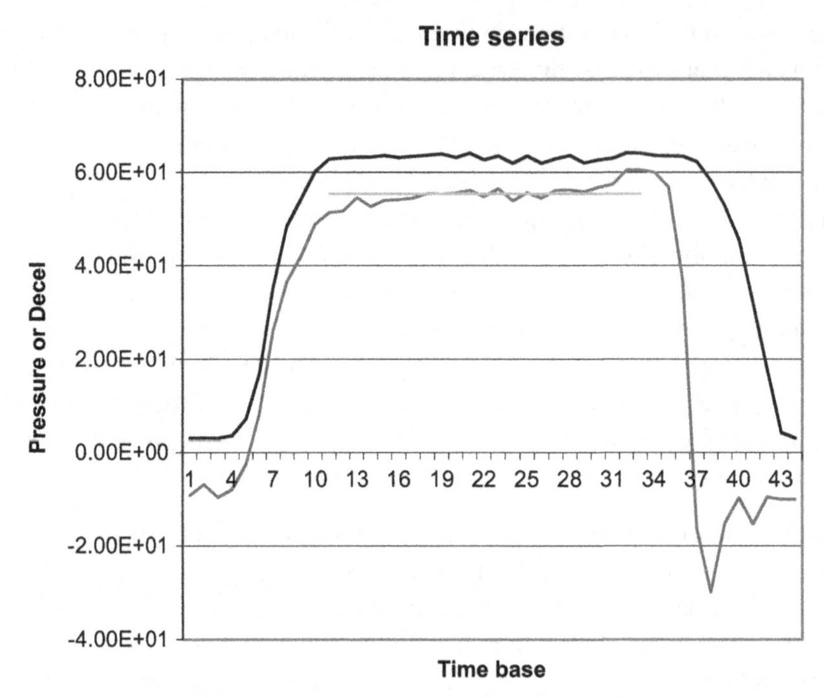

Figure 9.6: Example of Brake Pressure and Vehicle Deceleration Data vs. Time.

3. When the vehicle comes to rest (or nearly to rest), immediately release the brake actuation pressure and accelerate gently to a predefined road speed to facilitate brake cooling.
4. The vehicle should then be driven at constant speed without the use of the brakes until the indicated temperature has declined to reach the specified temperature.
5. Repeat the procedure (steps 1–4) at different settings (speed, temperature, deceleration, actuation pressure) until the test is complete.

Some typical 'raw data' test results are shown in Figures 9.6 and 9.7 from a basic brake performance test on a small car with hydraulically actuated brakes. Actuation pressure and deceleration are plotted as time series in Figure 9.6, from which the points below are highlighted:

1. There is a slight zero offset for both the pressure and the deceleration transducers, which would normally be corrected in the data processing.
2. Both the deceleration and the pressure traces show an initial 'ramp-up' section and a final 'ramp-down' section. The deceleration reaches zero before the actuation pressure (brake pedal) is released, and the deceleration indicates a reversal after the vehicle has come to rest. This is caused by torque reversals on the wheels and is the main reason why some testers only bring the vehicle 'nearly to rest'.

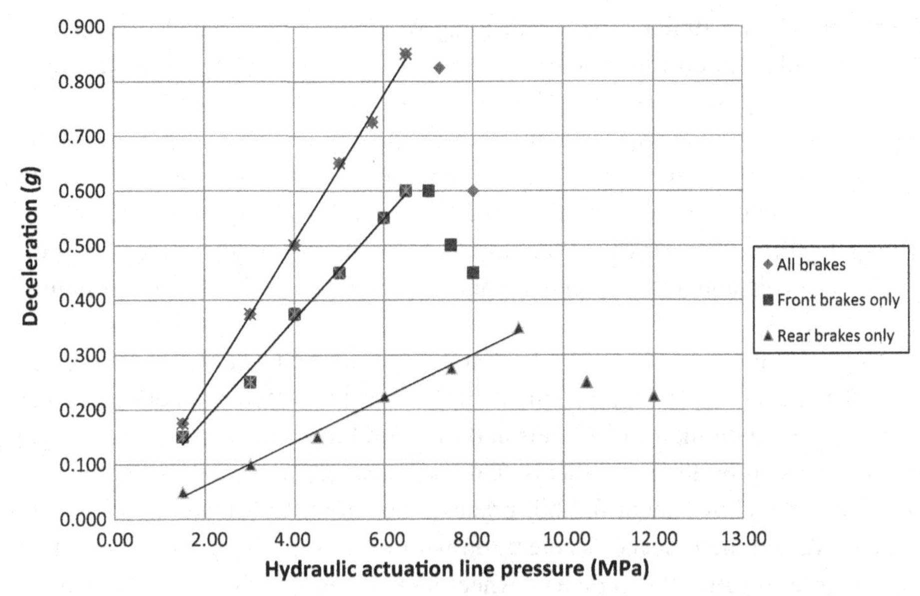

Figure 9.7: Brake Performance Graph (Data from Marshall, 2008).

3. The purpose of this particular test was to hold the deceleration constant and evaluate the 'in-stop' performance of the friction pair from the variation in actuation pressure. An estimate of the average deceleration over the 'steady' part of the brake application is indicated, which has been calculated as the arithmetic mean over the central portion. The selection of this 'central portion' can be difficult, especially if there is substantial in-stop variation caused either by an inexperienced test driver or by variable friction. There are various definitions of 'trigger' points for the start and end of averaging, which can include for the start pressure/deceleration above zero, 95% of fully developed deceleration, and for the end road speed within 5% of start speed, deceleration reduces to 90% of fully developed deceleration, or actuation pressure is reduced to zero. None of these is foolproof, but consistency in the data helps all of them to be utilised effectively if required.

When the brake performance test is completed, each dataset (Figure 9.6) generates a single value of, in this case, mean deceleration for each braking event, and the mean of at least three datasets at the same test conditions should be calculated to provide a single datapoint for the brake performance graph shown in Figure 9.7. If a significant effect of an assignable cause such as direction is identified, e.g. because of wind, an even number of datasets from consecutive braking events in opposite directions should always be used. The data in the brake performance graph (Figure 9.7) is from a different vehicle, and illustrates the following points. (These data are used to demonstrate design validation in Chapter 12.)

1. The data points fit a straight line, confirming Amontons' law (see also point 5 below).
2. 'All brakes' refers to all four brakes operating as designed. ABS was disabled here.
3. 'Front brakes only' refers to 'split' brake testing, where the rear brakes have been disabled by a valve in the system (interlocked and key operated to prevent the vehicle's being used illegally on the public road). From the gradient, the brake factor of the front (disc) brakes can be verified using the brake system design equation (6.26). Similarly the brake factor of the rear (disc) brakes can be verified from the 'rear brakes only' gradient. The calculated design performance datapoints are also shown in Figure 9.7 and are close.
4. The 'threshold' pressure for 'all brakes' and for 'front brakes only' is very low, while that for the rear brakes is higher. This demonstrates the difference between a standard caliper disc brake on the front wheels and the 'combined' disc brake (service plus parking actuation) on the rear wheels. This is discussed in more detail below.
5. Rates of braking above about 0.8 (all brakes), 0.6 (fronts only) and 0.35 (rears only) are not achieved, and the reduced deceleration with increased actuation pressure that can be seen in Figure 9.7 after this indicates wheel lock. From this the tyre/road adhesion coefficient can be estimated.

Figure 9.8 highlights the effect of threshold pressure and rolling resistance around the origin of the graph. Rolling resistance generates deceleration without actuation pressure, which is represented on the ordinate, while threshold pressure is the offset on the abscissa (actuation pressure but no deceleration). The straight-line fit of brake performance data should not be expected to pass through the origin and a common mistake in data analysis for system verification is to use individual datapoints or a line fit forced through the origin. This will give incorrect results and waste a lot of time chasing reasons for an apparently declining brake factor with actuation pressure!

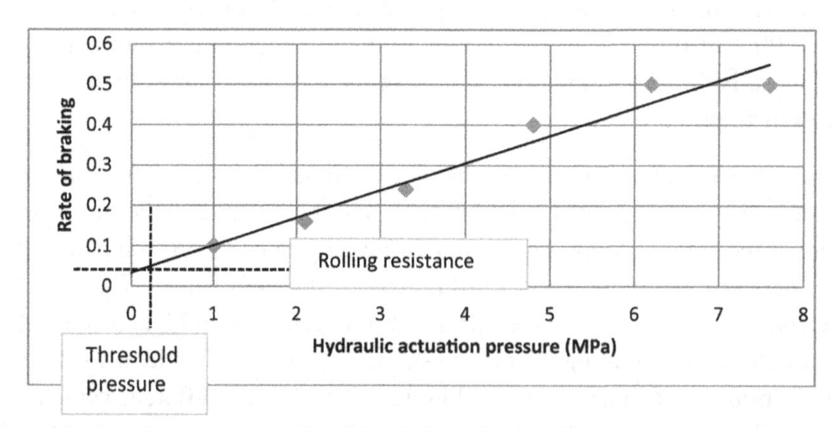

Figure 9.8: Brake Performance Graph Highlighting Threshold Pressure and Rolling Resistance at The Origin.

Wear Test Procedures

The ultimate test of wear and durability of a road vehicle braking system is through 'normal' service operation on actual vehicles driven by their owners or users. As previously explained, this approach is generally unacceptable because of the time and cost so accelerated (shorter duration) tests must be adopted to replicate real-world usage to estimate the likely wear life of the friction pair at the design stage, if possible using laboratory-based rather than vehicle-based methods. With any modern resin-bonded composite friction material there will be regions of operation and duty where extremes of high and low wear occur, so a basic wear test might consist of a series of repetitive brake applications from a given speed and at a given work rate, commencing at the lower end of the operating temperature range. The brake applications continue until a measurable amount of wear is achieved, which is measured either by mass or volume (thickness) change, and the next series is then commenced with increased temperature. This simplistic approach to wear testing can provide a basis for comparison between different friction pairs under the same conditions, but the extrapolation of this data to real vehicle applications is unlikely to be very accurate. Fundamentally, the parameter that affects wear most significantly is the work done or amount of energy dissipated, and so all wear test procedures should be based on constant torque/deceleration principles so that differences in nominal friction level and fade resistance do not affect the results. Durability needs to be evaluated on the rotor and other brake components, e.g. wear of the rotor, thermal cracking of the disc or drum, or catastrophic failure.

Standardised Test Procedures

Many different brake performance tests following standardised test procedures can be carried out on vehicles and on dynamometers; an excellent overview is given by Agudelo and Ferro (2005), who summarise the publicly available test procedures used by the automotive industry worldwide to assess brake and vehicle braking performance. These are:

- Federal Motor Vehicle Safety Standards (FMVSS) and the National Highway Traffic Safety Administration (NHTSA) Regulations
- Standards or Recommended Practices from the Society of Automotive Engineers (SAE)
- International Standards from the International Organisation for Standardization (ISO)
- Regulations from the United Nations Economic Commission for Europe (UNECE).

An example is the self-certification test procedure for passenger cars, FMVSS Standard 135 (US FMVSS135), which is carried out on the road (i.e. on a test track). This procedure specifies the requirements for the hydraulic service and parking brake systems of a passenger car to ensure stable braking performance during normal and emergency driving conditions. A summary of the procedure is shown in Table 9.2. Before any brake test, checks must always be made to confirm that the experimental set-up (especially the instrumentation and datalogging) is functioning correctly, and often this is formally included in the test procedure. The sections of the test procedure are largely self-explanatory, e.g. the 'wheel lock sequence' is to confirm that premature rear wheel lock does not occur in the extremes of loading condition as specified in the relevant legislation (see Chapters 3, 4, 6 and 8). 'Stops' are defined as brake applications to bring the vehicle to rest, while 'snubs' are defined as brake applications to reduce the vehicle's road speed from an initial value to a required 'end of application' speed, which is non-zero. 'Cold' effectiveness can define the 'baseline' performance, which can then be used to compare the performance in other parts of the test procedure, e.g. at higher temperature.

SAE J2522, summarised in Table 9.3, is an example of an inertia dynamometer standardised test procedure for motor vehicles fitted with hydraulic brake actuation to assess the effectiveness and behaviour of a friction material with respect to pressure, temperature and speed. It was developed in Europe by a partnership of vehicle, brake and friction material manufacturers (as the 'AK Master' test procedure) and is not a certification test. It has also been used to develop tests for cold temperature performance, wet effectiveness and parking brake evaluation ramp applications, and has been found to be useful for the evaluation of friction behaviour such as 'green' effectiveness, speed sensitivity, fade resistance, friction recovery and friction stability.

The FMVSS/TP 121D standard specifies performance and equipment requirements for pneumatic braking systems for commercial vehicles and trailers, and public service vehicles including buses and coaches. It includes vehicle and dynamometer tests with a test procedure for each; the inertia dynamometer test procedure is summarised in Table 9.4 for comparison with that shown in Table 9.3 for vehicles with hydraulic brake systems (passenger cars, vans and light trucks). The number of brake applications for the bedding/burnishing section is increased (doubled) for commercial vehicle brakes (Table 9.4), and higher temperatures are used, reflecting the higher durability of commercial vehicle friction materials.

The introduction of standards for replacement friction materials (see Chapter 8) has required appropriate experimental test procedures to be developed to confirm their suitability in comparison with the original equipment (OE) materials. The SAE J2430 dynamometer effectiveness characterisation test for passenger cars and light truck brake replacement friction materials forms the basis of an automotive industry-wide brake evaluation procedure. It uses parameters measured from an actual vehicle and aims to replicate the key sections of the FMVSS 105/135 procedure, measuring speed sensitivity and performance stability for comparison with a baseline friction material or an OE friction material. The procedure is summarised in Table 9.5. Experimental test procedures

Table 9.2: Summary of FMVSS 135 Vehicle Test Procedure (Agudelo and Ferro, 2005)

Section	Number of Stops/Snubs	Brake-Release Speed (km/h)	Control	Initial Brake Temperature (°C) or Cycle Time	Performance Requirement
Burnish at GVW	200 stops	80–0	3 m/s^2	<100°C or 2 km	–
Wheel lock sequence at LLVW/GVW 50km/h	3–6 stops	50–0	Pedal force ramp up to 1000 N or lock-up in 0.5–1.5 s	<100°C	Wheel lock-up 0.5–1.5 s from 0.15–0.8 adhesion coefficient
Wheel lock sequence at LLVW/GVW 100 km/h	3–6 stops	100–0	Pedal force ramp up to 1000 N or lock-up in 0.5–1.5 s	<100°C	Wheel lock-up 0.5–1.5 s from 0.15–0.8 adhesion coefficient
Adhesion utilisation (torque wheel method) at LLVW/GVW	10 stops	100–0 and 50–0	Pedal force ramp 100–200 N/s until lock-up or 1000 N	<100°C	No rear lock-up first 0.15–0.8 *g*
Cold effectiveness at GVW	6 stops	100–0	65–500 N pedal force	65–100°C	70 m
High speed effectiveness at GVW	6 stops	80% of V_{max} but not >160	65–500 N pedal force	65–100°C	153 m
Stops with engine off at GVW	6 stops	100–0	65–500 N pedal force	65–100°C	70 m
Cold effectiveness at LLVW	6 stops	100–0	65–500 N pedal force	65–100°C	70 m
High speed effectiveness at LLVW	6 stops 80% of V_{max} but not >160	65–500 N pedal force	65–100°C	153 m	
Failed antilock at LLVW	6 stops	100–0	65–500 N pedal force	65–100°C	85 m
Failed proportioning valve at LLVW	6 stops	100–0	65–500 N pedal force	65–100°C	110 m

Continued

Table 9.2: Summary of FMVSS 135 Vehicle Test Procedure (Agudelo and Ferro, 2005)—cont'd

Section	Number of Stops/Snubs	Brake-Release Speed (km/h)	Control	Initial Brake Temperature (°C) or Cycle Time	Performance Requirement
Hydraulic circuit failure at LLVW	4 stops	100–0	65–500 N pedal force	65–100°C	168 m
Hydraulic circuit failure at GVW	4 stops	100–0	65–500 N pedal-force	65–100°C	168 m
Failed antilock at GVW	6 stops	100–0	65–500 N pedal force	65–100°C	85 m
Failed proportioning valve at GVW	6 stops	100–0	65–500 N pedal force	65–100°C	110 m
Power brake unit failure at GVW	6 stops	100–0	65–500 N pedal force	65–100°C	168 m
Parking brake at GVW	Up to 2 uphill and 2 downhill	0	500 N pedal force or 400 N hand force	65–100°C	Hold vehicle stationary >5 min
Fade heating snubs at GVW	15 snubs	120–60 typical	3m/s^2	55–65°C for first then 45 s	–
Hot performance at GVW	2 stops	100–0	#1 pedal force shortest cold effectiveness #2500 N pedal force	#1 immediately #2 1.5 km after #1	Stop #1: deceleration 60% of shortest cold effectiveness Stop #1 or 2: 89 m
Cooling stops at GVW	4 stops	50–0	3 m/s^2	1.5 km after previous stops	–
Recovery at GVW	2 stops	100–0 pedal force shortest cold effectiveness 1.5 km after previous stops	deceleration 70–150% of shortest cold effectiveness		
Final inspection	Check corner assembly, brake system and indicators			Comply with S.1.17	

V_{max} is the maximum attainable speed within 3.2 km from 0 km/h initial speed at maximum acceleration rate and LLVW. LLVW refers to 'lightly loaded vehicle weight' and GVW refers to 'gross vehicle weight' as previously defined.

Table 9.3: Summary of SAE J2522 Inertia Dynamometer Global Brake Effectiveness Test Procedure (Agudelo and Ferro, 2005)

Section	Number of Stops/Snubs	Braking-release Speed (km/h)	Control	Initial brake Temperature (°C)
Green effectiveness	30	80–30	30 bar	100
Burnish/bedding	192	80–30	Various pressures	100
Characteristic check	6	80–30	30 bar	100
Speed/pressure sensitivity	8	40–<5	10, 20, ..., 80 bar	100
Speed/pressure sensitivity	8	80–40	10, 20, ..., 80 bar	100
Speed/pressure sensitivity	8	120–80	10, 20, ..., 80 bar	100
Speed/pressure sensitivity	8	160–130	10, 20, ..., 80 bar	100
Speed/pressure sensitivity	8	200–170	10, 20, ..., 80 bar	100
Characteristic check	6	80–30	30 bar	100
Cold braking check	1	40–<5	30 bar	40
Motorway braking check #1	1	100–5	0.6 g	50
Motorway braking check #2	1	0.9–0.5 V_{max}	0.6 g	50
Characteristic check	6	80–30	30 bar	100
First fade (maximum 160 bar)	15	100–<5	0.4 g	100–500 disc 100–300 drum
Recovery	18	80–30	30 bar	100
Pressure sensitivity	8	80–30	10, 20, ..., 80 bar	100
Increasing temperature sensitivity (500°C/300°C)	9	80–30	30 bar	100,150, ..., 500
Pressure sensitivity (500°C)	8	80–30	10, 20, ..., 80 bar	500
Recovery 2	18	80–30	30 bar	100
Second fade (maximum 160 bar)	15	100–<5	0.4 g	100–500 disc 100–300 drum
Characteristic check	18	80–30	30 bar	100

for brakes and braking systems continue to be developed and the ISO 15484 standard for the product definition and assurance of road vehicle brake lining friction materials represents a global standardisation of test procedures focused on the quality control of friction components for passenger cars and commercial vehicles. The standard covers friction component testing relating to product definition and assurance, visual inspection, physical characteristics, friction and wear, corrosion ageing, and noise and vibration.

The purpose of a fade test is to evaluate brake performance as the temperature increases, which can be achieved either by repeated snub applications as in Tables 9.2–9.4, or by

Table 9.4: Summary of FMVSS/TP-121D-01 Inertia Dynamometer Test Procedure (Agudelo and Ferro, 2005)

Section	Number of Stops/Snubs	Brake apply and Release Speed (km/h)	Control	Initial Brake Temperature (°C)/Cycle Time (s)	Performance Requirement
177°C (350°F) Burnish	200	64–1	3.05 m/s^2	177°C (350°F)	–
260°C (500°F) Burnish	200	64–1	3.05 m/s^2	260°C (500°F)	–
Brake power check stops	2	80–1	2.74 m/s^2	<177°C	–
Hot stop check stops	2	32–1	4.27 m/s^2	<177°C	–
Brake recovery check stops	2	32–1	3.66 m/s^2	<177°C	–
Brake retardation	7	80–1	1.38, 2.07, 2.76, 3.45, 4.14, 4.83 and 5.52 bar (20–80 psi)	77°C	Minimum retardation 0.02, 0.04, 0.05, 0.08, 0.09, 0.11 and 0.12 m/s^2 respectively
Brake power	10	80–24	2.74 m/s^2	66°C for first stop then every 72 s	<6.89 bar (100 psi)
Hot stop	1	32–1	4.27 m/s^2	60 s from end of last brake power snub	–
Brake recovery	20	32–1	3.66 m/s^2	1st stop 120 s from end of hot stop 2nd–20th every 60 s	1.38–5.86 bar (20–85 psi) for non-ABS 0.83–5.86 bar (12–85 psi) for ABS <5.86 bar for front axle of bus/truck N/A for front axle truck-tractor
End of test. final inspection					

Table 9.5: SAE 2430 Dynamometer Effectiveness Characterisation Test for Passenger Cars and Light Truck Brake Replacement Friction Materials (Agudelo and Ferro, 2005)

Section	Number of Stops/Snubs	Braking-Release Speed (km/h)	Control	Initial brake Temperature (°C)	Cycle Time (s)
Instrument check 50 km/h torque control	5	50–3	Torque @ 0.31 g	<100	
Instrument check 100 km/h torque control	5	100–3	Torque @ 0.31 g	100	
Instrument check pressure control	3	50–3	Pressure @ 75 N pedal force	100	
Instrument check 50 km/h ramp	5	50–[0.8 g, 1000 N or 3 km/h]	135 N/s pedal apply rate	100	
Instrument check 100 km/h ramp	5	100–[0.8 g, 1000 N or 3 km/h]	135 N/s pedal apply rate	100	
Instrument check cooling curve 80 km/h	18	80–80	Within cooling band	200 for front 150 for rear	15
Burnish	200	80–3	Torque @ 0.31 g	100°C first stop 100°C or 97 s	
Effect #1 post-burnish 50 km/h ramp	5	50–[0.8 g, 1000 N or 3 km/h]	135 N/s pedal apply rate	100	
Effect #1 post-burnish 100 km/h ramp	5	100–[0.8 g, 1000 N or 3 km/h]	135 N/s pedal apply rate	100	
Post-burnish cold effectiveness	6	100–3	Torque @ 0.65 g	100	
Fade heating cycles	15	120–56	Torque @ 0.31 g	55 first snub	45

Continued

Table 9.5: SAE 2430 Dynamometer Effectiveness Characterisation Test for Passenger Cars and Light Truck Brake Replacement Friction Materials (Agudelo and Ferro, 2005)—cont'd

Section	Number of Stops/Snubs	Braking-Release Speed (km/h)	Control	Initial brake Temperature (°C)	Cycle Time (s)
Hot performance effectiveness	2	100–3	1st at minimum pressure from cold effectiveness test 2nd at pressure @ 500 N pedal force for 135 test, 667 N for 105 test	–	1st at 35 2nd at 30
Cooling cycles	4	50–3	Torque @ 0.31 g	–	120
Recovery 100 km/h ramp	2	100–[0.8 g, 1000 N or 3 km/h]	135 N/s pedal apply rate	–	60
Reburnish	35	80–3	Torque @ 0.31 g	100°C first stop 100°C or 97 s	
Final effectiveness 50 km/h ramp	5	50–[0.8 g, 1000 N or 3 km/h]	135 N/s pedal apply rate	100	
Final effectiveness 100 km/h ramp	5	100–[0.8 g, 1000 N or 3 km/h]	135 N/s pedal apply rate	100	
Final effectiveness 160 km/h ramp	5	160–[0.8 g, 1000N or 3 kph]	135 N/s pedal apply rate	100	
Post-test cooling curve 80 km/h	18	80–80	Within cooling band	200 for front 150 for rear	15
Post-test cooling curve 112 km/h	18	112–112	Within cooling band	200 for front 150 for rear	15

drag braking, e.g. on a long downhill gradient. In the fade test sections in the test procedures illustrated above, a series of higher duty brake applications are made at defined time intervals, which heats up the brake. After the fade test, the 'recovery' test section aims to identify how quickly a material can 'recover' to baseline performance in a series of (generally) lighter duty brake applications at defined intervals, which allow the brake to cool down. Friction materials typically lose performance as temperature increases and should regain performance as the friction pair cools down. As well as the 'fade' test sections in the standard procedures shown above, most vehicle and brake manufacturers have their own in-house fade test procedures. Fade testing on a vehicle incorporates repeated brake applications on the vehicle in the laden condition from high road speed without allowing the brake to cool significantly in between, thereby maximising the thermal loading on the brakes. The test may be carried out on a time cycle, e.g. apply the brakes to decelerate the vehicle at a high rate to the finish speed (which may be zero), then accelerate the vehicle back up to the test speed, then drive at that constant speed until 1 minute has elapsed, and then apply the brakes to start the next cycle. Alternatively, the cycle time may be determined by the performance of the vehicle so that after decelerating it is accelerated at the maximum achievable rate until the required speed is reached and then the brakes are applied again. This type of test usually applies the brakes to create maximum deceleration without ABS intervention and is extremely punishing. The 'AMS' test for passenger cars is this kind of fade test; it was created to evaluate extreme brake duty on high-performance cars and has become an EU industry standard for passenger cars. It is not a legislative test but is claimed to represent consumer expectations. The AMS test schedule is, in summary, 10 stops from 100 km/h at maximum deceleration (without ABS intervention) with maximum acceleration back up to 100 km/h between each stop. The performance requirement is a stopping distance less than 40 m with minimal deterioration between the 10 stops; for vehicle manufacturers this typically means that the derived μ must not fade to below 0.3. Often two cycles are run to eliminate the first (uncharacteristic) fade from new brake pads. An example of measured temperature–time during a fade test is shown in Figure 9.9 (McLellan, 2012).

Brake Test Data Interpretation and Analysis

The universal use of computer-based data acquisition systems in brake experimental testing provides very large amounts of data, which requires close and careful analysis. Since the most critical wearing component in a conventional braking system is the friction material, which is a highly complex and constantly changing composite, established experimental procedures that recognise this inherent variation in performance must always be followed. Fundamentally this means understanding the statistical basis of variation, and testing a large enough number of samples from a suitable population. The alternative is

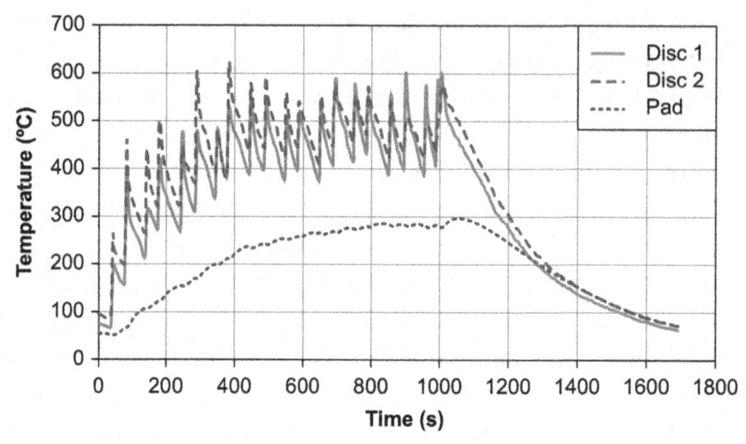

Figure 9.9: Example of Temperature–Time During a Fade Test (McLellan, 2012).

wide confidence limits on the information generated. Textbooks on statistical methods in research and development (Caulcutt, 1991) are readily available and provide easily accessible guidance for the analysis of experimental brake test data, as well as for the design of the experimental procedures.

The correct and meaningful interpretation of brake test data requires experience, knowledge and understanding of test procedures, vehicle set-up, instrumentation and sensors, external factors such as ambient temperature and wind, and most importantly the characteristic behaviour of the brake friction pair. It is always good practice to examine and check the test results as they comc, and not simply download them on to some recording media for subsequent processing, by which time it is too late to correct any mistakes or failures in any part of the set-up and operation of the tests. Many commercially available brake test systems, dynamometers, etc., record the data and process it into standard format reports (e.g. Figure 9.10). Whilst such reports are useful for comparison purposes, there are always other characteristics or features especially relating to the friction material and its tribological interaction with the rotor material that need to be considered.

Chapter Summary

- The role of testing in the design and verification of brakes and braking systems for modern road vehicles is explained and discussed. There is a continuing trend away from experimental testing, especially that which involves the road or track testing of actual vehicles, because of the complexity, cost and time involved, and towards computer simulation and 'virtual' testing. Nevertheless experimental testing is essential for design verification and for the provision of accurate data for design predictions.

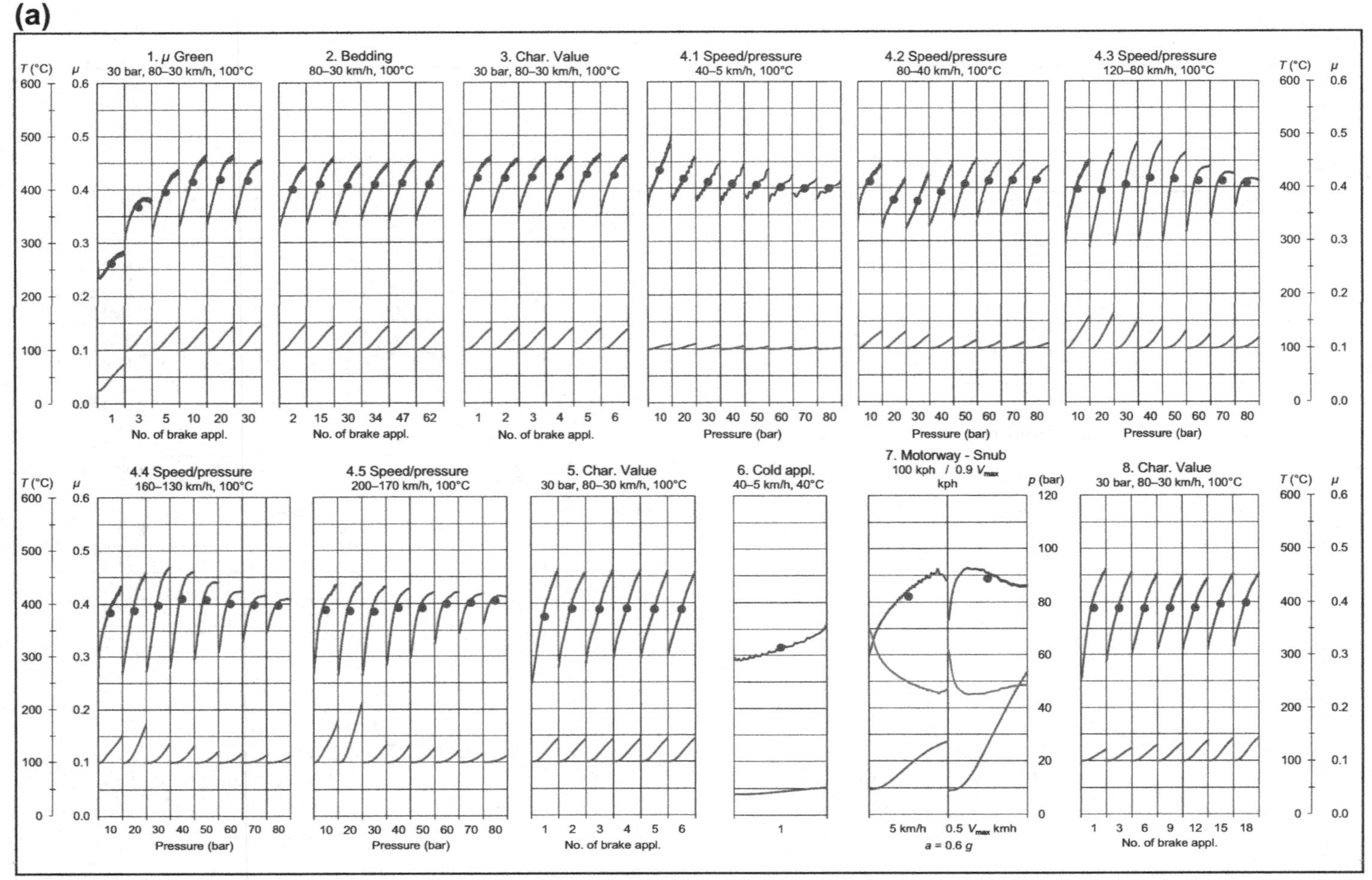

Figure 9.10: (a) Standard Report from SAE J2522 (Table 9.3) Part 1. (b) Standard Report from SAE J2522 (Table 9.3) Part 2. (Courtesy Federal Mogul)

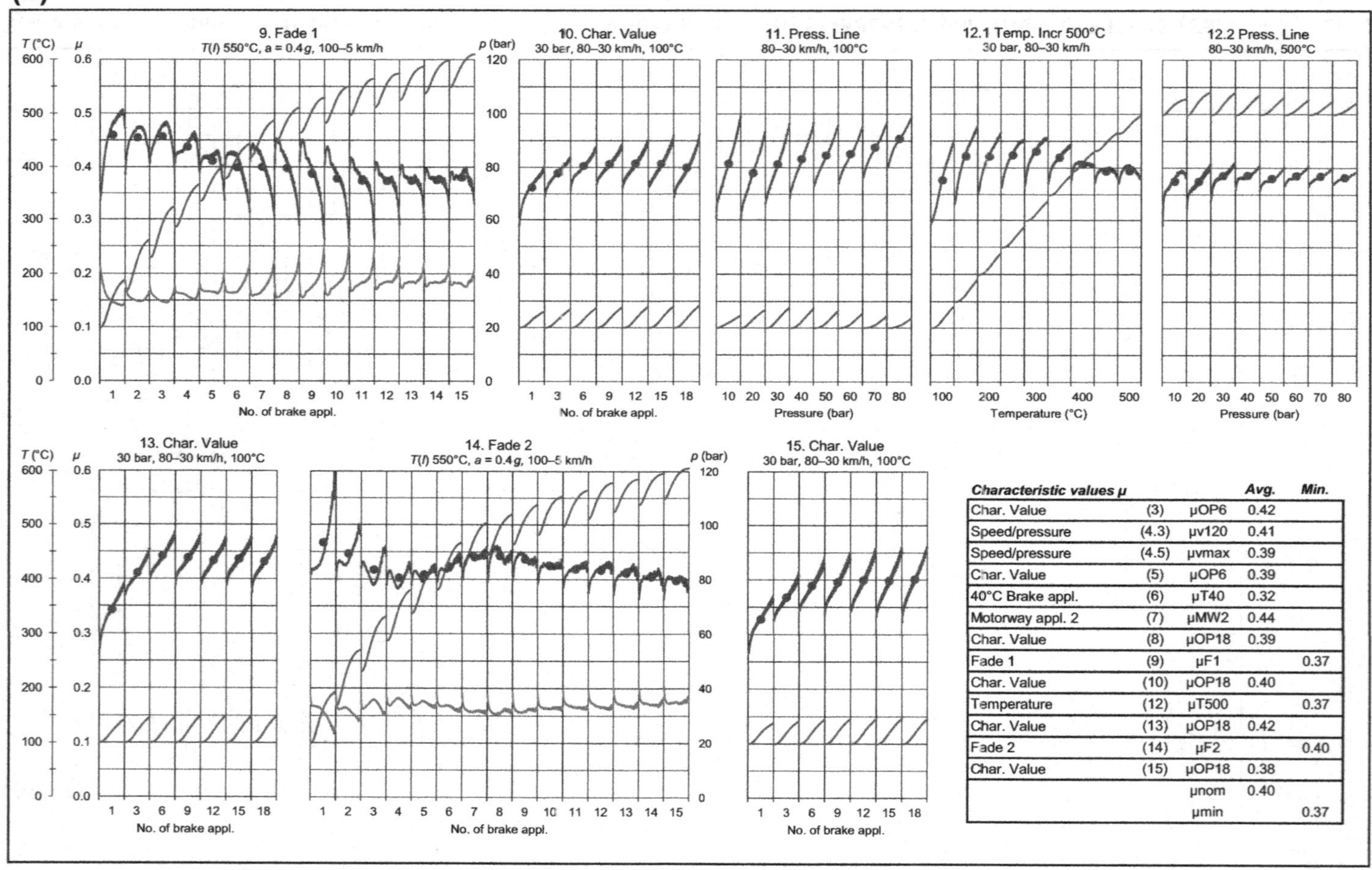

Characteristic values μ			Avg.	Min.
Char. Value	(3)	μOP6	0.42	
Speed/pressure	(4.3)	μv120	0.41	
Speed/pressure	(4.5)	μvmax	0.39	
Char. Value	(5)	μOP6	0.39	
40°C Brake appl.	(6)	μT40	0.32	
Motorway appl. 2	(7)	μMW2	0.44	
Char. Value	(8)	μOP18	0.39	
Fade 1	(9)	μF1		0.37
Char. Value	(10)	μOP18	0.40	
Temperature	(12)	μT500		0.37
Char. Value	(13)	μOP18	0.42	
Fade 2	(14)	μF2		0.40
Char. Value	(15)	μOP18	0.38	
		μnom	0.40	
		μmin		0.37

Figure 9.10
cont'd

- Brake experimental testing can be carried out on actual vehicles on test tracks and under laboratory conditions, e.g. on a 'rolling road' chassis dynamometer. Vehicles may also be set up to record 'real-world' user data on public roads, provided that any modifications to the vehicles are not dangerous and do not contravene legal requirements, the instrumentation and data acquisition do not interfere with the driver's control of the vehicle, and the vehicle is driven safely. All types of brake testing are potentially dangerous and hazardous, and some basic rules for safety are introduced.
- Parameters that can be measured in experimental brake testing are explained, and examples of the types of instrumentation and sensors used to measure these parameters are briefly described. Data acquisition and datalogging are explained.
- It is important to define and agree the purpose of experimental brake testing before commencing any test programme. Some aspects of experimental design for brake testing are explained in the context of a boundary diagram and a 'p-diagram', and procedures for experimental brake testing are introduced. The importance of standardisation of test equipment and procedures across different manufacturers and different countries is emphasised, so that consistency of braking system performance can be achieved.
- Different types of hardware for brake testing are described and explained, including vehicles, dynamometers and test rigs. The advantages and disadvantages of each are considered, from 'fleet' testing of passenger cars through 'performance' or 'effectiveness' testing on dynamometers, to 'small-sample' testing on scale friction rigs. The variability that is always present in any form of vehicle testing is discussed, and ways to either reduce the amount of variability by good test definition, preparation and control, or to take account of the variability in subsequent data interpretation and analysis, are discussed.
- The importance of preparing the brake friction pair through the processes of bedding and burnishing is emphasised, and a procedure to achieve bedding-in is explained. Other essential preparations include validating the test set-up, confirming brake actuation and control, health and safety provisions including risk assessments, checking and calibration of all sensors, transducers and instrumentation, and confirmation of the correct functioning of signal conditioning and datalogging equipment, and many brake test procedures include 'instrument checks' in the initial stages for these purposes.
- A generic brake performance test procedure for actual vehicles is described and discussed, and some example data are presented. Accelerated wear testing is briefly discussed. Standardised brake test procedures are discussed, using as examples procedures from the automotive industry that now have acceptance in many countries worldwide. These examples cover passenger car braking systems (hydraulic) and commercial vehicle braking systems (pneumatic), and relate to actual vehicle testing and inertia dynamometer testing.

- Because friction materials typically lose performance as temperature increases and regain it as the brake cools down, fade test procedures to evaluate brake performance as the temperature increases are important. This type of test can utilise repeated 'snub' applications or 'drag' braking, at defined time intervals that heat up the brake. After a fade test, a 'recovery' test aims to identify how quickly a material can 'recover' to baseline performance in a series of (generally) lighter duty brake applications at defined intervals that allow the brake to cool down. Fade testing on a vehicle incorporates repeated brake applications from high road speed without allowing the brake to cool significantly in between, thereby maximising the thermal loading on the brakes. The test may be carried out on a set time cycle, or the cycle time may be determined by the acceleration performance of the vehicle. This latter type of test usually applies the brakes to create maximum deceleration without ABS intervention and is extremely punishing.
- Brake test data interpretation and analysis are briefly discussed.

References

Agudelo, C.E., Ferro, E., 2005. Technical overview of brake performance testing for original equipment and aftermarket industries in the US and European markets, Link Technical Report FEV205–01, Link Testing Laboratories, Inc.

Caulcutt, R., 1991. Statistics in Research and Development, 2nd edn. Chapman & Hall. ISBN 0 412 35800-5.

Marshall, P.H., 2008. Case study, Braking of Road Vehicles 2008. University of Bradford.

McLellan, R.G., 2012. Brake testing, Braking of Road Vehicles 2012. University of Bradford. ISBN 978 1 85143 269 1.

Rhee, S.K., Schwartz, H.W., 1979. Test methods for automotive and truck friction materials, Wear tests for plastics, selection and use, a symposium. In: Bayer, R.G. (Ed.), Wear tests for plastics, selection and use, a symposium. ASTM, 1979.

Tumbrink, H., Bock, H., Noack, J., 2000. Comparison of Fatigue Testing and Fatigue Calculation for an Al Alloy Brake Caliper Housing, SAE Technical Paper 2000-01-2752, 2000. http://dx.doi.org/10.4271/2000-01-2752.

US Department of Transportation, Federal Motor Carrier Safety Administration, Rules and Regulations Standard No. 135; Light vehicle brake systems (FMVSS 135).

CHAPTER 10

Brake Noise and Judder

Introduction

Noise and judder from the braking systems of road vehicles are of major concern to manufacturers and users. Audible noise causes environmental pollution, especially in urban areas, and is regarded by vehicle owners and users as symptomatic of a fault in the braking system, i.e. an 'error state' (see Figure 9.3). Whilst judder may also create audible noise, the main problem from brake judder is the mechanical vibration that is sensed by the vehicle users and is also regarded as a fault. This mechanical vibration may be an irritant, e.g. steering wheel or brake pedal vibration, or may lead to premature failure of components through vibrational fatigue or even to loss of control of the vehicle. A road vehicle that demonstrates even a low level of brake noise or judder is very likely to be returned for attention as a complaint, and thus from the point of view of the manufacturer, the error state becomes a failure mode, in that corrective action is required under the terms of the vehicle's warranty. The cost of correcting brake noise and/or judder is a significant proportion of the annual warranty costs to vehicle manufacturers worldwide, and especially in the USA and Europe, where a high level of sophistication in vehicle systems and performance is expected by the customer. So vehicle manufacturers continue to search for methods to design quiet brakes and braking systems to reduce their warranty costs. This continues to be a challenge; noise and judder have been associated with friction brakes since the earliest days of motor vehicles and despite the huge amount of research that has been carried out on the subject, there is still no completely effective predictive method for the design of a quiet brake or a judder-free brake. Consequently there remains an expectation in the automotive industry that the friction brakes on any new vehicle will have a significant likelihood of the occurrence of noise and/or judder at the design verification stages or even in the customer domain. However, research and experience have increased the level of understanding of the root causes of brake noise and judder to the extent that if best practice design rules are applied at the system design stage, the propensity for noise or judder from the brakes is considerably reduced.

'Brake noise' is a generic term that may be defined as the audible noise emanating from a disc brake or a drum brake during vehicle use, which occurs at definite frequencies that are independent of usage conditions such as rotor speed. The term 'during vehicle use'

Braking of Road Vehicles.

does not limit the occurrence of brake noise to when actuation pressure or force is applied; brake noise can and does occur when the brakes are not being used, termed ‘off-brake’ noise. Brake noise occurs over a wide range of frequencies, typically from 100 Hz to above 20 kHz (which is the upper limit of human audibility), and the classic brake ‘squeal’ noise is often in the region of 1–6 kHz, where the human ear is most sensitive. Brake noise can occur at many distinct frequencies for any given brake and at each frequency the harmonic content can vary, which contributes to the characteristic of the sound. Most brake noise is fugitive in the sense that the same operating conditions of actuation force, speed, temperature and other parameters may generate noise during one brake application but not the next one. The noise may also be intermittent within a single wheel rotation due to the small changes in conditions caused by the rotor runout. In comparison, brake judder is a mechanical instability with a frequency related directly to rotor speed.

Within the broad definitions of brake noise and judder, there are many sub-categories of brake noise and judder, which were defined by North (1976), and subsequently by Lang and Smales (1983) and several more researchers. In ascending order of frequency some of the original sub-categories of brake noise are listed below:

- Judder is a speed-related non-resonant vibration with a low frequency (typically around 10 Hz), which is a multiple of the rotor rotational speed. It creates wheel and brake torsional vibration about the suspension or chassis, and vibration in the actuation system that can be transmitted back to the driver, e.g. via the brake pedal on passenger cars with hydraulic braking systems. The actual judder frequency is a multiple of the rotor speed and becomes intrusive when the frequency coincides with resonant frequencies in the vehicle’s suspension, steering or structure. There are two distinct categories of brake judder, namely cold judder and hot judder (sometimes termed ‘high-speed’ judder). Cold judder is associated with the mechanical effects of non-uniformity of the rotor rubbing path for which there can be several causes, e.g. non-circular or eccentric brake drums, runout or swash of brake discs, disc thickness variation (DTV) caused by off-brake wear, or circumferential friction variation, which might occur with uneven coatings or transfer films on the rotor friction surfaces. Hot judder is also a speed-related non-resonant vibration usually of a higher frequency (typically 200 Hz) because of the greater circumferential non-uniformity of the rotor surface, which is caused by thermomechanical interactions between the friction pair. These result in thermal distortion and deformation, and can also cause temperature-related phase changes of the rotor material. Such phase changes often create ‘blue spots’on the rotor friction surface where the cast iron has changed phase from a predominantly pearlitic microstructure to a martensitic microstructure, which has a higher specific volume. These regions initially protrude into the rotor rubbing path and are thus subjected to higher localised heat flux input, temperature and wear, and subsequently (on cooling) they may become concave. High-speed judder can produce a noise that is often referred to as 'drone' or 'rumble'.

- Groan is a semi-resonant vibration, typically <100 Hz, which is most prevalent at lower speeds and is usually considered to be caused by 'stick-slip'arising from a negative $\mu-v$ characteristic (i.e. μ decreases with increasing sliding speed). Groan usually displays a vibration mode of rigid body axial rotation of the disc brake caliper and the adjacent suspension components.
- Hum is a resonant sinusoidal vibration that generally occurs under 'off-brake' conditions, is independent of rotor speed, and occurs at frequencies between 100 and 400 Hz. It is associated with disc brakes that have a low caliper mounting torsional stiffness.
- Moan is a higher frequency vibration often associated with disc brake caliper whole body movement, which usually occurs at frequencies of 600–700 Hz.
- 'Low-frequency' squeal is the most intrusive type of brake noise that occurs with disc and drum brakes and usually has a frequency between 1 and 4 kHz, which is in the part of the audible spectrum to which the human ear is particularly sensitive. This frequency range is below the fundamental bending frequency of the pad assembly, which is significant because in disc brakes the squeal modes of vibration are diametral (two, three or four nodal diameters) with nodal spacing greater than the pad length. Squeal noise is particularly difficult to suppress because circumferential symmetry in a brake disc or drum allows vibrational energy to be continually transferred between two closely spaced modal frequencies ('pair modes') in a 'binary flutter' instability, as proposed by North (1972, 1976) and Millner (1978).
- High-frequency squeak in disc brakes involves higher order disc vibrational modes in the frequency range 4–15 kHz, which have typically five to 10 nodal diameters. The nodal spacing is therefore less than the pad length and the squeak natural frequencies are now higher than the fundamental pad bending mode. Squeak is a major problem for single-piston sliding calipers with high aspect ratio pads.
- Squelch in disc brakes is an amplitude modulated version of squeak noise with a 'beating' effect caused by general asymmetry in the disc.
- Wire brush is a non-resonant noise at a very high frequency (up to 20 kHz) with random amplitude modulation, and is often heard immediately prior to the development of a true unstable squeak.

It is widely accepted that most brake noise is caused by frictionally induced dynamic instabilities in the brake that create vibrations of the brake components such as the disc or drum, which then radiate the sound that is heard. The mechanism of brake noise can therefore be considered as having two parts: a 'source' that creates a vibrating signal input to the brake system, and a 'response' that amplifies the signal and radiates it as sound pressure variations or waves. This suggests two approaches to reducing noise propensity: one is to address the root cause, i.e. the source of the vibration, and the other is to address the effect, i.e. the way the vibration is amplified and radiated as sound waves. Published research on the subject of brake noise covers both, although in recent years the tendency appears to have been to concentrate on the latter.

This chapter explains the features and principles of brake noise and judder, summarises some of the approaches to modelling and analysing the physical mechanisms, and suggests best practice design rules to minimise brake noise and judder propensity. In this context 'noise propensity' and/or 'judder propensity' are terms used to describe the likelihood that a given brake design will suffer from noise or judder problems. For example, it is suggested from published work and experience that the following approaches can be beneficial in influencing brake noise propensity; however, some seem to conflict with others.

- Practical methods for reducing brake noise propensity depend on the individual nature of the noise, e.g. frequency, likely instability mechanism and the type of brake involved.
- Brake noise propensity tends to increase with coefficient of friction (μ), so the noise propensity of a particular brake may be reduced by the substitution of a lower μ friction material.
- Brake noise propensity tends to increase on cooling after heavy-duty and/or repeated brake applications.
- The contact between the friction material and the rotor, e.g. heel and toe contact on the leading shoe of drum brakes, and leading contact on disc brake pads, often increases noise propensity:
  - Drum brake noise can often be reduced by reducing the leading shoe factor, by grinding the lining to give crown contact, or shortening or chamfering the linings.
  - Disc brake squeal can often be eliminated by moving the contact between the piston and the pad backplate towards the trailing end, e.g. by a shim.
  - In disc brake calipers with two or more actuation pistons, stability can be increased (and noise propensity reduced) by utilising piston seal damping in the pad rotational mode, e.g. using shims to encourage contact between the pistons and the backplate extremities.
  - Squeal in disc brakes that have leading and trailing abutments can be reduced by locating the abutment at the leading end.
- Decoupling the piston and pad motions in the tangential direction by grease or a low friction coating and using the piston as a mass damper can reduce squeal noise propensity.
- Damping of a brake drum can be increased by fitting a band lined with friction material around its mouth, although the band tension is critical, and allowance must be made for drum expansion and centrifugal effects. The addition of friction damping to the rotor with a loose peripheral ring may similarly help to reduce brake noise propensity, but all such devices are adversely affected by corrosion in use.

- Damping the pad bending mode, e.g. by removing slots in the friction material, incorporating a high loss interlayer between the friction material and backplate, or applying a damping layer to the back of the pad, can reduce certain types of brake noise propensity. A thin viscoelastic layer between the backplate and a separate steel shim helps by converting small bending strains into large shear strains in the viscoelastic layer.

A huge amount of research and development effort has been and still is directed towards the solution of brake noise and judder problems, and the number of publications on the subject increases every day. From all this effort, the knowledge and understanding of brake noise and judder does advance, but it is not easy to see clearly how advances are made because often there are conflicting results arising from subtle differences in the conditions of investigation. Even the statements listed above would be enthusiastically debated by experts who would identify situations where they do not apply, or argue about the fundamental reasons for such relationships. So this chapter starts with a very brief review of brake noise based on four published reviews over a 40-year period.

Brake Noise Review

There is an enormous body of knowledge associated with brake noise that has been developed over the last 60 years, and gaining an understanding of the significant advances is a daunting prospect. As a starting point, some conclusions and observations from four significant reviews of brake noise published in the last 40 years are summarised here. These show a development of approach from understanding the nature and characteristic features of vibration associated with brake noise (frequencies, mode shapes), through lumped-parameter models, mathematical models, stability criteria, to modelling and simulation, all supported by increasingly sophisticated experimental measurement and research techniques such as laser holographic interferometry. Early work on brake noise includes Mills (1938), who reported the earliest structured investigation of drum brake noise, and identified many characteristics and possibilities, including 'trailing action' in the friction forces on the brake linings and the effects of friction material, both of which could be thought of as source parameters, and the potential for noise control using mass damping and friction damping devices, which could represent response parameters. Fosberry and Holubecki (1955) suggested that the source of drum brake squeal was a periodic excitation arising from μ variation with sliding speed (in particular, higher static μ than dynamic μ, creating 'stick-slip'). Basford and Twiss (1958) concluded from experimental studies of brake linings that stick-slip motion could not be responsible for vibration during high-speed operation of brakes. Spurr (1961) noted that brake squeal was excited by 'vibrations in friction', which could be caused by deflections in the exciter of a 'sprag-slip' mechanism (See Figure 10.1). He demonstrated the theory using a double cantilever source model, which indicated how a time-varying friction force could be generated even with a constant coefficient of friction through periodic variations in the normal force that were dependent on the design and mechanical properties of the actuation system. This was one of several 'kinematic constraint' source models that have been researched extensively because they demonstrated how a time-varying friction force could be generated even with a constant coefficient of friction.

North was a pioneer in investigating and modelling brake noise. He presented an eight degree of freedom lumped parameter rigid body mass/spring/damper dynamic model of a disc brake and was probably the first to correlate his predictions against experimental results on an actual squealing brake system. This (linear) model was novel because the friction forces between the disc and pads were modelled as 'follower' forces. He was also the first researcher to consider brake squeal as a self-excited vibration induced by friction forces with constant μ, and to attribute the mechanism of disc brake squeal to a type of 'flutter' instability. He adopted the criterion for brake squeal based on the onset of instability, which is basically the same as used today in complex eigenvalue analysis. In anticipation of this work, his review of published work on vibrations in braking systems (North, 1969) noted the characteristic features associated with drum brake squeal, which are listed below:

- The frequency of squeal is independent of the rubbing speed.
- Brake torque increases while the brake is squealing.
- Squeal tendency increases with increased actuation pressure.
- Drum vibration is 'bell-like' with six to 14 modes measured.
- Squeal is most likely to occur after high-temperature and high-pressure operation, which could be a friction interface contact effect or a recovery effect in the friction material.
- Squeal is most prevalent at temperatures below 100°C.
- Squeal is most likely to occur when the brake linings have high μ, e.g. with cold brake usage (North's term was 'early morning sharpness').

For disc brakes, North noted the following:

- The disc vibrates with an even number of radial nodes, the spacing of which is widest at the position of the pads and approximately equal over the rest of the disc.
- The pads are positioned approximately on an antinode and vibrate with the disc.
- An increase in temperature causes a decrease in squeal frequency.
- Squeal frequency is independent of disc speed.
- Squeal is again most likely after high-temperature and high-pressure operation.
- Pad thickness has no effect on squeal frequency or the likelihood of squeal.
- Squeal is again likely as a result of high μ (early morning sharpness).

Crolla and Lang's (1991) review of brake noise and vibration noted that the key features of a self-excited oscillation or dynamic instability in a friction brake were that the source and the response are (or become) in-phase, positive feedback occurs, the system is unstable and the amplitude of vibration increases until 'limit cycle' operation is reached ('limit cycle' describes the vibrational response of the system in the unstable condition, and may be large-amplitude vibrations or loud noise limited by kinematic constraints or damping in the system, such as occurs with resonance). The limitations of sprag-type source models developed in the form of pin-on-disc models were identified as being geometrically unable to relate to real brake design. The binary flutter mechanism (North, 1972) was proposed for brake squeal noise, which led to 'stability modelling', where the squeal propensity was predicted from the rate of divergence of unstable roots. Murukami et al. (1984) presented a multiple degree of freedom lumped parameter model of a disc brake to predict squeal propensity, which was correlated with an experimental brake to provide some qualitative design guidelines based on, for example, μ characteristics (higher μ and decreasing μ–speed both increase squeal propensity) and caliper/actuation geometry. Crolla and Lang noted the use of finite element (FE) analysis and commented that modal analysis of individual components correlated well but did not help the understanding of the causes (source) of brake noise because the necessary interactions between different parts of the brake could not be modelled. Their review concluded that brake noise was a systemic problem that could not be solved by investigating individual components such as the disc, pad, caliper and mountings, because in a real brake they all interact with each other. The key to further progress lay in higher fidelity computer modelling and better ways of representing frictional coupling at the rotor/stator interface, both of which have come about with the increased availability of sophisticated modelling and simulation software and computer power, as indicated in the fourth review by Ouyang et al. (2005) considered later.

The review of disc brake squeal by Kinkaid et al. (2003) presents a thorough analysis of a large number of significant publications covering every aspect of the brake noise problem, from which they summarised the conclusions relating to brake squeal listed below:

- A braking system can generate squeal noise at several distinct frequencies.
- Higher μ and variation of μ with speed and actuation pressure may all be associated with higher squeal noise propensity (the effect of any decreasing μ–speed characteristic on squeal noise occurrence is still not agreed; see recent work by Eriksson et al. (2002) and Yuan (1995)).
- Pad/disc contact in a disc brake is important in squeal noise propensity.
- Vibration amplitudes during brake squeal noise occurrence are very small (microns).
- Knowledge of the modal characteristics of individual uncoupled disc brake system components does not necessarily indicate squeal noise propensity.
- Squeal noise is generated by coupled vibrations of a disc brake system.
- Squeal noise frequency is not necessarily close to any resonant frequency of the brake rotor.

- It is likely that the contributions of different brake components to squeal noise generation depend on the squeal noise frequency.
- Out-of-plane and in-plane vibrations of the brake disc are generated with brake squeal noise, and the coupling between them is important.
- The vibration of a squealing disc brake assembly rotating at constant speed may be a standing wave or a travelling wave, and if a travelling wave, its speed is not the same as the squeal noise frequency.
- During the generation of squeal noise, disc brake pads vibrate with various bending and torsional modes, and disc mode shapes are based on an even number of radial nodes.
- The amount of squeal noise increases when the natural frequencies of the pads, caliper and disc are close together.
- Detailed tribological phenomena, e.g. the size of 'contact plateaux' at the brake friction interface (Eriksson et al., 2002), may also contribute to the generation of brake squeal noise.

The review notes that sprag-slip and similar theories for source models were defined as 'geometrically induced' or 'kinematic constraint' models (Jarvis and Mills, 1963; Crisp et al., 1963), and although these have been investigated in research configurations by many researchers, they agreed with Crolla and Lang that they appear to be quite difficult to simulate computationally. A mechanical impact model termed 'hammering' has been proposed as a source mechanism (Rhee et al., 1989), which seems to represent many features associated with brake noise generation and has been used in recent research by Alasadi et al. (2013). Turning from the source to the response, evidence that brake squeal noise was associated with not only transverse vibrations of a brake disc but also of in-plane vibrations was presented by Matsuzaki and Izumihara (1993) in the context of high-frequency squeal. Subsequently, in-plane disc vibrations have been considered very important in brake squeal noise generation, and have been experimentally demonstrated, e.g. by Talbot et al. (2002) using laser holographic interferometry. The role of coupling between the in-plane and out-of-plane modes was described as the 'key to produce squeal' by Chen et al. (2002), who also observed that the squeal frequency was close to the disc in-plane mode frequency and not the out-of-plane mode frequency in the coupled mode of vibration.

The review also considered methods to eliminate brake squeal noise. Some of these methods are often referred to as 'fixes', indicating that they are applied post-design, usually to solve brake noise problems that have occurred in the customer domain (the 'field'). Commonly used fixes include:

- Grease or viscous layer between the pad backplates and caliper
- 'Shims' between the pad backplate and the actuation piston or mechanism
- Chamfers or slots in the brake pad material

- 'Refreshing' the disc friction surfaces, which may take the form of cleaning, lightly abrading or re-machining
- Lubrication of moving parts, e.g. the pin slides of a sliding caliper.

The rationale behind these fixes is that they change the interconnection and interaction characteristics of the various parts of the brake. For example, the interposition of a shim (which may have a viscous layer incorporated) between the pad backplate and the actuation piston or tappet can alter the coupling stiffness and damping between the two, and is found to be effective for high-frequency noises, e.g. above 4 kHz. Lubricating moving parts changes the contact friction characteristics at the interfaces. Chamfers change the contact between the pad and the disc, and slots change the flexural characteristics of the pad assembly while at the same time influencing the tribological behaviour of the friction material, which would also be a result of refreshing the disc friction surface. In practice it has been found over many years that the effectiveness of these fixes decreases over time, and at best can offer only short-term control of the problem. Another noticeable effect is that manufacturers are reluctant to delete a fix on subsequent designs, so they tend to continue to be specified even when no longer appropriate, often masking the true nature of the noise problem. Other methods to eliminate brake squeal noise mentioned by Kinkaid et al. include active control ('dither') and fundamental design features such as the reduction of the circumferential symmetry of rotors to separate doublet (also called pair) modes. Most published work over the last 25 years has focused on linear analysis to predict the onset of instability, which is then correlated to the onset of squeal; the use of FE analysis was first reported as a model for disc brake squeal in by Liles (1989), and the challenge now seems to be understanding and predicting what happens after instability has occurred, which needs a non-linear modelling approach.

Ouyang et al. (2005) reviewed the state of the art of CAE simulation and analysis for disc brake squeal but emphasised the importance of experimental methods as well as theoretical methods in the understanding of brake noise. The complex eigenvalue analysis approach is a linearised stability analysis that is considered to be 'a good approximation if it is linearised properly in the vicinity of an equilibrium point like the steady sliding position' and is widely used in industry for FE-based brake design and noise propensity analysis. These authors explained how, when the method is used in FE analysis, geometric coupling can be used to incorporate friction at the brake friction interface, e.g. via a spring element linking nodes on the pad and rotor surfaces with tangential friction forces; the stiffness matrix of this element is asymmetric. Mathematically this is not entirely straightforward but accurate computational methods have been developed and commercial FE systems now include this type of formulation. An alternative is to include friction as a 'follower force' in the brake system model (a 'follower' force can be defined as a force that depends upon the displacement or movement of the body on which it acts). A non-linear transient analysis can include the effect of time-varying forces in the brake system, e.g. from start-up to steady state, as well as stability analysis, and FE-based studies have been published demonstrating the procedure and good experimental correlation.

The review discusses the advantages and disadvantages of complex eigenvalue analysis and transient analysis. The complex roots can all be obtained from an eigenvalue analysis in one run, and they reveal the unstable vibration modes allowing design modifications to be made, e.g. to change modal frequencies or add damping to make the mode stable. However, they do not predict the real magnitude of a vibrational motion or provide any information about a resulting noise. In a transient analysis, many runs are required until a limit-cycle motion is found, and it is therefore recommended that a complex eigenvalue analysis is always completed before a transient analysis to indicate where unstable motion might exist. The main limitation of the complex eigenvalue analysis is that it is only accurate around steady-state sliding. Away from steady-state sliding conditions the effect of non-linearities may be significant. Other disadvantages are that features such as variable material properties cannot be included. Despite these limitations, the complex eigenvalue approach is widely used to inform structural modifications to prevent brake noise at the design stage, although there is an increasing realisation that this is insufficient to design out brake noise. Transient analysis in theory is not limited by all the assumptions associated with complex eigenvalue analysis and can include changing friction contact area, non-linear friction, moving loads and variable material properties. The main disadvantage is the very large computing demands. Because of the high frequencies considered, time steps must be very small for explicit integration, typically 10^{-5} s. Using an implicit integration algorithm allows a much larger time step, but the system equations must be solved at each step. It is expected (Mahajan et al., 1999) that a variety of analytical methods for brake noise modelling and prediction will be used to improve design methods for quiet brakes.

'Outstanding issues' are discussed in the conclusion of Ouyang et al.'s review. These include knowing which aspects of brake design dominate particular features of brake noise generation; e.g. the follower force effect should be addressed by increasing the bending stiffness of the rotor, mode coupling should be addressed by brake component and interconnection design, and friction coefficient variation with speed and pressure should be reduced, possibly also adding more damping into the system. Contact and tribological effects at the friction interface, and contact effects at other interfaces such as the actuator and pad backplate on a disc brake, are under-represented in brake noise modelling, as are thermal effects. The incorporation of damping in the models is also under-researched. Finally, variation in many parameters associated with friction brakes, ranging from geometric dimensions and tolerances to uncertainties in the braking friction process, present a great challenge to the verification of noise predictions. It has already been stated that brake noise is fugitive in the sense that sometimes it occurs and sometimes it does not, even under nominally the same conditions, and a study of variability and uncertainty in brake noise phenomena is seen as a worthwhile endeavour.

The Source of Brake Noise

Frictionally induced dynamic instabilities in the brake may be generated by asperity contact and the physical mechanisms of friction as introduced in Chapter 2, which in turn create time-varying frictional forces at the friction interface to act as the source function for brake noise. As previously explained, stick-slip or the μ–speed characteristic of the friction material was thought to contribute to the source instability of brake noise and although they can be the source of certain types of brake noise (e.g. groan) they are not the source of most types of brake noise, especially squeal. If μ is assumed to decrease linearly with sliding velocity, the equations of motion of the pad will be as indicated in Equation (10.1), which can be rewritten in the form of Equation (10.2):

$$M\ddot{x} + c\dot{x} + kx = \mu_D P = [\mu_S - \alpha(V - \dot{x})]P \tag{10.1}$$

$$M\ddot{x} + (c - \alpha P)\dot{x} + kx = (\mu_S - \alpha V)P \tag{10.2}$$

The damping term $(c - \alpha P)$ would be negative, representing instability, if $\alpha P > c$ and the amplitude of the pad vibration would then increase, although the instability could be controlled by reducing α or increasing the damping coefficient (c). This μ–speed friction material characteristic theory has been shown to have limited applicability as most brake noises occur with friction materials having no μ–speed dependence or even a positive μ–speed characteristic, but a negative μ–speed dependence is known to exacerbate noise caused by other mechanisms at low speeds.

Spurr's sprag-slip model (Spurr, 1961) proposed a mechanical basis for time-varying friction force variation. Figure 10.1(a) shows a rigid strut O′P pivoted at O′, which is the free end of a cantilever O′–O″, where O″ meets an abutment. Taking moments about O″ the friction drag force F is calculated from:

$$F = \frac{\mu L}{(1 - \mu \tan\theta)} \tag{10.3}$$

The strut 'sprags', i.e. F tends to infinity when $1 - \mu\tan\theta = 0$, which Spurr associated with the onset of squeal noise when the 'sprag angle' (θ) was close to the limit indicated in Equation (10.2).

$$\theta = \cot^{-1}\mu \tag{10.4}$$

The extension of this and other similar theories to an actual disc brake considers the position of the centre of pressure of the pad and assumes the double cantilever system to

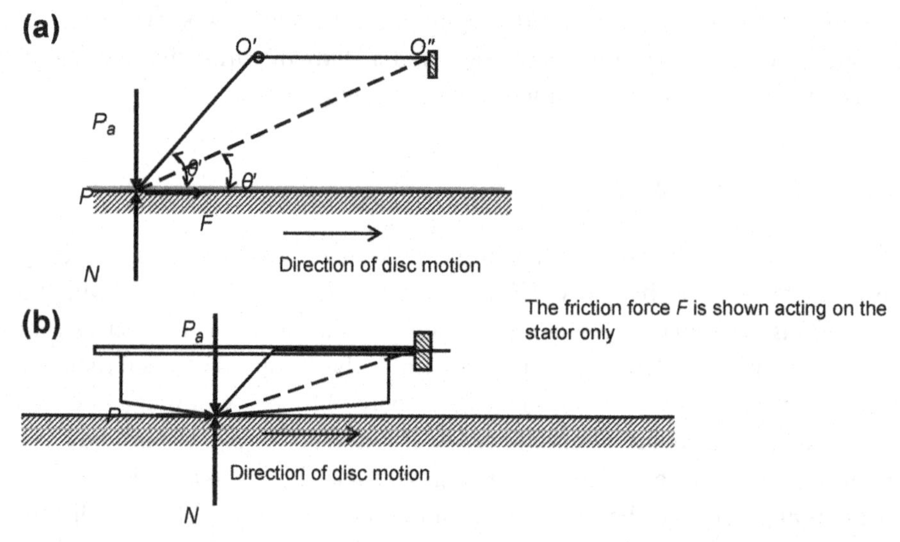

Figure 10.1: (a) 'Sprag-Slip' Model (Spurr, 1961) (Upper). (b) 'Sprag-Slip' Model in a Disc Brake Scenario (Lower).

lie within the pad material, as indicated in Figure 10.1(b). For $\mu = 0.4$, the sprag angle would be 68°, and Fieldhouse et al. (2011) consider the direction of this resultant force vector with respect to the caliper mounting rather than the pad/caliper abutment, which is discussed later.

Fieldhouse et al. (2011) carried out experiments to examine the noise propensity of a passenger car disc brake with a varying centre of pressure (CoP) offset by controlling the position of contact between the actuation piston and the backplate. They found that with a leading offset between 12 and 15 mm, noise was generally generated at all operating temperatures, and noise propensity was least with a zero or trailing contact offset, or a large leading contact offset, and thus a region of high noise propensity could exist between them. They also observed that as the temperature decreased and μ increased, the critical offset changed from 12 to 15 mm, and concluded that a leading offset increased noise propensity. Using a 'co-planar' analysis based on Equations (5.10) and (5.11) – see Figure 5.6 in Chapter 5 – they calculated the offset between the actuation force and the CoP as defined by the position of the normal reaction and predicted where the resultant force vector (normal force and friction force) passed with respect to the caliper mounts (not the backplate abutments or actuation piston). Their theory is illustrated in Figure 10.2, which shows first a leading CoP with which the resultant force causes the mounting to rotate clockwise, and second a reduced leading CoP where the resultant force causes the mounting to rotate counterclockwise. Their suggestion was that if the CoP hunts between these two extremes it will pass a point where the resultant force will cause only axial movement, as shown in Figure 10.2(c), which corresponds to the highest normal reaction

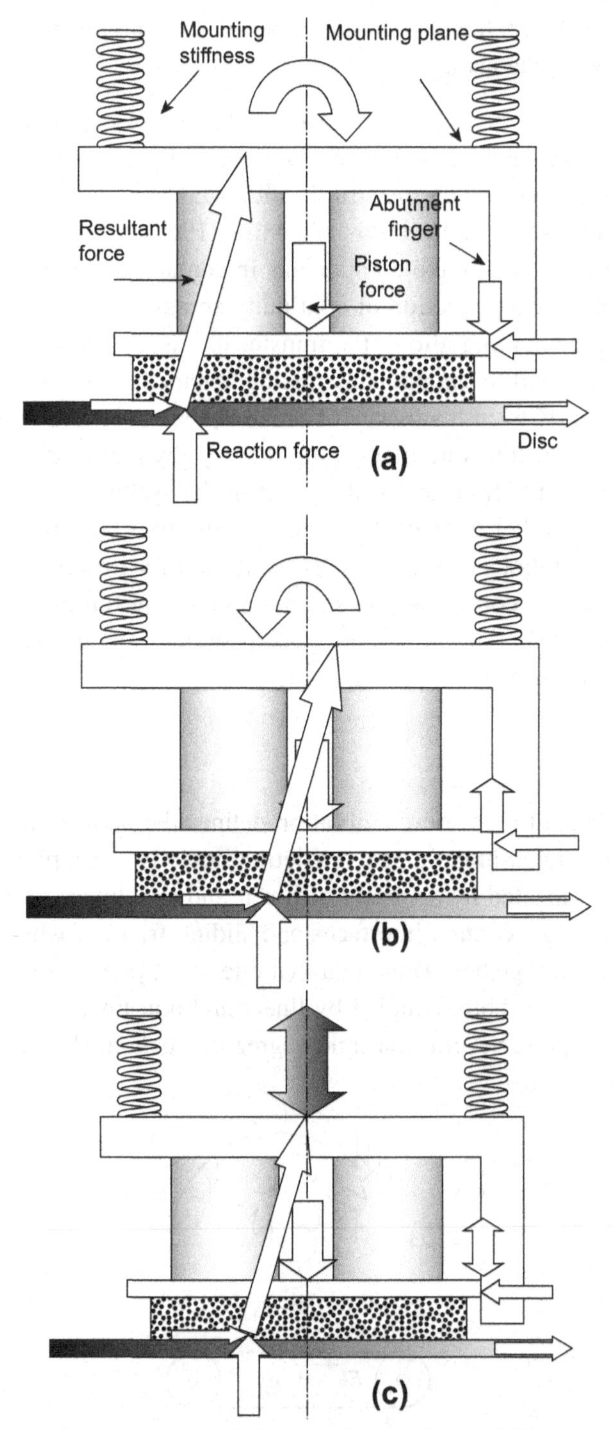

Figure 10.2: Sprag Theory Applied to a Disc Brake (Fieldhouse et al., 2011).

and hence friction drag force, thus generating a periodic excitation force that could be a noise vibration source mechanism.

The 'kinematic constraint' type of source model for brake noise couples one mode of vibration (e.g. in the strut mechanism in Spurr's model) to that of another part of the system (e.g. the body against which the strut is sliding). Geometric coupling was identified as being important in brake noise by Jarvis and Mills (1963), who explained how the normal vibration modes of two brake components in contact, e.g. a pad and a disc, may be mathematically coupled by an equation of continuity across the surface, resulting in a non-linear oscillation. Mode coupling allows the transfer of energy between different modes of vibration and is fundamental to the phenomenon of 'binary flutter', which was first incorporated by North (1976) in his theory of brake squeal. Alternative models to sprag-slip have been proposed to illustrate mode coupling, e.g. by Hoffman et al. (2002) – see Figure 10.3; energy from the frictional sliding part of the system is transferred to vibrational energy in the other parts of the system by the friction force, which links the vertical and horizontal motions of the mass. The criterion for instability is that the coupling force balances the corresponding structural coupling force of the system (Hoffman et al., 2002). Potentially this represents a 'source' model that could be incorporated into a non-linear modelling approach.

System Response

The response of the brake to the source vibration defines the nature (modes and frequencies) of the generated brake noise. A friction brake is a complicated assembly of components that are connected by a variety of linear and non-linear devices and physical phenomena such as springs, contact interfaces and sliding friction, which couple various parts of the brake system together. Thus it can vibrate as a system of rigid bodies or as a system of flexible bodies, or both, coupled by linear and non-linear contact interfaces, forces, springs and dampers. Several multiple degree of freedom (DoF) models have been

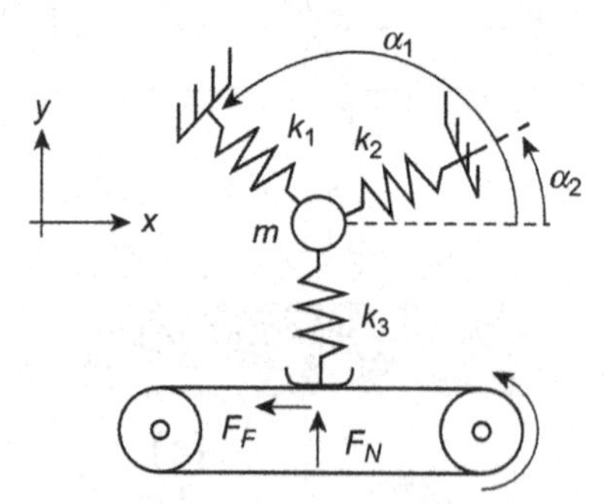

Figure 10.3: A Physical Mechanism Visualising Mode-Coupling Instability of Self-Excited Friction-Induced Oscillations (Hoffman et al., 2002).

proposed over the last 40 years, which have given valuable insight into the parameters and factors that influence the vibration response of the brake.

Eigenvalue stability analysis has been widely reported in the published literature relating to brake noise, and can be demonstrated on simple lumped parameter models, e.g. the dynamic response of a single degree of freedom (DoF) mass/spring/damper system subjected to a sinusoidal input ($f = f_o \sin(\omega t)$), as illustrated in Figure 10.4. The frequency response can be calculated from the system transfer function:

$$\frac{X}{f_o} = \frac{1}{-\omega^2 m + i\omega c + k} \tag{10.5}$$

The system stability can be analysed by calculating the roots or eigenvalues of the characteristic equation (10.6); a negative real part indicates stability.

$$-\omega^2 m + i\omega c + k = 0 \tag{10.6}$$

Extending this to a 2DoF coupled system as illustrated in Figure 10.5, where one mass has two coupled vibration modes of bounce and pitch, shows the effect of damping on stability.

The frequency response is calculated from the transfer function for each DoF, which can be done by evaluating the determinant of the nominator and denominator of the matrix equations (10.5):

$$\frac{X_1}{f_o} = \frac{\begin{bmatrix} 1 & -\left[\left(k_f a - k_r b\right) + \left(c_f a - c_r b\right) i\omega\right] \\ 0 & \left[\left(k_f a^2 + k_r b^2\right) - \omega^2 I + \left(c_f a^2 + c_r b^2\right) i\omega\right] \end{bmatrix}}{\begin{bmatrix} \left[\left(k_f + k_r\right) - \omega^2 m + \left(c_f + c_r\right) i\omega\right] & -\left[\left(k_f a - k_r b\right) + \left(c_f a - c_r b\right) i\omega\right] \\ -\left[\left(k_f a - k_r b\right) + \left(c_f a - c_r b\right) i\omega\right] & \left[\left(k_f a^2 + k_r b^2\right) - \omega^2 I + \left(c_f a^2 + c_r b^2\right) i\omega\right] \end{bmatrix}} \tag{10.5a}$$

$$\frac{X_2}{f_o} = \frac{\begin{bmatrix} \left[\left(k_f + k_r\right) - \omega^2 m + \left(c_f + c_r\right) i\omega\right] & 1 \\ -\left[\left(k_f a - k_r b\right) + \left(c_f a - c_r b\right) i\omega\right] & 0 \end{bmatrix}}{\begin{bmatrix} \left[\left(k_f + k_r\right) - \omega^2 m + \left(c_f + c_r\right) i\omega\right] & -\left[\left(k_f a - k_r b\right) + \left(c_f a - c_r b\right) i\omega\right] \\ -\left[\left(k_f a - k_r b\right) + \left(c_f a - c_r b\right) i\omega\right] & \left[\left(k_f a^2 + k_r b^2\right) - \omega^2 I + \left(c_f a^2 + c_r b^2\right) i\omega\right] \end{bmatrix}} \tag{10.5b}$$

The modal frequencies for zero damping are shown in Figure 10.6 and the effect of damping at each of the degrees of freedom is illustrated in Figures 10.7 and 10.8 for the two system natural frequencies ω_1 and ω_2. The mode shapes are shown in Figure 10.9; both modes represent a combined translation and rotation. For this 2DoF system the stability can be assessed as shown in Table 10.1 for three cases of zero damping and high damping at each of the two masses. This indicates that this simple example system is

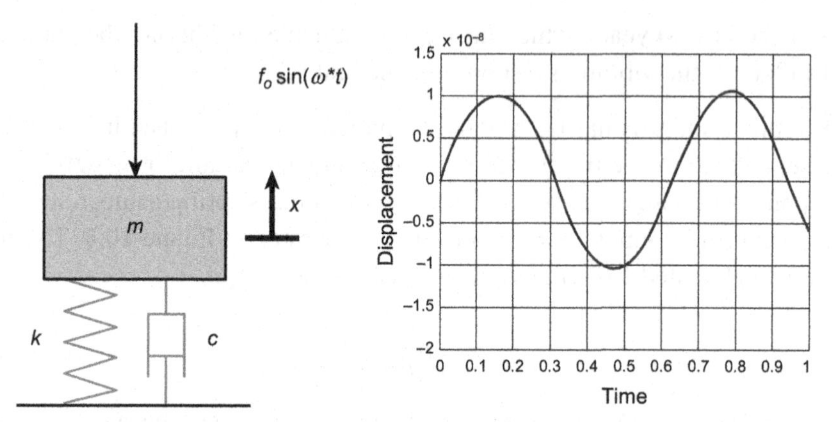

Figure 10.4: Single Degree of Freedom System Vibrational Response (Displacement vs. Time).

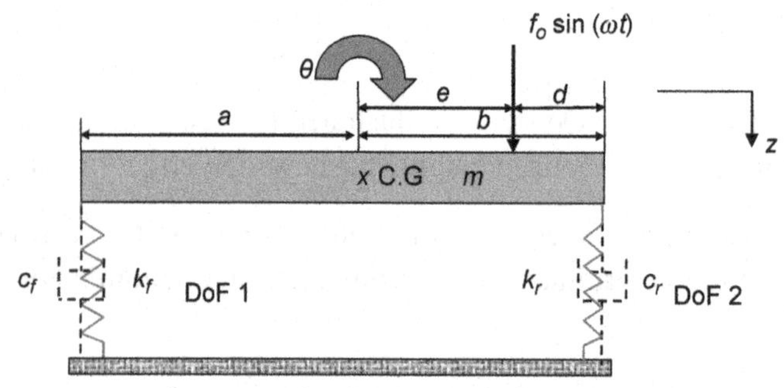

Figure 10.5: 2DoF Coupled System.

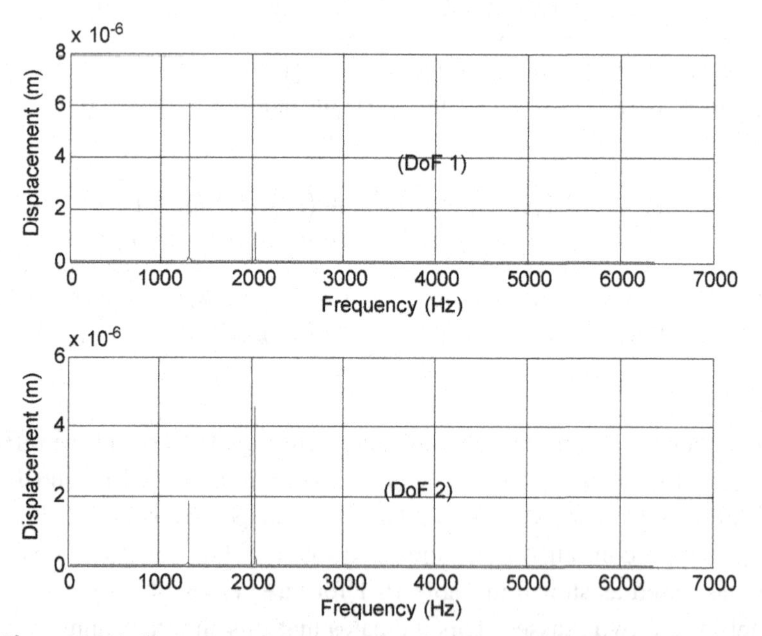

Figure 10.6: Frequency Response for Zero Damping (DoF 1 and 2).

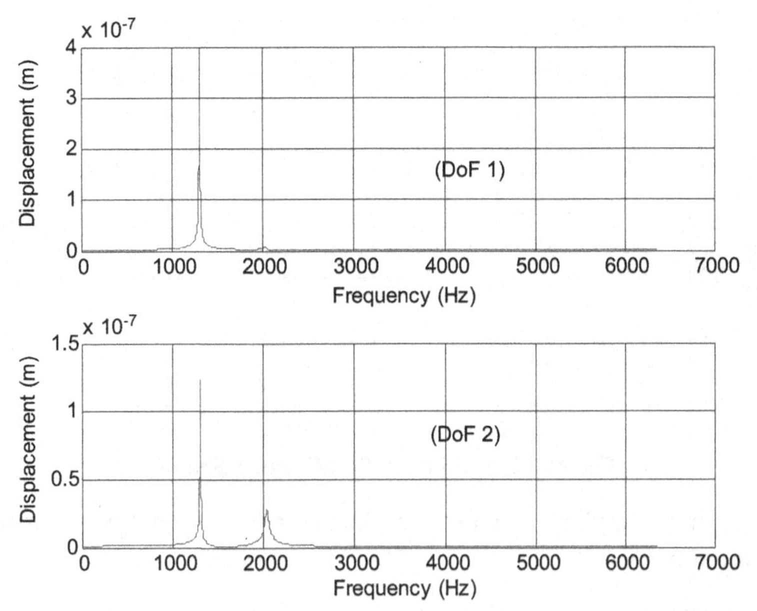

Figure 10.7: Frequency Response With High Damping At DoF 1 (DoF 1 and 2).

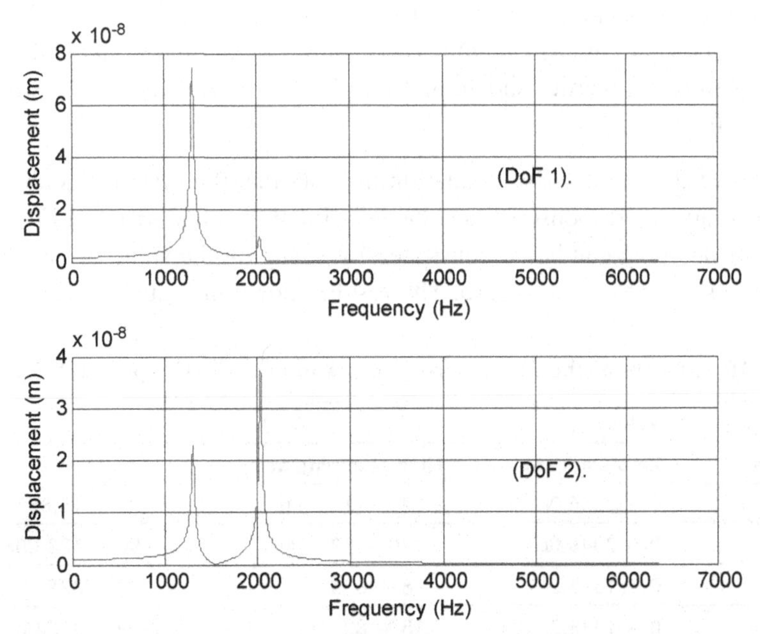

Figure 10.8: Frequency Response with High Damping at DoF 2.

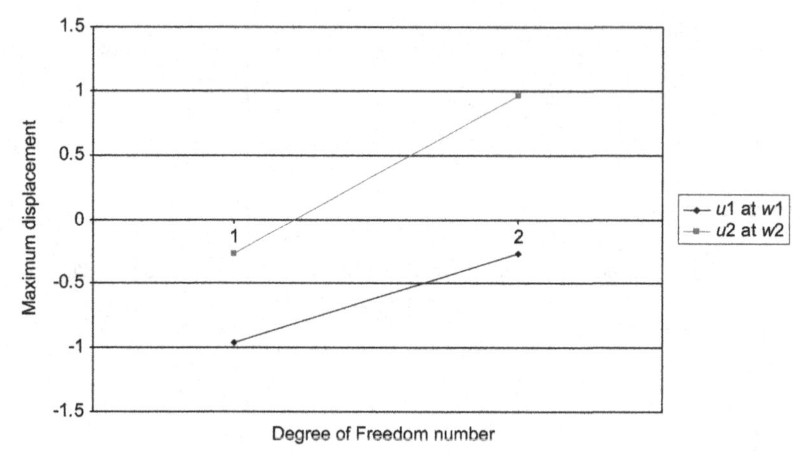

Figure 10.9: Coupled 2DoF Mode Shapes.

stable when damping is applied, and also that the damping does not necessarily suppress particular vibration modes.

The principles demonstrated in these simple 1DoF and 2DoF system models are only intended to help understand some basics of the dynamic behaviour of vibrating mechanical systems. Real vibrating systems have distributed properties such as mass, stiffness and damping, which can be linear and/or non-linear. They can be modelled by decomposing them into degrees of freedom representing the key elements and interconnections, which was the approach taken by researchers including North (1976) and Millner (1976, 1978). The analytical solution of the equations of motion for such systems is in principle possible but computer methods are quicker. Figure 10.10 shows a multiple (four) DoF linear model, which for the example properties shown in Table 10.2 generates the results summarised in Figures 10.11–10.13.

The eigenvalues of the characteristic equation indicate that the system is stable, as expected. The frequency response for a sinusoidal input is summarised in Figure 10.11, and the effect of damping can be demonstrated by changing the value of one damping coefficient while setting the rest to zero. The results shown in Figure 10.12(a) illustrate

Table 10.1: Roots of the Characteristic Equation for the Coupled 2DoF System

	Values		
Roots	Zero Damping	High Damping at #1	High Damping at #2
s_1	$0 + 2046.0i$	$-120 + 12{,}854i$	$-90 + 12{,}853.0i$
s_2	$0 - 2046.0i$	$-120 - 12{,}854i$	$-90 - 1285.0i$
s_3	$0 + 1313.2i$	$-18 + 8251i$	$-99 + 8252.0i$
s_4	$0 - 1313.2i$	$-18 - 8251i$	$-99 - 8252.0i$

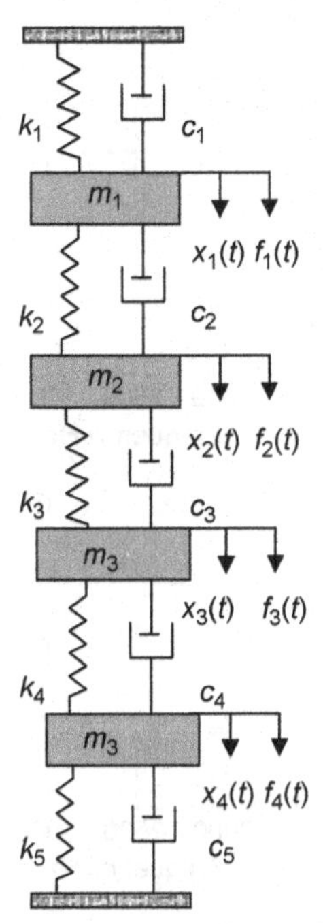

Figure 10.10: Example Multiple (Four) Degree of Freedom Lumped Parameter Mechanical Model.

Table 10.2: Example Properties for Model Parameters in Figure 10.11

Parameter	Values
k_1 (N/m)	10^8
k_2 (N/m)	10^7
k_3 (N/m)	10^8
k_4 (N/m)	3.5×10^7
k_5 (N/m)	5×10^7
m_1 (kg)	12.3
m_2 (kg)	0.4
m_3 (kg)	0.26
m_4 (kg)	8.8
c_1 (kg/s)	0
c_2 (kg/s)	0
c_3 (kg/s)	0
c_4 (kg/s)	0
c_5 (kg/s)	0
f_1 (N)	$f_o\sin(\omega t)$
f_2, f_3 and f_4	0

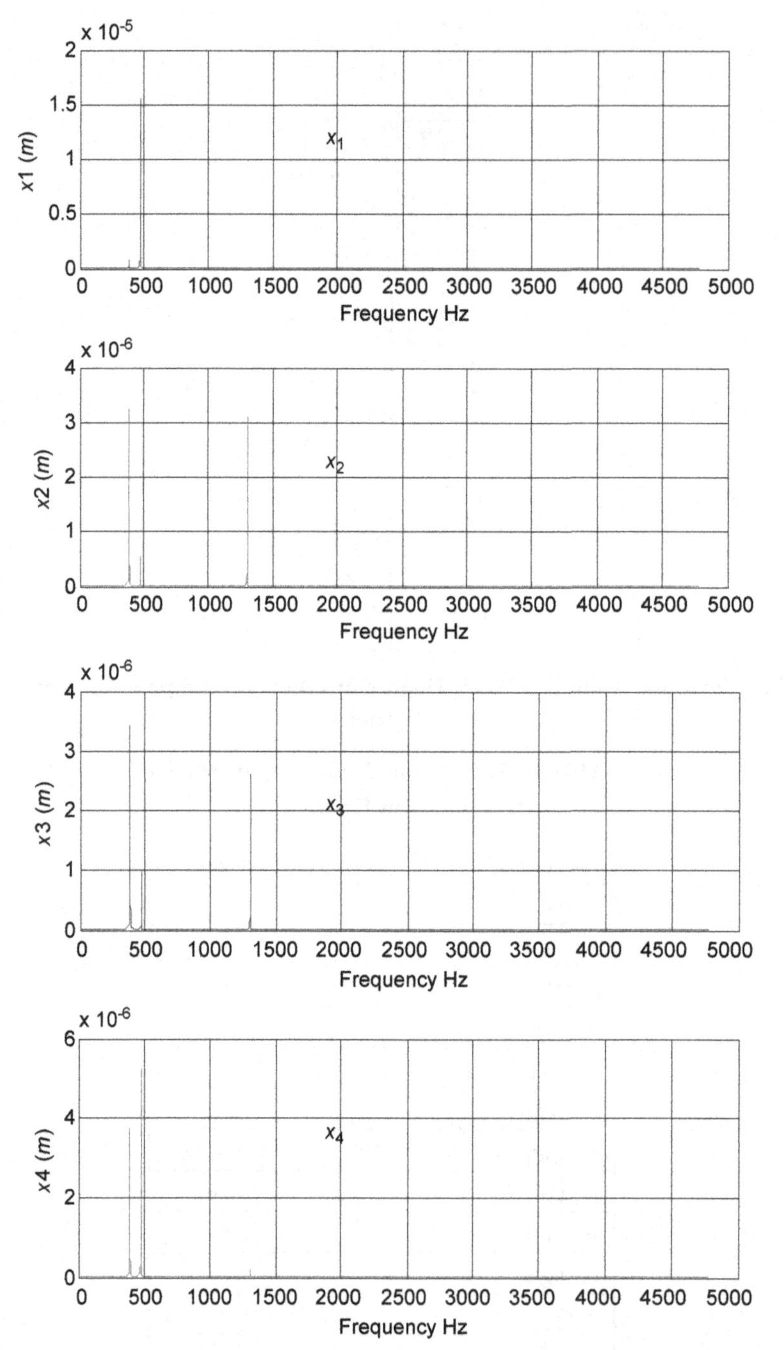

Figure 10.11: Frequency Response of the System for Damping Coefficients (c_1–c_5) = 0.

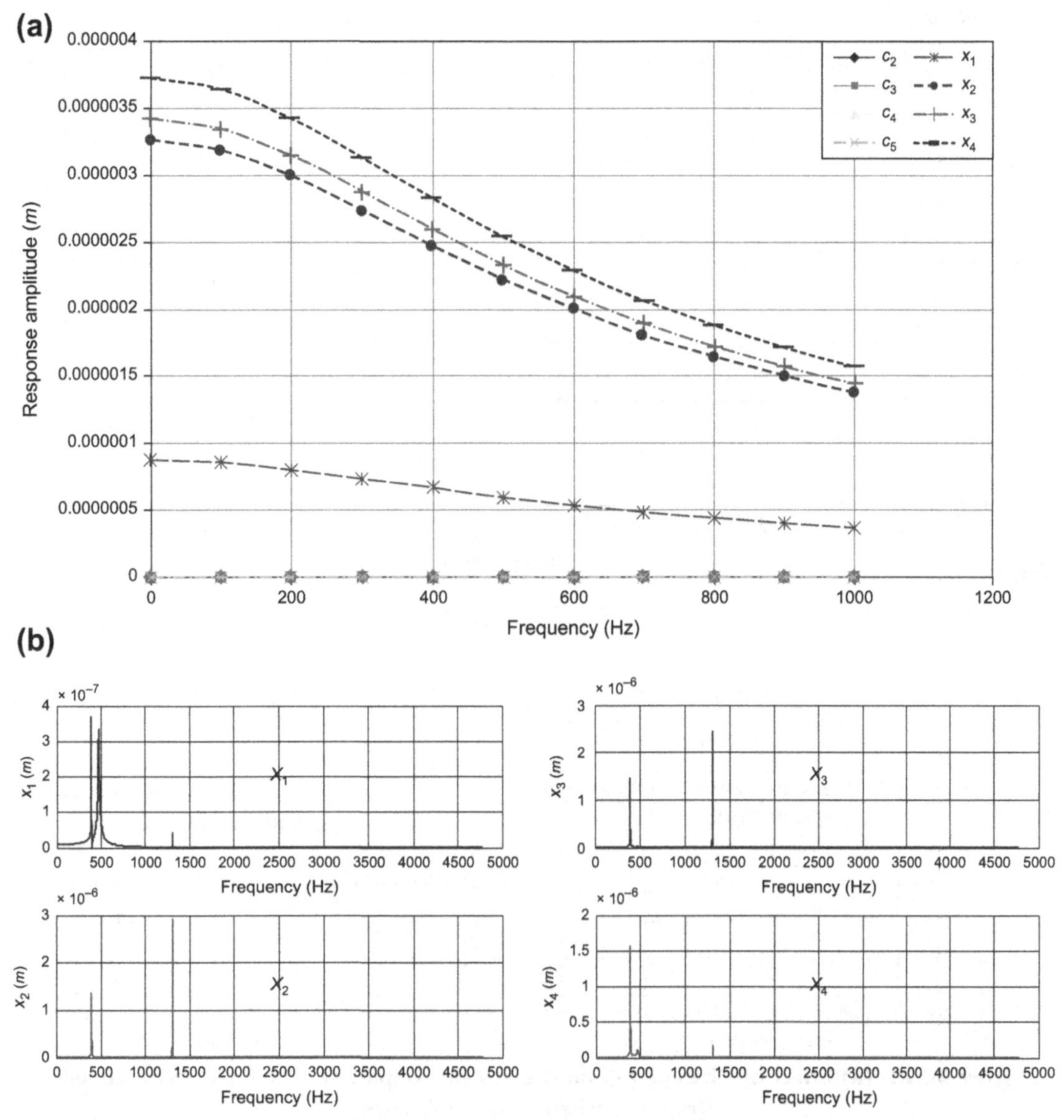

Figure 10.12: (a) Effect of Changing c_1 on the System Amplitude Response. (b) Frequency Response when $c_1 = 1000$ kg/s.

that increasing the damping coefficient c_1 affects the system response (amplitude of vibration is reduced as c_1 is increased), and Figure 10.12(b) (with damping c_1) can be compared with the undamped amplitudes shown in Figure 10.11. In comparison, increasing the damping coefficient c_4 has very little effect on the system response, as illustrated in Figure 10.13(a), and comparing Figure 10.13(b) with Figure 10.11. The point of this is to demonstrate that the application of damping to a vibrating mechanical system may apparently have little or no effect because of where it is being applied to the system.

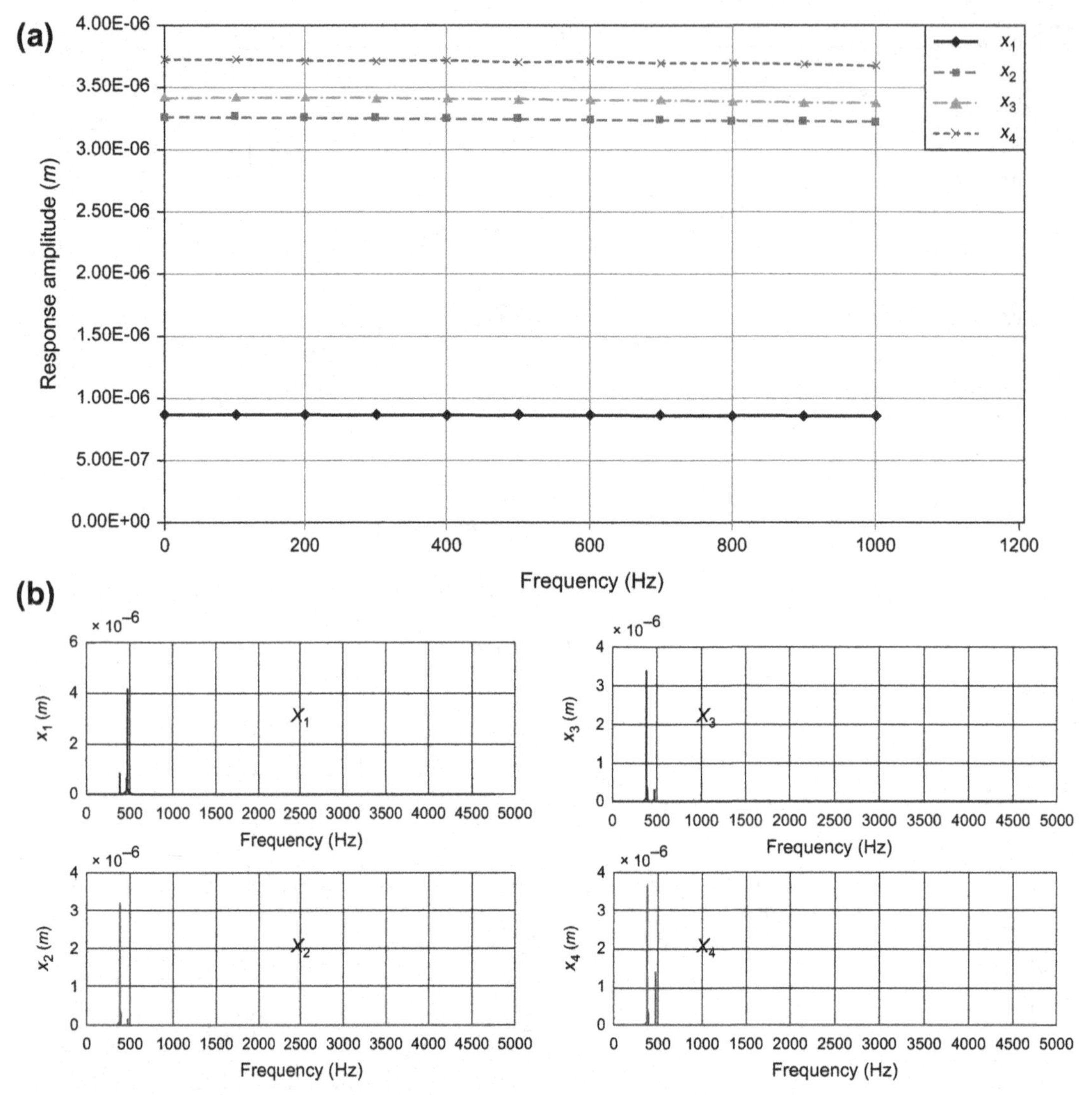

Figure 10.13: (a) Effect of Changing c_4 on the System Amplitude Response. (b) Frequency Response when $c_4 = 1000$ kg/s.

This possibly explains why sometimes damping appears to affect brake noise problems, and sometimes not; damping needs to be applied to the appropriate part of the system in order to produce the required effect. The difficulty for brakes is identifying where to apply the damping to affect the relevant DoF.

A multiple DoF lumped parameter model for low-frequency 'hum' noise on a disc brake (200–400 Hz) with constant μ, which occurs under 'off-brake' conditions with only one pad touching the disc, is illustrated in Figure 10.14 (Lang and Smales, 1983).

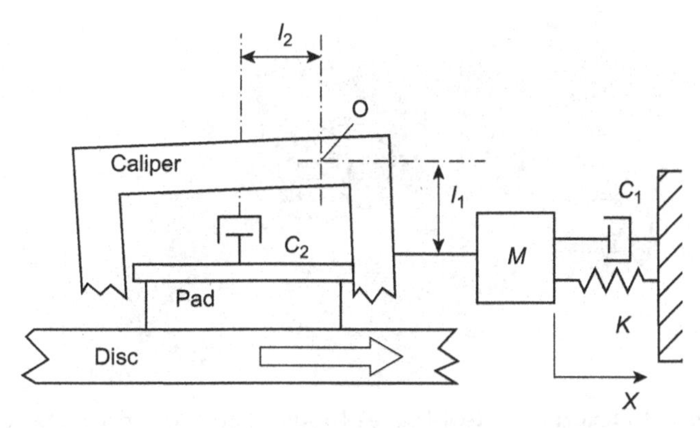

Figure 10.14: Lumped Parameter Disc Brake 'Hum' Noise Model (Lang and Smales, 1983).

The caliper oscillates as a rigid body on its mounting of stiffness K about a node at O, with damping c_1. The actuator (piston in the cylinder bore) provides further damping c_2 attributed to the piston seal, and the disc does not vibrate because the frequency of interest is below its first flexural mode so that the piston is effectively stationary. An initial pad velocity of $\dot{x}$ in the direction of disc rotation produces a relative velocity in the damper c_2 of $(l_2/l_1)\dot{x}$ and hence a normal force at the friction face of $c_2(l_2/l_1)\dot{x}$. The resulting friction force of $\mu c_2(l_2/l_1)\dot{x}$ is a damping force acting in the same direction as $\dot{x}$, and acts as a negative damping force leading to instability. The equation of motion is:

$$M\ddot{x} + \left[c_1 + c_2\left(\frac{l_2}{l_1}\right)\left(\frac{l_2}{l_1} - \mu\right)\right]\dot{x} + kx = 0 \tag{10.7}$$

and hence unstable oscillation occurs when:

$$\mu > \frac{l_2}{l_1} - \frac{c_1 l_1}{c_2 l_2} \tag{10.8}$$

The stability of this model can be increased by decreasing μ and/or increasing the caliper mounting damping c_1, which was verified experimentally with a mass damper fitted on the suspension. Equation (10.8) also indicates that instability does not occur if l_2/l_1 is negative, i.e. if the caliper mounting is designed with O positioned towards the leading end of the brake. Lang and Smales also found experimentally that the hum noise was eliminated if the caliper mounting were stiffened asymmetrically to decrease l_2, and that calipers with lower values of c_2 did not produce hum noise.

Lumped parameter models where the components of the brake are either not modelled separately or are modelled as rigid bodies, do not adequately represent the mechanisms of brake squeal noises that occur at higher frequencies (e.g. above 1 kHz). Squeal noises are always associated with complex modes of vibration of the brake and its components, which cannot be usefully modelled using lumped parameter models, as the vibration of the flexible

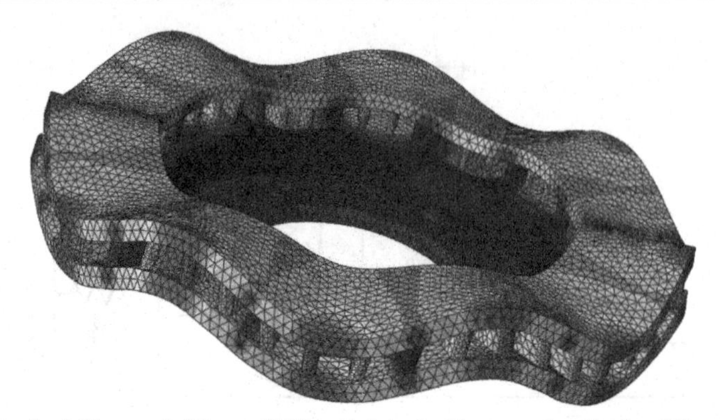

Figure 10.15: Typical Flexural (Out-of-Plane) Mode Shape of a Brake Disc (Bryant, 2013).

components is important in defining the system response. In particular, flexural vibrations of the rotor occur and the natural frequencies of the disc or drum appear to have some influence on the noise frequency. The 1 kHz lower frequency for squeal noise is defined by the frequency of the fundamental mode of vibration of discs and drums of passenger cars, although this can be lower for large commercial vehicle brake drums so squeal noise on CV brakes can be lower than 1 kHz. A typical flexural (out-of-plane) mode of a brake disc is shown in Figure 10.15, and if the brake pad is short compared with the wavelength of the disc mode, its contact with the disc may be considered to be approximately flat so that the brake pad assembly behaves as a rigid beam having 2DoF associated with two different modes of vibration, transverse and rotational (see Figure 10.5). The coupling of the two modes of vibration by the friction force can then produce an instability known as 'binary flutter' due to its similarity with aircraft wing flutter, as proposed by North (1976).

A lumped parameter model of binary flutter is shown in Figure 10.16, comprising a 2DoF disc section interfacing with a spring representation of the friction material with the backplate fixed. The two equations of motion are:

$$M\ddot{y} + \left(k_d + k_p\right)y + F\theta = 0 \tag{10.9}$$

$$I\ddot{\theta} - \mu h k_p y + \left(S_d + k_p\frac{l^2}{3}\right)\theta = 0 \tag{10.10}$$

These can be written in matrix form (see Equations (10.5)):

$$[M]\{\ddot{X}\} + [K]\{X\} = 0 \tag{10.11}$$

The solutions are of the form $y = y_0 e^{\lambda t}$ and $\theta = \theta_0 e^{\lambda t}$ and hence $\{\ddot{X}\} = \lambda^2\{X\}$, which on substitution into the equations of motion gives the eigenvalue equation:

$$\lambda^2\{X\} + [A]\{X\} = 0 \tag{10.12}$$

where $[A] = [M]^{-1}[K]$.

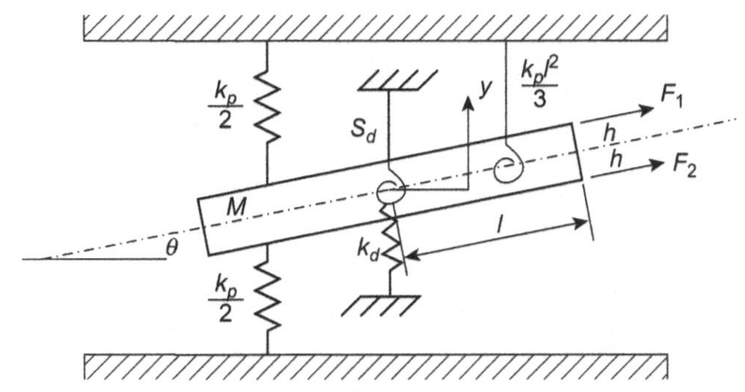

Figure 10.16: Binary Flutter Model (North, 1976).

If the associated characteristic equation in the eigenvalue λ has complex conjugate solutions $(p + iq)$ the solution for y and θ is oscillatory $(y = y_0 e^{pt}\cos(qt))$. The stability criterion can be applied again; as previously explained a positive value of the real part of λ indicates instability (the imaginary part indicates the frequency of oscillation), which gives the condition for instability as:

$$\frac{4MIF\mu h}{\left(I - M\frac{l^2}{3}\right)^2} > k_p > 0 \qquad (10.13)$$

This type of analysis has been extended (North, 1972; Millner, 1978; Murakami et al., 1984) to include the effects of additional degrees of freedom representing the pads, pistons and caliper in order to obtain better correlation with the real noise problem. Millner's (1978) 6DoF model showed that the position of the contact between the caliper actuation piston and the pad could influence stability and this explained the basis for shims, which are fitted as a 'fix' between the pad backplate and the actuation piston. All these models predict that higher μ destabilises the system, but the effect of other parameters such as stiffness and mass are more difficult to quantify in terms of actual brake design.

Noises such as 'squeak' occur at frequencies higher than those of squeal (but are often also termed squeal or high-frequency squeal), typically above 5 kHz. The disc flexural mode is a higher order with a shorter wavelength, which causes dynamic changes in the contact and pressure distribution between the pad and the disc. The fundamental bending frequency of a disc brake pad is in the range 2–5 kHz) so the contact and pressure distribution may also be affected by pad flexure. Experimental examination of the mode shapes of brake components during high-frequency squeak noise has indicated a very complicated range of dynamic deformations of the disc and pad, which cannot be included in the lumped parameter models described above. Other noises that are thought to be related to squeak include 'squelch', which is an amplitude modulated 'squeak', and 'wire brush', which appears to be the result of decaying oscillations at squeak frequency when full instability does not occur (Fieldhouse, 2013).

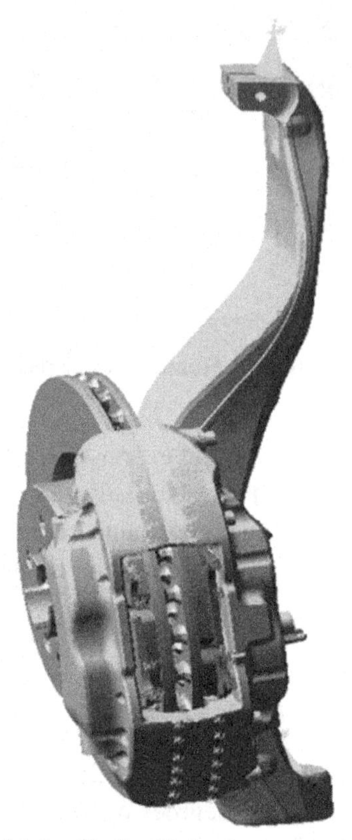

Figure 10.17: MBDS Model for Brake Noise Investigation (Alasadi et al., 2013).

FE analysis for brake noise tends to concentrate on the time domain, while multi-body dynamics (MBD) computer simulation can predict the vibration behaviour of the brake parts in the time domain. Recent work on brake noise modelling using MBD has presented a full dynamic model of an opposed piston passenger car disc brake that includes 3D flexible body modelling of the disc, pads, pistons and caliper, plus the hub and steering knuckle as shown in Figure 10.17 (Alasadi et al., 2013). Sliding frictional contact between the friction material and the rotating disc was modelled using a nonlinear impact contact algorithm, which aligns with the 'hammering' mechanism (Rhee et al., 1989). Vibrational modes of the brake pads were predicted that showed considerable similarities to published experimental observations (Fieldhouse et al., 2011), and confirm that a vibrational response from the brake components at frequencies that were known to be associated with brake noise could be generated with a constant coefficient of friction. The predominant frequency identified from the simulation varied between 2.6 and 2.8 kHz over a range of different actuation pressures and friction coefficients. Other frequencies were also predicted, in particular the onset of 5.7 kHz over 15–30 bar hydraulic actuation pressure and with a friction coefficient of 0.55. The predicted vibrational response of the brake disc, caliper, and the inner and outer brake pads is illustrated in Figure 10.18 and shows that 2.7 kHz was the predominant

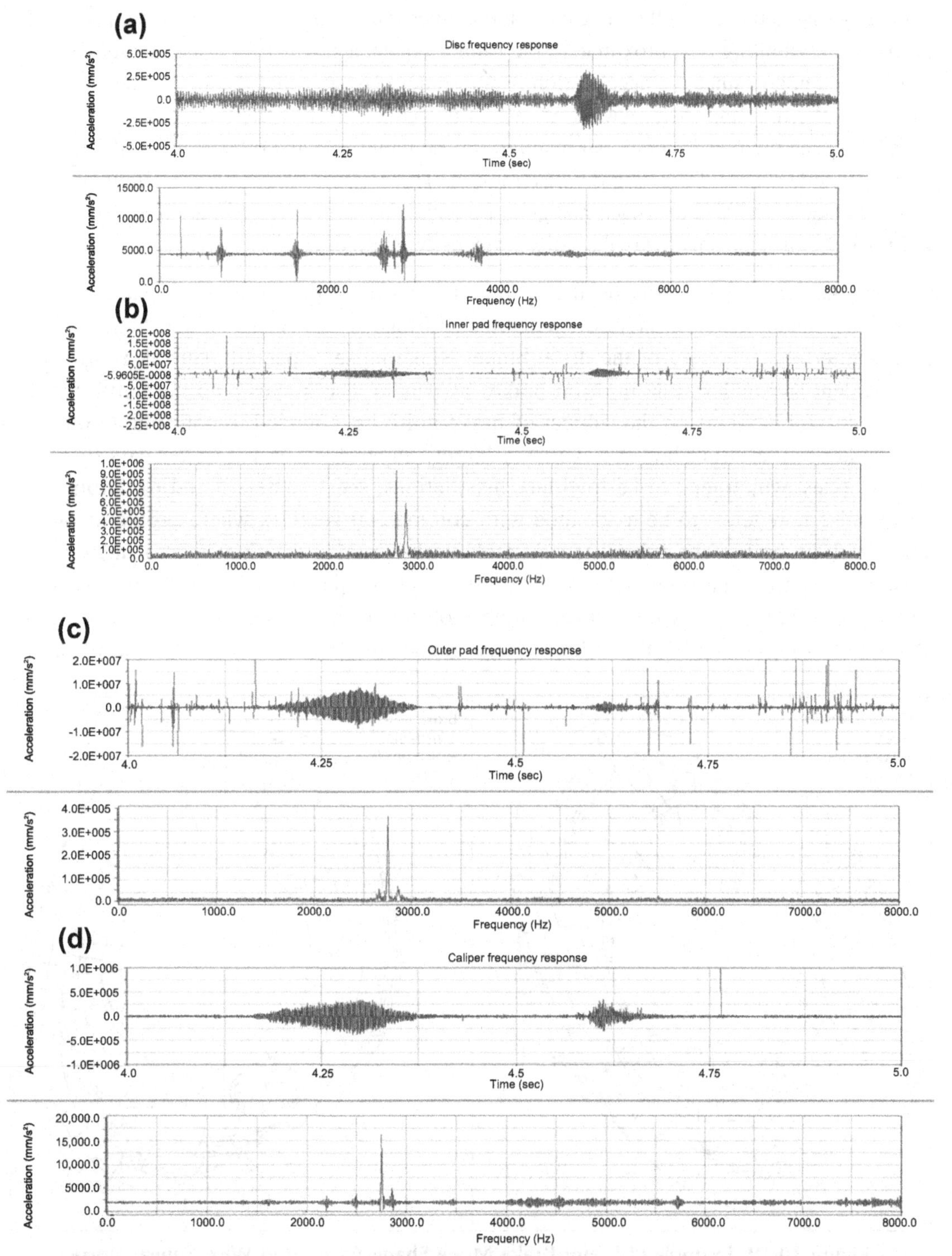

Figure 10.18: (a) Predicted Disc Frequency Response. (b) Predicted Inner Pad Frequency Response. (c) Predicted Outer Pad Frequency Response. (d) Predicted Caliper Frequency Response (Alasadi et al., 2013).

frequency that appeared on the disc, pads and caliper. These indicate that the disc, pads and caliper were vibrating at approximately the same frequencies, which indicated coupling between them, and since the acceleration response demonstrates that the amplitude varies with time, the coupling varies with time, which is characteristic of the binary flutter instability. The predicted vibrational response of the pads indicated combined flexible and torsional mode shapes, following the disc motion. This type of modelling and simulation indicates that the MBDS approach has great potential for investigating brake noise source and response mechanisms for brake noise propensity predictions.

So far brake noise has mostly been considered in terms of disc brakes that have a (nominally) plane friction interface geometry and thus are easier to model than the curved interface geometry of the drum brake. Brake noise, especially squeal, is still just as prevalent and undesirable in drum brakes as it is in disc brakes, and the only reason that recent publications on the subject of brake noise have tended to concentrate on disc brakes is that drum brakes have been largely superseded on passenger cars (except on rear axles, which tend to be low-duty installations; see Chapters 5 and 6). Drum brake noise now tends to be associated with commercial vehicles where problems of noise in drum brakes are still all too evident. Brake drum mode shapes associated with noise have similar characteristics to brake disc mode shapes in that they form a wave around the rubbing path with an even number of nodal lines across the rubbing path, as illustrated in Figure 10.19. Drum brake noise models must include deformation of the

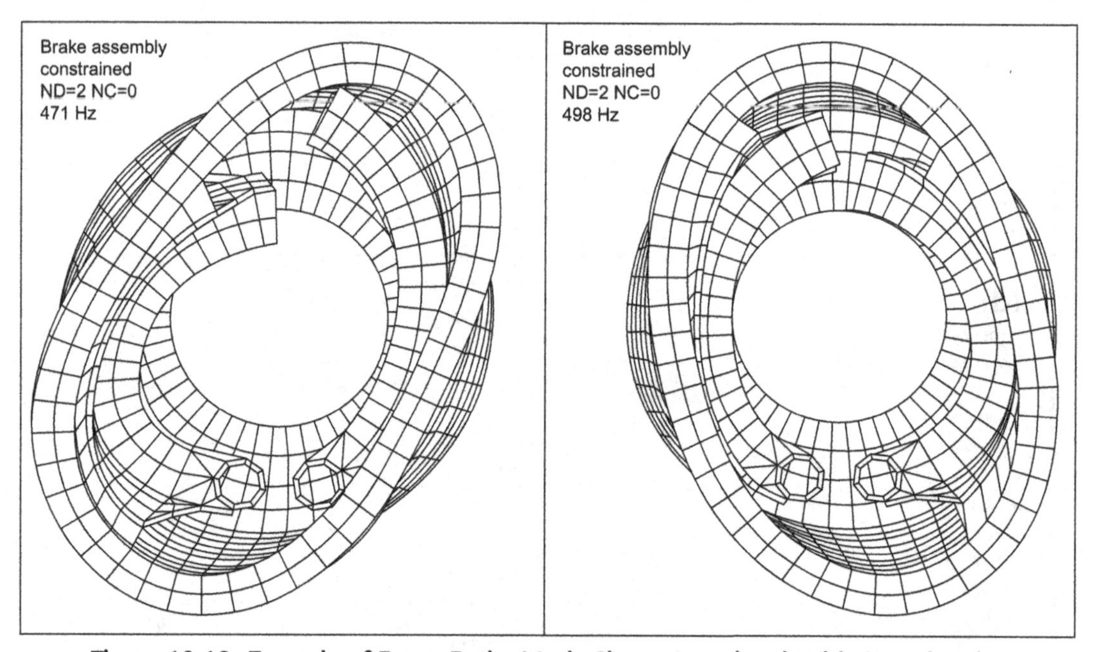

Figure 10.19: Example of Drum Brake Mode Shape Associated With Squeal Noise (Day and Kim, 1996).

brake drum and the brake shoes over the friction contact area, as was first presented by Millner (1976) in his theory of drum brake squeal. He predicted that squeal noise propensity increased with μ, heel-and-toe lining contact was destabilising compared with crown contact, and there was no simple relationship between stability and type of friction material, and shoe or drum stiffness and mass. In most modes he found that damping of the drum tended to reduce squeal noise propensity while damping of the shoe did the opposite.

Modal Analysis in Brake Noise

Modal analysis studies the mode shapes of vibrating bodies and systems when they are vibrating at a resonant frequency, i.e. the response to a time-varying forcing vibration is significant. A mode shape can be defined as a graph that shows the deflected form of the body or the system when the displacements are at a maximum, and there is one mode shape for each mode of vibration (e.g. Figure 10.9). There are two approaches to modal analysis, namely experimental modal analysis where mode shapes are measured using sensors and transducers for, say, acceleration and displacement, and theoretical modal analysis where the mode shapes are predicted, e.g. by FE-based eigenvalue extraction, or MBD analysis of the vibration behaviour of the brake assembly. Modal analysis can quickly evaluate the effect of changes to the component or system in terms of the dynamic behaviour. Experimental testing for brake noise and vibration was briefly introduced in Chapter 9; experimental modal analysis is usually used for verification of predictive methods and the investigation of operational noise problems in which it may focus on individual components or complete brake installations. Often experimental techniques cannot reliably be applied to a complete brake installation on a vehicle in the field so either it needs to be brought to a laboratory, e.g. to operate on a chassis dynamometer, or the brake needs to be installed on an appropriate dynamometer for controlled investigation. Generally, brake noise above 1 kHz can be investigated on the foundation brake alone, while noise above 3 kHz can be investigated by concentrating on the friction pair with the caliper and abutments in position to provide the correct disposition of the pads to the disc. Below 1 kHz brake noise investigations require the foundation brake with all the associated components such as the hub, wheel, tyre, axle, suspension arms, etc., to be included in the experimental set-up. Experimental investigation of brake noise usually starts with the audio recording of the noise, which is then analysed by fast Fourier transform (FFT) to identify the frequencies involved in the form of a frequency response function (FRF). It is also important to note the detail of the operational conditions under which the noise occurs, e.g. speed, deceleration, temperature, loading and usage.

All experimental modal parameters are obtained from measured operating deflection shapes (ODSs) while artificially exciting the body or system. An ODS describes the forced motion of two or more points on a body. The most widely used experimental modal analysis method is impact testing, which excites the body or system by an impact. An example FRF from a brake disc impact test is shown in Figure 10.20. For impact testing the body is usually supported in the 'free–free' condition, e.g. by suspending it on low-stiffness 'bungee' cords, and the equipment listed below is required:

- An impact hammer with a load cell attached to its striking face ('head') to measure the input force.
- An accelerometer to measure the response; this may be achieved by transducers mounted on the body or a laser accelerometer device.
- An FFT analyser to compute the frequency response functions (FRFs).
- Post-processing software for identifying modal parameters and displaying the mode shapes.

Forced vibration excitation can be achieved by 'shaker' testing; an electrodynamic or hydraulic shaker can be attached to the body or system often using a long slender rod (high stiffness with low mass) or 'stinger'. The excitation force is applied only in the direction of the axis of the rod, and a load cell can be included to measure its magnitude and waveform. Figure 10.21 shows different vibration mode shapes in a brake disc with stinger attached; the mode shapes are indicated by grains of sand (the disc is horizontal), which agglomerate at the nodal lines. In this case Figure 10.21(a) indicates a two nodal diameter mode of vibration that is associated with a resonant frequency of about 700 Hz,

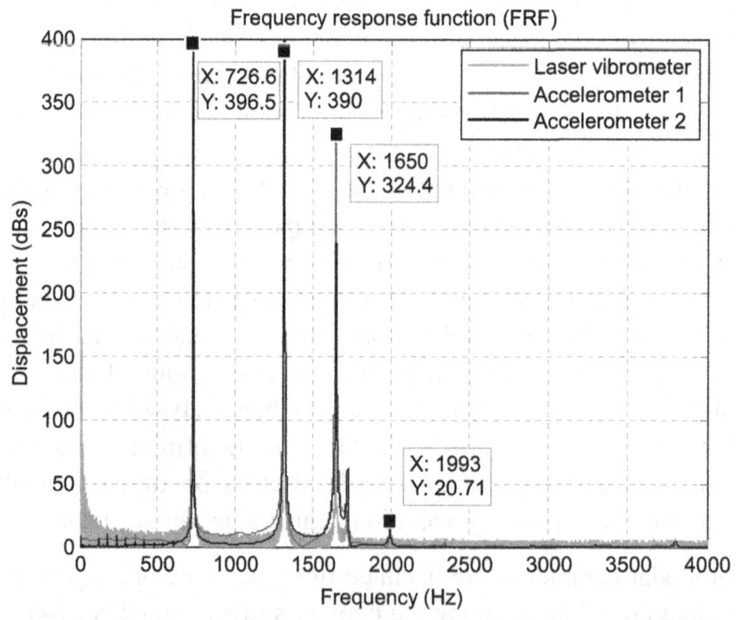

Figure 10.20: Example FRF from a Brake Disc Impact Test.

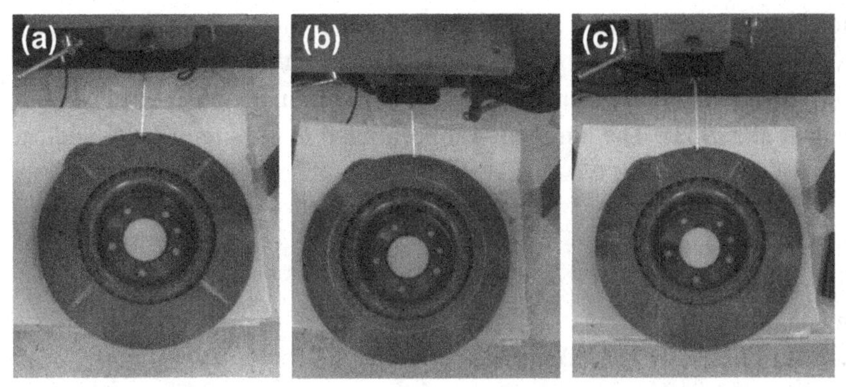

Figure 10.21: Example Brake Disc With Stinger Attached Showing Different Vibration Modes.

Figure 10.21(b) indicates a circumferential mode associated with a resonant frequency of about 1.3 kHz, and Figure 10.21(c) indicates a three nodal diameter mode with a resonant frequency of about 1.7 kHz. Again, accelerometers are used to measure the response. Shaker testing is useful when the system response from impact testing is too small, and can be used for the experimental modal analysis of the components of brake assemblies, usually while the rotor is stationary. To determine the mode shapes of brake components while the brake is operating, more sophisticated methods such as holographic interferometry are used. This particular technique requires specialised equipment in a research laboratory environment, and has been extensively described by Fieldhouse (2013), whose results have provided remarkable insight into the modes of vibration of noisy brakes.

While a disc brake is generating noise, experimental modal analysis has shown that the disc vibrates with a diametral mode of vibration that may be stationary or rotate around the disc axis at an angular velocity determined by the quotient of the frequency and the mode order. It has been shown (Fieldhouse, 2013) that there is a well-defined relationship between disc diametral vibration modes and frequency, which can be written in the form shown in Equation (10.14), where α is the node pitch angle and γ and β are constants for the particular disc design:

$$\text{Frequency (Hz)} = \gamma\alpha^{-\beta} \tag{10.14}$$

Similarly when a drum brake generates noise the drum vibrates with a radial mode of vibration, which may again be stationary or rotate around the drum axis. Both experimental and theoretical modal analyses have also confirmed the existence of 'doublet' or 'pair' modes, which is important because circumferential symmetry in a brake disc or drum allows vibrational energy to be continually transferred between two closely spaced modal frequencies or pair modes in a binary flutter instability. In practice there is always circumferential asymmetry in a brake rotor because of manufacturing imperfections and tolerances, and it has been found that increasing the circumferential asymmetry by, for example, adding equally spaced concentrated masses to a rotor periphery (while maintaining balance) can reduce or eliminate noise by separating the pair modes (Lang et al., 1983). The number of concentrated masses required is indicated by:

$$\frac{2M}{R} = \text{positive integer} \tag{10.15}$$

where:

M = mode order;
R = number of equally spaced masses.

This approach was investigated by (Day and Kim, 1996) in a parametric study of an S-cam drum brake using theoretical modal analysis (FE analysis) of the coupled brake drum and shoes, and an instability condition derived from Lang et al. (Lang et al. 1993) which indicated that bringing two brake drum normal modes together increased the instability propensity, while moving them apart decreased it. The parametric study showed that the parameters listed below increased the separation of pair modes of two nodal diameters, 471 and 498 Hz (see Figure 10.19), both frequencies close to an observed squeal noise of 480–490 Hz:

- Increasing coupling stiffness between the brake linings and the brake drum, e.g. by increasing lining stiffness or contact pressure (actuation force).
- Uniform or crown contact of the linings with the drum friction surface.

Brake drum and shoe stiffness appeared not to affect squeal noise propensity on this design of brake, but it was predicted that higher stiffness would increase the frequency of the squeal noise generated. The addition of four lumped masses on the drum periphery was found to have the largest effect on separating the pair modes. More information is given in Chapter 12.

The experimental set-up for double pulsed holographic interferometry investigations on brake assemblies (Felske and Happe, 1977) utilises mirrors so that images of the front, side and top view of a brake assembly are recorded simultaneously. Other mirrors enable other parts, e.g. the ends of the pads, also to be viewed. The basic holographic technique was extended by Fieldhouse (2013) to provide amplitude and direction data and their relative phase relationships. Results from many disc brake investigations indicated that the 'stationary' mode shape may actually travel slowly round the disc, and many other features of brake and component vibration were used to interpret causal mechanisms and suggest ways to avoid noise or reduce noise propensity in brake design. For example, Talbot and Fieldhouse's experimental modal analysis using laser holographic interferometry has confirmed the presence of in-plane vibration modes. Referring to Figure 10.22 (Talbot and Fieldhouse, 2002), diagonal fringes can be seen along the edge

Figure 10.22: Photograph of Mode Shapes Using Laser Holographic Interferometry (Fieldhouse, 2013).

of the brake disc (upper part of the image) that represent combined rotational motion, in-plane vibration and out-of-plane vibration. Vibration of the pad assembly was shown to be particularly complex, usually a combination of bending and torsional modes over a range of frequencies above the first flexural mode, usually above 2.5 kHz. Talbot and Fieldhouse also observed that the trailing end of the pad backplate in a 'tension' abutment design tended to vibrate in all cases of audible noise and compared pad vibration to the associated disc vibration, suggesting that the pad abutment was possibly critical in 'triggering' noise instability. They also explained how the clamp force generated by the pads of a disc brake tended to change the antinode pitch in the disc mode; increased actuation pressure 'squashed' the disc vibration as it passed the pads, influencing the frequency. This effect could be controlled by careful design of the pad effective length. They summarised five basic criteria associated with this effect:

- The system tries to fit a whole number of antinodes along the pad friction interface.
- The pad effective length can be between 80% and 100% of its design length.
- The antinode will generally compress to fit under the pad friction interface.
- Antinode compression will be a maximum of 60% of the original free mode angle.
- For frequency to be considered a problem the free mode antinode subtended angle must be greater than that of the pad.

Talbot and Fieldhouse proposed an approach to predict potential disc brake noise frequencies based on pad–disc interaction starting with determining the free modes of vibration of the brake disc, and explained how chamfers, which are often introduced at the ends of disc brake pads as a noise fix, work by adjusting the pad effective length. They concluded that although the principles do not in themselves help to predict brake noise propensity, they may be applied to new disc brake designs to predict the frequencies of brake noise, if the brake does generate noise. And if it does, changes may be proposed that are informed by this theory to, for example, pad profile, positions of actuation pistons (or fingers) and disc stiffness.

Variability in Brake Noise

One of the most interesting (and frustrating) features of brake noise is its fugitive nature; it is well known that even across different vehicles of the same make and model with the same brake installation, some exhibit brake noise, while others do not. Given the complexity of the underlying physical principles of brake noise generation that have been outlined above and the emphasis on instability, it appears that relatively small changes in one or more of the many design and operational parameters associated with friction brakes can tip a brake installation from quiet to noisy. Operational parameters such as temperature, speed, actuation pressure and humidity have all been mentioned, and many vehicle and brake manufacturers have studied the possible influence of variation arising from manufacturing tolerances. The role of the pad assembly in disc brake noise is becoming increasingly well understood, and the nature of the abutment contact is influenced by the dimensional tolerances between the pad backplate and the caliper abutments, which may change over time as a result of wear, plastic deformation, and the build-up of debris and corrosion. Brake pads are generally closely toleranced in terms of thermophysical properties during and immediately after manufacture, but once they enter the customer domain, variability increases. The manufacture of brake discs and drums is closely controlled but achieving and maintaining close tolerances on the internal geometry of ventilated brake discs, for example, is costly. Examination by the author of samples of ventilated brake discs within the same batch and across different batches has indicated a 3σ variation of $\pm 5\%$ in resonant frequencies up to 6 kHz, with pair modes separated by amounts ranging from 10 to 60 Hz. The close examination of components from examples of brakes that exhibit noise, and comparison with nominal dimensions and values, is thus a useful exercise.

Brake Judder

As previously described, brake judder is a mechanical instability that creates a non-resonant forced vibration usually with a low frequency (typically up to 200 Hz) related directly to rotor speed. It may be caused by circumferential variations or imperfections in the rubbing path of the brake rotor, which cause wheel and brake torsional vibration at a frequency that is a multiple of the rotor speed. The vibration can be transmitted through the vehicle suspension, brake actuation system, steering and the structure, so that it can be felt during braking by the driver via the steering wheel as steering 'shimmy' (sometimes also known as 'nibble') and the brake pedal as vibration, and by the driver and passengers through the vehicle body and the seats, e.g. cab vibration on a commercial vehicle (Hussain et al., 2007). These may be irritating and tiresome when the amplitude of vibration is small, but become a problem when their frequency coincides with resonant frequencies in the vehicle's suspension, steering or structure, and large-amplitude low-frequency resonant vibration occurs.

The two distinct categories of brake judder are cold judder and hot (or high-speed) judder. Cold judder typically creates very low frequency vibration (around 10 Hz), which is felt by the vehicle occupants rather than heard. Cold judder is associated with mechanical non-uniformity of the rotor rubbing path for which the causes can include those listed below. As a result of improvements in manufacturing techniques and good understanding of the causes of cold judder, it is seldom a major problem on modern road vehicles.

- Non-circular or eccentric brake drums. Brake drums, especially on commercial vehicles, can become non-circular over time with use because of thermoplastic deformation and circumferential variations in the original casting. On the smaller passenger car brake drums, eccentricity was a problem, often caused by poor location of the brake drum on the hub via the wheel studs or 'bolt-up' distortion. This is avoided by the modern practice of using combined hubs and brake drums.
- Runout or swash of brake discs. Brake disc runout, e.g. where the disc swashes from side to side during running due to poor manufacturing practice, bolt-up distortion or hub flange contamination, causes intermittent contact between the brake pads and the disc in the off-brake condition. Runout can increase with time and good practice is to minimise it from new, ideally to less than 80 μm, although some manufacturers aim for 10 μm, by maintaining close tolerances and clean hub flange faces. Disc brake pad retraction also affects disc runout; poor pad retraction, perhaps due to corroded pistons or aged seals, may cause intermittent off-brake pad/disc contact.
- Disc thickness variation (DTV) caused by poor manufacturing practice or in-use by 'off-brake wear'. At the positions where the intermittent pad/disc contact occurs, the disc will wear, increasing DTV. The amount of wear depends on the number of disc rotations; therefore, high-speed driving over long distances (such as motorway driving) can create substantial levels of DTV. The critical threshold for DTV to generate noticeable cold judder has been identified as 10–15 μm (De Vries and Wagner, 1992; Jacobsson, 2003). Increasing piston seal roll-back can help to reduce cold judder but at the expense of increasing pad/disc clearance and brake actuation response.
- Circumferential μ variation generated by uneven coatings or transfer films on the rotor friction surfaces, or corrosion. Transfer films on the disc rubbing surface can become patchy over time, especially as a result of intermittent vehicle usage, when the area

under the pads while the vehicle is stationary can start to adhere to the pad material through localised corrosion, a phenomenon known as 'pad imprinting'.

Hot judder is more problematic because it is fundamentally a thermally induced phenomenon, and it occurs in disc and drum brakes, though it is particularly a problem in large heavy-duty disc brakes, e.g. on commercial vehicles and certain types of large high-performance passenger cars. Permanent thermoplastic rotor deformation results from localised yielding of cast iron under thermal cycling, the same mechanism that causes thermal cracks (see Chapter 7) and is also relevant to hot judder. The main cause of hot judder though is considered to be 'hot spotting' or 'blue spotting' (see Figure 10.23), which is a phenomenon associated with localised phase change in the cast-iron material of the rotor friction surface from pearlitic to martensitic as a result of localised high temperatures, as explained in Chapter 7. The higher specific volume of the martensite causes these regions to 'stand proud' of the rest of the surface as 'blue spots' (so called because of their characteristic oxidised blue colour) and these may eventually be worn away, creating local hollows in the friction surface. In turn these may be replaced by other 'blue spots' in other regions of the friction surface. The uneven rubbing surface caused by hot spots, together with their associated different coefficient of friction, creates cyclic variations in brake torque leading to judder. Vibrations from hot judder are higher frequency than with cold judder and are associated with a characteristic speed-related noise called 'drone' or 'rumble', which is a drumming noise of frequency typically between 100 and 200 Hz.

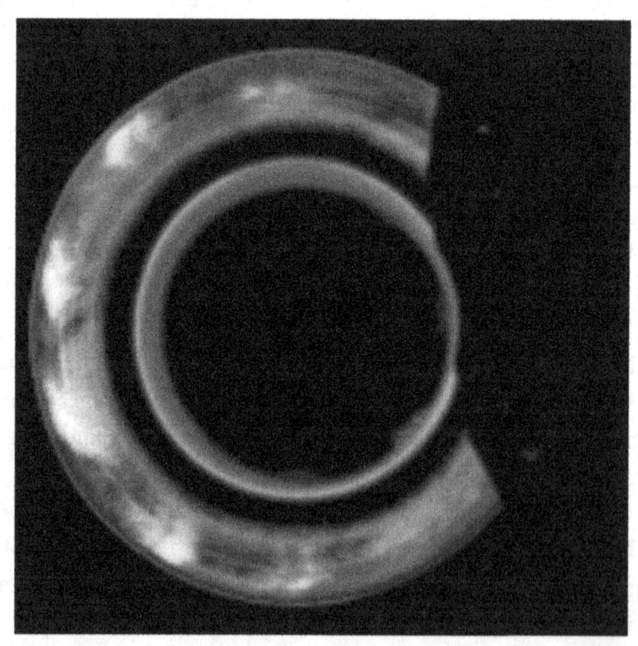

Figure 10.23: Hot Spots on a Brake Disc Rubbing Surface (Bryant, 2010).

The generation of hot spots starts in regions of localised high rotor friction surface temperature, which may be caused by hot circumferential banding across the rubbing path as a result of radial interface contact and pressure distributions in disc brakes, and axial interface contact and pressure distributions in drum brakes. Where the rotor is continuous, e.g. in a conventional brake drum or solid disc, these hot bands on the rotor surface try to expand but are constrained by the neighbouring cooler material. This induces compressive stress within the rotor rubbing path and the only way that thermal expansion can be accommodated is for the hot band to buckle, creating a circumferential waveform on the friction surface. The number of hot spots and their radial positions are possibly determined by the need to accommodate an integer number of hot spots around the rubbing path. These hot spots also appear to move around to other positions in the rubbing path as they wear away and then re-establish. Panier et al. (2004) used an IR camera to investigate hot spot formation on a brake disc and concluded that the energy flow into the disc brake pads was significant in their formation of MHS. They identified five types of hot spot formation on the brake disc, as illustrated in Figure 10.24:

- Type 1: the asperity type hotspot with only small areas of rapid temperature rise.
- Type 2: temperature gradients along a hot band that create small contact areas from a single rubbing path.
- Type 3: hot bands then forming from pad expansion causing larger circumferential contact patches in the plane.
- Type 4: buckling effects creating large, distinct hot spots.
- Type 5: the type 4 hot spots coalescing to form hot contact areas over most of the rubbing path, usually evident at the end of a braking application.

Yi et al. (2000) developed mathematical models to predict hot spots and thermoelastic instability (TEI) on brake rotors using an eigenvalue approach. They suggested that circumferential frictional variations could be caused by an instability, possibly flutter related, at a frequency close to the fundamental resonant frequency of the system, which could be a way of relating hot judder to TEI. On solid brake discs, type 4 hot spots often occur spaced uniformly around the rubbing path and are antisymmetric, i.e. the

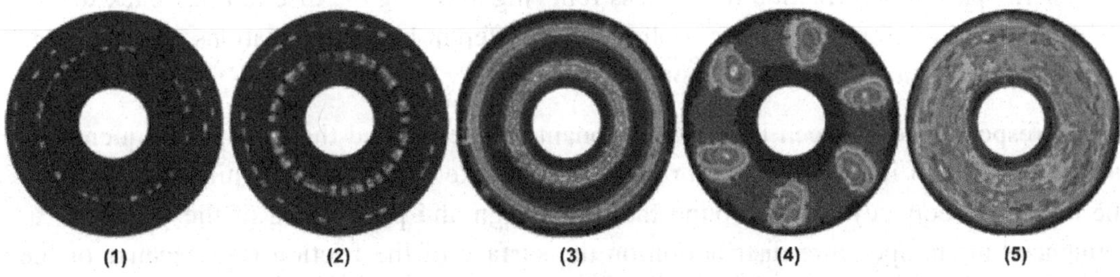

Figure 10.24: Hot Spot Formation Types (Panier et al., 2004).

spots on one side are positioned circumferentially in between those on the other side, which further supports the buckling theory. Hot spot formation on ventilated brake discs is affected by the positioning and design of the vents because of the localised conduction path and stiffening provided by the vanes, and also shows antisymmetric disposition of the hot spots on one side of the disc compared to the other. Kubota et al. (1998) recommended four design rules to minimise disc brake drone noise (and the underlying hot judder):

- Design the rotor to minimise thermal deformation and maintain uniform pad/disc contact.
- Use a low-compressibility friction material to ensure more uniform pad/disc contact.
- Specify high thermal conductivity rotor materials to minimise surface temperatures and circumferential temperature variations.
- Design discs with a narrow rubbing path width.

Bryant (2010) measured pulsation of the hydraulic actuation pressure of a passenger car disc brake (ventilated disc), which exhibited a small amount of hot judder. His measurements identified a brake pressure variation of 0.5 bar and a DTV of 5 μm, which was below the critical threshold of 10–15 μm. He demonstrated how judder developed over a sequence of 30 vehicle brake applications, reaching a brake pressure variation of 1.7 bar when the judder was distinctly noticeable by the vehicle occupants. The brake pressure variation had changed from start to finish of the sequence of 30 brake applications from a relatively well-defined second order to a more complex sixth order, indicating the development of an increased number of circumferential irregularities, but since there was no evidence of hot spots this was ascribed to thermoelastic deformation. He was also able to correlate the brake pressure variation with the measured tangential acceleration of the brake pads. The factors that Bryant ascribed to the development of judder included:

- Thermoelastic deformation due to rapid heat dissipation through the brake disc, causing buckling in the rubbing path.
- The clamping action of the brake pads causing the buckling form to change as it passes between them.
- Thermoplastic deformation from stress relieving allowing the disc to relax back to a state of cold deformation after cooling; after different brake applications, the states of cold deformation after cooling could be different.

No correspondence between brake discresonant frequency and the hot spot frequency was found, which indicated that vibration of the brake disc had no significant effect on the hot spot frequency. Bryant found that the design and positioning of the vanes influenced the temperature distribution on the surface of the friction ring because of the increased local thermal mass, which allowed the disc to conduct heat from the surface to a vane faster than to the air flowing through the vanes. This did appear to influence the number and positioning of hot spots to coincide with groups of vanes. He concluded that minimising thermoelastic deformation will reduce brake judder and can be achieved by adopting a disc design with:

- Higher buckling stiffness to resist circumferential buckling
- Lower buckling stiffness to promote higher order circumferential buckling
- Reduced thermal gradients to allow greater thermal expansion to avoid buckling
- More uniform distribution of the heat to reduce maximum temperature (and temperature gradients).

Another theoretically possible option would be to reduce the thermal loading on the brake disc by reducing the vehicle mass, but cost, mass and packaging constraints would usually preclude this approach. Bryant's preferred approach was to suggest a novel vent design with a false vane at the midpoint between two adjacent vanes with the aim of achieving more uniform heat and temperature distributions.

Chapter Summary

- The problems of brake noise and vibration have been described in terms of how they affect vehicle drivers and occupants, vehicle manufacturers and the manufacturers of brake component parts. The two distinct problem areas are brake noise and brake judder; the former includes a range of road speed-independent brake noises over a frequency range from a few hundred hertz to 10 kHz or more, while the latter is categorised into cold judder generating road speed-related vibration at low frequencies (around 10 Hz), which are felt rather than heard, and hot judder at road speed-related frequencies typically 100–200 Hz. Descriptive names for different types of brake noise are briefly explained.
- Brake noise is considered first, starting with an explanation of 'source' mechanisms, which create a vibrating signal input to the brake system, and then moving on to the 'response' of the system, which amplifies the signal and radiates it as sound pressure variations or 'waves'.
- Brake noise and vibration have been extensively researched over many years and a short review of published work on the subject is given based on four significant reviews by notable researchers in the field over the last 40 years. These show a development of approach from understanding the nature and characteristic features of vibration associated with brake noise (frequencies, mode shapes), through lumped-parameter models, mathematical models, stability criteria, to modelling and simulation, all supported by increasingly sophisticated experimental measurement and research techniques such as laser holographic interferometry. Some recommendations and conclusions relating to the reduction of brake noise propensity (described as the likelihood that a given brake design will suffer from noise problems) and the solution of brake noise problems are

summarised, some of which, e.g. the effect of the friction coefficient μ, have been well known for over 50 years.

- The source of brake noise creating vibration is explained as frictionally induced dynamic instabilities in the brake that are generated by asperity contact and the physical mechanisms of friction to create time-varying frictional forces at the friction interface. Various mechanical models are described and their usefulness explained, e.g. the kinematic constraint family of models such as the 'sprag-slip' model (Spurr, 1961), which is brought up to date by a brief summary of Fieldhouse et al.'s (2011) recent work on centre of pressure and caliper mounting geometry.
- The response of the brake system and components to the source vibration is studied and simple lumped parameter models are used to explain some basic theoretical concepts, including how the system stability can be analysed by calculating the roots or eigenvalues of the characteristic equation; a negative real part indicates stability. This method is widely used in predicting brake noise propensity for design purposes.
- Brake noises that occur at higher frequencies (e.g. above 1 kHz for passenger car brakes but below this for larger commercial vehicle brakes) where the human ear is most sensitive cannot be adequately modelled by lumped parameter modelling. These are often referred to as 'squeal', and are always associated with complex modes of vibration of the brake and its components as the vibration of the flexible components is important in defining the system response. In particular, flexural vibrations of the rotor occur and the natural frequencies of the disc or drum appear to have some influence on the noise frequency. It is explained that if the brake pad is short compared with the wavelength of the disc mode, its contact with the disc may be considered to be approximately flat so that the brake pad assembly behaves as a rigid beam having two degrees of freedom associated with two different modes of vibration. An important development in brake noise analysis was the coupling of these two modes of vibration by the friction force, which can produce an instability known as 'binary flutter'. Higher frequency noises such as 'squeak' occur at higher frequencies, typically above 5 kHz, and the disc flexural mode is a higher order with a shorter wavelength, which causes dynamic changes in the contact and pressure distribution between the pad and the disc.
- Recent work on brake noise modelling using 'multi-body dynamics' computer simulation (MBDS) is described. This can predict a wide range of vibrational responses for a complete disc brake installation, which includes 3D flexible body modelling of the disc, pads, pistons and caliper, plus the hub and steering knuckle (Alasadi et al., 2013). Sliding frictional contact between the friction material and the rotating disc was modelled using a non-linear impact contact algorithm that aligns with the 'hammering' mechanism (Rhee et al., 1989), and the predictions show considerable similarities to published experimental observations (Fieldhouse et al., 2011), and confirm that a

vibrational response could be obtained from computer predictions using a constant coefficient of friction μ.

- The use of experimental modal analysis in understanding brake noise is explained, including impact testing and shaker testing. Results from more advanced experimental techniques such as laser holographic interferometry are described, and the substantial contribution made to the knowledge and understanding of brake noise by this method is noted. An example of controlling brake squeal by using FE modal analysis in the form of eigenvalue analysis of a large commercial vehicle drum brake is presented showing how the approach of separating 'pair' modes that facilitate binary flutter instability in brake noise can be applied.
- The role of variability in actual component geometry and manufacturing tolerances is briefly discussed.
- Cold and hot judder are explained; brake judder differs from brake noise in that the frequency of noise and/or vibrations caused by judder are road speed (or rotor rotational speed) dependent, while those of brake noise are independent of road speed. The causes of cold judder in both disc brakes and drum brakes are well understood, and attention to a few basic design and manufacturing rules enables cold judder to be avoided by vehicle manufacturers.
- Hot judder is more problematic because it is fundamentally a thermally induced phenomenon that occurs in both disc and drum brakes, though it is particularly a problem in large heavy-duty disc brakes, e.g. on commercial vehicles and certain types of large high-performance passenger cars. Thermal cycling of the cast-iron material of brake rotors creates permanent thermoplastic rotor deformation from localised yielding. 'Hot spotting' or 'blue spotting' results from high localised temperatures and resulting phase change in the cast-iron rotor friction surface, which causes these spots to 'stand proud' of the rest of the surface. Together with a change in μ these create circumferential variations leading to judder. The vibrations from hot judder are associated with 'drone' or 'rumble' noise, of frequency typically between 100 and 200 Hz.

References

Alasadi, A., Bannister, P., Bryant, D., Day, A., Hussain, K., 2013. The prediction of pad vibration in a noisy disc brake using multibody dynamics analysis, Eurobrake 2013, paper EB2013-NVH-003.

Basford, P.R., Twiss, S.B., 1958. Properties of Friction Materials I - Experiments on Variables Affecting Noise.

Bryant, D., 2010. Thermo-elastic deformation of a vented brake disc. PhD thesis. University of Huddersfield, UK.

Chen, F., Chern, J., Swayze, J., 2002. Modal coupling and its effect on brake squeal. SAE, 2002-01-0922.

Crisp, J.D.C., Fosberry, R.A.C., Holubecki, Z., Gough, V.E., Hales, F.D., Lanchester, G.H., Martland, L., Spurr, R.T., Wilson, A.J., 1963. Communications on Jarvis and Mills' paper. Proc. IMechE. 178 (32), 857–864.

Crolla, D.A., Lang, A.M., 1991. Brake noise and vibration – the state of the art. Vehicle Tribol. 18, 165–174.

Day, A.J., Kim, S.Y., 1996. Noise and vibration analysis of an S-cam drum brake. Proc. IMechE. 210, 35–43.

De Vries, A., Wagner, M., 1992. The brake judder phenomenon. SAE, 920554.

Eriksson, M., Bergman, F., Jacobson, S., 2002. On the nature of tribological contact in automotive brakes. Wear 252 (1–2), 26–36.
Felske, A., Happe, A., 1977. Vibration analysis by double pulsed laser holography. SAE Tech. Paper 770030, 1977. http://dx.doi.org/10.4271/770030.
Fieldhouse, J.D., 2013. Brake noise and vibration, Braking of Road Vehicles. University of Bradford, 2013.
Fieldhouse, J.D., Bryant, D., Talbot, C.J., 2011. The influence of pad abutment on the generation of brake noise. Int. J. Vehicle Struct. Syst. 3 (1), 45–57. Print ISSN 0975-3060; Online ISSN 0975-3540, 2011.
Fosberry, R.A.C., Holubecki, Z., 1955. An investigation of the cause and nature of brake squeal, MIRA report 1955/2.
Hoffman, N., Fischer, M., Allgaier, R., Gaul, L., 2002. A minimal model for studying properties of the mode-coupling type instability in friction induced oscillations, Pergamon. Mech. Res. Commun. 29, 197–205.
Hussain, K., Yang, S.H., Day, A.J., 2007. A study of commercial vehicle brake judder transmission using multi-body dynamic analysis. Proc. IMechE. Part K. J. Multi-body Dyn. 221, 311–318.
Jacobsson, H., 2003. Aspects of disc brake judder. Proc. IMechE. vol. 217. Part D: J. Auto. Eng., 419–430.
Jarvis, R.P., Mills, B., 1963. Vibrations induced by dry friction. IMechE. Proc. 178 (Pt. 1, No. 32), 847–857.
Kinkaid, N.M., O'Reilly, O.M., Papadopoulos, P., 2003. Automotive disc brake squeal – review. J. Sound Vib. 267, 105–166.
Kubota, M., Suenaga, T., Doi, K., 1998. A study of the mechanism causing high speed brake judder. SAE, 980594.
Lang, A.M., Smales, H., 1983. An approach to the solution of disc brake vibration problems. Paper C37/83, IMechE. Conference on Braking of Road Vehicles, 1983.
Lang, A.M., Schafer, D.R., Newcomb, T.P., Brooks, P., 1993. Brake squeal - The influence of rotor geometry. In: IMechE International Conference on the Braking of Road Vehicles. IMechE, pp. 161–172.
Liles, G.D., 1989. Analysis of disc brake squeal using finite element methods. SAE, 891150.
Mahajan, S.K., Hu, Y., Zhang, K., 1999. Vehicle disc brake squeal simulation and experiences. SAE, 1999-01-1738.
Matsuzaki, M., Izumihara, T., 1993. Brake noise caused by longitudinal vibration of the disc rotor. SAE, 930804.
Millner, N.A., 1976. Theory of drum brake squeal. Proc. Auto. Div. IMechE. (C39/76), 177–181.
Millner, N., 1978. An analysis of disc brake squeal. SAE, 780332.
Mills, H.R., 1938. Brake squeak, Automobile Research Committee.
Murakami, H., Tsunada, N., Kitamura, T., 1984. A study concerned with a mechanism of disc brake squeal. SAE, 841233.
North, M.R., 1969. A survey of published work on vibrations in braking systems. MIRA.
North, M.R., 1972. Frictionally induced self-excited vibrations in a disc brake system. PhD Thesis. Loughborough University of Technology.
North, M.R., 1976. Disc brake squeal. Paper C38/76, IMechE. Conference on Braking of Road Vehicles.
Ouyang, H., Nack, W., Yuan, Y., Chen, F., 2005. Numerical analysis of automotive disc brake squeal – a review. Int. J. Vehicle Noise Vib. 1 (Nos. 3/4), 207–231.
Panier, S., Dufrénoy, P., Weichert, D., 2004. An experimental investigation of hot spots in railway disc brakes. Wear 256 (7–8), 764–773.
Rhee, S.K., Tsang, P.H.S., Wang, Y.S., 1989. Friction-induced noise and vibration of disc brakes. Wear 133, 39–45.
Spurr, R.T., 1961. A theory of brake squeal. Proc. Auto. Div. Inst. Mech. Engrs. 1961–1962 (1), 33–52.
Talbot, C., Fieldhouse, J., 2002. Investigations of In-Plane Disc Vibration Using Laser Holography. SAE Technical Paper 2002-01-2607.
Yi, Y., Barber, J.R., Zagrodzki, P., 2000. Eigenvalue solution of thermoelastic instability problems using Fourier reduction. Proc. Roy. Soc. vol. A456, 2799–2821.
Yuan, Y., 1995. A study of the effects of negative friction-speed slope on brake squeal. In: Proceedings of DETC'95, DE, vol. 84-1. ASME, New York, 1153–1162.

CHAPTER 11

Electronic Braking Systems

Introduction

The term 'electronic braking systems' used in this chapter covers the use of electrical, electronic and computer technology in the actuation and control of road vehicle braking systems, and includes aspects of regenerative braking that depend heavily on electronic control technology for its effective implementation. Electronic braking systems have made it possible to utilise the brakes to enhance road vehicle safety through technologies such as antilock braking (ABS) and Electronic Stability Control (ESC), and in this way the brakes of a road vehicle have become 'active safety' devices. Two distinct aims associated with road safety enhancement have been addressed through electronic braking: first, improving the braking performance of road vehicles in terms of both deceleration and stability to avoid or mitigate a collision, and secondly providing the means to intervene in the dynamic control of the vehicle to assist the driver to avoid instability or loss of control. The former utilises electronic and computer control to achieve optimum braking under all conditions, and the latter represents (see Chapter 8) automatically commanded braking designed to detect an emergency or critical situation and activate the vehicle braking system to decelerate and/or control the vehicle with the purpose of avoiding wherever possible a dangerous situation or event, such as a collision or a rollover. Intelligent electronic control of the brakes to enhance the vehicle's 'handling' safety can be summarised by the generic term ESC (Electronic Stability Control), which is the term used in this book, although there are many other terms that effectively mean the same, e.g. VSC (Vehicle Stability Control). ABS was the first electronically controlled braking system (Bosch/Mercedes, 1978), and was followed in 1987 by the first Traction Control System (TCS) on a production vehicle (Mercedes and BMW), and then in 1995 by the first ESC application on a production Mercedes car. Mercedes also produced the first passenger car with a braking system including part electrical actuation in 2001. The first production road vehicle with roll stability control was a Volvo passenger car in 2003, although at about that time rollover stability control was also available as an option for articulated commercial vehicle combinations (tractor and semi-trailer). The ability to control actively the distribution of brake force at individual wheels to optimise adhesion utilisation has become standard fitment to many road vehicles (passenger cars and commercial vehicles) in the last 10 years. Now, the term 'ESC' (or similar) can include all the functionality described above in a single system that monitors the vehicle's dynamic behaviour and makes control interventions where considered necessary to stabilise it in critical manoeuvres, preventing undesirable and potentially dangerous responses such as excessive understeer, oversteer, skidding or rollover.

Vehicle 'handling' can be defined in terms of whether a road vehicle maintains a path that accurately reflects the driver's intent (as defined, for example, by the steering wheel angle) whilst at the same time remaining stable, i.e. it responds as the driver expects it to. Stable handling can only be achieved if the vehicle can consistently maintain tyre/road adhesion when a braking, acceleration or steering input (or a combination of these) is applied by the driver. The vehicle's course will then follow closely the driver's intent (steering angle) as long as the tyre/road forces remain below the limit defined by the coefficient of adhesion between the tyres and the road. When the driving conditions are such that a tyre approaches the limit of adhesion, ultimately adhesion may be lost and the tyre will skid when the required tyre/road force exceeds the available force from the tyre/road adhesion, as explained earlier in this book. ESC incorporates 'closed loop' control of both vehicle yaw rate and tyre slip angle because these have been found to be effective indicators of the onset of vehicle dynamic instability resulting from the loss of tyre adhesion, and Electronic Brakeforce Distribution (EBD) utilises independent wheel brake control to assist the vehicle to avoid excessive slip angles and/or yaw rate. In this chapter stability control system technologies and functions are outlined and explained in terms of a range of control and monitoring mechanisms that utilise the braking system to provide active safety systems on modern road vehicles, and aspects of regenerative braking technology are briefly explained and discussed.

Antilock Braking System (ABS)

ABS was the first active safety technology to utilise the vehicle's braking system. It aims to detect the potential occurrence of wheel lock, e.g. by comparing the wheel speed with vehicle speed, and cyclically releasing and reapplying the brake so that the wheel continues to rotate and thus provide maximum braking force (based on adhesion coefficient rather than sliding tyre/road friction) and directional stability. The first road vehicle antilock braking system to be put into production in Europe was the Dunlop Maxaret 'anti-skid' device, a mechanical system fitted to a Jensen car in 1966. The first electronic antilock braking system was a Bosch system fitted to a Mercedes production car in 1978, and with this the term 'ABS' (which comes from the German term 'Anti Blockier System') came into general use to mean an antilock braking system. ABS was first required as a mandatory fitment in the USA in the 1970s, but this requirement was subsequently withdrawn as the technology was not robust at that time. In Europe ABS became standard specification on heavy commercial vehicles and coaches in the mid

1980s, and this was followed by a mandatory requirement in 1991. Modern ABS control systems and actuators are usually combined in one unit (the ABS module), which is an integral part of any road vehicle ESC system.

The operation of ABS in cyclically releasing and reapplying brake actuation pressure may cause the time-averaged braking force developed at the road surface to be reduced below the maximum that might otherwise be achievable by a skilled driver. But usually, under conditions of braking where wheel lock may occur, such as emergency braking or braking on low adhesion road surfaces (wet or icy), a road vehicle fitted with ABS will achieve shorter stopping distances than the same vehicle without ABS, because the braking forces at each wheel utilise the maximum tyre/road adhesion coefficient rather than the lower sliding coefficient of friction between the tyre and the road when the wheel is locked and skidding.

ABS can adversely affect the handling performance of the vehicle because, when activated, it forces the wheels to run at a higher average braking slip than under normal braking conditions. This reduces the cornering potential of the affected wheels, because they have to run at greater slip angles to maintain the required lateral forces. An ABS system should therefore not be used to cover up a poor braking system design with an inadequate braking distribution, which is why the principles of braking system design covered in earlier chapters of this book are so important. Vehicle and braking system design should try to achieve the best compromises for adhesion utilisation and braking compatibility, and only rely upon ABS for emergency situations. Because of this, many vehicle manufacturers continue to fit load-sensing valves and other devices to improve distribution and compatibility in addition to ABS. For most road vehicles, especially commercial vehicle combinations (see Chapter 4) where the difference between unladen and laden states is large, without load sensing to control braking distribution the braking efficiency of the solo vehicle could be low, and compatibility between a towing vehicle and its trailer would be poor. The result could be excessive brake wear on over-braked axles or 'glazing' (see Chapter 2) of the friction material as a result of under-braking, both of which would have a deleterious effect on the overall performance of the brakes, and hence the deceleration and stopping distance capabilities of the vehicle.

Despite the fact that ABS intervention can reduce stopping distances under some conditions of operation, the most important benefit of ABS is that it improves vehicle safety and stability by allowing steering control under all braking conditions because the wheels are not allowed to lock and skid. It achieves this by detecting incipient wheel lock, i.e. measuring some parameter(s) associated with the rotation of the wheel being controlled, e.g. wheel deceleration, and comparing this with some preset value (such values have to be set by 'tuning' the system, which can be difficult and time-consuming) to ascertain whether wheel lock is likely to occur. Control action can then be taken to avoid actual wheel lock. Another way of achieving the aims of ABS is to design the controller to focus on the slip (defined in Equation (3.7)) of the road wheels, maintaining it at a level so that the available tyre/road adhesion is maximised and a high level of lateral adhesion is maintained. This is known as 'slip control' ABS, and for each wheel that indicates abnormally high slip, the brake actuation pressure is modulated (reduced) to keep the wheel rotating while maintaining as much braking force as possible as requested by the driver, until the tyre/road adhesion increases to a level where it is no longer required. When excessive braking slip is detected by the ABS controller it releases and reapplies the brake, allowing the wheel to accelerate back up to a speed where the slip is consistent with peak adhesion (point 'C' in Figure 3.12). A diagram of the forces and torques on a road wheel 'i' which is rotationally accelerating while the brake is 'off' under ABS cycling is shown in Figure 11.1.

T_{wi} = Braking force at axle 'i' (N);
τ_{wiacc} = Torque to accelerate rotationally the wheel and brake rotor assembly (Nm);
P_{wi} = Weight of the wheel and associated rotating parts ('unsprung' weight) (N);
ω_i = Angular (rotational) velocity of the wheel (Rad/s).
$\dot{\omega}_i$ = Angular (rotational) acceleration of the wheel (Rad/s^2).
I_{zzi} = Polar moment of inertia of the wheel and hub assembly (kgm^2).

If the coefficient of adhesion between the tyre and the road is k, the maximum force (T_i) that can be generated at the tyre/road interface of the wheel is $T_{wi} = k \cdot N_i$ (Equation 3.9, Chapter 3). Equation [11.1] demonstrates that the wheel can accelerate rotationally only if $T_{wi} \cdot r_r > \tau_{wiacc}$.

$$T_{wi} \cdot r_r - \tau_{wiacc} = I_{zzi} \cdot \dot{\omega}_i \tag{11.1}$$

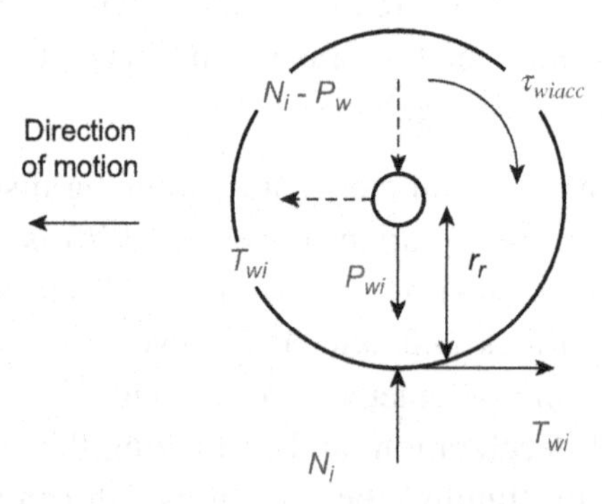

Figure 11.1: Forces and Torques Acting on a Road Wheel Under Braking.

This demonstrates how the ABS controller must take account of the acceleration of the wheel back up to speed. Sufficient wheel torque is needed and thus the coefficient of adhesion (k) must be high enough to achieve this within the ABS cycle time, but in the case where the coefficient of adhesion (k) is very low (see Figure 11.2), e.g. on snow or ice, or the wheel dynamic normal reaction is very small, this may be a problem. Because of issues like this, most ABS systems are 'tuned' on the vehicle. The level of wheel slip generated beyond that at which peak adhesion is reached is dependent on several factors, including the brake type, level of maintenance and braking system response. High levels of brake hysteresis or slow system response will allow a significant increase in wheel slip with the potential for the loss of vehicle stability. However, the actual time a wheel is in the unstable slip zone (C to D in Figure 3.12) is relatively short and as a result there is usually insufficient time for the vehicle to become unstable before the wheel returns to the stable zone (A to C in Figure 3.12). The exception to this is when a vehicle is on a low-adhesion surface, where wheels with high inertia may be slow to recover as there is limited adhesion to accelerate the wheel(s). This is particularly relevant with heavy commercial vehicles, which have high transmission inertia associated with the driven wheels, and are equipped with endurance brakes that further increase the inertia of the drivetrain.

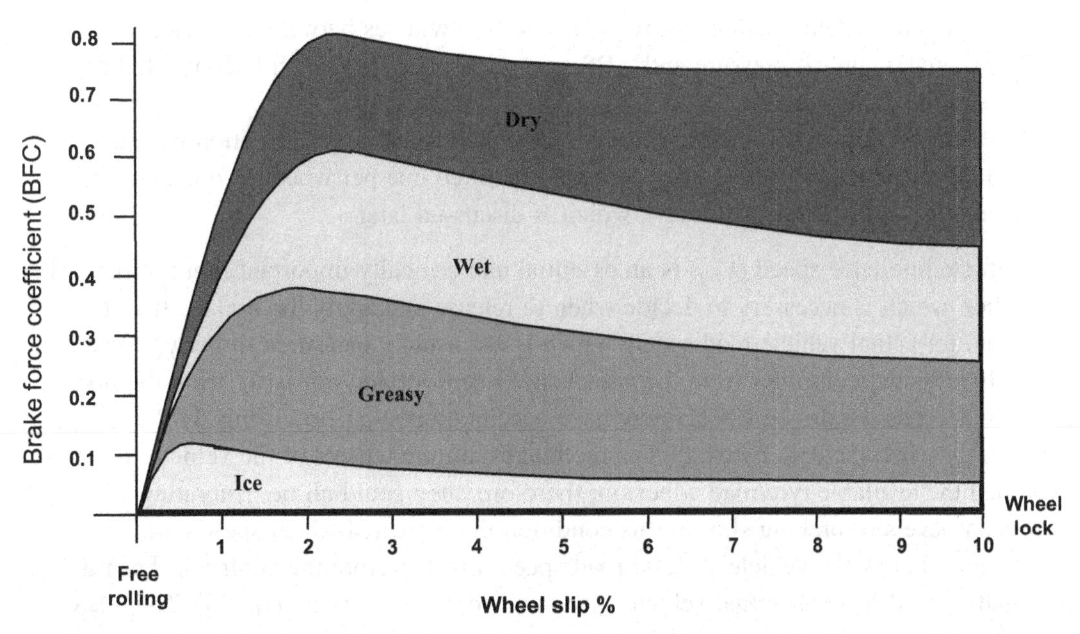

Figure 11.2: Brake Force Coefficient (BFC) vs. Wheel Slip: Example Characteristics for Different Road Surfaces (Ross, 2013).

ABS Technology and Control Strategy Options

The basic components of an ABS control system include the following components and technology (refer to Figures 11.3 and 6.19):

- With hydraulically actuated braking systems as fitted to passenger cars and similar vehicles, an ABS pump is required to maintain hydraulic pressure during ABS operation. In a pneumatically actuated braking system, actuation pressure is already provided by the compressor.
- Wheel speed sensors provide wheel rotational speed information to the ABS controller at the required cycling speed; all ABS control decisions are based on this information. All two-axle rigid vehicles have one per wheel but other types of vehicle may not.
- The main functions of the ABS electronic controller are:
  - Monitor wheel speed data
  - Calculate the vehicle reference speed (V_{ref})
  - Make decisions when to release and reapply the brakes
  - Control the pressure modulators and cycling valves
  - Monitor general system integrity.
- There is usually one pressure control valve per wheel on two-axle rigid vehicles, although this can depend on the installation and the ABS control strategy, particularly for multi-axle commercial vehicle combinations. In pneumatic systems each pressure control valve can deliver states of pressure reduction, pressure hold or pressure increase. In hydraulic systems each pressure control valve switches between the driver ('muscular') applied pressure and ABS generated pressure, which is controlled not to exceed the driver intent.
- Pressure cycling (ABS) valves: these are designed to cycle the actuation pressure on and off at the required frequency, and may be fitted one per wheel or one per axle depending on the control strategy, which is discussed later.

The vehicle reference speed (V_{ref}) is an essential and critically important input to the ABS controller, which is necessary to decide when to release and apply the brakes. It is a substitute for actual vehicle road speed, which is not usually measured directly; it can be determined by the controller from the wheel speed sensor data, primarily from the non-driven wheels as the driven wheels may have acceleration slip. But during ABS cycling, particularly when the driver first applies the brakes, all the wheels of the vehicle may have exceeded the available tyre/road adhesion; therefore, they could all be generating significant levels of braking slip. In this condition the indicated wheel speeds are significantly below the vehicle's actual road speed, and therefore the controller has no information to determine actual vehicle speed and so an estimate is required. The ABS controller has no information to indicate the prevailing tyre/road adhesion so it assumes that V_{ref} is reducing at a rate greater than if the vehicle were being braked on a high-adhesion surface. The actual rate utilised within the controller depends on the vehicle characteristics, particularly its actual braking capability. If V_{ref} is incorrect two problems may ensue:

- Over-braking, where the controller estimates the vehicle deceleration to be higher than the actual, which results in ABS control ceasing too soon, i.e. while the vehicle is still travelling too fast.
- Under-braking, where the controller estimates the vehicle deceleration to be lower than the actual, which results in the available adhesion being under-utilised and increased stopping distance.

When the wheel begins to accelerate the controller can start to estimate the rate of vehicle deceleration again. It is especially useful with 'slip control' ABS (see below) to allow the wheel to accelerate to zero slip to verify that the calculated V_{ref} aligns with actual vehicle speed. Determining V_{ref} on all wheel drive vehicles is also difficult because all the wheels may have significant levels of acceleration slip, and thus may be rotating at a speed higher than the true vehicle speed. As a result the calculated V_{ref} would be higher than the actual vehicle speed, which can have two effects: first that ABS cycling occurs prematurely and secondly that an accelerating wheel can never reach V_{ref}, i.e. the speed when brake reapply should take place. In both cases the result is that the vehicle is under-braked and, to avoid this, additional information is needed for the more accurate calculation of V_{ref}, e.g. vehicle acceleration (provided by a longitudinal acceleration sensor).

ABS should utilise as much of the available tyre/road adhesion as possible, while at the same time limiting wheel slip to ensure vehicle stability, and several different control strategies can be used to achieve this. Most ABS systems now use an individual wheel control strategy, which requires sensors and control on each wheel. Earlier generation ABS configurations used strategies that simultaneously controlled both wheels on an axle using only one pressure modulator (ABS valve) per axle. This 'axle control' had wheel speed sensors at each wheel and the pressure modulator controlled both brakes on the axle as determined by the ABS controller. The actuation pressure at each brake was therefore equal, and the control logic options for ABS action for this arrangement are therefore as follows:

- 'Select low', where ABS control commences when the first of the two wheels exceeds the defined deceleration or slip thresholds. This has the advantage of enhanced stability as one wheel will be under-braked (less wheel slip) but at the expense of increased stopping distance.
- 'Select high', where ABS control commences when the second of the two wheels exceeds the defined deceleration or slip thresholds. This control logic reduces stopping distance at the expense of stability as one wheel will be locked.

The former strategy could limit the generated axle braking force to below what it could have been, while the latter could allow one wheel to lock while controlling the other, and thus both could reduce braking efficiency under ABS operation. On passenger cars and light vans individual control with wheel speed sensors and associated pressure modulators for each brake on an axle allows the adhesion utilisation at each wheel to be individually optimised, and is now almost universally used. On commercial vehicles with pneumatically actuated braking systems individual control can be difficult to tune to avoid excessive yaw on 'split-μ' road surfaces because of low ABS cycling speed (influenced by actuation system inertia and hysteresis in the pneumatic actuation system). 'Modified independent control' is the same as independent control except that under certain conditions the pressure difference between left and right wheels is limited, and this is used on heavy commercial vehicles and buses, where because of the steering geometry design, differential brake torques produce a steering moment that is transmitted to the driver's steering wheel. If large enough, the differential can result in instability. Road surface conditions are the biggest single variable in tyre/road adhesion as illustrated in Figure 11.2, but some loose surfaces such as snow and loose gravel may exhibit completely different adhesion characteristics where the peak adhesion is generated when a wheel is locked. To take account of such conditions, off-road vehicles may have a switch to disable the ABS control or have an off-road mode that permits greater wheel slip to be developed before ABS is activated and during ABS cycling. Systems do exist that combine 'select low' and 'select high' strategies and automatically switch between the two depending on the prevailing tyre/road adhesion conditions.

When a road wheel reaches peak adhesion it will rapidly decelerate towards lock (100% slip), so the wheel speed sensor will indicate an unusually high wheel deceleration provided that the deceleration threshold is higher than the deceleration the vehicle is capable of achieving. The Bosch ABS control cycle for passenger cars in high-adhesion conditions (Bosch, 1993) uses four preset thresholds:

- λ_1, known as the 'slip-switching threshold', which relates to the difference between the speed of the wheel and the 'reference speed' of the vehicle.
- $-a$, $+a$ and $+A$, which relate to defined features of the wheel angular deceleration and acceleration during ABS control.

Brake actuation pressure is increased as the driver applies the brakes until the wheel deceleration decreases below the '$-a$' value when the actuation pressure is held constant. If the wheel speed then reduces to below that of the reference speed by more than the slip-switching threshold, the actuation pressure is reduced until the acceleration increases above the '$+a$' value when the actuation pressure is held constant. When the wheel acceleration reaches the '$+A$' value, the wheel is considered to be entering the stable region of the μ–slip curve and actuation pressure can be increased again according to the driver's demand. As an alternative to (or in addition to) the wheel deceleration criterion as described in the control cycle above, a wheel slip criterion may be employed. While it is possible for ABS to function using wheel acceleration/deceleration detection only, there are conditions when a wheel can decelerate at a rate below the detection threshold and continue to lock, e.g. on vehicles with high-inertia drivetrains on a low-adhesion road surface combined with low driver demand. To overcome this problem a second detection threshold can be used based on the magnitude of the slip generated by the wheel.

The operation of ABS on a passenger car with hydraulic actuation is illustrated in Figure 11.3. Under normal braking with the ABS inactive (Figure 11.3(a)) the hydraulic actuation pressure is transmitted through open pressure control valves to the wheel cylinders and the ABS valves are closed. Under ABS operation, Figure 11.3(b) illustrates how the pressure control valves are closed and the ABS valves are opened, allowing the ABS pump to provide hydraulic actuation pressure to the individual wheel cylinders, which can be modulated as required.

The pneumatically actuated braking systems fitted to most commercial vehicles (see Figure 6.19) rely upon ABS operation through the control line, where pneumatic valves and logic have recently been replaced with electronic control, which has significantly improved the control and distribution of braking on commercial vehicles. Closed-loop control is a fundamental requirement for ABS, which requires measurement of the actuation pneumatic pressure via transducers at each brake. Previous generation pneumatic control ABS systems were open-loop as actuation pressure was not measured, and decisions on pressure control were made based on the reaction of the braked wheels and the known response characteristics of the actuation pressure modulators. With an electronically controlled braking system the pressure modulator valves are able to provide all states required, enabling full ABS control in terms of controlled pressure rise, hold and decrease. As a result ABS control for pneumatically braked commercial vehicles is now a function within the electronic braking system where the additional information available from the system pressure transducers may be used to improve the ABS pressure control to enable greater braking efficiency through improved adhesion utilisation. As mentioned above, pneumatic brake actuation is slower than hydraulic brake actuation, so commercial vehicle ABS cycles at a lower frequency than for passenger cars (1 Hz compared with 6–8 Hz), and this also makes it more difficult for slip control to be implemented on HGVs (Miller and Cebon, 2010).

All brakes have hysteresis, the amount of which is dependent on the brake type, its design and how well the brake is maintained. Brake hysteresis can be defined as the difference in the input (actuation) torque required to produce either an increase or decrease in generated brake torque when changing state from increasing brake demand to reducing brake demand, and vice versa. Figure 11.4 illustrates this effect for commercial

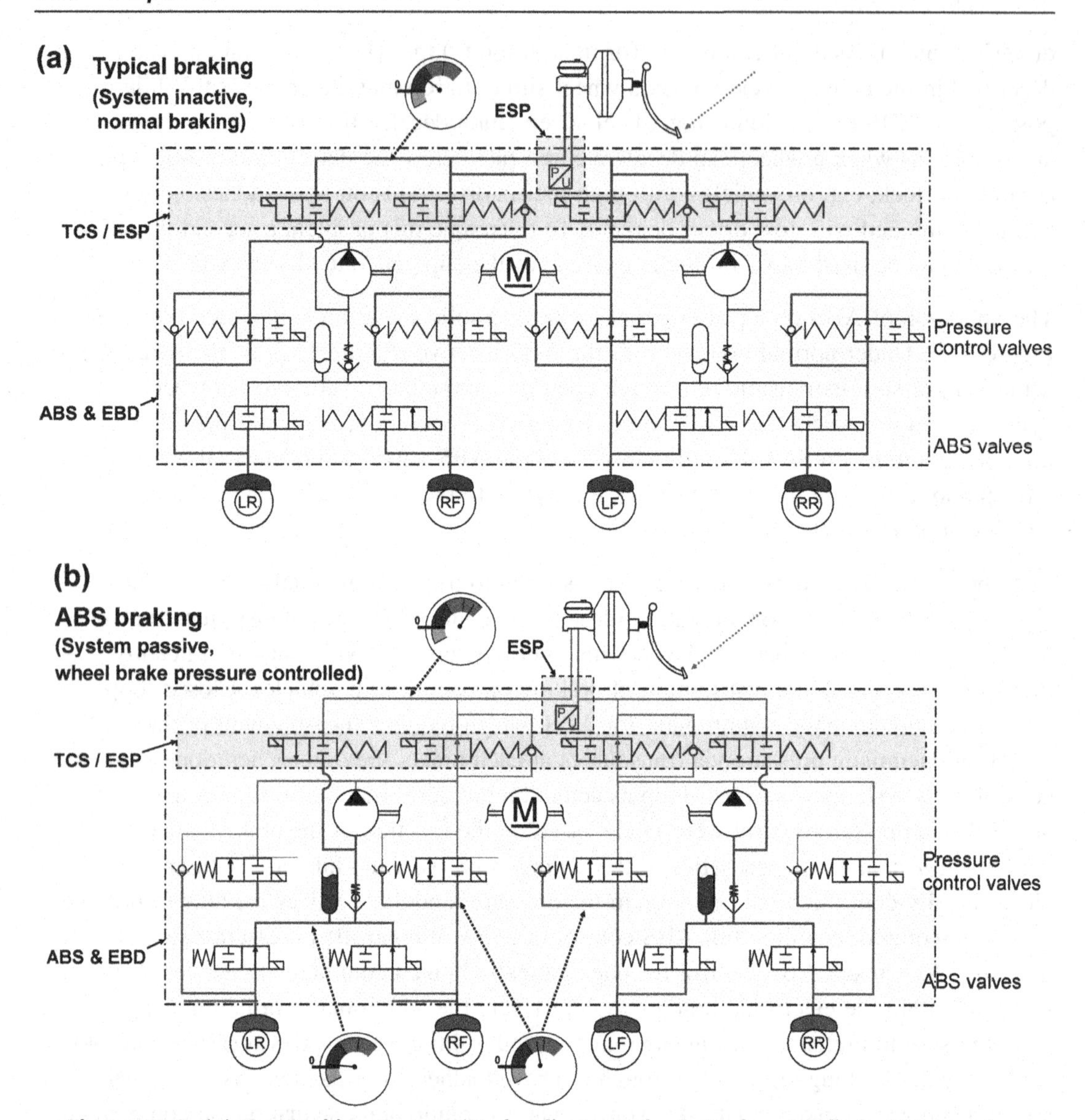

Figure 11.3: (a) Car Braking System (Hydraulic Actuation); Brakes Applied, ABS Inactive. (b) Car Braking System (Hydraulic Actuation); Brakes Applied, ABS Active (Moore, 2013).

vehicle (air-actuated) systems. When brake pressure increases there is a corresponding increase in brake torque generated. When ABS cycling commences (point A), actuation pressure is reduced; however, the brake torque generated remains the same until a reduction from 5 to 2.8 bar (point A to B) has been produced, which is equivalent to 44% hysteresis. Equally when the brakes are reapplied, the applied brake pressure must increase from 1.6 to 3.1 bar (point C to D) before there is an increase in the generated

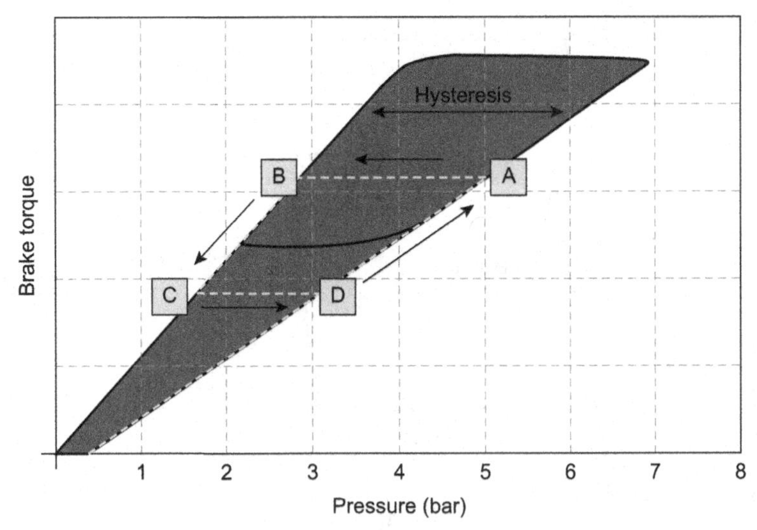

Figure 11.4: Hysteresis in Pneumatic Brake Operation on a Commercial Vehicle (Ross, 2013).

brake torque. It is essential to cross the hysteresis loop as quickly as possible, otherwise there are detrimental effects on stability (controlled wheels continue to decelerate during pressure release) and reduced adhesion utilisation (increased stopping distance because there is no increase in brake torque generated).

To minimise the effect of brake hysteresis in ABS operation, actuation pressure modulators must be designed for high flow, especially those used in pneumatic braking systems because air is compressible. Also, to ensure that the flow capability of the pressure modulator is fully utilised the flow from the pressure modulator(s) to the brake actuator(s) must be maximised. During brake reapply with pneumatic systems, one way of allowing for system hysteresis is a rapid pressure rise (primary rise) followed by a slower pressure rise (secondary rise) to move through the optimum slip band relatively slowly, as illustrated in Figure 11.5. Disc brakes have lower hysteresis than drum brakes, so especially on commercial vehicles this enables the wheels to be controlled over a narrower slip band. The pressure control used in conjunction with slip control is much less defined and often includes multiple small pressure reductions and increases.

Measurement of ABS Efficiency

The braking performance of a road vehicle is usually defined in terms of minimum stopping distances and/or minimum mean fully developed deceleration (MFDD), but neither of these criteria is suitable to define the braking performance of a road vehicle when the ABS is operating. This is because both stopping distance and MFDD assume that the tyre/road adhesion is sufficient to generate the required rate of braking and thus

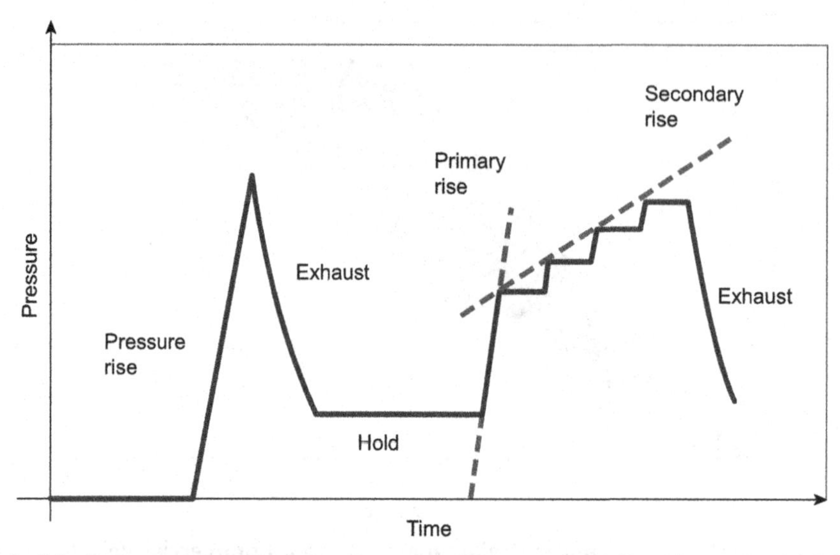

Figure 11.5: 'Ideal' Pneumatic Actuation Pressure Characteristic Showing a Rapid Primary Rise Followed by a Slower Secondary Rise (Ross, 2013).

the vehicle's braking performance depends upon the vehicle braking system design and specification. But when the ABS is operating, the adhesion limit has by definition been exceeded, and the braking performance therefore depends upon the tyre/road adhesion. Two parameters that can affect ABS operation during a brake performance test are the non-constant tyre/road adhesion coefficient (k) (which may be caused by variations in the road surface, environmental conditions, or weather), and different tyres on driven and non-driven wheels.

The relationship between tyre/road adhesion (in the form of brake force coefficient, BFC) and braking slip was introduced in Chapter 3; Figure 3.12 illustrates a typical tyre/road adhesion characteristic. For each individual curve there is a point where the maximum BFC is developed, which is known as the 'peak adhesion', and one way of quantifying the efficiency of the ABS would be to compare the rate of braking of the vehicle with ABS cycling to the maximum rate of braking that can be achieved by the vehicle without ABS and without wheel lock, i.e. with all wheels at peak adhesion. Because the tyre/road adhesion is variable, the maximum braking force generated by a wheel at peak adhesion is also variable as it may occur at different levels of slip from wheel to wheel. Additionally, the maximum braking force at each individual road wheel depends on the dynamic normal force at that wheel (Equation (3.9)), which varies both laterally and longitudinally due to dynamic load transfer. Therefore, evaluating the maximum achievable rate of braking without ABS is very difficult because all the wheels are unlikely to be at peak adhesion at the same time.

Comparing the braking performance of a vehicle under ABS operation with the maximum rate of braking achievable under ABS disabled but without wheel lock could result in the ABS operation performance apparently exceeding the maximum rate of braking under ABS disabled, which is not physically possible. This can occur because one or more road wheels are likely to be operating below peak adhesion in the ABS disabled condition, whereas with ABS operation all the wheels are operating at (or very close to) peak adhesion. To avoid this, the test procedures specified in UN Regulation 13 for assessing the performance of an ABS equipped vehicle require that the ABS efficiency is measured on both high- and low-adhesion surfaces for both laden and unladen conditions. For the high-adhesion road surface the peak adhesion must be about 0.8, and the peak adhesion of the low-adhesion surface must be a maximum of 0.3 (additional requirements apply to ensure that it is not too low, that the peak adhesion is well defined, and that the adhesion coefficient is not speed dependent). The only other requirement is that the ratio of the peak adhesion to the adhesion at 100% wheel slip (μ_t) is between 1 and 2. A basalt surface is usually used for low-adhesion ABS tests; this requires water to reduce the adhesion but near 100% wheel slip the adhesion increases as illustrated in Figure 11.6. As a result the ratio of the peak adhesion to the adhesion at 100% wheel slip is often very close to the minimum value of 1.

Because ABS operates by releasing brake actuation force when excessive wheel slip or imminent lock is detected, it is by definition not possible to achieve 100% adhesion utilisation when the ABS is operating, and therefore it is accepted that the actual braking distance with ABS operating will be greater than the theoretical minimum. UN Regulation 13 requires that on low- and high-adhesion surfaces the 'adhesion utilised' (ε) must be ≥ 0.75 (laden and unladen). In this case, the adhesion utilised is defined in Equation (11.2), where z_{AL} is the maximum measured rate of braking with the ABS operating and

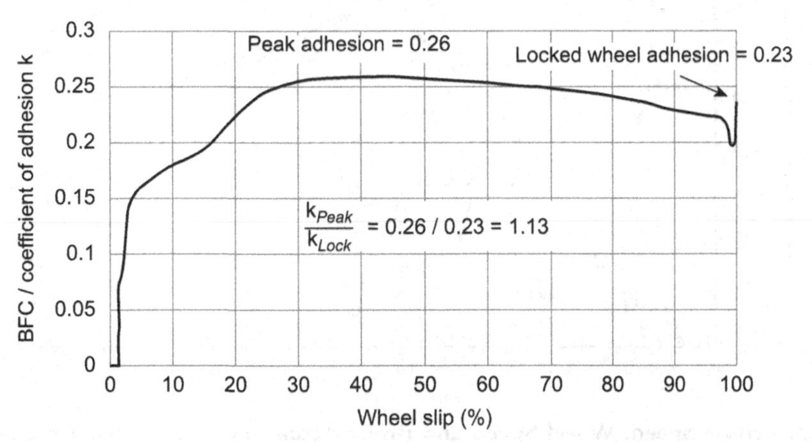

Figure 11.6: Tyre/Road Characteristics of Wet Basalt (Ross, 2013).

k_M is the tyre/road coefficient of adhesion. For example, in Figure 11.6 the peak adhesion is 0.26; therefore, the rate of braking achieved with the ABS cycling must be $z_{AL} \geq 0.195$ $(= 0.26 \times 0.75)$.

$$\varepsilon = \frac{z_{AL}}{k_M} \tag{11.2}$$

A methodology for measuring the ABS efficiency of any road vehicle is presented here which has two parts: measurement of the ABS braking performance of the road vehicle, and measurement of the peak adhesion value. When measuring the ABS performance, only that part of the braking event when the ABS is in control of the wheels must be considered; when the brakes are first applied the ABS is not in control because the actuation pressure rise at the start of the brake application is very rapid, which results in high wheel deceleration and slip. After the ABS has reacted to the condition and started to control the wheels, assessment of the system performance can be commenced. Similarly, when the vehicle road speed approaches zero, wheel slip cannot be accurately evaluated (see Figure 11.7), and so this part of the braking event should also not be included.

The test procedure for measuring ABS performance is summarised below, with the set speed parameters listed in Table 11.1. Three measurements on each test surface for each load condition are required, and the average time to decelerate between the specified

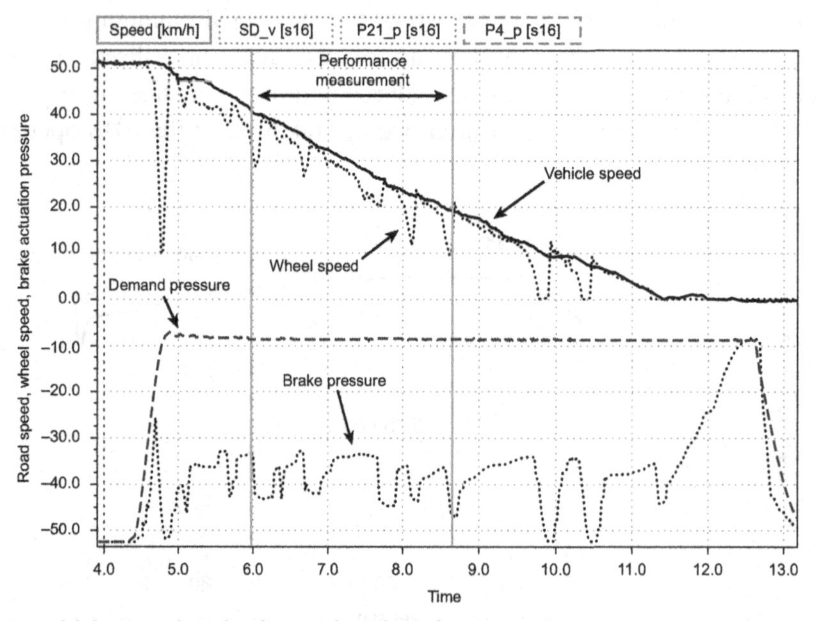

Figure 11.7: Vehicle Speed, Wheel Speed and Brake Actuation Pressure vs. Time During ABS Operation (Ross, 2013).

Table 11.1: Speed Parameters for ABS Performance Measurement

ABS Performance; Measured Speed Parameters	Solo or Towing Vehicles	Trailers*
Speed at which the brakes are first applied (km/h)	55	50
Speed at which ABS performance measurement commences (km/h)	45	40
Speed at which ABS performance measurement ceases (km/h)	15	20

*Trailers have different test measurement speeds as only the trailer is braked and must also decelerate the unbraked mass of the towing vehicle.

speeds is used to calculate the braking rate and dynamic axle loads, from which the peak adhesion for the whole vehicle is calculated as described below. The level of deceleration produced must allow the ABS to cycle fully, e.g. a front axle may commence ABS cycling but as dynamic load transfer occurs the load on that axle increases and slip is reduced.

- Approach the test area at a speed above the brake apply speed.
- Disengage the engine (select neutral gear) and coast to the required brake apply speed (Table 11.1).
- Apply the brakes.

The data to be recorded include:

- Vehicle speed
- Wheel rotational speed (all wheels)
- Driver demand (e.g. hydraulic or pneumatic actuation system pressure at the control)
- Actuation pressure at each brake actuator.

ABS performance tests and the tests to determine the peak adhesion should be completed as close together as possible and it is preferable to complete the ABS tests first to avoid any effect of temperature differentials between the brakes. The determination of peak adhesion can produce high brake temperatures and since each axle is tested independently, significant temperature differentials between the axles can affect the braking performance of the vehicle and hence the ABS.

Measurement of the peak adhesion value for the vehicle (k_M) starts with measuring the peak adhesion for the front and rear axles independently. The actuation pressure to the braked axle is gradually increased until the time to decelerate between two speeds (as shown in Table 11.2) without wheel lock is minimised. The adhesion utilisation at each wheel on any axle will always be different because of side-to-side differences in the adhesion coefficient (k), the actual brake performance and the dynamic wheel loads, so for

Table 11.2: Speed Parameters for Peak Adhesion Measurement

ABS Performance Measurement Speed Parameters	Solo or Towing Vehicles	Trailers
Speed at which the brakes are first applied (km/h)	50	50
Speed at which peak adhesion measurement commences (km/h)	40	40
Speed at which peak adhesion measurement ceases (km/h)	20	20

the ABS performance measurement individual actuation pressure adjustment to each brake on the axle is needed. The test procedure is:

- For all vehicle types apply the brakes at an initial speed of 50 km/h.
- Record the time to decelerate the vehicle from 40 to 20 km/h without wheel lock.

The peak adhesion is then calculated from the minimum deceleration time (t_{min}). UN Regulation 13 requires three values of t_{min} to be obtained within $1.05t_{min}$, but if this is not achievable it is also permitted to utilise the lowest deceleration time t_{min}. Using the lowest value produces the maximum braking rate and is therefore more representative of the peak adhesion.

An example of the type of data recorded during peak adhesion measurement on a commercial vehicle fitted with pneumatic brake actuation is illustrated in Figure 11.8. This particular test utilised a controlled brake application combined with the ABS remaining operational (which is permitted by UN Regulation 13) to help avoid problems of wheel reaction at the initial stage of the brake application, when the rapid rise in actuation pressure can result in a wheel exceeding peak adhesion and reaching lock. Figure 11.8 indicates that ABS cycling starts soon after 50 km/h but has ceased by 40 km/h when the peak adhesion measurement commences, and never restarts. In comparison, Figure 11.9 illustrates a situation where ABS cycling restarts below 20 km/h, but again the ABS is not operating during the 20–40 km/h speed measuring range. There are cases when it is not possible to prevent wheel lock or ABS cycling occurring within the 20–40 km/h speed measuring range without reducing the brake actuation pressure to a level that affects the peak adhesion deceleration, which would artificially increase the ABS efficiency. This can be resolved by deviating from the prescribed procedure and not recording the actual time to decelerate from 40 to 20 km/h but by extrapolating a section of the vehicle speed slope to give a more realistic deceleration time. This is most likely to be needed during low-adhesion testing. Once the peak adhesion has been obtained from one axle the procedure is repeated for the other axle.

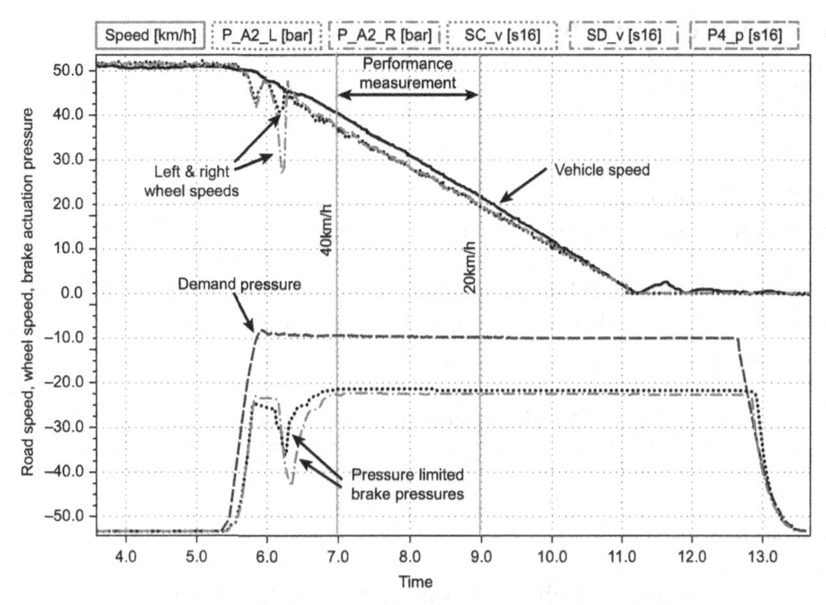

Figure 11.8: ABS Operates to Control Initial Wheel Deceleration on Brake Application but is not Operating Between the Speeds ABS Performance is Measured (Ross, 2013).

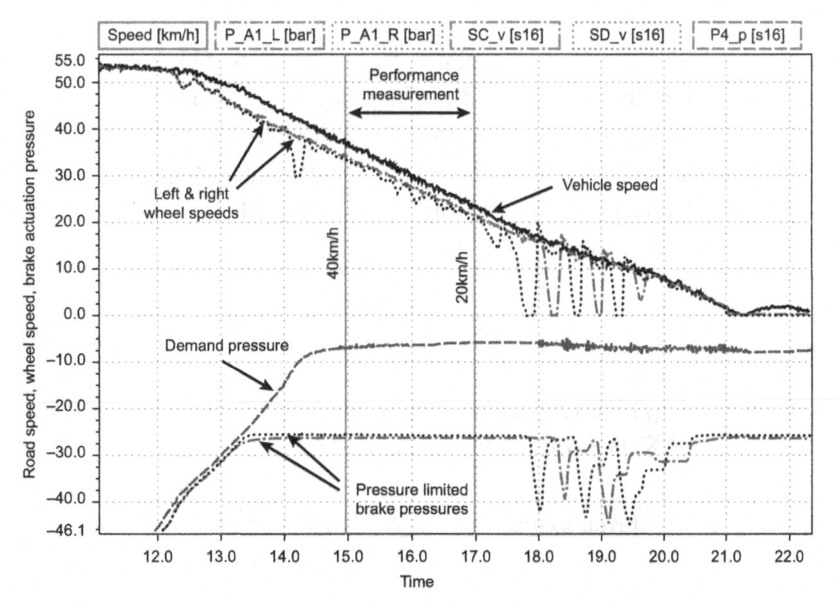

Figure 11.9: ABS Operates Below 20 km/h but is not Operating Between the Speeds ABS Performance Is Measured (Ross, 2013).

The peak adhesion deceleration for each axle can be calculated as follows:

$$z_m = 0.566/t_m \tag{11.3}$$

where:

z_m is the rate of braking of the vehicle;
t_m is the minimum deceleration time measured between 40 and 20 km/h.

To determine the peak adhesion rate of braking from the vehicle rate of braking (z_m) the rolling resistance of the unbraked wheels and the dynamic load transfer to or from the axle under test must be taken into account. For a front axle:

$$k_f = \frac{\left[z_m Mg - (\text{Rolling resistance of unbraked rear axle} \cdot P_2)\right.}{\left[P_1 + \left(\frac{h}{E}\right) z_m Mg\right]} \tag{11.4}$$

And for a rear axle:

$$k_r = \frac{\left[z_m Mg - (\text{Rolling resistance of unbraked front axle} \cdot P_1)\right.}{\left[P_2 - \left(\frac{h}{E}\right) z_m Mg\right]} \tag{11.5}$$

The rolling resistance is specified as the rate of braking (z) of the vehicle attributed to each axle, which is defined in UN Regulation 13 as:

- Driven axle = 0.015
- Non-driven axle = 0.010.

The respective peak adhesion values calculated from Equations (11.4) and (11.5) are not directly used to determine the vehicle's ABS efficiency as these must be weighted according to the dynamic axle loads associated with the tests when the ABS is in operation throughout the stop. To determine the ABS efficiency, the times from three tests for the vehicle to decelerate from 45 to 15 km/h are averaged (t_m) and the average rate of braking (z_{AL}) is calculated from:

$$z_{AL} = \left(\frac{0.849}{t_m}\right) \tag{11.6}$$

Using the braking rate (z_{AL}) the respective dynamic axle loads are calculated using Equations (11.7) and (11.8) (see Equations (3.2) and (3.3)):

Front axle: $$N_1 = P_1 + P z_{AL} \frac{h}{E} \tag{11.7}$$

Rear axle: $$N_2 = P_2 - P z_{AL} \frac{h}{E} \tag{11.8}$$

The peak adhesion braking rates for the front axle (k_f) and rear axle (k_r) are then weighted according to the dynamic load distribution using Equation (11.9) to produce a peak adhesion braking rate for the whole vehicle (k_M):

$$k_M = \frac{k_f N_1 + k_r N_2}{P} \tag{11.9}$$

The ABS braking efficiency (η) is then determined from Equation (11.10), and must be at least 0.75 (laden and unladen) – see Equation (11.2).

$$\eta = \frac{z_{AL}}{k_M} \tag{11.10}$$

Traction Control System (TCS)

Traction control (TCS) is a stability control feature closely associated with ABS. It allows brake actuation pressure to be generated to brake a driven wheel to prevent it spinning when peak adhesion has been lost during acceleration, and thus maintain traction by transferring the drive torque to another drive wheel that does have full adhesion. Communication with the engine control unit is essential, so the ESC controller also requests engine torque reduction during a traction control event; thus TCS control is more complicated than ABS control. The advantage of TCS is that it allows optimised vehicle acceleration, but the disadvantage is that it increases brake usage, which can in turn increase friction material and rotor wear, reducing brake life. The operation of TCS is illustrated in Figure 11.10 for a front wheel drive (FWD) car; in the event of wheel spin the pressure control valves to the front brakes are closed and the ABS valves are opened, allowing the ABS pump to provide hydraulic actuation pressure to the front brake wheel cylinders, which can be modulated as required.

Wheel spin that occurs during acceleration may result in understeer in a front wheel drive vehicle and oversteer in a rear wheel drive vehicle. The wheel speed is monitored continuously by the ABS; therefore determining whether a wheel is spinning is easily achieved on a 2WD vehicle by comparing the speed of the driven wheels with that of the non-driven wheels. When the wheels of the driven axle are on a road surface with uniform adhesion and either or both of the wheels tends to spin, traction control can be achieved using only the engine torque control. Once the spinning wheel has been brought under control, engine torque must be increased to accelerate the vehicle while at the same time relying on the TCS to control the wheel slip. As with ABS the TCS can only utilise the available adhesion but just at the driven wheels, whereas braking force is transmitted via every wheel of the vehicle. Where the coefficient of adhesion is different each side of the vehicle (on a split-μ road surface) the first wheel to lose adhesion will always be that on the low-adhesion surface. In this case reducing the

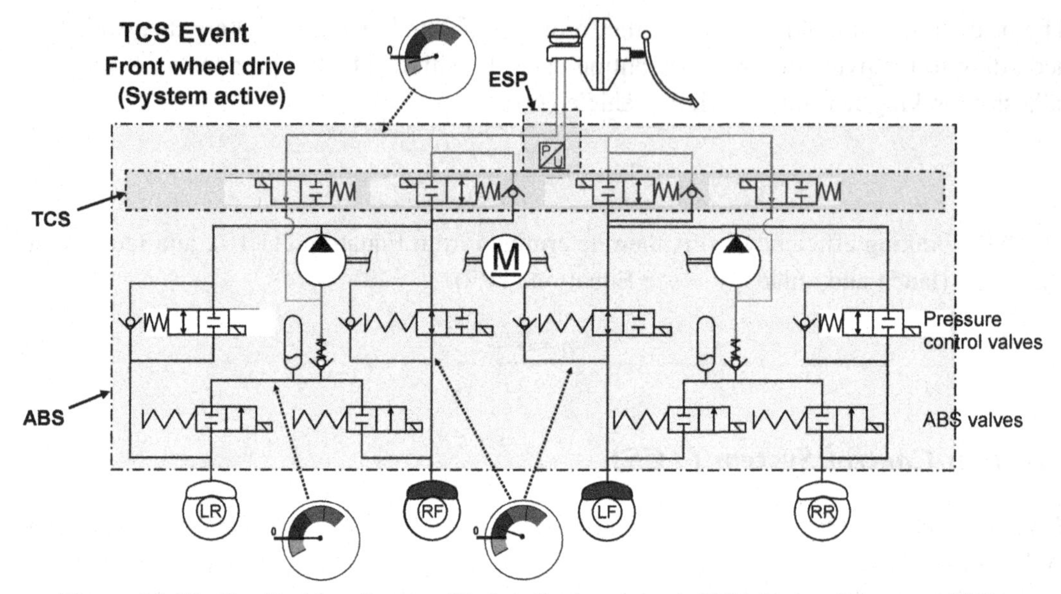

Figure 11.10: Car Braking System (Hydraulic Actuation), TCS Active (Moore, 2013).

engine power to limit the torque transmitted by the low-adhesion wheel restricts that which may be transmitted by the other wheel, i.e. it is under-utilising the adhesion that is available. Here it is possible to utilise some of the higher adhesion by applying a brake force to the spinning wheel that, via the differential, transmits the drive torque to the non-spinning wheel. Using a combination of brake and engine control it is then possible for the vehicle to accelerate at a higher rate than would have been possible (without the use of a differential lock). The basic TCS logic to control wheel spin is therefore as follows:

- Single spinning wheel (split-μ surface):
  - Apply brake to spinning wheel to bring wheel under control
  - Use a combination of engine control and brake to prevent either wheel from spinning.
- Single spinning wheel (uniform adhesion surface):
  - Control wheel slip by engine control only (applying the brake to the spinning wheel simply causes the opposite wheel to spin).
- Two spinning wheels (homogeneous surface):
  - Control wheel slip by engine control only.

ABS only reacts when imminent wheel deceleration with associated increased slip is detected, which is generated by the application of the brakes by the driver. When the vehicle drivetrain is transmitting power from the engine, the brakes are not being applied, and therefore for traction control to function the ABS must be modified to enable an automatic brake application of the driven wheels to occur. This is achieved by the addition of additional valves, labelled 'TCS' in Figure 11.10, which allow pressure to be applied to the driven wheel brakes as necessary. To ensure that brake pressure is only applied to the spinning wheel requires the brake of the non-spinning wheel to be isolated, using the ABS pressure control valve to prevent actuation pressure being generated at that wheel. It is then possible to apply actuation pressure only to the wheel that is spinning to bring it under control. Automatic application of the brakes is generally speed limited so the brakes cannot be applied at road speeds above a defined threshold, which is vehicle dependent but is typically 60–80 km/h.

Electronic Stability Control (ESC)

Following recommendations of many studies and statistics, stability control has become a standard feature in Europe, North America and Canada. The North American FMVSS126 legislation mandated ESC by September 2011 for all new vehicles with a mass less than 4.5 tonnes, and in Europe standard requirements for stability control will be fully implemented in 2014 (see Table 8.1). Many passenger cars have already had some form of ESC or 'roll stability control' (RSC) fitted as standard to improve their dynamic stability, especially to prevent rollover of high centre of gravity vehicles such as MPVs and SUVs. Commercial vehicles are particularly vulnerable to rollover, having a high centre of gravity especially when laden, and rollover prevention systems that utilise autonomous brake intervention have been developed to the extent that they are now included in European legislation (see Chapter 8). The articulated commercial vehicle tractor and semi-trailer combination is one of the most unstable vehicles on the road because the towing vehicle (tractor) usually has a short wheelbase and high centre of gravity plus the additional braking duty and load transfer imposed by the trailer. As well as being prone to rollover, these vehicles are very dangerous when unstable; e.g. the 'jack-knife' instability that occurs when the rear wheels of the tractor exceed the available lateral adhesion and the resulting high yaw rate of the tractor mean that the driver cannot intervene with any meaningful corrective action. The result is that the tractor folds around the trailer as the trailer attempts to continue in its original direction.

ESC utilises the vehicle's braking system to create yaw forces via the wheels to steer the vehicle. When the ESC becomes active the brakes can still be applied by the driver but the ESC adjusts the priorities that govern the braking system from the basic function of decelerating the vehicle to one of keeping the vehicle stable and on course. Specific braking intervention under ESC is directed at individual wheels; e.g. in a left turn, the left rear wheel is braked to counter understeer or the right front wheel is braked to counter oversteer. Additionally, ESC may not only initiate braking intervention but can also interface with the engine control system to accelerate the driven wheels. More recently 'torque vectoring' has become available to increase the effectiveness of engine intervention. Thus, there is a 'discriminatory' control concept based on these two individual intervention strategies: ESC can brake selected wheels (selective braking) or accelerate the driven wheels (selective acceleration) to keep the vehicle on the road and reduce the risk of collision and rollover. So ESC combines ABS, EBD and TCS, and also can decelerate a single wheel to generate a yaw moment in order to compensate oversteer or understeer. The advantages include a very high level of vehicle stability under all conditions (governed by physical limits), but the disadvantages include reduced feedback about road conditions to the driver.

In terms of vehicle handling, specifically directional control, if a road vehicle does not respond to the commands of the driver then at least one of the following parameters has been exceeded:

- High wheel slip that indicates that the braking force has exceeded the tyre/road adhesion.
- High wheel slip that indicates that the traction force has exceeded the tyre/road adhesion.
- High slip angle that indicates that the sideways forces have exceeded the available lateral adhesion.

The first two can be controlled by ABS and TCS respectively but the third requires stability control. The maximum lateral and longitudinal forces that can be generated at any tyre/road interface are limited by the total adhesion as illustrated in Figure 11.11

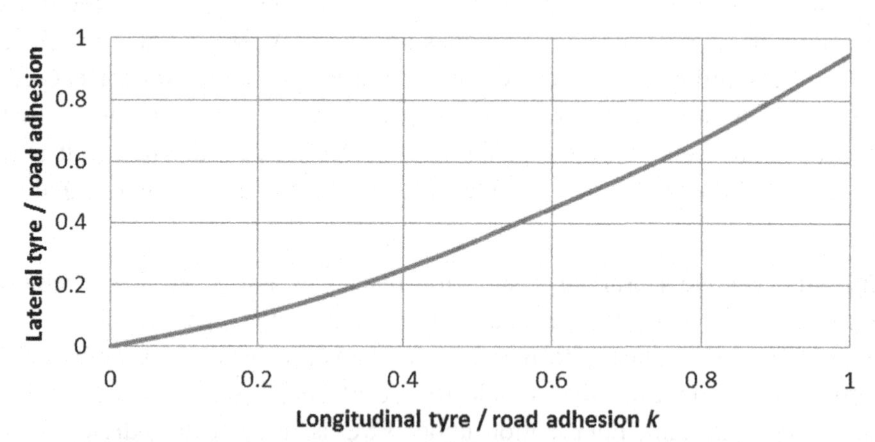

Figure 11.11: Illustration of the Relationship Between Maximum Lateral and Longitudinal Tyre/Road Adhesion.

(refer to Figures 3.12, 3.14 and 3.15 in Chapter 3). When a manoeuvre is carried out the vehicle will remain stable and follow the direction intended by the driver until the steering forces exceed the available lateral tyre/road adhesion, when the vehicle will no longer respond to the control inputs of the driver and instability will result. When the steering forces exceed the lateral adhesion of the steered axle the vehicle will 'understeer' and will follow a wider arc than intended. During a manoeuvre when the steering forces do not exceed the lateral adhesion the vehicle will follow the direction intended by the driver up to the point when the lateral adhesion of the rear axle is exceeded. In this case the rear axle will slide out of line in the direction of the destabilising forces and result in 'oversteer' and the vehicle will now follow a tighter arc than intended, rotating about the front axle (spin). In semi-trailer combinations the 'oversteer' condition would result in a 'jack-knife' as the trailer would be pushing the tractor, thereby increasing the destabilising forces.

Figure 11.12 illustrates how an ESC system on a two-axle rigid vehicle addresses understeer; as the vehicle drifts outwards while cornering, the brake on the inside rear wheel is actuated to provide an opposing yaw moment, which returns the vehicle to the desired trajectory. If the ESC system detects oversteer, i.e. the rear of the car drifts outwards while cornering, the brake on the outside front wheel is actuated to provide an opposing yaw torque that turns the vehicle in the desired direction.

ESC can create brake actuation pressure without driver effort and requires all the ABS components described earlier plus a steering wheel angle sensor and inertial measurement sensors. The former provides information about the driver's steering intent, i.e. the intended direction of the vehicle, which is compared to the actual vehicle movement

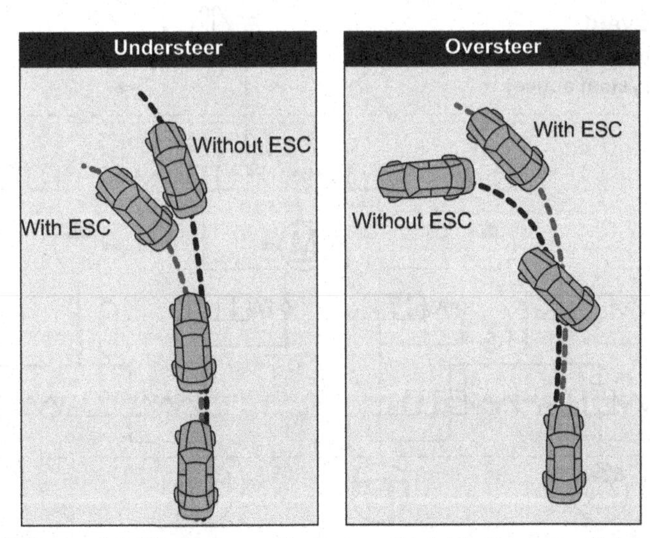

Figure 11.12: ESC Operation To Correct Oversteer And Understeer. *(Adapted from Continental)*

derived from the inertial measurement sensors. At least two inertial measurement sensors are required by an ESC system: lateral acceleration, and yaw rate. In addition, a longitudinal sensor is required for 4WD vehicles as previously explained for V_{ref} and for features such as 'Hill Start Assist' that need to determine road gradient. A roll rate sensor is required for a full 'Roll Stability Control' (RSC) system. On a passenger car the sensors are usually located in a separate module ideally packaged near the centre of gravity of the vehicle, and the latest systems integrate them into either the brake Electronic Control Unit (ECU) or the Restraint (airbag) controller. Additional capabilities that can be provided by the ESC function include:

- Emergency brake assist: increasing brake pressure when emergency actuation is detected.
- Adaptive cruise control interface: smooth pressure modulation according to cruise control requirements.
- Failed booster support/cold start support/overboost by pressure increase by ESC in the cases of insufficient or no vacuum, recognised for example by a vacuum sensor or differential travel sensor.

The operation of ESC on a passenger car with a hydraulic braking system is illustrated in Figure 11.13. This illustrates how the pressure control valves to the brakes are open to allow the driver to apply the brakes, but in the case illustrated no actuation pressure is generated by the driver's muscular effort. However, the ECU has detected that a brake intervention is required (to the left rear brake) and the hydraulic pump is activated, which creates the required hydraulic pressure.

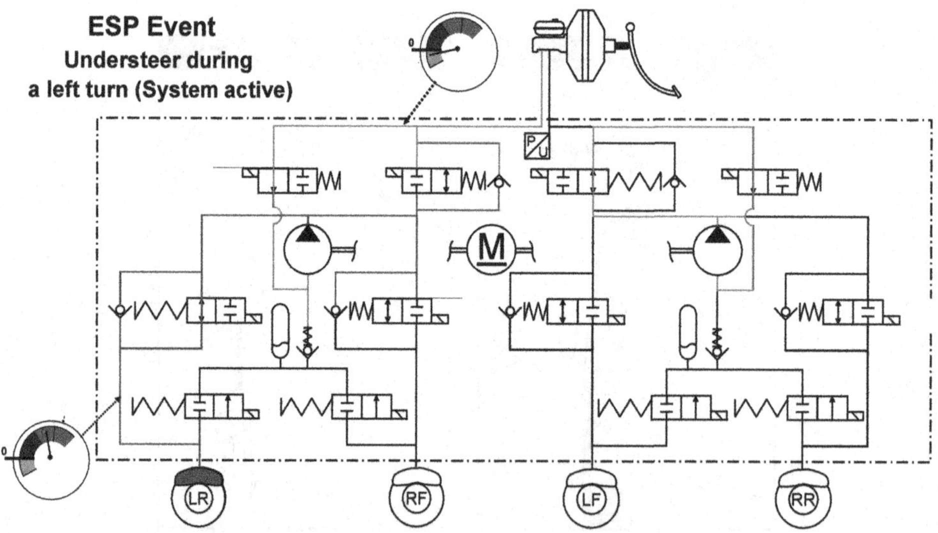

Figure 11.13: Car Braking System (Hydraulic Actuation), ESC Active (Moore, 2013).

On commercial vehicles with pneumatic brake actuation and electronic control, an ESC system (often termed 'Electronic Stability Program' or ESP on commercial vehicles) has a central ECU, which receives the driver demand from a footbrake module as illustrated in Figure 11.14. This controls, via a controller area network (CAN) databus, the 'brake apply' modules at each axle, which perform the following functions:

- Ensure air pressure is available to actuate the brakes (from the pressure supply/reservoir/storage system)
- Specify and generate the required actuation brake pressure at the selected brake
- Hold or exhaust actuation pressure at the selected brake
- Measure brake pressures and wheel speeds.

The correction of the loss (or imminent/anticipated loss) of directional control by the driver of a commercial vehicle is similar to the ESC control actions outlined previously for a passenger car. Oversteer and understeer correction in a commercial vehicle tractor/semi-trailer articulated combination is illustrated in Figures 11.15 and 11.16.

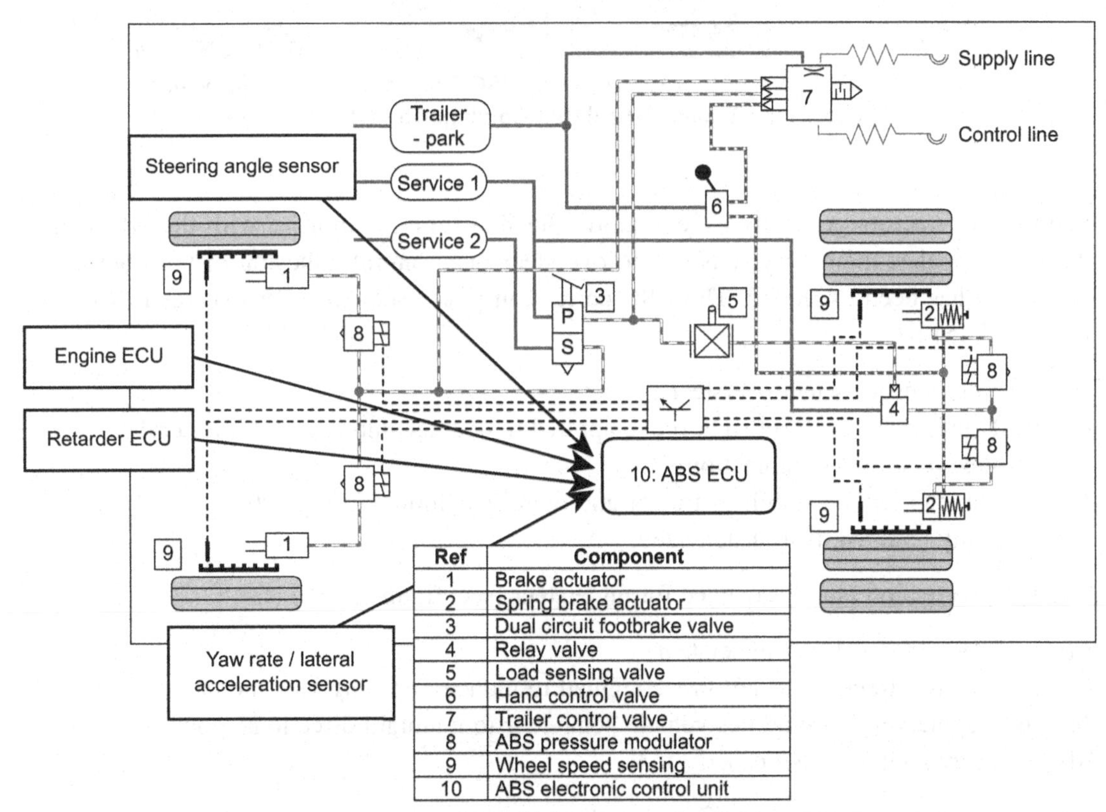

Ref	Component
1	Brake actuator
2	Spring brake actuator
3	Dual circuit footbrake valve
4	Relay valve
5	Load sensing valve
6	Hand control valve
7	Trailer control valve
8	ABS pressure modulator
9	Wheel speed sensing
10	ABS electronic control unit

Figure 11.14: Commercial Vehicle Pneumatic Braking ESC System (Compare with Figure 6.19(a)) (Ross, 2013).

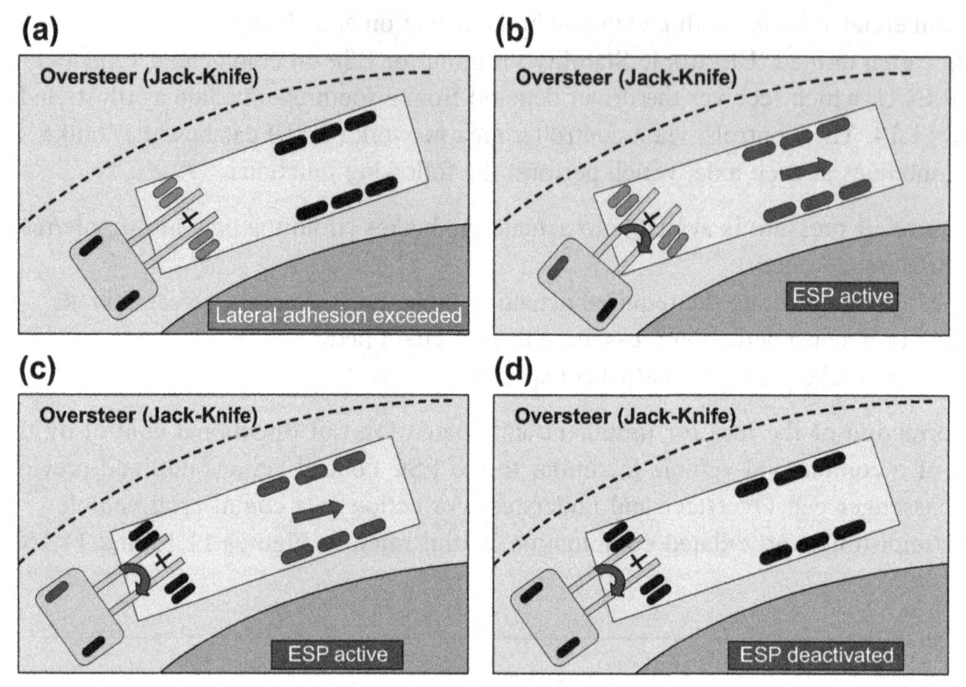

Figure 11.15: Correction of Oversteer by ESC/ESP in a Commercial Vehicle Tractor/Semi-Trailer Articulated Combination (Ross, 2013).

Semi-trailer tractors generally have a short wheelbase that, combined with the effect of the trailer, makes them very sensitive to oversteer once lateral adhesion of the tractor drive axle has been exceeded. The ESC correction phase sequence for oversteer (see Figure 11.15) is:

(a) Vehicle sensors detect oversteer
(b) ESP active, tractor front right brake applied to counter initial yaw, trailer brakes applied to stretch combination
(c) Brake intervention stabilises tractor and reduces combination speed
(d) Stability retained, ESP deactivated.

The ESC correction phase sequence for understeer (see Figure 11.16) is:

(a) Vehicle sensors detect understeer
(b) ESP active, tractor rear left brake applied to induce opposing yaw moment
(c) Brake intervention modifies vehicle behaviour to maintain directional control
(d) Stability retained, ESP deactivated.

Although with understeer the ESC intervention modifies the vehicle behaviour to enable directional control to be maintained, road speed reduction is also desirable.

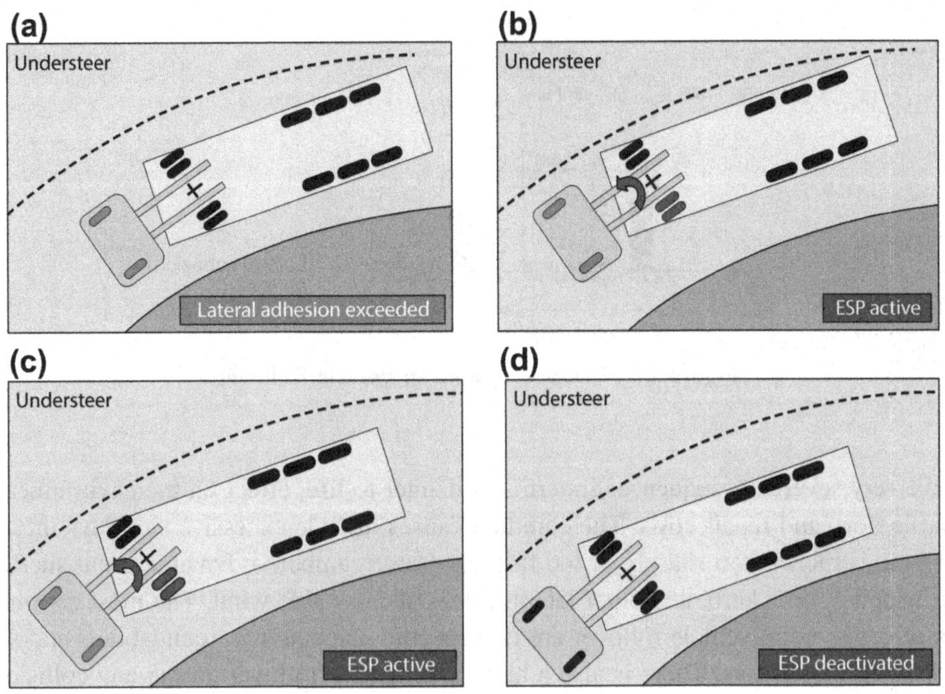

Figure 11.16: Correction of Understeer by ESC/ESP in a Commercial Vehicle Tractor/Semi-Trailer Articulated Combination (Ross, 2013).

Rollover Stability Control (RSC)

ESC sensors can also be used to recognise conditions that might lead to vehicle rollover, and initiate brake interventions to prevent it as far as possible, thus providing rollover protection. This feature is known by various names including Roll Stability Control (RSC), Roll Stability Program (RSP), Active Rollover Mitigation (ARM) and Rollover Mitigation (ROM). The purpose is to detect a potential vehicle rollover condition and reduce the risk of rollover occurring by slowing down the vehicle and generating understeer via brake intervention to reduce the vehicle's radial acceleration and hence lateral forces, e.g. by applying the brakes on the outside front wheel. For passenger cars and similar vehicles with a high centre of gravity, RSC systems require additional sensing, e.g. a roll sensor to provide a more accurate estimate of vehicle roll, vehicle pitch and heave.

Commercial vehicles with a high centre of gravity are particularly susceptible to rollover, and semi-trailer combinations are most at risk of rollover initiated from the trailer. Additionally, as in the case of semi-trailer combinations, the driver can be remote from what is happening at the trailer and therefore unlikely to become aware of impending instability until it is too late. Rollover of a heavy commercial vehicle is very serious and

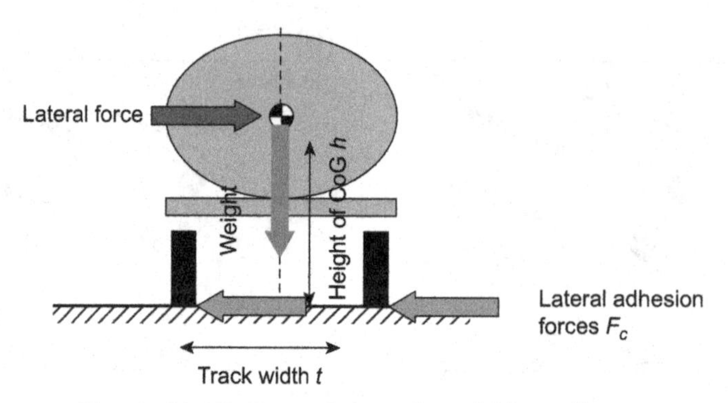

Figure 11.17: Force System in Vehicle Rollover.

can have very severe consequences in terms of danger to life, effect on the environment and traffic flow, and repair costs. The common causes include excessive sideways inertia forces from cornering too sharply or too fast, excessive camber, a dynamic event such as a wheel 'tripping' on a kerb, and other lateral forces such as side wind. The most common causes of commercial vehicle rollover are two specific manoeuvres: roundabouts and slip roads leaving motorways. There is also a high potential for rollover during any collision avoidance manoeuvre in the form of a single or double lane change. These types of manoeuvres generate sideways (lateral) inertia forces that depend primarily on vehicle road speed and the radius of the turn, as indicated in Chapter 3. Since the lateral force while cornering is $F_c = mv^2/r$, where v is the vehicle speed (m/s) and r is the radius of the turn (m), speed is a particularly critical causal parameter in rollover.

Sideways inertia forces act through the vehicle's centre of gravity and are resisted by the lateral adhesion forces at the tyre/road interfaces, as illustrated in Figure 11.17 (see also Figure 3.6). From Chapter 3, Equations (3.6), it is clear that the vehicle rollover about the outer tyre/road contact point will be initiated if the normal reaction (N_i) between the tyre and the road at the inner wheel becomes zero, as indicated in Equation (11.11):

$$\frac{P}{2} = \frac{hF}{t} \tag{11.11}$$

If the lateral force exceeds the maximum lateral adhesion force at the wheels, the vehicle will skid rather than roll. The rollover susceptibility therefore depends on three design parameters: height of the centre of gravity, longitudinal position of the centre of gravity (position of the load centre of gravity) and the 'effective track'. Because semi-trailers are coupled to the tractor via the coupling (kingpin or fifth wheel), the effective track depends upon the position of the trailer centre of gravity, as illustrated in Figure 11.18. The most stable condition to reduce the possibility of rollover is for the CoG to be over the rear of the trailer with no load on the coupling, but this is an undesirable arrangement that could

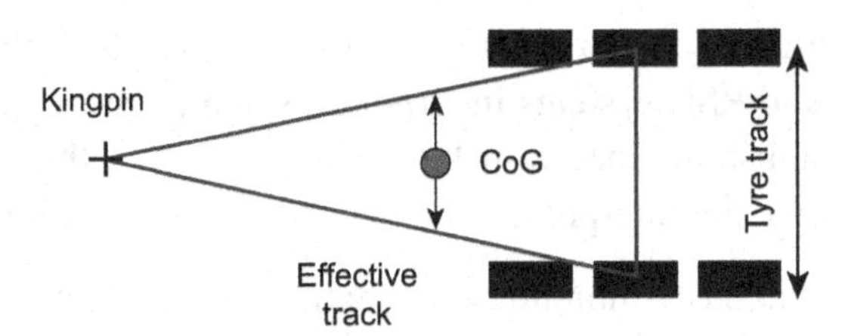

Figure 11.18: Effective Track of a Semi-Trailer.

have an adverse effect on the dynamic handling and stability of the combination during normal driving and braking because of the increased possibility of yaw oscillations and jack-knife. Other factors that affect the susceptibility of articulated semi-trailer vehicle combinations to rollover include:

- Road camber
- Coupling position and design
- Suspension roll stiffness
- Chassis torsional stiffness
- Tyre pressures
- Load shift
- Roll centre position.

The tractor/semi-trailer coupling permits relative pitch and yaw movement, and is designed to provide some (limited) lateral or roll support between the tractor and the semi-trailer. When the articulation angle between the tractor and semi-trailer is not 180° (i.e. when they are not in a straight line), the roll support provided by the coupling reduces and becomes zero when the articulation angle is 90°.

On an articulated commercial vehicle (tractor and semi-trailer combination), rollover usually starts at the semi-trailer. If impending rollover were detected by the tractor's ESC using the sensors to indicate the reaction of the trailer on the tractor, the first useful control action would be to reduce the vehicle's road speed. But additional rollover protection can be achieved using RSC fitted only to the semi-trailer, because the brakes of the semi-trailer can be automatically applied (independently of the driver) to achieve increased stability by selective braking. A lateral acceleration sensor and controller are required on the semi-trailer but no information or control signals need to be passed from the towing vehicle, so provided that it is equipped with an ISO7638 connector to permanently power the trailer braking system, the trailer RSC function can always operate. The best system, however, is RSC which includes both the towing vehicle and the trailer in the sensor and control logic.

Any automatically commanded brake control system must have at least one threshold to pass before it can become active. If the threshold is set too low the system will intervene prematurely and cause driver dissatisfaction, and if it is set too high the intervention may be too late. Many ESC and RSC systems incorporate some learning function in their control logic, e.g. to establish the load distribution or position of the centre of gravity, in order to be able to improve the interpretation of the sensor data, as illustrated below:

- System passive: continuously monitors information from vehicle sensors, checks for potential rollover and compares with predefined threshold.
- Detect potential rollover: the indicated lateral acceleration has exceeded the threshold value.
- Automatic brake application: automatically generates a low actuation pressure in the brakes.
- Check wheel response: ABS wheel speed sensors provide information to decide whether the vehicle is approaching a critical rollover state, involving either of the following:
    - No wheel response: the inside wheels are not decelerating, which indicates that the trailer is not in a critical rollover condition. The detection threshold is modified and the system returns to passive mode.
    - Wheel response: the inside wheel(s) decelerate due to lateral load transfer, reducing the inside wheel load, indicating imminent rollover.
- Brake application: the brakes of the tractor and/or trailer are automatically applied. In the case of motor vehicle systems engine power is also reduced.
- Speed reduction: the reduction in speed due to the application of the brakes reduces the lateral acceleration.
- Stability maintained: the reduction in lateral acceleration results in the stability of the vehicle/trailer being maintained.
- System passive: after the brake intervention and reduction in lateral acceleration the brakes that have been applied are released and the system returns to a passive mode.

Such a learning process takes time and often the rate of change of lateral acceleration is so high that there is no time to complete the low-pressure brake application before the rollover has occurred. The learning process would then be bypassed, allowing the system to automatically apply the brakes. Usually when an automatic brake application has been initiated as a result of the low-pressure brake application and wheel speed response there is a high probability of imminent rollover, and therefore the braking force generated is likely to be relatively high to ensure that the vehicle speed is reduced as quickly as possible. It is important to achieve the right duration of any brake intervention in this type of RSC control. A system that detected too early and repeatedly applied the brakes, or that left the brakes applied for too long, would cause driver discomfort by continually having to respond to unnecessary speed reduction. Alternatively, it would be very easy for a driver to drive the vehicle too fast and rely on the stability system to get out of trouble, and this clearly defeats the objective of such systems. So when the system detects a potential rollover and the brake intervention to stabilise the vehicle is high, the duration of the brake application should be such that a given manoeuvre can always be completed more rapidly by driving at a sensible speed for the conditions than relying on the system to correct for bad driving.

Electronic Brakeforce Distribution (EBD)

EBD systems control the balance between the brake forces on different axles to optimise braking efficiency across all vehicle loading and driving conditions. It has the advantage of being able to sense and control the amount of braking applied to an axle more accurately in proportion to the dynamic axle load. Wheel speed sensor information is used to detect circumferential tyre slip on the front and rear axles, and when a defined rear tyre slip level is exceeded, the actuation pressure to the rear wheels is limited or reduced. This has the advantage of providing optimum braking force at all the wheels of the vehicle regardless of loading conditions prior to ABS intervention. It also distributes brake thermal loading more evenly around the vehicle by ideal utilisation of all the brakes, which improves brake performance, especially under high-duty conditions.

The way EBD works can be explained as follows for the simplest case of a two-axle rigid vehicle. If the rotational speeds of the front and rear wheels are ω_1 and ω_2 respectively, the differential slip (dQ) can be calculated from:

$$dQ = (\omega_1 - \omega_2)/\omega_1 \tag{11.12}$$

A braking control system based on differential slip would then individually control the actuation pressure (pneumatic or hydraulic) applied to the brakes on each axle to minimise the differential slip, thereby ensuring that the wheels are operating in the same region of the adhesion/slip curve; i.e. they are being braked in proportion to the dynamic normal forces on the wheels. As explained in Chapter 4, for towing vehicle and trailer combinations, the towing vehicle may not directly control the braking system of the trailer but supplies a pneumatic and electric signal to the trailer from which, based on the load carried by the trailer, its braking system reacts to deliver actuation pressure to the trailer brakes according to the setting of the load sensing on the trailer.

Wheel speed information is fundamentally important to ABS, ESC and EBD. Other vehicle electronic systems can also generate information that can be communicated to the EBS via the vehicle CAN bus. For example, when the vehicle is accelerating, knowledge of the engine torque derived from the instantaneous fuel consumption and the engine speed, and the gearbox transmission ratios, together with the vehicle acceleration derived from the wheel speeds, enables the total mass of the vehicle combination to be calculated. With the known or learned braking force/braking pressure relationship for the towing vehicle thus available, the braking characteristics of the trailer can then be deduced. A coupling force control (CFC) system on the towing vehicle can use this information to adjust the pneumatic and electric signals at the coupling head to improve the compatibility. This applies irrespective of whether the trailer is equipped with EBS or a conventional pneumatic braking system.

Emergency Brake Assist (EBA)

As discussed above with RSC, reducing the vehicle speed as quickly as possible has many advantages in avoiding or reducing the severity of dynamic instabilities in motor vehicles; if nothing else it will reduce the kinetic energy prior to a collision. Brake interventions initiated by an ESC system operate at a much faster response time than can be achieved by a human driver, and Emergency Brake Assist (EBA) is a generic name for the way in which some of the features of ESC can be utilised to help the driver brake faster and more effectively in an emergency. These include reducing system delays in generating brake pressure, e.g. by recognising certain conditions that might indicate imminent brake application, such as when the throttle pedal is quickly released or when a stability index increases beyond a preset limit, and 'pre-filling' or 'pre-pressurising' the brake actuation to bring the pads to touch the discs. This is normally associated with hydraulically actuated braking systems on passenger cars and similar vehicles, and can be achieved by briefly running the ABS pump to build up sufficient brake fluid pressure to close the gap between the pads and the discs.

The requirement for brake assist is a part of the European pedestrian protection legislation, and recognises that often in a panic situation the driver may brake quickly but insufficiently (see Figure 11.19). Applying the brakes quickly can be recognised by the

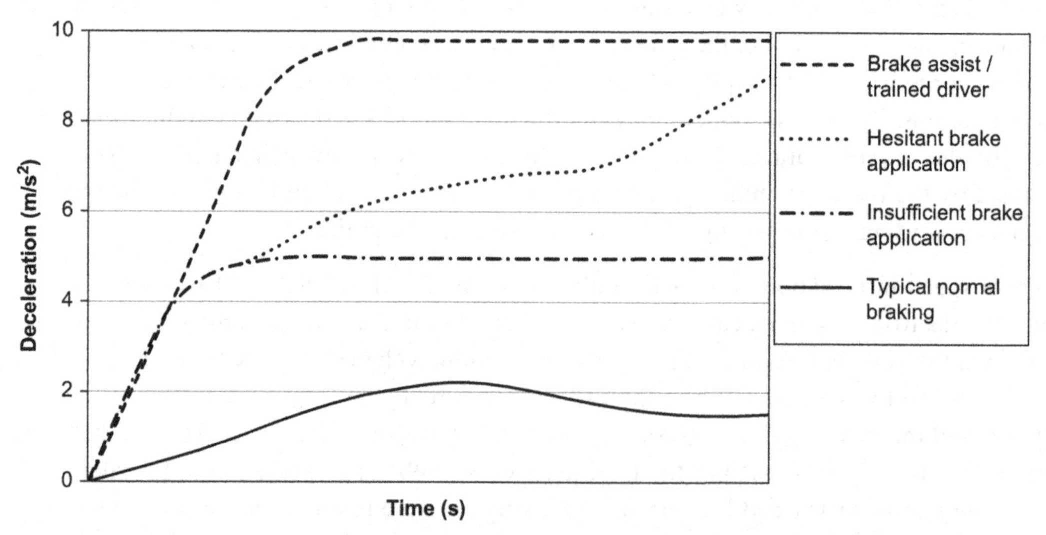

Figure 11.19: Effectiveness of Emergency Brake Assist (Moore, 2013).

EBA function, which can use the ABS/ESC system to increase the brake actuation pressure to maximise adhesion utilisation and force ABS operation; if ABS is active then the vehicle is also steerable. Alternatively, and especially for vehicles with ABS but not ESC, a spring mechanism in the brake master cylinder can trigger the required increase in brake pressure.

Adaptive Cruise Control (ACC)

Adaptive Cruise Control (ACC) combines constant speed cruise control with the automatic positioning of a road vehicle with respect to another vehicle in front. It uses the vehicle's engine and transmission to maintain the set speed, and the vehicle's braking system as necessary to maintain a safe distance behind the vehicle in front. The primary components of ACC are a microprocessor controller and a range finding sensor, which scans the road ahead and reports the location and rate of approach of a vehicle or obstacle in front. The controller then determines the safest course of action for the vehicle and the driver and implements it. The possible responses include:

- No intervention
- Request a vehicle speed and/or engine torque decrease
- Request the brakes to be applied
- Warn the driver of a potentially critical stop situation.

An ACC system can request a vehicle deceleration to match the speed of the preceding vehicle and then will resume the preset speed and distance once the road in front of the vehicle is clear. Requests to reduce vehicle speed or engine torque, or to apply the brakes, are passed to the engine controller, and to the ESC module if the requested vehicle speed reduction cannot be achieved via engine/transmission braking alone. The ESC module then applies an appropriate brake response to achieve the desired vehicle deceleration. The ACC system design includes driver safety features such as audible and visual warnings and signals to the driver. Normally an ACC system operates at road speeds above 30 km/h, and would require additional technology for safe operation at lower speeds.

Collision Mitigation by Braking (CMbB)

As explained in Chapter 8, a 'Collision Mitigation Braking System' (CMBS) is a type of 'Advanced Emergency Braking System' (AEBS) that can autonomously apply emergency braking in order to mitigate the severity of a collision that has become unavoidable. In comparison, a 'Collision Avoidance Braking System' (CABS) can autonomously apply emergency braking in order to avoid a collision. AEBS is a generic name for any system that can apply automatically commanded braking, and is designed to detect automatically an emergency situation and activate the vehicle braking system to decelerate the vehicle with the purpose of avoiding or mitigating a collision. An ACC forward sensing system automatically controls the vehicle to decelerate the vehicle gently to match the vehicle's speed with that of a preceding vehicle if slower than the set cruise speed. The same system can also detect a potential collision if the relationship between the vehicle closing speed and the distance predicts it, and take appropriate control action including a brake pre-fill and emergency braking typically up to 8 m/s^2.

Electric Parking Brake (EPB) Systems and Hill Start Assist (HSA)

Current EPB systems take the form of either a cable-operated device consisting of a single electromechanical actuator with an electronic controller (part of the brake ECU), which actuates the parking brakes via bowden cables, or an electric motor directly attached to the disc brake caliper. The latter type of EPB system saves package space but the caliper may have a weight penalty, which may not be acceptable because of the increased unsprung mass. The caliper may also be pre-actuated using hydraulic actuation pressure via the ESC module and then locked electromechanically. Combining an EPB system with ESC can provide the following optional additional functions:

- Hydraulic support for the parking brake (e.g. on a grade)
- Hill Start Assist
- Automatic Vehicle Hold.

Hill Start Assist (HSA) prevents the vehicle rolling back when the service brake is released without the need to apply the parking brake. Typically the HSA system holds the brake actuation pressure set by the driver's pedal force input for a defined time interval (typically up to 2 seconds) after it is released without the vehicle immediately rolling back. The pressure hold duration is critical, since the driver must remain aware of this temporary assistance. Automatic Vehicle Hold (AVH) is an extension of HSA, whereby the brakes may be held for a longer period of time than HSA allows via the EPB; the vehicle can be safely held stationary, e.g. if the driver has left the vehicle.

Trailer Sway Control (TSC)

Trailer sway or 'snake' is a dynamic instability where the trailer (usually centre axle types, especially caravans) sways from side to side with increasing amplitude while travelling, and can be initiated by side winds, track ruts, fast steering movements or a badly configured trailer load (drivers should always ensure that there is a small downward tow hitch force to minimise the potential for instability). It is extremely difficult to recover from trailer sway once it has started, and the end result is usually an overturned trailer and possibly the towing vehicle as well. TSC detects the onset of snaking and can counteract the resulting yaw movement by generating a stabilising opposing yaw torque with alternating brake interventions, while at the same time reducing the engine torque and increasing the pressure in all wheel brakes to reduce the road speed.

Torque Vectoring by Braking (TVbB)

Torque vectoring is a feature that is now often installed on high-performance vehicles to actively direct the drive torque at the road wheels to maximise adhesion utilisation during acceleration. Torque Vectoring by Braking (TVbB) provides some of the benefits of TVbD through the use of the vehicle's brakes. Brake torque is applied to one driven wheel on a drive axle, which directs ('vectors') drive torque to the opposite wheel via the differential. Excessive wheel spin is prevented and more importantly yaw rate during manoeuvres can be controlled before stability control thresholds are reached and TCS or ESC intervenes.

Engine Drag Control (EDC)

On some very low friction surfaces (e.g. ice) a vehicle may become unstable when the throttle is closed and the residual drag torque from the engine prevents the driven wheels from rotating sufficiently. To avoid this phenomenon the brake control system requests a positive torque (torque increase) from the powertrain control system to compensate for the engine drag and powertrain losses during specific driving manoeuvres. This feature is of particular benefit for RWD vehicles, but can also be usefully installed on FWD vehicles.

ESC Mode Switching

It can sometimes be beneficial for the driver to have more than one stability control setting. Some passenger car manufacturers fit an on/off ESC switch while others choose not to; such a switch may be a physical switch or a software switch. ESC is fitted to vehicles to assist with the control of their dynamic stability in use, and there can be situations where ESC is not required (e.g. driver choice), tyre/road adhesion settings need to be changed (e.g. driving on grass), or its operation aggravates problems that can occur in certain types of driving conditions, e.g. deep snow. Some of the options that are currently available on passenger cars are summarised in Table 11.3.

Driving a vehicle out of deep snow can be helped if parts of the ESC function are disabled, e.g. engine traction control only. In this case stability control, brake traction control, ABS and EBD may remain active with their standard parameter settings, and the vehicle can be rocked to and fro out of the snow. Vehicles designed with off-road capability may require a number of ESC settings to allow the driver to select the appropriate tyre/road adhesion profile for driving on sand, mud or grass, etc. Typically

Table 11.3: Examples of ESC Switch Modes

ESC Modes	No ESC Switch	ESC Switch with 'Get out of Deep Snow' Function	Cars with a 'Sport Heritage'	Cars Designed also for the 'Racetrack'
ABS, EBD, BAS	Active	Active	Active	Active
ACC interface	Active	Active	Off	Off
CMbB interface	Active	Active	Off	Off
Engine TCS	Active	Off	Extended tuning	Off
Brake TCS	Active	Active	Extended tuning	Off
Stability control	Active	Active	Extended side slip	Off

each setting would alter the limits for wheel slip in ABS mode and for traction control but leave the standard stability control settings. Hill Descent Control (HDC) is intended to assist the driver when driving the vehicle down a steep slope. Integrating with the cruise control system for fine control of low vehicle speed, the ESC unit generates brake pressure in a similar way to the ACC interface. Once the vehicle is off the slope the HDC function may automatically disable, e.g. above a road speed of 15 km/h.

Regenerative Braking

The increasing functionality of electronic braking control systems to enhance road safety for road vehicles has also been associated with alternatives to the use of friction brakes for retarding the vehicle. For many years it has been accepted that a vehicle's kinetic energy dissipated during braking is lost to the environment. Even the provision of retarders (or endurance brakes) to augment the braking power of the friction (foundation) brakes on a road vehicle has not changed the accepted practice that the heat energy from braking is generally dissipated to the environment. The reason seems to be cost-effectiveness; many technologies have been developed and successfully applied, but until the recent market acceptance of hybrid drive vehicles, particularly passenger cars in the form of hybrid electric vehicles (HEVs) or full electric vehicles (FEVs), the additional cost of the technology was prohibitive. The EU has an agreement with the European Automobile Manufacturers Association (ACEA) to reduce the average CO_2 emissions per manufacturer to 130 g/km by 2015 for all new cars in the EU (Euro 6). This represents a significant shift towards a more sustainable personal transport infrastructure, and as a result road vehicle manufacturers are considering ways of reducing fuel consumption (and hence CO_2) by overall vehicle energy control and management, i.e. not just by increasing the efficiency of the internal combustion engine, but that of the whole vehicle, including storing and reusing energy where possible.

Many researchers have described how regenerative braking can be applied to road vehicle powertrains. For electrical regenerative braking systems, high battery current demands during acceleration and braking limit the amount of regenerative braking that can be used, and energy conversion rates are a challenge for all regenerative braking system technologies. The retarding torque generated by most motor/generator technologies is speed dependent; in particular the torque decreases with speed, and thus no system of regenerative braking is effective at low speeds and therefore friction brakes are likely to remain essential in order to provide full road vehicle braking capability. Additionally, no current form of regenerative braking is able to generate sufficiently high braking force for emergency braking from relatively high speed because of the high power dissipation and the rate of energy transfer required. Friction braking is the only retardation technology that can be practically implemented on road vehicles to provide reliable braking at all road speeds, and dissipate large amounts of energy very quickly. Therefore, a combination of friction brakes and regenerative braking is likely to be required for the foreseeable future, and this presents a challenge for the management of braking effort generated by each system at each wheel. Challenges for road vehicle regenerative braking systems therefore include:

- The provision of two types of braking system, namely friction and regenerative, necessary to meet legislative requirements under all operating conditions.
- Transition between and distribution of regenerative and friction braking.
- Stability and response time under low road adhesion conditions and emergency braking.

Brake system technologies that have been described in this book and that could be suitable for vehicles with regenerative braking include:

- Electrohydraulic brake (EHB)
- Electromechanical brake (EMB)
- Conventional brake with active booster and electric vacuum pump
- Electromechanical rear and conventional front brake.

Regenerative braking systems aim to recover, store and reuse some of the vehicle's braking energy to improve fuel efficiency or boost the range of electric and hybrid vehicles (FEV/HEV). Energy storage media include electric batteries and/or ultracapacitors, flywheels and hydraulic accumulators. Some form of motor/generator augments the friction braking (from the foundation brakes) where possible; as the driver applies the brakes through a conventional pedal, the motor/generator creates braking torque that may provide sufficient retarding force to meet the driver demand, or may supplement the friction braking. An example layout for a passenger car is shown in Figure 11.20.

There are different levels of 'hybrid' road vehicles:

- Micro hybrid. Some form of alternative power source in the vehicle provides low levels of energy conversion and reuse, e.g. providing regenerative braking by using the vehicle alternator to charge the vehicle battery, or drive ancillary equipment only during vehicle deceleration.

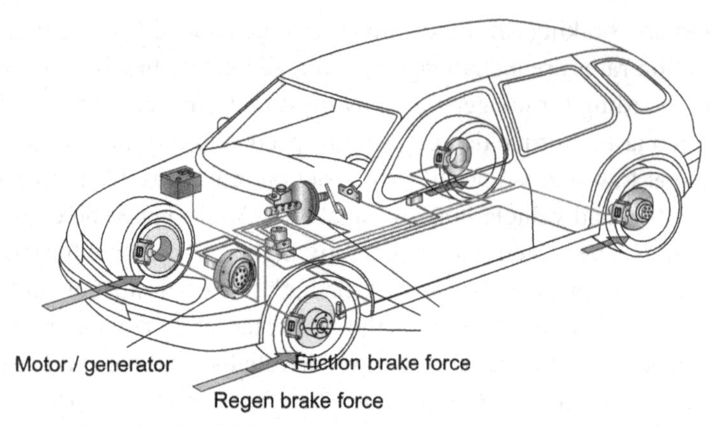

Figure 11.20: Hybrid Vehicle Regenerative Braking System (Continental).

- Mild hybrid. Some form of alternative power source in the vehicle provides sufficient power to drive the vehicle to some extent, e.g. by augmenting the internal combustion engine (ICE) during acceleration.
- Full hybrid. Some form of alternative power source in the vehicle provides sufficient power to drive the vehicle without assistance from the ICE.
- Plug-in hybrid. The vehicle's batteries can be recharged by connecting an external power source to further decrease the reliance on the ICE.

While a road vehicle with a hybrid powertrain will always incorporate regenerative braking, this does not mean that regenerative braking can only be fitted to hybrid vehicles. On all hybrid designs, regenerative braking will charge the energy storage units (e.g. batteries); however, the cost and extra mass of the hybrid powertrain with regenerative braking may more than offset any fuel economy gains, so there needs to be a trade-off between the efficiency gains from energy recovery, storage and subsequent reuse, and the extra power required to propel a significantly heavier vehicle. 'Micro-hybrid' powertrain technology harvests low amounts of braking energy through regenerative braking e.g. to charge the vehicle's battery and power ancillary features, and has the advantage of low cost, since the system is designed to store only a relatively small proportion of the braking energy, and only for a short period of time. The amount of energy that could be recovered, stored and reused via regenerative braking depends upon the design of the powertrain and the way the brakes are used. Brake usage varies with the type of vehicle, the type of road, the terrain and the driver, and experience has indicated that in passenger cars the distribution of service brake applications tends to peak around $0.12 \leq z \leq 0.17$, and high rates of braking ($z > 0.5$) are only employed occasionally, e.g. 1% of service brake applications.

The prediction of energy dissipated during vehicle braking is essential to design an effective regenerative braking system so that fuel and energy efficiency can be predicted.

Standard legislative drive cycles are now used extensively by vehicle manufacturers to define vehicle operation in the operating domain. These are intended more for evaluating engine usage and fuel consumption, and as a result their applicability to braking events is very limited. If such cycles are used to evaluate regenerative braking systems, the indicated results may not be representative of real-world usage, performance and benefits. For this reason regenerative braking systems should be evaluated based on field driving data rather than legislative drive cycles. On flat roads it has been possible to define the work done by the brakes in terms of a duty factor D_k as shown in Equation (11.13) (Meise, 2009). D_k describes the energy dissipated during braking per unit time after allowing a correction factor for drag losses, and can be related to the mean journey road speed $\overline{V}$ and a parameter V_0 that relates to the route and represents the maximum speed that the test route could be driven without applying the brakes (and without engine braking). Its magnitude was found to depend upon a number of factors, including the traffic density and the handling of the vehicle, but varied only by $\pm 15\%$. The most significant effect came from the type of road used, e.g. narrow and twisty or open main road.

$$D_k = A\left[\frac{\overline{V}}{V_0^{0.75}}\right]^B \tag{11.13}$$

Values for the constants A and B were found to be 9.61×10^{-6} and 6.3 respectively. The mean journey speed can be related to the maximum road speed V_{max} of the vehicle used, as shown in Equation (11.14):

$$\overline{V} = \beta V_0^{0.62} V_{max}^{0.45} \tag{11.14}$$

The value of the parameter β depends upon the manner in which the vehicle is driven: 0.93 for normal driving and 1.17 for fast driving.

Electrical regenerative braking systems are the most common in hybrid and electric passenger cars and light vans because of the high efficiency and controllability of electric motor/generators and energy storage units. The power to weight ratio of hydraulic systems has tended to limit regenerative braking to heavy vehicles, but recent advances have made their use possible for passenger cars. The two main designs are a series system where the ICE drives a hydraulic pump at a constant speed with a very high efficiency, and a parallel system where the ICE drives the wheels via a clutch and the hydraulic pump is connected to the drive shaft of the axle. In both cases, the hydraulic energy drives hydraulic motors on each axle or wheel and when the vehicle is in regenerative braking mode the ICE is disconnected from the system and the energy is stored in hydropneumatic accumulators. Mechanical (kinetic) energy systems based on flywheels have a good power to weight ratio and use environmentally friendly materials. Concerns over vehicle handling problems associated with gyroscopic effects can be solved, and the main disadvantage relates to risk and safety; if a critical component failed and released the flywheel while rotating, or the energy had to be 'dumped' quickly, e.g. in vehicle recovery after an accident, this could present a serious hazard. Pneumatic energy systems for in-vehicle energy storage can utilise a piston compressor or the ICE (as in a retarder) to generate a retardation torque on the drive axle(s) and store the energy as compressed air in a pressure vessel. Their advantage is cost-effectiveness since no expensive electrical components are needed, but a disadvantage may be the size and weight of the pressure storage vessel.

The design and implementation of regenerative braking system control is quite difficult; the basic strategy can be to split the braking energy between both systems, prioritising the use of the regenerative braking over the use of the friction brakes, which are used to complete the braking demand. The simplest control of a retarder was a single level of operation (on or off), but now retarders usually have several levels of operation; e.g. where the retarder operates during all service brake applications, an initial movement of the brake pedal can introduce the retarder prior to friction braking. For regenerative braking such a simple control would not be satisfactory for anything more than a residual level of regenerative braking. The operation of the regenerative braking system has to be 'blended' with that of the friction braking system to avoid braking balance and vehicle stability problems, and to provide an acceptable control interface (brake pedal characteristic, in particular 'feel') for the driver. The challenges of blending regenerative braking with friction brakes so that brake operation is completely transparent to the driver are considerable; ultimately the regenerative braking control system would have to integrate with all other electronic braking control systems including ABS and ESC.

The legislative aspects of the control and performance of 'blended' brake systems on vehicles with regenerative braking are covered in Chapter 8 of this book. As explained there, in EU braking legislation regenerative braking systems are separated into two categories:

- Category A: a regenerative braking system that is not part of the service braking system; typically regenerative braking is introduced when the accelerator pedal is released.
- Category B: a regenerative braking system that is part of the service braking system.

For vehicles with category B regenerative braking systems the service braking system can only have one control device. Vehicles fitted with both categories of regenerative braking (A and B) must not cause the action of the service braking control to reduce the electric regenerative braking effect generated by the release of accelerator control. This represents an important control strategy consideration – whether to apply regenerative braking on lift-off of the accelerator pedal to simulate or augment engine braking, or to apply it all on the brake pedal, or a combination of the two. An example of a brake system blending profile required during a brake application is illustrated in Figure 11.21.

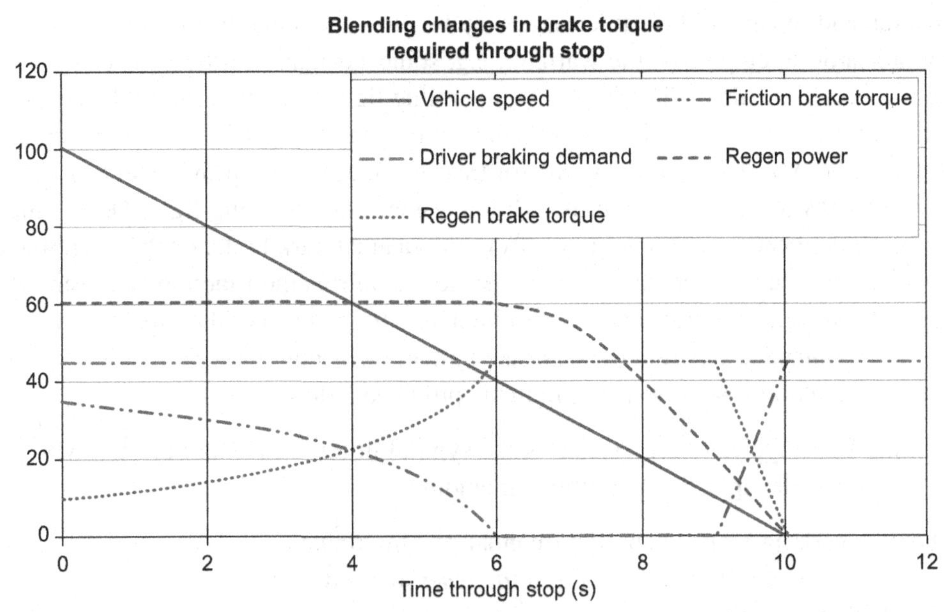

Figure 11.21: Example of a Brake System Blending Profile During a Brake Application (Curry, 2013).

Whichever control strategy is selected, the use of regenerative braking in conjunction with the friction (service) brakes can potentially significantly bias the front to rear braking distribution of the vehicle, which may promote wheel lock and thus affect vehicle stability. Although the vehicle's stability control systems might be expected to intervene, there is great opportunity for conflict between these different systems and an overall control strategy for regenerative braking with stability control is essential. One current solution is to disable regenerative braking when ABS or ESC is operating. An example of the basic design of a braking system for a passenger car with regenerative braking is included in Chapter 12.

System Warnings and Driver Interfaces with Electronic Braking

The modern vehicle braking system has many electronic functions and capabilities and it is essential that the driver is aware firstly of which functions are being utilised, and secondly if any particular function has failed or is not working properly. The requirements for braking system warnings and indicators are included in the UN Regulations 13 and 13H, which means that requirements and symbols are well standardised across different vehicles. Visible interfaces in the form of warning lights indicate the status of braking system functions to the driver, who may be obliged to stop the vehicle immediately (in a safe fashion) or be able to continue driving in a controlled and safe manner to a place of investigation and repair. Audible interfaces may be provided alone or in conjunction with a visual indicator, to emphasise the warning; e.g. some car manufacturers provide an audible warning to the driver in addition to a warning light if the parking brake is applied while the vehicle is driving. Noise from the actuation can also provide an indication of its function, e.g. the ABS/ESC pump and motor that, when activated, provide a clear indication of ABS or ESC operation in addition to the visual warning light. During the parking brake application, a ratchet noise from the manual park brake or the operation of motors/cables in an electronic parking brake system confirms the function has been set. Most other brake-related noises can be classified as error states, i.e. they indicate an undesired or unintended function of the brake system or its components. A brief description of braking system warning lights is provided below.

Red warning lights typically represent a brake symbol in the instrument cluster, which must illuminate under specified conditions including:

- When the parking brake is applied (manual parking brake or EPB)
- When a low level of the brake fluid in the reservoir is detected
- When EBD is disabled due to a system failure.

Amber lights are used to indicate a failure or undesired function of the electronic braking system. They must be continuously illuminated until the fault is no longer detected, either because it has been repaired or because the brake controller self-test diagnosis in that ignition cycle has detected that it has corrected itself. It is extremely beneficial for a Diagnostic Trouble Code (DTC) to be logged in the memory of the brake controller for each individual failure to facilitate detection and repair. The same amber warning light can be used to warn the driver that a stability and/or safety system is operating; e.g. an amber flashing 'skidding car' icon could indicate a driving situation when ESC or traction control mode is active.

Chapter Summary

- The role of intelligent electronic control of a road vehicle's braking system in enhancing road safety is introduced and explained for both passenger cars and commercial vehicles. Stability control functions are outlined and explained in terms of a range of control and monitoring mechanisms that utilise the braking system to provide active safety systems on modern road vehicles.
- Antilock braking (ABS) principles and technology are explained with the emphasis on controlling the 'slip' of the road wheels so that the available adhesion is maximised and a high level of lateral adhesion is maintained. The three main components of ABS, namely wheel speed sensors, electronic controller (ECU) and pressure modulators, are listed and the importance of determining the vehicle reference speed is explained.

- An overview of the way ABS works is presented, explaining how the system releases and reapplies the brake as excessive slip is sensed. Control logic options for ABS action are summarised; these include 'select low' and 'select high'. The difficulties arising from low-adhesion surfaces, brake hysteresis and wheel/rotor/drivetrain inertia are discussed.
- ABS efficiency is discussed in detail and a practical procedure to measure it based on the UN Regulation 13 Annex 13 requirements is presented and explained.
- Traction control (TCS) is introduced as a stability control feature associated with ABS, which allows brake actuation pressure to be generated to brake a driven wheel to prevent it spinning when peak adhesion has been lost during acceleration. It is explained how traction is maintained by transferring the drive torque to another drive wheel that does have adhesion, and the basic TCS logic to control wheel spin is summarised.
- The development and widespread implementation of Electronic Stability Control (ESC) on passenger cars and commercial vehicles is explained. ESC utilises the vehicle's braking system to create yaw forces via the wheels to steer the vehicle. For optimal implementation ESC not only initiates braking intervention, but can also interface with the engine control system to accelerate the driven wheels.
- ESC technology includes all the sensors required for ABS plus a minimum of two inertial measurement sensors: one measuring lateral acceleration and the other measuring yaw rate. The ESC correction phase sequence for oversteer and understeer in passenger cars and commercial vehicles is explained.
- The susceptibility of different types of road vehicle to rollover is discussed along with the way in which RSC can recognise conditions that might lead to vehicle rollover, and initiate brake interventions to prevent it as far as possible to provide rollover protection. Different approaches are summarised, and the system used on commercial vehicles is described.
- Other intelligent electronic brake system control features are outlined, including Emergency Brake Assist (EBA), Adaptive Cruise Control (ACC), Collision Mitigation by Braking (CMbB), Electric Parking Brake systems (EPB) and Hill Start Assist (HSA), Trailer Sway Control (TSC), Torque Vectoring by Braking (TVbB), Engine Drag Control (EDC), Hill Descent Control (HDC) and ESC mode switching.
- An overview of the principles and technologies of regenerative braking is presented, and the continuing need for friction brakes on road vehicles especially for emergency braking is explained. Brake system technologies that could be suitable for vehicles with regenerative braking are listed.
- An example of how the work done by the brakes of a road vehicle can be predicted on flat roads is briefly described. Strategies for the control of non-friction braking (including retarders) are discussed in the context of the need to 'blend' the operation of

two different types of braking systems. The need to prioritise the regenerative braking over the friction braking to maximise energy recovery, while maintaining braking stability, is explained. The importance of creating a blended braking system so that the driver is satisfied in terms of brake system response and pedal feel is discussed. The importance of cooperation between regenerative braking control systems and stability control systems such as ABS and ESC is highlighted.

- Finally, the important function of system warnings and driver interfaces with electronic braking is described, with examples of visual and audible display signals.

References

Bosch, 1993. Automotive Handbook, 3rd edn. SAE. ISBN 1 56091 372 X.
Curry, E., 2013. Braking of Road Vehicles 2013. University of Bradford. ISBN 978 1 85143 272 1.
Meise, P., 2009. Energy flow in regenerative braking for road vehicles, PhD thesis. University of Bradford.
Miller, J.I., Cebon, D., 2010. A high performance pneumatic braking system for heavy vehicles. Vehicle System Dynamics: Int. J. Vehicle Mechanics Mobility 48, 373–392.
Moore, I., 2013. Braking of Road Vehicles 2013. University of Bradford. ISBN 978 1 85143 272 1.
Ross, C.F., 2013. Braking of Road Vehicles 2013. University of Bradford. ISBN 978 1 85143 272 1.

CHAPTER 12

Case Studies in the Braking of Road Vehicles

Introduction

The chapters in this book have presented some of the basic theory and practice of road vehicle braking, and some examples of design calculations have been included to demonstrate how the theory is applied. The purpose of this chapter is to present and discuss some examples that illustrate firstly an experimental verification of a braking system design by comparison of measured braking performance (in terms of deceleration vs. actuation hydraulic line pressure for the basic braking system of an example passenger car), and subsequently how and why the actual performance of brakes and braking systems can vary from the design performance. As explained in previous chapters and referring particularly to Figure 9.3, the input to a vehicle braking system is the driver effort, i.e. the force applied to the brake pedal, and the desired response is retardation force to decelerate the vehicle. Drivers expect consistent and reliable performance from the braking system, therefore variations from the design performance can be regarded as failure modes, and as indicated in Figure 9.3 these can include:

- Retardation force too low
- Retardation force too high
- Variable retardation force
- Unplanned or unintended retardation force.

Failure modes that result from actuation system failure, e.g. an air or hydraulic system leak, are legislated for in the system partial failure design performance and are not considered further here.

Brake System Design Verification

In Chapters 3 and 4 the theory relating to the braking dynamics of rigid vehicles with two or more axles and vehicle/trailer combinations with more than two axles was presented, and examples were shown to illustrate braking performance parameters such as efficiency and adhesion utilisation. In Chapter 6 braking system design and actuation layout was explained and examples were presented for a passenger car (specification in Table 6.2) and a commercial vehicle (specification in Table 4.6). In Chapter 9 methods of experimental

Braking of Road Vehicles.

Table 12.1: Passenger Car Design Specification (Marshall, 2007)

Design Parameter	Specification	
Loading Condition:	**DoW**	**Test Weight (Partially Laden)**
Wheelbase, E (mm)	2754	2754
Vehicle mass (kg)	1486	1710
Vehicle weight (N)	14,575	16,780
Dynamic radius of tyres (mm)	307	307
Centre of gravity height (mm)	565	570
Position of centre of gravity behind the front axle, L_1 (mm)	1132	1243
Position of centre of gravity in front of the rear axle, L_2 (mm)	1622	1511
Front axle weight, N_1 (N)	8585	9205
Rear axle weight, N_2 (N)	5990	7575
GVW (N)	18,394	18,394
Vehicle maximum speed (km/h)	210	210

brake testing were presented, describing how the braking performance of a road vehicle could be evaluated. An example is presented here (Marshall, 2007) to illustrate the verification of a passenger car braking system design by comparing the calculated (predicted) braking performance with that measured experimentally on the vehicle. The vehicle design specification is summarised in Table 12.1 and the brake system design data are summarised in Table 12.2.

The results from brake performance tests carried out on the vehicle are summarised in Table 12.3 (Marshall, 2007). The test procedure was as described in Chapter 9, and the

Table 12.2: Brake System Design Data (Marshall, 2007)

	Specification	
Design Parameter	**Front – Single Piston Sliding Caliper Disc Brake**	**Rear – Single Piston Sliding Combined Caliper Disc Brake**
Caliper piston diameter (mm)	57	38
Threshold pressure (bar)	0.75	1.5
Pad friction coefficient	0.4	0.38
Effective radius of rotor (mm)	125	119
Actuation system efficiency (η)	0.95	0.95

Table 12.3: Results from Brake Performance Tests Carried Out on the Vehicle Compared with Calculated Values Based on Nominal μ

Actuation Line Pressure (MPa)	Rate of Braking (Measured)	Rate of Braking (Calculated)	Comments
All Brakes (ABS/EBD Disabled)			
1.50	0.175	0.195	
3.00	0.375	0.403	
4.00	0.500	0.542	
5.00	0.650	0.681	
5.75	0.725	0.785	
6.50	0.850	0.889	Rear wheels locked just before the end
7.25	0.825	0.994	Rear wheels locked immediately, front wheels then locked
8.00	0.600	1.098	All wheels locked for the entire application
Front Brakes Only (Rear Brakes and ABS/EBD Disabled)			
1.50	0.150	0.142	
3.00	0.250	0.290	
4.00	0.375	0.389	
5.00	0.450	0.488	
6.00	0.550	0.587	
6.50	0.600	0.637	
7.00	0.600	0.686	One front wheel locked just before the end
7.50	0.500	0.736	Both front wheels locked early in the application
8.00	0.450	0.785	Both front wheels locked throughout
Rear Brakes Only (Front Brakes and ABS/EBD Disabled)			
1.50	0.050	0.054	
3.00	0.100	0.113	
4.50	0.150	0.173	
6.00	0.225	0.233	
7.50	0.275	0.293	
9.00	0.350	0.352	Both rear wheels locked just before the end
10.50	0.250	0.412	Both rear wheels locked early in the application
12.00	0.225	0.472	Both rear wheels locked throughout

measured brake performance of this vehicle was presented in Figure 9.6. The measured brake performance, although close, is slightly lower than the calculated values, as illustrated in Figure 12.1 in the region where wheel lock does not occur. The design calculation does not include the tyre/road coefficient of adhesion so the effect of wheel lock has not been covered. The reasons for the difference between the calculated and measured values could include:

- The assumed actuation system efficiency is too high (estimated at 95%).
- The coefficient of friction μ between the brake pads and discs may be different (lower) than the design values (0.4 front, 0.38 rear) because of heat generation and temperature.
- There may be experimental error, e.g. arising from wind or incline (which cannot always be excluded even with repeat testing in opposite directions).
- Calibration of the instrumentation may be imperfect.

These experimental data and analyses can also be used to estimate the tyre/road coefficient of adhesion at the time of the test by assuming that the maximum rate of braking when the brakes on one axle only are operating occurs at the limit of adhesion. Figure 12.1 indicates that front wheel lock (FWL) occurred at 7 MPa and because this data point lies well away from the linear characteristic it can be assumed that the maximum deceleration occurred at

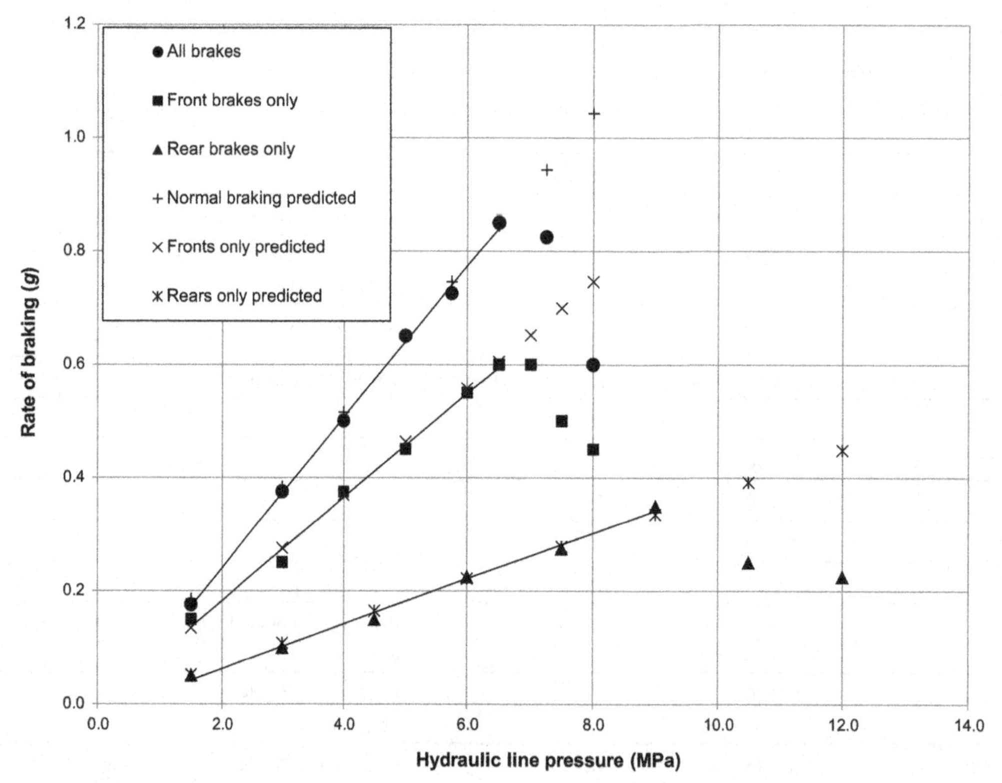

Figure 12.1: Comparison of Measured and Calculated Brake Performance (see Table 12.3).

about 6.5 MPa, when the predicted rate of braking is $z = 0.605$. Similarly rear wheel lock (RWL) occurred at 9 MPa but this data point lies on the linear characteristic; therefore it can be assumed that the maximum deceleration occurred at this actuation pressure, for which the predicted rate of braking is $z = 0.335$. The calculated values of adhesion utilisation are $f_1 = 0.90$ (at the front wheels) and $f_2 = 0.87$ (at the front wheels), which therefore indicate a tyre/road coefficient of adhesion of around 0.9, assuming it is the same for the front and rear wheels (which is not necessarily the case). This result is consistent with good quality conventional passenger car tyres on a good quality dry road surface, which was representative of the actual test conditions.

On this vehicle the actuation system design data was:

- Pedal ratio = 3.26:1
- Master cylinder piston diameter = 23.89 mm
- Boost ratio = 4.5:1.

Using Equation (6.21c), the driver effort required to achieve $z = 0.25$ in the 'servo-fail' condition at GVW (= 18,394 N) would be approximately 300 N.

Braking Performance Variation

In the previous section the measured performance of a vehicle braking system was shown to be close to the design performance. The measurement of the vehicle's braking performance was made under carefully controlled test conditions in the 'as new' condition. It is good practice to quantify the magnitude of any such variation and evaluate its effects on the vehicle's braking performance at the design stage so that top and bottom limits to the design operation can be set. Working to a $\pm 10\%$ tolerance on friction materials (see Chapter 2) the braking performance range is illustrated in Figure 12.2. In

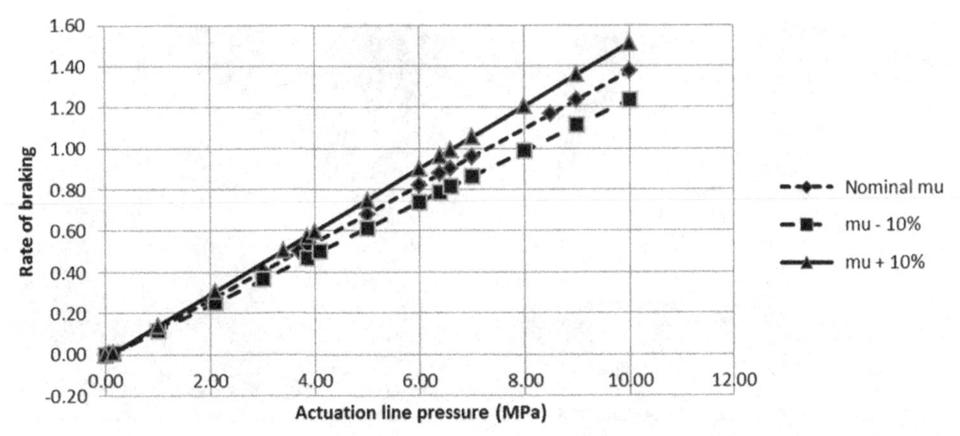

Figure 12.2: Braking Performance Range at 10% μ.

comparison with the data shown in Table 12.3 and Figure 12.1, the hydraulic actuation line pressure required to generate $z = 0.5$ (at the vehicle test weight) ranges from 3.4 to 4.1 MPa, while the pedal effort ranges from 110 to 130 N. At GVW, the driver effort required to achieve $z = 0.25$ in the 'servo-fail' condition with μ at the lower limit would be approximately 330 N.

There are many reasons why the braking performance of a road vehicle may change over time, and any deterioration would be recognised by the driver as a reduction in the perceived rate of braking for the same pedal effort on a hydraulically actuated system. On a commercial vehicle with pneumatic brake actuation, the driver is more likely to notice an increase in stopping distance for the same pedal operation. Changes in the braking performance over time may be caused by an increase in threshold forces, perhaps also associated with a decrease in actuation efficiency, caused by ageing and contamination of seals, and corrosion and contamination at interfaces such as disc brake pad abutments, drum brake shoe pivots and mechanical actuators. Changes in braking performance are often attributed to the friction material; as explained in Chapter 2, the coefficient of friction (μ) between a friction material and a rotor does vary, especially with respect to temperature, but most modern resin-bonded composite friction materials demonstrate fairly consistent performance in operation unless duty levels and the associated operating temperatures are high for a significant proportion of the vehicle usage. The following case studies are presented to illustrate the complexity of possible causes of brake performance variation.

Figure 12.3 shows photographic images of one of a pair of badly corroded brake discs taken from the rear axle of a sports utility vehicle (SUV) on which a thick layer of rust had built up over a region towards the inner radius of the rubbing path on the inner face of the brake disc. This layer was so thick that the contact between the brake pad and the disc was concentrated on the rust layer with the cumulative deleterious effects of reducing the

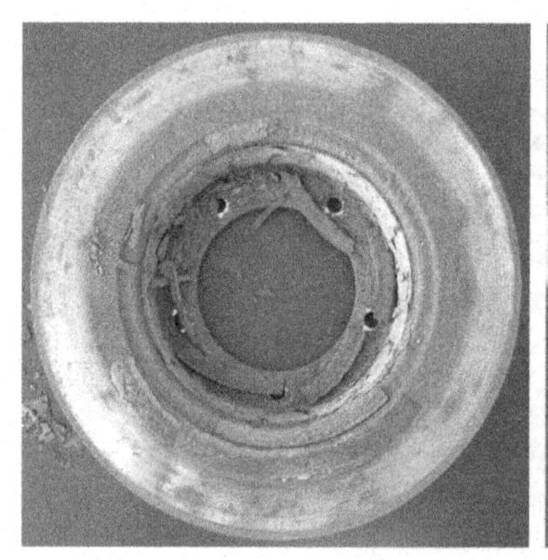

Figure 12.3: Corroded Brake Disc.

effective radius of the disc and reducing the effective value of μ because the friction material was operating against rust rather than the designed disc material. High brake pad temperatures would also have been generated because of the reduced heat transfer into the disc through the substantial thermal barrier presented by the rust layer. The reported effect was a long-term reduction in the vehicle braking performance, excessive wear of the front brake pads and the recent occurrence of brake judder. The problem was attributed to the quality of a new set of front brake pads fitted to replace those that were worn out; after their fitment the poor braking performance continued and judder developed. A full investigation of the vehicle's brakes identified the bad condition of the inner face of the rear brake discs, which was hidden by the dust shields; the outer rubbing paths of the rear discs were in reasonably good condition and the pads were not worn to the extent that it had been considered necessary to replace them. As can be seen, the rust layer had started to detach causing the judder, and the increased wear of the front brake pads was the result of the increased demand on the front brakes because of the reduced torque generated by the rear brakes. This example demonstrates the importance of checking all parts of the braking system diligently, as well as not jumping to conclusions about the cause of the problem.

Another example of brake performance variation over time was reported by Day and Harding (1983) relating to an S-cam drum brake as used on an articulated vehicle semi-trailer. Figure 12.4 shows the specific torque of a 420 mm × 180 mm S-cam drum brake measured on an inertia dynamometer over nearly 4000 brake applications, to investigate the cause of brake performance variation on semi-trailers reported by operators in the field (4000 brake applications were estimated as possibly representing as much as 32,000 km on long-distance haulage operation). The first 800 brake applications were bedding-in, and

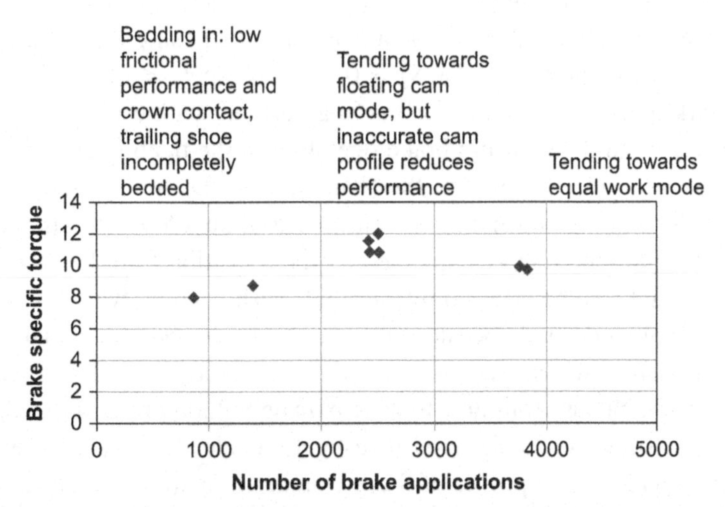

Figure 12.4: Measured S-cam Drum Brake Specific Torque vs. Time (Day and Harding, 1983).

it was noted that at the end of the 4000 brake applications, bedding of the trailing shoe was still incomplete suggesting that the brake performance had still not reached steady-state operation.

The specific torque of an S-cam brake was defined in Equation (5.30) as brake torque/actuation torque ($\tau_s = \tau/\tau_a$). The design features and operation of an S-cam drum brake, and the calculation of the brake factor and specific torque of an S-cam drum brake, were explained in Chapter 5, and Table 5.5 demonstrated how the specific torque could vary depending upon the frictional losses in the rotation of the actuator mechanism, i.e. the camshaft. As explained in Chapter 5, the contact and pressure distribution between a brake lining mounted on a brake shoe and the brake drum affects the shoe factor; 'crown' contact (when the rubbing radius of the lining is marginally less than that of the drum) reduces the shoe factor, and 'heel-and-toe' contact (when the lining rubbing radius is marginally greater than that of the drum) increases the shoe factor. The wear of the friction material always tends towards a condition of uniform pressure over the friction surface, although different actuation forces and operating temperatures mean that there is always variation about the uniform pressure condition, which is one source of torque variation in drum brakes. However, when a drum brake is new, it is good practice for the brake linings to be machined to give a condition of slight 'crown' contact (typically no more than 1 mm in 200 mm radius) so that shoe flexure and lining compression tend to create a near uniform pressure distribution to avoid localised high pressure and heat generation, which can lead to thermal damage to the drum.

As also explained in Chapter 5, S-cam drum brakes operate in a 'floating cam' mode when new, and tend towards an 'equal work' mode as bedding-in progresses. Following the derivation of specific torque (τ_s) for the 'equal work' mode of operation for this brake discussed in Chapter 5, the 'floating cam' specific torque where the actuation force applied to both shoes is the same ($P_{a1} = P_{a2} = P_a$) can be calculated using Equation (5.38) for the camshaft actuation torque and $S_1 = 1.46$, $S_2 = 0.45$, yielding $\tau_s = 15.3$. This compares with the equal work mode value of 11.0 and is in agreement with established knowledge that the brake factor of an S-cam drum brake operating in the floating cam mode is higher than in the equal work mode. The part of the characteristic shown in Figure 12.4 that demonstrates that the brake performance decreases with time after approximately 2000 brake applications could therefore be explained by the transition from floating cam to equal work mode. This would also be affected by camshaft bush friction, which was most significant in the equal work mode because of the unbalanced shoe tip actuation forces, and air actuator efficiency, which was not included in the predictions. However, the initial low performance of the brake, from new to approximately 2000 brake applications, was not explained by this transition, so the authors' investigation extended to the S-cam actuation mechanism. They found that the profile of the S-cam actuator was inaccurate over the initial part of its rotation (equivalent to a change from $r_b = 14.32$ mm to $r_b = 13.10$ as designed), which meant that τ_s was predicted as 13.3 in the 'new' condition. Whilst this does not completely explain the initial low performance of this brake, it does demonstrate how brake performance can be affected by several different parameters. In the original published work (Day and Harding, 1983), it was stated 'in practice, therefore, a combination of initial crown contact and lower unbedded friction is likely to obscure the expected high performance of the brake operating in the floating cam mode before bedding-in is complete.'

One of the basic design checks that must be completed for drum brakes is for 'sprag' of the leading shoe (see Chapter 5, Equation (5.23a)). The occurrence of sprag mentioned in Chapter 5 in the context of drum brake lining contact and pressure distributions was one of two instabilities noted by Harding and Day (1981), both of which were associated with the disposition of the friction material linings and the parking brake mechanism of a typical simplex drum brake fitted to the rear axle of a passenger car. It was a sliding parallel abutment design on which the friction material linings were disposed towards the trailing ends of the brake shoes in the forward direction of rotation to achieve a more uniform wear pattern and increase the life of the linings. Forward instability could occur when the parking brake was actuated while the vehicle was moving; the conditions of contact between the lining and the brake drum appeared to cause the leading end of the trailing shoe to lift off the abutment, turning it into a leading shoe with the parking brake actuator strut acting as the abutment. Because the strut was connected to the other (leading) brake shoe, the effect was to switch the brake into a duo-servo mode, with consequent high performance. Releasing the parking brake was found to release the brake except where weaker pull-off springs had been fitted, when the brake could not be released without reversing, and this constituted full sprag where the brake could not be released by removing the actuation force. The transition from normal operation to sprag-type operation was associated with uncontrollable rear wheel lock and a force impulse that was characterised by plastic deformation of the brake shoe at the point of contact between it and the parking brake actuator, and was often associated with permanent damage to the brake abutments and backplate.

A reverse instability was also studied, which although also associated with the parking brake mechanism, was found to have a different cause although the end result was very similar in terms of a locked brake. The lining arc length on the leading shoe had been reduced from 110° to 90° arc length (perhaps to reduce cost because on this design of brake the lining wear can be tapered towards the trailing end and the wear at the leading end can be negligible). The disposition of the 90° arc length lining towards the trailing end of the leading shoe created over-performance in reverse rotation, which displayed the characteristics of sprag and occurred more frequently with time. This instability was found to be initiated by a wedging effect as the shoe slid across the abutments under reverse rotation, being constrained only by a small amount of friction material, which was subject to very high compressive and shear forces that eventually caused it to crack away. At this point the movement of the brake shoe was constrained only by mechanical contact between the parking brake actuator strut and the hydraulic cylinder.

It was mentioned in Chapter 5 that a disc brake pad backplate must always present a clearance fit between the caliper abutment features under all operating conditions. If this clearance fit is too tight, the pad assembly will stick between the caliper abutments, reduce or even prevent actuation of the pad, and possibly exert a significant normal force on the disc even when the actuation force is released. The resulting residual torque from a dragging brake will continue to heat up the pad assembly, expand the backplate and jam it between the caliper abutments more firmly, and the problem may be further exacerbated by corrosion and contamination. There is potential for problems of tolerance stack-up between the pad backplate and the caliper abutment gap; too much clearance will permit impact noise, e.g. associated with direction reversals, while too little can lead to restricted movement. Problems have occurred in disc brakes if the pads cannot move freely perpendicular to the disc; these are characterised by erratic (and usually reduced) performance, and are often associated with the occurrence of brake noise. Because there is no positive retraction of the pads (only the seal roll-back and 'knock-back' from any disc runout, which is usually very small), residual brake torque can be generated that heats up the brake even though the driver is not actuating it. The problem can be masked if the vehicle is driven at high speed, e.g. on a motorway, because of the much greater brake cooling at high speed, and can be most noticeable in low-speed driving, e.g. in urban environments where brake cooling is low.

The residual torque from the front disc brake of a passenger car exhibiting the symptoms of a dragging brake was measured under quasi-static conditions at ambient temperature, and found to be approximately 30 Nm. An FE thermomechanical analysis was conducted in which a circumferentially uniform steady-state heat flux input equivalent to a constant torque of 30 Nm at a vehicle speed of 50 km/h was applied to the rubbing surfaces of the friction ring of a simple FE model. A surface heat transfer boundary condition was applied to the exposed surfaces of the disc using values of heat transfer coefficient derived for the specific brake installation from brake cooling curves on an actual vehicle. The calculated steady-state temperature distribution is shown in Figure 12.5. This indicates that very high temperatures, in this case around 600°C, could be reached even under low-torque continuous drag and were confirmed by experimental measurement. The same FE model was used to predict the thermal distortion of the brake disc and indicated that the brake disc coning distortion was 0.5 mm (outwards). This result was verified using test dial indicators on a stationary vehicle with the drive wheels jacked off the ground and driven against a brake torque of 30 Nm. This investigation demonstrated that if the brake pad movement perpendicular to the disc rubbing path were restricted, e.g. by binding between the caliper abutments, even a small amount of residual torque could result in the

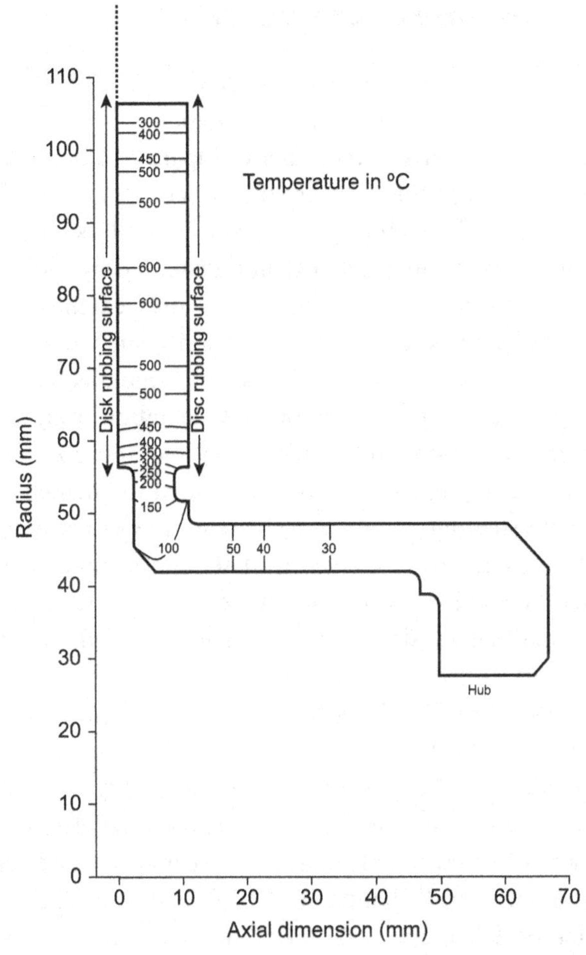

Figure 12.5: Temperature Distribution of a Brake Disc Subject to a Constant 30 Nm Drag Torque at 50 km/h.

generation of very high temperatures and consequent thermal expansion and distortion of the pad backplate and the brake disc that could exacerbate the effect still further. This was a real problem reported by customers in terms of high brake temperatures (with an associated smell of burning), erratic braking performance, excessive friction material degradation and wear, and loss of braking function from brake fluid vaporisation. All the drivers who reported the problem claimed that their use of the car was limited to local journeys in city-type traffic. The immediate solution (containment action) was to control the tolerances relating to the abutment aperture in the caliper and the length of the pad backplates. Longer term solutions included improved disc design to reduce thermal coning distortion and better cooling of the car's front brakes at low speed.

Interaction Between the Brakes and the Vehicle

There are many interactions that occur between a vehicle and its brakes, either by structural connections, e.g. the suspension, or the actuation system, or by the transmission of heat and noise to the vehicle's driver and other occupants. Brake judder, for example, has been shown to be more of a problem where the effect is transmitted to the driver (and passengers), e.g. by steering shake, cabin shake or noise, and less where the judder at the brake is just the same but is attenuated rather than transmitted by the vehicle's structure. The consistency of a vehicle's braking performance not only extends to longitudinal braking performance, but also to lateral stability and consistency. One overriding expectation that is common to all drivers of all types of vehicle is that they expect to be able to achieve any required deceleration without any deviation from their intended path of travel. A significant lateral deviation while braking that requires considerable corrective action is usually termed 'brake pull' and is often attributed to unequal side-to-side braking forces, which interact with steering geometry to increase the yaw moment instead of decreasing it (this is the reason why negative offset steering geometry is preferred). A small amount of lateral deviation during braking, e.g. when the driver's application of the brakes causes very small lateral deviations that may vary in direction during the braking event, is referred to as 'steering drift'. Brake pull is potentially dangerous and is seldom a problem on modern road vehicles unless associated with a significant failure. Minor deviations associated with steering drift during braking are still known to occur but remain unacceptable by today's standards of vehicle driveability. The causes of 'brake pull' are well understood, and include side-to-side brake torque differences caused by, for example, poor adjustment of operating clearances between stator and rotor (automatic adjusters are mandatory), large brake temperature differentials (e.g. if a brake is dragging), contamination, corrosion or damage to one brake, and even different manufacturing techniques or materials (e.g. trace elements such as titanium and vanadium in the rotor cast iron (Chapman and Hatch, 1976).

Four parameters associated with any road vehicle's steering geometry that can affect steering drift during braking are 'toe', 'camber', 'caster' and 'steering offset' (also known as 'scrub radius'). Poor adjustment of the toe setting can cause unstable straight-line driveability and a steering reaction is generated if the toe steer angles are different left to right. The amount of toe-in or toe-out can also depend on the suspension position relative to the steering gear and in particular can change in opposite directions during body roll, thereby affecting the vehicle's directional stability. Wheel camber generates camber steer: positive camber will make the wheels toe-out while negative camber will make the wheels toe-in. Caster provides stability by the self-alignment of the steered wheels; decreasing caster will reduce directional stability and vice versa. Steering offset is the geometric distance between the centre line of the tyre contact patch and the steering axis at the ground plane (see Figure 5.2). It is known that dynamic changes to these parameters during operation can result from the forces generated by the application of the brakes, and thus may affect a vehicle's 'straight-ahead' directional stability unless the design is robust to the changes. Tyres also contribute to road vehicle stability during braking; lateral force and self-aligning torque arising from pneumatic trail determine the effective slip angle and total axle lateral forces. A different lateral stiffness between left and right tyres will lead to a permanent steering drift, independent of braking. Camber of the road surface creates a sideways gravitational force component, which also leads to a permanent steering drift.

The kinematics of vehicle suspension systems can introduce 'compliance steer' because lateral or longitudinal forces at the tyre contact patch deflect the suspension bushings and change the camber and toe angles (Momoiyama and Miyazaki, 1993). Compliance exists in all road vehicle suspension systems because they may include elastomeric suspension pivot bushes, cross-member mounts, steering rackmounts, and the elastic deformation under the application of braking forces of components such as suspension and steering links, and subframes. Compliance in the suspension system is viewed as essential to achieve a good ride characteristic, but compliance steer is considered to be one of the biggest contributors to straight-line stability during braking. As an example, when the steering control arms move because of elastomeric bush deflection a change in steering angle can result, and if this is different side to side, the vehicle can steer to one side while braking.

Klaps and Day (2003) reported a series of experiments to investigate steering drift during braking on a 1.2 km straight test track. The car had a MacPherson strut design of front suspension, in which a lower suspension arm provided lateral and longitudinal location (Figure 12.6), and was known to demonstrate a small amount of steering drift during braking. The following parameters were investigated:

A. Tyre and brake temperature
B. Suspension geometry toe steer curve
C. Steering gear housing/engine subframe reinforcement cover

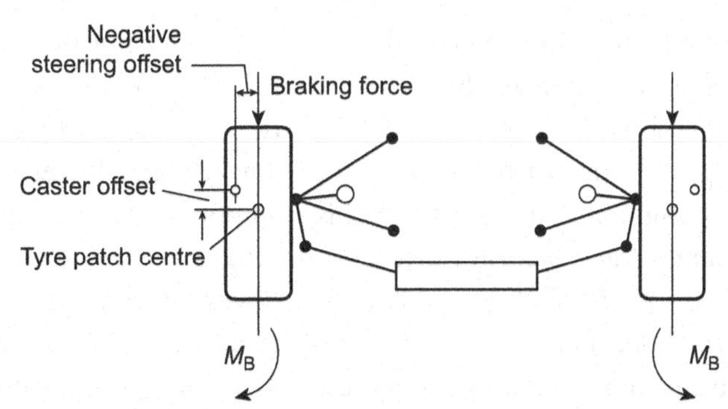

Figure 12.6: Test Car Steering Geometry (Klaps and Day, 2005).

D. Front suspension lower wishbone rear bush stiffness
E. Wheel offset (y-direction).

Side-to-side variation of braking force (e.g. from different brake discs or friction materials) was not included since this was a known effect. The measured response was the lateral displacement, measured by the number of carriageway lanes moved from the straight-ahead position during the braking manoeuvre. The test procedure comprised brake applications to decelerate the vehicle to rest in neutral gear from 100 km/h at a set deceleration with the steering wheel 'free' (hands off). The deviation from straight-line braking was measured, and the procedure was repeated five times. The results are summarised in Table 12.4.

Having established that the suspension compliance and steering offset had a significant effect on steering drift during braking, further vehicle tests were carried out to investigate how steering drift during braking could be reduced or eliminated. Steering drift during braking to the left occurred as shown in Figure 12.7; under fixed control (the driver held the steering wheel in the straight-ahead position) the yaw velocity initially increased and then decreased towards the end of the deceleration, but remained positive throughout, indicating a continuous drift to the left. Under free control (hands off the steering wheel) the yaw velocity characteristic also remained positive throughout, showing an initial increase, then a sharp decrease and then an increase before decreasing towards the end of the deceleration. Again this represented a drift to the left, but was less continuous. The brake pressures were measured and found to be higher at the left front wheel than at the right front wheel, but when the brake pipe connections were swapped from left to right, the steering drift to the left remained.

As indicated in Table 12.4, the effect of the front suspension lower wishbone rear bush stiffness was found to be most significant so a stiffer bush was designed, manufactured and fitted, which generated significantly reduced steering drift during braking. Figure 12.8(a) and (b) illustrate the improvement in terms of toe steer changes during the braking event from the original to the new design.

From the work reported, the major cause of steering drift during braking was found to be side-to-side dynamic variation in the deformation and deflection of suspension and steering components and not any form of side-to-side variation in braking forces. The most significant effect was the stiffness of the rear bush in the lower suspension arm. The deflections and compliance arising from this component generated large changes in the side-to-side camber and steering offset, changing toe steer angles during braking, which actually reversed during the braking event. Changing to a stiffer rear bush in the lower wishbone minimised the deflection and controlled the wheel orientation better during braking. It was concluded that the most effective means of

Table 12.4: Results of Initial Investigation of Steering Drift During Braking (Klaps and Day, 2003)

Parameter		Effect
A	Tyre and brake temperature	Insignificant
B	Suspension geometry toe steer curve	Insignificant
C	Steering gear housing/engine subframe reinforced cover	Significant negative effect. Best suitable level is + (with perimeter frame reinforcement cover)
D	Front suspension lower wishbone rear bush stiffness	Most significant effect. D+ (voided rear bush) gave smallest steering drift
E	Wheel offset	Second most significant effect
A × B	Tyre and brake temperature/suspension geometry toe steer curve interaction	Insignificant
A × C	Tyre and brake temperature/steering gear housing/engine subframe reinforced cover interaction	Insignificant
A × D	Tyre and brake temperature/front suspension lower wishbone rear bush stiffness interaction	Minor interaction effect. When 'D' and 'A' were +, A × D became + as well, which meant a reduction of the steering drift
A × E	Tyre and brake temperature/wheel offset interaction	Insignificant
B × C	Suspension geometry toe steer curve/engine subframe reinforced cover interaction	Insignificant
B × D	Suspension geometry toe steer curve/front suspension lower wishbone rear bush stiffness interaction	Insignificant
B × E	Suspension geometry toe steer curve/wheel offset interaction	Insignificant
C × D	Steering gear housing/engine subframe reinforced cover/front suspension lower wishbone rear bush stiffness interaction	Insignificant
C × E	Steering gear housing engine subframe reinforced cover/wheel offset interaction	Insignificant
D × E	Front suspension lower wishbone rear bush stiffness/wheel offset interaction	Largest significant interaction effect. When D is + and E is +, then the interaction also becomes +. This has a negative effect on the response (greater steering drift)

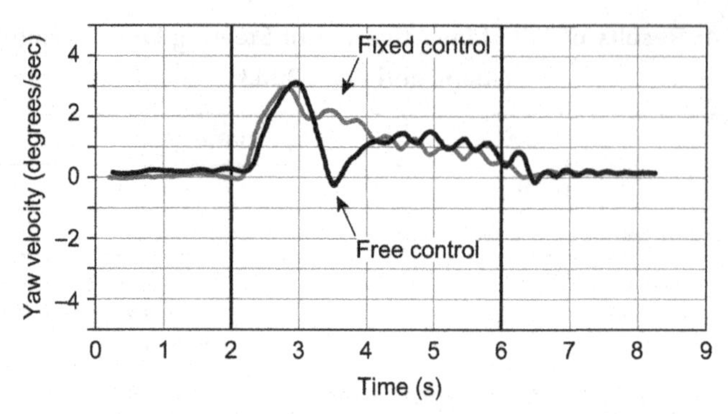

Figure 12.7: Typical Yaw Velocity Measurements, Fixed and Free Control (Klaps and Day, 2005).

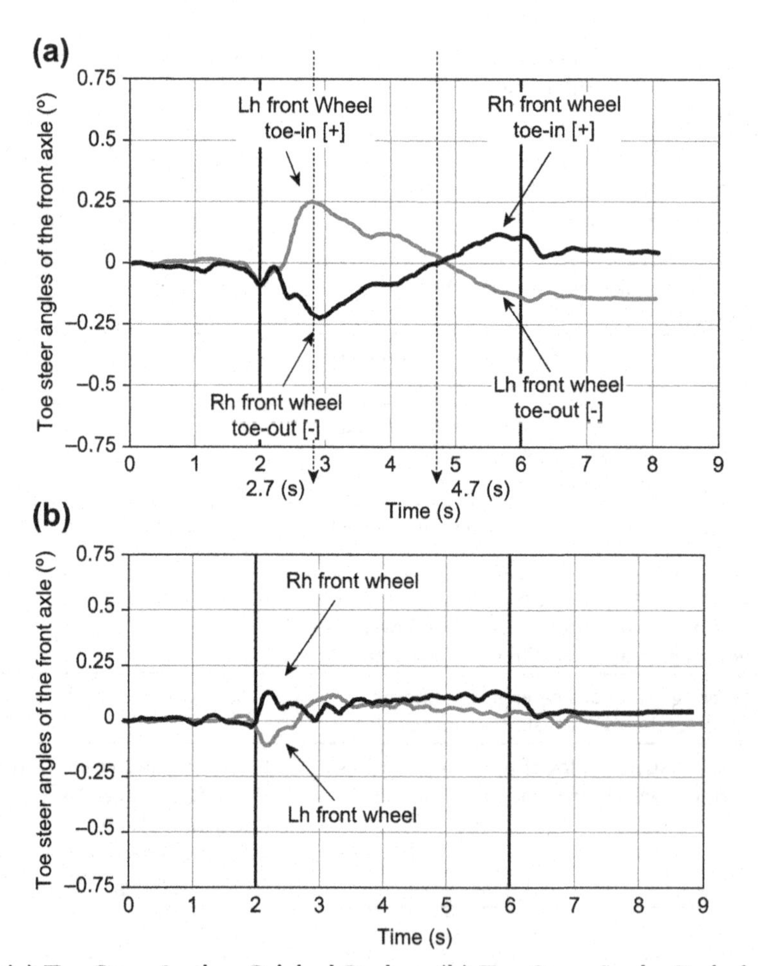

Figure 12.8: (a) Toe Steer Angles, Original Bushes. (b) Toe Steer Angle, Redesigned (Stiffer) Bushes (Klaps and Day, 2005).

controlling any tendency towards steering drift during braking is to ensure, in the design of the steering and suspension system, that there is minimum side-to-side variation in suspension deflection and body deformation both statically and dynamically.

Mixed-Mode Braking Systems

Road vehicles that are fitted with other types of braking systems such as endurance braking (retarders) and regenerative braking in addition to conventional friction foundation brakes are called 'mixed-mode' braking systems. They need to have actuation control systems that integrate or blend the function of each type to provide consistent and reliable braking under all conditions of operation. As stated in Chapter 10, the challenges of blending regenerative braking with friction brakes so that brake operation is completely transparent to the driver are considerable. The basic design of a mixed-mode braking system needs to start with the same considerations of adhesion utilisation and braking distribution between the axles of the vehicle, as were introduced in Chapters 3 and 4 of this book, and an example based on a passenger car vehicle with regenerative braking is introduced here.

The total braking force on a two-axle rigid road vehicle is defined by Equation (3.16), assuming no lateral variation:

$$Pz = T_1 + T_2 \tag{3.16}$$

In a mixed-mode braking system the brake forces may comprise a friction brake force component and a 'second mode' brake force component, e.g. from a retarder or from regenerative braking, so Equation (3.16) can be rewritten as:

$$Pz = \left(T_{1fb} + T_{1rb}\right) + \left(T_{2fb} + T_{2rb}\right) \tag{12.1}$$

where the subscript '$_{fb}$' refers to friction brake force and '$_{rb}$' refers to regenerative brake force. As explained in Chapter 3, the ratio of the braking force generated by the front wheels to the braking force generated by the rear wheels of a two-axle rigid road vehicle is defined in Equation (3.19a) as the ratio X_1/X_2, where X_1 and X_2 are the proportion of the vehicle's total braking force generated at the front and rear axles respectively. Therefore:

$$\left(T_{1fb} + T_{1rb}\right)/\left(T_{2fb} + T_{2rb}\right) = X_1/X_2, \text{ and } X_1 + X_2 = 1 \tag{12.2}$$

In many designs of hybrid (HEV) or electric (FEV) vehicles, regenerative braking operates only on the drive axle, so for an FWD car, only friction braking applies at the rear wheels, i.e. $T_{2rb} = 0$. Despite having two drive axles, some 4WD vehicles also apply regenerative braking only on one axle, e.g. the rear axle, in which case only friction braking applies at the front wheels, i.e. $T_{1rb} = 0$.

From a basic braking stability point of view, the limiting rate of braking (z) for a specified adhesion coefficient (k) is still defined by the vehicle's dimensions as in Equations (3.26) and (3.27), and adhesion utilisation (Equation (3.40)) is defined in the same way. Thus, the design of a road vehicle's braking system can still follow the procedure set out in Chapters 3 and 4, which uses considerations of adhesion utilisation to ensure that a good basic design is achieved, provided that all modes of braking (as appropriate) are included in the calculation of brake forces at each axle.

$$f_i = \left(T_{ifb} + T_{irb}\right)/N_i \tag{3.40}$$

A major challenge with regenerative braking is that the brake force developed by the regenerative braking system is highly variable, e.g. it depends upon road speed and the capacity of the energy storage system to receive and hold the recuperated energy (e.g. the state of charge of the batteries of an HEV). Therefore, in operation, the vehicle's brake control system must blend the brake force from the regenerative braking system with the brake force from the friction brakes so that the total brake force at each wheel creates the total vehicle braking force, which matches the driver's demand. The detail of brake blending control systems is beyond the scope of this book, but the principle is illustrated in the following example of an HEV as specified in Table 12.5. This braking system has been designed to provide full braking capability without regenerative braking, thereby meeting the requirements of UN Regulation 13H (see Chapter 8). The design value of z_{crit} in the unladen (driver-only) condition is approximately 0.63.

This vehicle has regenerative braking on the rear axle only, i.e. $T_{1rb} = 0$, provided by the 60 kW electric motor/generator, and for the purposes of this example 60 kW can be assumed to be the maximum available regenerative braking power $\dot{Q}_{\rm rb}$ (ignoring losses and inefficiencies in the system). The regenerative braking torque at the rear axle is calculated from Equation (12.3), and the torque–road speed characteristic is shown in Figure 12.9.

$$\tau_{2rb} = \dot{Q}_{rb}/\omega \tag{12.3}$$

The brake torque shown in Figure 12.9 represents the maximum regenerative braking torque (ignoring losses and inefficiencies in the system) available at any road speed. It is clear that the regenerative brake torque and hence brake force is lowest at high road speeds and greatest at low road speeds, although regenerative braking is normally disabled at very low speeds. In practice, electric motor/generators are torque limited and speed limited to avoid damage, and their operational power rating is usually specified in terms of transient, intermittent and continuous power modes, the highest being transient mode. The actual regenerative brake torque that can be applied by the blended braking system controller depends upon the state of charge of the electric power system batteries, and the braking stability of the vehicle. If the batteries can accept no further charge, the regenerative brake

Table 12.5: Vehicle and Braking System Data

Electric Vehicle and Braking System Data	
Wheelbase	2700 mm
Position of centre of gravity (behind front axle)	1444 mm (driver only) 1512 mm (GVW)
Height of centre of gravity	690 mm (driver only) 750 mm (GVW)
Effective radius of tyres	355 mm
Unladen mass	2192 kg
Front axle static weight	10,000 N (driver only) 11,000 N (GVW)
Rear axle static weight	11,200 N (driver only) 14,000 N (GVW)
Maximum speed	140 km/h
Brake data	
Front Disc Brakes – Single Piston Sliding Caliper	
Piston diameter	59 mm
Threshold pressure	0.5 bar
Pad friction coefficient	0.40
Effective radius of rotor	160 mm
Rear Disc Brakes – Single Piston Sliding Combined Caliper	
Piston diameter	47 mm
Threshold pressure	0.8 bar
Pad friction coefficient	0.38
Effective radius of rotor	160 mm

force will be zero and the friction brakes have to provide 100% of the required braking force. In practice there may be control strategies that avoid this, e.g. providing a means of dumping excess recuperated energy, but such strategies are not covered here.

At road speeds of 40 and 140 km/h (maximum speed), two extremes of mixed-mode braking system operation have been examined on this example vehicle: no regenerative braking and maximum regenerative braking. The purpose is to illustrate the effect on the vehicle braking performance and the difficulties involved. Using Equations (3.38) and (3.39), the ideal braking ratio (X_1/X_2) to achieve optimum braking (100% braking efficiency) can be shown to vary from 47/53 at $z = 0.1$ to 70/30 at $z = 1$ for the vehicle fully laden (GVW). Using Equation (6.27) the hydraulic actuation pressure required at the

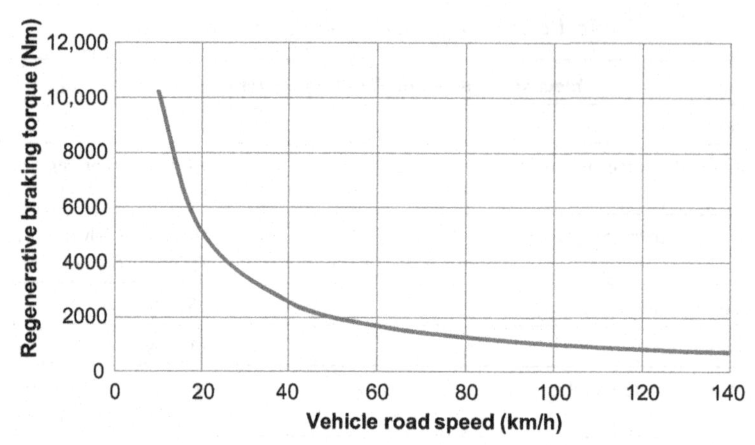

Figure 12.9: Regenerative Brake Torque vs. Road Speed for the Example Vehicle (Rear Axle Only).

front and rear brakes is shown in Figure 12.10, which represents the condition with no regenerative braking. The other extreme (maximum regenerative braking) could be achieved in several ways, but two are considered here to illustrate the options.

First, the regenerative braking on the rear axle of the vehicle could take precedence over front axle braking until the limit of adhesion were approached. At this point the front brakes would be actuated. This would have the advantage of maximising braking energy recovery, but operating the rear wheels at significant levels of braking slip and the front wheels at zero braking slip with the potential for a sudden change being required could

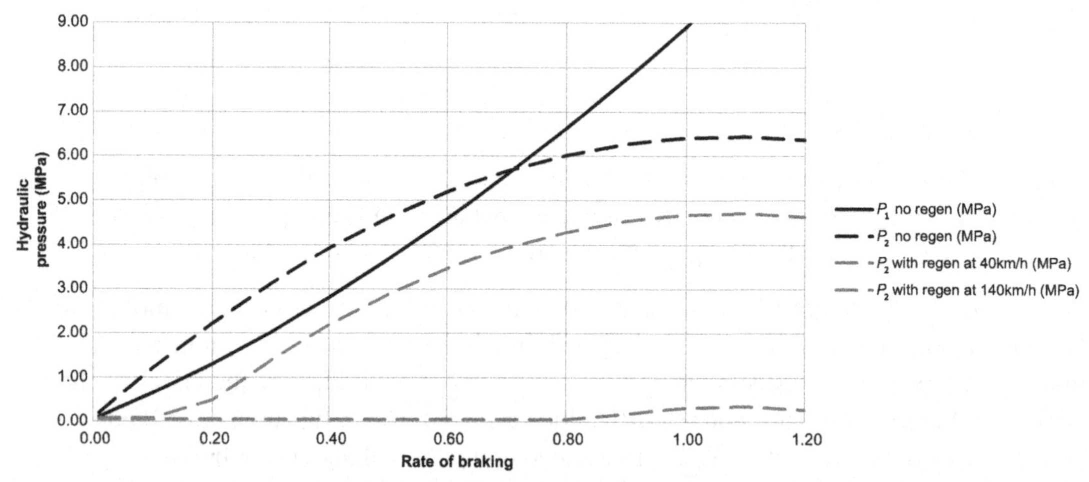

Figure 12.10: Required Hydraulic Actuation Pressure to Achieve Optimum Adhesion Utilisation vs. Rate of Braking (Vehicle at GVW).

introduce handling peculiarities that would probably be unacceptable. The second approach would be to use the regenerative braking to achieve ideal braking, which is also illustrated in Figure 12.10. The regenerative braking contributions at two speeds, 40 km/h (representing low-speed operation) and the vehicle's maximum speed of 140 km/h to represent high-speed operation, are shown. This illustrates that at low speed the friction brakes are not required below a rate of braking of about 0.8, and even at high speed the front brakes define the actuation pressure requirement. Also, because the actuation pressure to the rear brakes is reduced due to the regenerative braking, benefits in terms of fluid consumption and pedal feel may ensue.

It is important to remember always that the limiting factor in all brake system design is the adhesion between the tyre and the road, and the introduction of mixed-mode braking in the form of additional braking technology that operates through the road wheels does not change this in any way. Although this example has been presented in terms of a two-axle passenger car, the same operational and physical principles apply to all road vehicles, including commercial vehicles with endurance braking systems. The main difference between regenerative braking systems and endurance braking systems lies in the way they are designed to be used; the former are designed for intermittent braking duty while the latter are intended for continuous (drag) braking operation. If a regenerative braking system or an endurance braking system were able to take a significant proportion of braking duty from the foundation brakes, then there could be other benefits, e.g. weight reduction by downsizing the friction brakes. However, any attempt to do this has to take into account the braking duty requirements in the event of mixed-mode braking system deactivation (e.g. under ABS or ESC intervention) or failure for any reason. The fundamental requirement of braking safety must never be compromised; road vehicles must be fitted with braking systems that are able to decelerate the vehicle safely, controllably and sufficiently under all conditions of use.

Chapter Summary

- In this chapter some case studies have been presented that aim to illustrate how the theory is applied in actual braking situations, and also provide some discussion of the types of problems that brake and vehicle designers may encounter. The focus is on design verification and performance variation; the latter can be regarded as a 'failure mode' relating to the generated retardation force being too high or low, variable, unplanned, etc. Failure modes that occur as a result of system or part system failure are covered in the earlier chapters as part of the design process and legislation.
- An example of braking system design verification is shown based on a medium-sized passenger car. Braking performance data from experimental tests on the vehicle are presented and compared with the calculated performance using the theory presented in

this book (Chapters 3, 5 and 6). Close agreement is demonstrated, and reasons for the difference between the calculated and measured values are discussed. It is also shown how an estimate of tyre/road adhesion can be made from vehicle brake test data.

- The reasons why the braking performance of a road vehicle may change over time are discussed; any deterioration would be recognised by the driver as a reduction in the perceived rate of braking or an increase in stopping distance for the same pedal operation. In a disc brake installation, variation in the effective radius (R_e) can sometimes be interpreted as a change in the frictional performance of the brake pad friction material. An example is discussed in which heavy corrosion of the inner rubbing path of the rear axle brake discs of a two-axle vehicle reduced the braking torque generated by these rear brakes partly because of a reduction in R_e, and partly because of the presence of a rust layer between the friction material and the cast-iron disc material. The 'knock-on' effect was increased wear of the front brakes and judder.
- In Chapter 5, the effect of contact and pressure distributions between the brake linings and the brake drum of a commercial vehicle S-cam drum brake was explained. Using the S-cam brake theory from Chapter 5, the change in the performance (in terms of the brake factor/specific torque) with use (time) is examined. It is demonstrated that part of the observed variation can be attributed to 'mechanical' effects relating to the distribution of work between the two shoes of the brake, imperfections in the actuation mechanism (the S-cam), and a combination of contact and friction effects between the brake linings and the brake drum before bedding-in is complete.
- Drum brakes are still used on the rear axles of many passenger car and light van type vehicles. One of the basic design checks that must be completed for drum brakes is for 'sprag' of the leading shoe, and an example is presented that explains how a sprag-type failure mode could occur in forward and reverse directions when the brake geometry (specifically the lining arc disposition on the leading and trailing brake shoes) is changed.
- The importance of the correct clearance at the working interfaces of brakes is demonstrated by an example relating to the clearances between a disc pad backplate and the caliper abutments. Tolerance stack-up in the abutment clearances meant that the pad backplate movement under actuation was restricted, resulting in inadequate retraction and residual drag in the 'off-brake' condition. The temperatures generated in the brake disc caused it to cone excessively, exacerbating the problem, resulting in customer complaints of high brake temperatures, erratic braking performance, excessive friction material degradation and wear, and loss of braking function from brake fluid vaporisation.
- The interaction between a vehicle's braking performance and its handling performance is discussed in terms of steering drift during braking. 'Drift' refers to a small amount of lateral deviation during braking (as opposed to 'pull' where the lateral deviation is

large) that can occur on many road vehicles and can be a source of dissatisfaction in customers. The major cause of steering drift during braking was found to be side-to-side dynamic variation in the deformation and deflection of suspension and steering components and not any form of side-to-side variation in braking forces. Changing to a stiffer rear bush in the lower suspension wishbone controlled the wheel orientation better during braking and it was concluded that the most effective means of controlling any tendency towards steering drift during braking is to ensure that there is minimum side-to-side variation in suspension deflection and body deformation both statically and dynamically.

- An example is presented that explains the basic principles and challenges of mixed-mode braking systems, i.e. friction brakes plus another type of braking system. In the example given, regenerative braking is briefly examined in terms of adhesion utilisation, ideal braking distribution and basic control strategies. The same principles apply to endurance braking systems, all of which operate through the road wheels. The example concludes by emphasising the importance of understanding fundamental operational and physical principles, which apply to all road vehicles to ensure that they are fitted with braking systems that are able to decelerate the vehicle safely, controllably and sufficiently under all conditions of use.

References

Chapman, B.J., Hatch, D., 1976. Cast Iron Brake Rotor Metallurgy, paper C35/76, Braking of Road Vehicles Conference, IMechE.

Day, A.J., Harding, P.R.J., 1983. Performance variation of cam-operated drum brakes, Conf. Braking of Road Vehicles. IMechE., 60–77, paper C10/83.

Harding, P.R.J., Day, A.J., 1981. Instability in the hand-brake performance of cars and vans. Proc. IMechE. 195 (No. 27), 325–333.

Klaps, J., Day, A.J., 2003. Steering drift and wheel movement during braking – parameter sensitivity studies. Proc. IMechE. Part D 217, 1107–1115.

Klaps, J., Day, A.J., 2005. Steering drift and wheel movement during braking: static and dynamic measurements. Proc. IMechE. Part D 219, 11–19.

Marshall, P.H., 2007. Car case study, Braking of Road Vehicles short course. University of Bradford, 2007.

Momoiyama, F., Miyazaki, K., 1993. Compliance steer and road holding of rigid rear axle for enhancing the running straightness of large sized vehicles. SAE, 933009.

Nomenclature and Glossary of Terms

The following list of nomenclature summarizes the main terms and symbols used in this book. Every effort has been made to be consistent but in some specific analyses certain symbols may have different meanings or interpretations to those in the list. Where this occurs, they are explained in the text alongside the analyses.

Nomenclature

$\mathbf{A}$	Area (m^2)
$\mathbf{A}_a$	Brake actuator area, i.e. of hydraulic pistons or pneumatic actuators which convert actuation system pressure into force applied to the stator (m^2)
$\mathbf{A}_{mc1}$, $\mathbf{A}_{mc2}$	Cross-sectional areas of the primary and secondary pistons respectively in a hydraulic master cylinder
$\mathbf{A}_N$	Apparent (or nominal) area of contact (indicated by the overall size of the contact interface) at any friction interface (m^2)
$\mathbf{A}_p$	Surface area of a disc brake pad (m^2)
$\mathbf{A}_R$	Real area of contact (based on the total surface areas of the microscopic asperities in contact) at any friction interface (m^2)
$\mathbf{A}_s$	Surface area (m^2)
$\mathbf{A}_{sr}$, $\mathbf{A}_{ss}$	Friction surface area of a brake rotor and stator respectively (m^2)
$\mathbf{a}_2$	Horizontal distance of the trailer centre of gravity behind its tow hitch (m)
$\mathbf{C^*}$	Alternative brake performance parameter to brake factor which refers to the 'Internal' brake ratio, i.e. the theoretical ratio of the total frictional force generated by the brake to the actuation force
$\mathbf{C}_p$	Specific heat at constant pressure (J/kgK)
$\mathbf{c}$	Horizontal distance from a towing vehicle rear axle to its tow hitch (m)
$\mathbf{c}$	Damping coefficient (kg/s or Ns/m)
$\mathbf{c}_{bf}$	Brake fluid compressibility (MPa^{-1})
$\mathbf{c}_{bp}$	Fluid consumption coefficient for hydraulic brake lines (per metre length) (mm^3/MPa.m)
$\mathbf{c}_{mc}$	Fluid consumption coefficient for a hydraulic master cylinder (mm^3/MPa)
$\mathbf{c}_{sc}$	Hydraulic slave cylinder bore and seal deformation fluid consumption coefficient (mm^3/MPa)
$\mathbf{D}$	Drawbar force, i.e. the coupling force between a towing vehicle and a chassis trailer (N)
$\mathbf{D}_k$	Duty factor for work done by brakes while driving
$\mathbf{D}_{mc}$	Hydraulic master cylinder bore diameter (m)

D_p	Brake actuation piston diameter (m)
d	Thickness of a disc brake pad backplate (m)
d_l	Semi-thickness of a brake rotor (m)
E	Vehicle wheelbase $= L_1 + L_2$ (m)
E, E_c	Young's modulus, compressive modulus (Pa)
E_2	Horizontal distance from the trailer hitch to its axle centre-line (m)
F	Friction force between two bodies in sliding contact (N)
F_c	Lateral or Cornering force (N)
F_d	Force applied by the driver to the brake pedal, also termed driver effort or control force (N)
F_{f1}, F_{f2}	Seal friction forces of the primary and secondary pistons respectively of a hydraulic master cylinder
F_i	Friction force between the tyres and the road when the wheels on axle 'i' are locked (N)
F_p	Brake pedal output force (N)
F_{yi}	Lateral dynamic friction force between the tyres and the road when the wheels on axle 'i' are skidding (N)
F_1, F_2	Friction drag force generated by the leading and trailing shoes respectively of a drum brake (N)
f	The specific friction force (the tangential friction force per unit real area of contact) between two regions in sliding contact (N/m^2)
f_i	Adhesion utilization at axle i, $= T_i / N_i$
G	Brake Gain of an over-run braking system, defined as the ratio of the braking force at the wheels of the trailer to the force in the coupling
g	Acceleration due to gravity (= 9.81 m/s^2)
H	Height (m)
h	Height of the centre of gravity of a vehicle above the road surface (m)
h	Heat transfer coefficient (W/m^{2}K)
h_{ab}	Height of a disc brake pad abutment reaction above the disc friction surface (m)
h_{am}	Height of the centre of gravity of the additional mass on the coupling of an articulated vehicle tractor and semi-trailer combination (m)
h_{cond}	Thermal conductance (W/m^{2}K)
h_2	Height of the centre of gravity of a trailer above the road surface (m)
h_3	Height of the tow hitch or coupling above the road surface (m)
Izz	Polar moment of inertia (kgm^2)
i	Number (i = 1 − n)
J	Deceleration of the vehicle (m/s^2)
J_m	Mean fully developed deceleration (also termed MFDD) (m/s^2)
J_{rr}	coast-down deceleration
k	Coefficient of thermal conductivity (W/mK)
k	Stiffness (N/m)
k	Theoretical coefficient of adhesion between the tyre and the road
L	Length or distance (m)
L_c	Critical length of a disc brake pad (m)
L_1	Horizontal distance from the front axle to the vehicle's centre of gravity (m)
L_2	Horizontal distance from the rear axle to the vehicle's centre of gravity (m)

l	Brake actuation lever length on a pneumatic brake actuation system (m)
M	Vehicle mass (kg)
M	Mode order
$\mathbf{M}_m$	Simulated tractor mass (= M + M_s) (kg)
$\mathbf{M}_s$	Additional mass for legislative testing (EU) to simulate the effect of a laden semi-trailer coupled to a tractor (kg)
$\mathbf{M}_{so}$	The difference between the maximum laden mass of a tractor (in a tractor and semi-trailer articulated combination) and its unladen mass, i.e. the maximum allowable static load on the coupling (kg)
$\mathbf{M}_t$	Trailer mass (kg)
m	Mass (kg)
$\mathbf{m}_i$	Equivalent mass associated with wheel 'i' of a vehicle (kg)
N	Normal force between two bodies in sliding contact (N)
$\mathbf{N}_c$	Clamp force on a disc brake pad (N)
$\mathbf{N}_i$	Dynamic normal reaction between the tyres and the road at axle 'i' under braking conditions (N)
$\mathbf{N}_{xi}$	Dynamic normal reaction at the road surface of the inside wheel of a vehicle axle while cornering (N)
$\mathbf{N}_{xo}$	Dynamic normal reaction at the road surface of the outside wheel of a vehicle axle while cornering (N)
$\mathbf{N}_4$	Tow hitch vertical load between a 2-axle motor vehicle and a single axle (or close-coupled axles) trailer (N)
P	Vehicle weight (N); ***this differs from the Regulations 13 & 13H where P is specified as the 'mass of the vehicle'. In this book the mass of the vehicle is denoted as 'M' (kg)***
$\mathbf{P}_a$	Actuation force applied to the brake stator (disc pad or drum brake shoe) (N)
$\mathbf{P}_{ac}$	Force generated by a pneumatic actuator (N)
$\mathbf{P}_{c1}$, $\mathbf{P}_{c2}$	Damping forces arising from the viscosity of the hydraulic fluid of a hydraulic master cylinder (N)
$\mathbf{P}_i$	Normal reaction between the tyres and the road at axle 'i' under static conditions (N)
$\mathbf{P}_{k1}$, $\mathbf{P}_{k2}$	Preload forces on the return springs of a hydraulic master cylinder (N)
$\mathbf{P}_m$	Simulated tractor weight (= M_m . g) (N)
$\mathbf{P}_{mc}$	Actuation force to a hydraulic master cylinder (N)
$\mathbf{P}_r$	Semi-trailer weight (N)
$\mathbf{P}_t$	Trailer weight (N)
$\mathbf{P}_{ti}$	Brake actuation threshold force (N)
$\mathbf{P}_w$	Weight of a road wheel and tyre assembly and its associated rotating parts ('unsprung' weight) (N)
p	Pressure (Pa)
$\mathbf{p}_m$	Mean or average pressure (Pa)
$\mathbf{p}_{mc}$	Hydraulic master cylinder pressure (MPa)
$\mathbf{p}_t$	Actuator threshold pressure (Pa)
$\mathbf{p}_1$, $\mathbf{p}_2$	Hydraulic pressure in the primary and secondary circuits of a hydraulic brake actuation system (Pa)
Q	Energy (J)

Q_a	Arrhenius activation energy (J)
q	Instantaneous energy (J)
$\dot{Q}$	Power (W)
$\dot{Q}_i$	Instantaneous power dissipated in a brake 'i' (W)
$\overline{\dot{Q}}_i$	Mean power dissipation (over a brake application) (W)
$\dot{q}$	Instantaneous power density (W/m^2)
R	Universal Gas Constant (8.314462J/mol.K)
R_b	Brake servo boost ratio
R_c	Reaction force on an S-cam brake camshaft (N)
R_p	Brake pedal lever ratio
R_{1x}, R_{1y}, R_{2x}, R_{2y}	Abutment reaction force components in drum brake analysis (N)
r	Radius (m)
r_b	Cam base circle radius (m)
r_e	Rotor effective radius (m)
r_i, r_o	Inner and outer radius of the disc rubbing path (m)
r_m	Brake disc rubbing path mean radius (m)
r_r	Tyre rolling (or dynamic) radius (m)
S_1, S_2	Drum brake leading (suffix 1) and trailing (suffix 2) shoe factors
s	Stopping distance (m)
T_i, T_{xi}	Braking force exerted by the brakes on axle 'i' under normal braking conditions on the road (N)
T_m	Total braking force on the tractor of an articulated vehicle combination (N)
T_r	Total braking force on the semi-trailer of an articulated vehicle combination (N)
T_{wi}	Braking force at 1 wheel of an axle 'i' (N)
T_{yi}	Sideways force on axle 'i' e.g. during cornering (N)
T_4	Horizontal drawbar force between a 2-axle motor vehicle and a single axle (or close-coupled axles) trailer (N)
t	Time (s)
t	Vehicle track width (m)
t	Thickness of friction material on a disc brake pad (m)
u_r	Radial expansion of a master cylinder bore (m)
V	Vehicle road speed (km/h)
V_{max}	Maximum vehicle road speed (km/h)
V_{mc}	Volume of brake fluid of the master cylinder (m^3)
V_{ref}	Vehicle reference speed (km/h)
v	Linear speed (m/s)
Δw	Wear (weight loss per unit area kg/m^2)
X, Y, Z	Cartesian axis set, for a road vehicle X is the vehicle longitudinal axis, Y is the lateral axis, and Z is the vertical axis
X_i	Proportion of the total vehicle braking force generated at axle 'i'
x	Distance (m)
x_p	Brake pedal travel (m)
Y_i	Proportion of braking provided by a towing vehicle and a trailer to decelerate the vehicle combination
z	Rate of braking of the vehicle = J/g
z_{crit}	Rate of braking with optimum braking distribution ($\eta = 1$) on a rigid solo vehicle

z_1	Rate of braking with optimum braking distribution ($\eta = 1$) on a commercial vehicle and trailer combination
α	Thermal diffusivity (m^2/s)
α	Inclination angle (Degrees)
γ	Coefficient of thermal expansion (K^{-1})
γ	Vehicle roll angle (rad)
Δ	Indicates an increment
δ	Angle subtended by the pad sector arc length (rad)
ϵ	Surface emissivity
η	Efficiency
ηC^*	Alternative brake performance parameter to brake factor which includes the efficiency of the internal actuation system
η_{mc}	Efficiency factor associated with hydraulic master cylinders and slave cylinders
θ	Temperature (°C)
θ_K	Absolute temperature (K)
μ	Dynamic (or sliding) coefficient of friction between the rotor and the stator components; may also be written μ_D when necessary to emphasis the differentiation between static and dynamic friction
μ_{ab}	Coefficient of friction between a disc brake pad and the caliper abutment
μ_s	Static coefficient of friction, when the bodies in contact have no relative motion between them
μ_t	Coefficient of sliding friction associated with a tyre on a locked wheel sliding on the road surface
ν	Poisson's ratio
ρ	Density (kg/m^2)
σ	Stefan–Boltzmann constant ($Wm^{-2}K^{-4}$)
τ	Torque (Nm)
τ_a	Actuation torque applied to a pneumatically actuated brake (Nm)
τ_{disc}	Torque generated by a disc brake (Nm)
τ_{drum}	Torque generated by a drum brake (Nm)
τ_s	Specific torque (alternative parameter to brake factor)
τ_{wi}	Brake torque reacted at 1 wheel of an axle 'i' (Nm)
ω	Angular velocity (rad/s)
$\dot{\omega}$	Angular acceleration (rad/s^2)

Glossary of Terms

ABS	Anti-lock braking system
ACC	Adaptive cruise control
AEBS	Advanced emergency braking system
ARM	Active rollover mitigation
AVH	Automatic vehicle hold
BBW	Brake by wire
BF	Brake factor
BFC	Braking force coefficient
C*	Equivalent or alternative parameter to brake factor

CABS	Collision avoidance braking system
CAD	Computer aided design
CAE	Computer-aided engineering
CFD	Computational fluid dynamics
CMbB	Collision mitigation by braking
CMBS	Collision mitigation braking system
CNSL	Cashew nut shell liquid
CoP	Centre of pressure
CSF	Combined shoe factor of a drum brake (= $S_1 + S_2$)
DoE	Design of experiments
DoF	Degree of freedom
DoW	Driver-only weight, i.e. the minimum driving vehicle weight which is the unladen weight plus the driver's weight (N)
DTV	Disc thickness variation
EBA	Emergency brake assist
EBD	Electronic brakeforce distribution
EDC	Engine drag control
EHB	Electro-hydraulic brake
EMB	Electro-mechanical brake
EPB	Electric parking brake
ESC	Electronic stability control
EU	European Union
EV	Electric vehicle
FEA	Finite element analysis
FEV	Full electric vehicle
FFT	Fast Fourier transform
FMVSS	Federal Motor Vehicle Safety Standards
FRF	Frequency response function
Fo	Fourier number
Foundation brake	The friction brake assembly at the road wheel
FWD	Front wheel drive
GVW	Gross vehicle weight, i.e. the maximum allowable vehicle weight fully laden (N)
HAS	Hill start assist
HEV	Hybrid electric vehicle
HGV	Heavy goods vehicle
ICE	Internal combustion engine
ISO	International Organization for Standardization
L & T	Leading and trailing (drum brake)
LVDT	Linear variable displacement transducer
MFDD	Mean fully developed deceleration (also termed d_m) (m/s^2)
NAO	Non-asbestos organic friction material
NHTSA	National highway traffic safety administration
NVH	Noise, vibration and harshness
OEM	Original equipment manufacturer
ROM	Rollover mitigation
RSC	Rollover stability control

RSP	Roll stability program
RWD	Rear wheel drive
SAE	Society of automotive engineers
SFC	Sideways force coefficient (also termed cornering coefficient in the US)
SSTR	Single stop temperature rise (°C)
TCS	Traction control system
TEI	Thermoelastic instability
TSC	Trailer sway control
TVbB	Torque Vectoring by Braking
UNECE	United Nations Economic Commission for Europe
US	United States (of America)
1D, 2D, 3D	1-dimensional, 2-dimensional, 3-dimensional
4WD	4-wheel drive

Index

Note: Page Numbers followed by "f" indicate figures; "t", tables; "b", boxes.

B

C

D

Printed in the United States
By Bookmasters